Lecture Notes in Computer Science 1027

Edited by G. Goos, J. Hartmanis and J. van Leeuwen

Advisory Board: W. Brauer   D. Gries   J. Stoer

Springer
Berlin
Heidelberg
New York
Barcelona
Budapest
Hong Kong
London
Milan
Paris
Santa Clara
Singapore
Tokyo

Franz J. Brandenburg (Ed.)

Graph Drawing

Symposium on Graph Drawing, GD '95
Passau, Germany, September 20-22, 1995
Proceedings

Springer

Series Editors

Gerhard Goos
Universität Karlsruhe
Vincenz-Priessnitz-Straße 3, D-76128 Karlsruhe, Germany

Juris Hartmanis
Department of Computer Science, Cornell University
4130 Upson Hall, Ithaca, NY 14853, USA

Jan van Leeuwen
Department of Computer Science,Utrecht University
Padualaan 14, 3584 CH Utrecht,The Netherlands

Volume Editor

Franz J. Brandenburg
Lehrstuhl für Informatik, Universität Passau
Innstraße 33, D-94030 Passau, Germany

Cataloging-in-Publication data applied for

Die Deutsche Bibliothek - CIP-Einheitsaufnahme

Graph drawing : proceedings / Symposium on Graph Drawing,
GD '95, Passau Germany, September 20 - 22, 1995. Franz J.
Brandenburg (ed.). - Berlin ; Heidelberg ; New York ;
Barcelona ; Budapest ; Hong Kong ; London ; Milan ; Paris ;
Santa Clara ; Singapore ; Tokyo : Springer, 1995
  (Lecture notes in computer science ; Vol. 1027)
  ISBN 3-540-60723-4
NE: Brandenburg, Franz J. [Hrsg.]; GD <1995, Passau>; GT

CR Subject Classification (1991): G.2.2, F.2.2, D.2.2, I.3.5, J.6, H.5.2

1991 Mathematics Subject Classification: 05Cxx, 68R10, 90C35, 94C15

ISBN 3-540-60723-4 Springer-Verlag Berlin Heidelberg New York

© Springer-Verlag Berlin Heidelberg 1996
Printed in Germany

Typesetting: Camera-ready by author
SPIN 10512376   06/3142 – 5 4 3 2 1 0   Printed on acid-free paper

Preface

Graph Drawing addresses the problem of visualizing structural information. More specifically, it is concerned with the construction of geometric representations of abstract graphs and networks. The automatic generation of drawings of graphs has important applications in key computer science technologies such as database design, software engineering, VLSI, network design, and visual interfaces, and also in engineering, production planning, chemistry, and biology. The range of issues being investigated in graph drawing includes algorithms, graph theory, order theory, graphic languages, applications, and practical systems. A great deal of research in graph drawing is motivated by applications to systems for viewing graphs and interacting with graphs. The interaction between theoretical advances and implemented solutions is an important part of the area of graph drawing.

Graph Drawing '95 (GD'95) was held in Passau, Germany, on September 20-22, 1995. It was preceeded by the GD'94 DIMACS International Workshop in Princeton, USA (1994), the GD'93 ALCOM Workshop in Paris, France (1993), and the GD'92 Work Meeting in Rome, Italy (1992).

In response to the call for papers and demos for GD'95, the program committee received 88 submissions. Six submissions applied for both categories. This indicates the strong and growing interest in the area. All but six submissions were received electronically. Each submission was refereed and evaluated by at least four program committee members, or their referees, with six members on the average. The program committee made its final selection via email, from July 10 to 14. It selected 40 papers and 12 demos.

The technical program of GD'95 consisted of 12 sessions and included the presentation of all papers, all but two demonstrations, and a poster gallery with 8 poster displays. Also, in conjunction with GD'95, there was a graph-drawing contest with prizes awarded at the workshop. This volume contains the accepted papers and descriptions of the demos of GD'95, and a report on the graph-drawing contest.

GD'95 gathered 127 participants from 14 countries from all over the world, again with a significant presence of participants with industrial affiliations. The presentations of the papers and the demos were excellent. All presentations were attended by almost all participants with intensive discussions immediately afterwards. There was a stimulating atmosphere at GD'95.

We wish to thank the many people who contributed to the success of the symposium: all authors of submitted papers, demos, and posters, the program committee members, and their referees, and the organizers of the graph-drawing contest. We are grateful to the sponsors of GD'95, particularly to the Deutsche Forschungsgemeinschaft, and the Bayerische Staatsministerium für Unterricht, Kultus, Wissenschaft und Kunst for providing financial support to the symposium, and to the Universität Passau for offering the use of their facilities. This helped to keep the conference fees low. Special thanks go to the symposium secretaries Susanne Lenz and Renate Schönberger for their excellent settlement of the registration and local arrangements, and to Andreas Stübinger for a perfect handling of all submissions, notifications, and final papers as the GD'95 postmaster.

Passau, October 1995   *Franz J. Brandenburg*

Organizer:
Franz J. Brandenburg (University of Passau, Germany)

Program Committee:
Franz J. Brandenburg, chair (University of Passau, Germany)
Peter Eades (University of Newcastle, Australia)
Jürgen Ebert (University of Koblenz, Germany)
Michael Kaufmann (University of Tübingen, Germany)
Jan Kratochvíl (Charles University of Prague, Czech Republic)
Anna Lubiw (University of Waterloo, Canada)
Joe Marks (Mitsubishi Electric Research Labs., Cambridge, USA)
Kozo Sugiyama (Fujitsu Labs., Shizuoka, Japan)
Roberto Tamassia (Brown University, Providence, USA)
Ioannis G. Tollis (University of Texas at Dallas, USA)

Graph-Drawing Contest Organizers:
Peter Eades (University of Newcastle, Australia)
Joe Marks (Mitsubishi Electric Research Labs., Cambridge, USA)

Local Arrangements:
Susanne Lenz
Renate Schönberger
Franz J. Brandenburg

Demo Arrangements:
Fred Dichtl
Frank Heyder
Andreas Stübinger

Supported by:
Universität Passau, Passau
Deutsche Forschungsgemeinschaft, Bonn
Bayerisches Staatsministerium für
Unterricht, Kultus, Wissenschaft und Kunst, München
Volksbank-Raiffeisenbank Passau-Freyung eG, Passau
SUN Microsystems GmbH, Grasbrunn
Bavaria Computer Systeme, Passau
AT&T Bell Laboratories, Murray Hill
Mitsubishi Electric Research Laboratories, Cambridge
Tom Sawyer Software, Berkeley
Silicon Graphics GmbH, Grasbrunn
Begegnungszentrum der Wissenschaft Schloß Neuburg, Neuburg
Fremdenverkehrsverein Passau, Passau
Buchhandlung Pustet, Passau

Contents

Quasi-Planar Graphs Have a Linear Number of Edges[*]

Pankaj K. Agarwal[1], Boris Aronov[2], János Pach[3,4,5], Richard Pollack[4], and Micha Sharir[4,6]

[1] Department of Computer Science, Duke University, Durham, NC 27708-0129 USA
[2] Computer and Information Science Department, Polytechnic University, Brooklyn, NY 11201-3840 USA
[3] Department of Computer Science, City College, CUNY, New York, NY, USA
[4] Courant Institute of Mathematical Sciences, New York University, New York, NY 10012, USA
[5] Hungarian Academy of Sciences, Budapest, Hungary
[6] School of Mathematical Sciences, Tel Aviv University, Tel Aviv 69978, Israel

Abstract. A graph is called *quasi-planar* if it can be drawn in the plane so that no three of its edges are pairwise crossing. It is shown that the maximum number of edges of a quasi-planar graph with n vertices is $O(n)$.

1 Introduction

We say that an undirected graph $G(V, E)$ without loops or parallel edges is *drawn* in the plane if each vertex $v \in V$ is represented by a distinct point and each edge $e \in E$ is represented by a Jordan arc connecting the points corresponding to endpoints of e. Throughout this paper, we assume that any two arcs of a drawing have at most one point in common, which is either a common endpoint or a common interior point where the two arcs cross each other. We do not make

---

[*] Work on this paper by Pankaj K. Agarwal has been supported by NSF Grant CCR-93-01259, an NYI award, and matching funds from Xerox Corporation. Work on this paper by Boris Aronov has been supported by NSF Grant CCR-92-11541 and by a Sloan Research Fellowship. Work on this paper by János Pach, Richard Pollack, and Micha Sharir has been supported by NSF Grants CCR-91-22103 and CCR-94-24398. Work by János Pach was also supported by Grant OTKA-4269 and by a CUNY Research Award. Work by Richard Pollack was also supported by NSF Grants CCR-94-02640 and DMS-94-00293. Work by Micha Sharir was also supported by NSF Grant CCR-93-11127, by a Max-Planck Research Award, and by grants from the U.S.-Israeli Binational Science Foundation, the Israel Science Fund administered by the Israeli Academy of Sciences, and the G.I.F., the German-Israeli Foundation for Scientific Research and Development. Work on this paper was done during the participation of the first four authors in the Special Semester on Computational and Combinatorial Geometry organized by the Mathematical Research Institute of Tel Aviv University, Spring 1995.

any notational distinction between vertices of G and the corresponding points in the plane, or between edges of G and the corresponding Jordan arcs.

A graph that can be drawn in the plane without crossing edges is planar. We call a graph *quasi-planar* if it can be drawn in the plane with no three pairwise crossing edges. The aim of this paper is to establish that the number of edges of any quasi-planar graph with n vertices is $O(n)$. This improves an earlier result of Pach et al. [6].

Theorem 1. *If $G(V, E)$ is a quasi-planar graph, then $|E| = O(|V|)$.*

We prove this theorem in Section 2, and in Section 3 we consider some related problems and generalizations.

2   Proof of Theorem 1

To simplify our presentation, we only prove the theorem in the special case when G has a straight-line drawing with no three pairwise crossing edges (straight line segments). Remarkably, the proof for the general case requires only minor modification.

The set of edges $E = E(G)$ defines a cell complex in the plane, whose 0-, 1-, and 2-dimensional cells will be called *nodes*, *segments*, and *faces*, respectively. This cell complex is known as the *arrangement* of the set of edges of G and is denoted by $\mathcal{A}(E)$. For example, *nodes* of $\mathcal{A}(E)$ are the endpoints and crossings of graph edges, and a *segment* of $\mathcal{A}(E)$ is a portion of a graph edge between two consecutive nodes. (To avoid ambiguity we hereafter refer to vertices and edges of G, and to the corresponding points and line segments in the plane, as "graph vertices" and "graph edges," respectively.) Let X be the set of crossings of graph edges, $N = V \cup X$ the set of all nodes of $\mathcal{A}(E)$, S the set of its segments, and F the set of its faces. For a face $f \in F$, the *complexity* $|f|$ of f is the number of segments of S on the boundary ∂f of f. As usual, if both sides of an edge are incident to the interior of f, then it contributes 2 to $|f|$. Let $t(E) = |\{f \in F : |f| = 3\}|$ be the number of triangular faces in F.

Lemma 2. *Let $G(V, E)$ be a graph drawn in the plane. Then the total complexity of all non-quadrilateral faces of the arrangement $\mathcal{A}(E)$ is at most $4t(E) + 20|V|$.*

Proof. It is sufficient to prove the lemma with the assumption that the planar graph (N, S) is connected and $|S| > 1$.

Recall the following familiar facts:

$$\sum_{f \in F} |f| = 2|S| \ ,$$

$$2|S| = \sum_{v \in N} \deg(v) = \sum_{v \in V} \deg(v) + \sum_{v \in X} \deg(v) \geq 2|E| + 4|X| \ ,$$

and

$$|V| + |X| + |F| = |N| + |F| = |S| + 2 \ .$$

The first two lines just express two different ways of counting the edges of the planar graph (N, S), as the sum of face complexities and of vertex degrees, respectively. The third line is Euler's relation. These easily yield

$$\sum_{f \in F} |f| \leq 4|V| + 4|F| - 2|E| - 8, \quad \text{hence} \quad \sum_{f \in F} (|f| - 4) \leq 4|V| .$$

Finally, since $|f| > 4$ implies $|f| \leq 5(|f| - 4)$, we have

$$\sum_{f \in F, |f| \neq 4} |f| = 3t(E) + 5 \sum_{f \in F} (|f| - 4) \leq 3t(E) + 20|V| .$$

□

Lemma 3. *Let $G(V, E)$ be a quasi-planar graph drawn in the plane. Then the overall complexity of all faces f of $\mathcal{A}(E)$, such that f is either a non-quadrilateral face or a quadrilateral face incident to at least one vertex of G, is $O(|V| + |E|)$.*

Proof. Note that $t(E) = O(|E|)$, as each triangular face f of $\mathcal{A}(E)$ must be incident to a vertex of G. For otherwise there would be three pairwise crossing edges. It is easy to check that the number of faces of $\mathcal{A}(E)$ incident to graph vertices is at most $2|E|$. In addition, this implies that the overall complexity of all quadrilateral faces of $\mathcal{A}(E)$ incident to a graph vertex in V is also $O(|E|)$. The lemma is now an immediate consequence of Lemma 2. □

Let $G(V, E)$ be a quasi-planar graph drawn in the plane with $n = |V|$ vertices. Returning to the proof of Theorem 1, we may assume without loss of generality that G is connected, as it suffices to establish a linear bound on the number of graph edges in each connected component of G. Let $G_0 = (V, E_0)$ be a spanning tree of G, so $|E_0| = n - 1$. Let $E^* = E \setminus E_0$. Note that each face of the arrangement $\mathcal{A}(E_0)$ is simply connected, for otherwise the union of nodes and segments of $\mathcal{A}(E_0)$ would not be connected, contradicting the connectedness of G_0. Moreover, by Lemma 3, the complexity of all faces of $\mathcal{A}(E_0)$, which are either non-quadrilaterals or quadrilaterals incident to a point in V, is $O(n)$. We refer to the remaining faces of $\mathcal{A}(E)$ as *crossing quadrilaterals*.

In the sequel, we use the following notion. A graph is called an *overlap graph* if its vertices can be represented by intervals on a line such that two vertices are adjacent if and only if the corresponding intervals overlap but neither contains the other [1]. Gyárfás [2] (see also [3]) has shown that every triangle-free overlap graph can be colored by a constant number, c, of colors, and Kostochka [4] proved that this is true with $c = 5$.

For each graph edge $e \in E^*$, let $\Xi(e)$ denote the set of segments of $\mathcal{A}(E_0 \cup \{e\})$ that are contained in e. In other words, it is the set of segments into which e is cut by the graph edges from E_0. By construction, each segment $s \in \Xi(e)$ is fully contained in a face of $f \in \mathcal{A}(E_0)$ and its two endpoints lie on the unique connected component of ∂f. For each face f of $\mathcal{A}(E_0)$, let $X(f)$ denote the set of all segments in $\bigcup_{e \in E^*} \Xi(e)$ that are contained in f, and let $H(f)$ denote the quasi-planar graph whose set of edges is $X(f)$. Since f is simply connected, any two segments in $X(f)$ cross each other if and only if their endpoints interleave along the boundary of f. By cutting the boundary of f so that it becomes an interval and associating with each segment in $X(f)$ the connected interval along the boundary of f between its endpoints, we obtain a collection of intervals with the property that two elements of $X(f)$ cross if and only if the corresponding intervals overlap and neither is contained in the other. This defines a triangle-free overlap graph on the vertex set $X(f)$. Therefore the segments of $X(f)$ can be colored by at most five colors, so that no two segments with the same color cross each other. (Note that, for a graph edge $e \in E^*$, several segments in $\Xi(e)$ may be contained in the same face f and thus belong to the same $X(f)$. These segments may be colored by different colors.)

Let f be a face of $\mathcal{A}(E_0)$ other than a crossing quadrilateral, and let $H_1(f)$, ..., $H_5(f)$ be the monochromatic subgraphs of $H(f)$ obtained by the above coloring. Fix one of these subgraphs, say $H_1(f)$, and re-interpret it as a graph whose vertices are the (relative interiors of the) edges of ∂f together with the elements of V on ∂f, and whose edges are the segments of $H_1(f)$. The resulting graph, $H_1^*(f)$, is clearly planar. We call a face of $H_1^*(f)$ a *digon* if it is bounded by exactly two edges, and we call an edge of $H_1^*(f)$ *shielded* if both of the faces incident to it are digons. The remaining edges of $H_1^*(f)$ are called *exposed*. Observe that, by Euler's formula, there are at most $O(n_f)$ exposed edges in $H_1^*(f)$, where n_f is the number of vertices of $H_1^*(f)$, which is at most $2|f|$.

We repeat this analysis for each of the other subgraphs $H_2(f), \ldots, H_5(f)$, and for all faces f of $\mathcal{A}(E_0)$ other than crossing quadrilaterals. It follows that the number of graph edges $e \in E^*$ containing at least one exposed segment (in the graph $H_i^*(f)$ containing it) is $O(\sum_f |f|)$, where the sum extends over all such faces f. By Lemma 3, this sum is $O(n)$.

It thus remains to bound the number of graph edges in E^* with no exposed subsegment; we call these edges *shielded*, borrowing the terminology used above. If e is a shielded graph edge, then, for each $s \in \Xi(e)$, either s lies in a crossing quadrilateral face of $\mathcal{A}(E_0)$, or else s is shielded in its subgraph. Note that no graph edge $e \in E^*$ can consist solely of segments passing through crossing quadrilaterals, as the first and last segments necessarily meet faces of $A(E_0)$ that have at least one graph vertex on their boundary, namely an endpoint of e.

Lemma 4. *There are no shielded edges.*

Proof. Suppose that $e \in E^*$ is shielded. Let a and b be the endpoints of e. We claim that there exists a graph edge $e^+ \in E^*$ such that (1) e^+ is a graph edge of E^* emanating from a next to e, and (2) for each segment $s \in \Xi(e)$, there is a corresponding segment $s^+ \in \Xi(e^+)$, such that s and s^+ connect the same pair of segments of $\mathcal{A}(E_0)$. Let $s_1, \ldots, s_k$ denote the segments in $\Xi(e)$, appearing along e in this order.

We prove, by induction on j, that the claim holds for $s_1, \ldots, s_j$. Consider first the case $j = 1$. Let a and b be the endpoints of e, so that s_1 is incident to a and s_k is incident to b. Then s_1 connects a with some edge τ_1 of $\mathcal{A}(E_0)$ (note that for a shielded graph edge e, $s_1 \neq e$). Since s_1 is shielded, there exists another graph edge $e^+ \in E^*$ with a subsegment $s_1^+ \in \Xi(e^+)$ that connects a to τ_1. Clearly, we can choose e^+ with these properties to be the graph edge emanating from a nearest to e, proving the claim for $j = 1$.

Suppose next that the assertion is true for $j - 1$ and e^+ is the graph edge satisfying the inductive assumption. Suppose that s_j connects two segments τ_{j-1} and τ_j of $\mathcal{A}(E_0)$ such that $u = \tau_{j-1} \cap e$ is the common endpoint of s_{j-1} and s_j, and $v = \tau_j \cap e$ is the other endpoint of s_j. (If $j = k$ then we take τ_j to be the other endpoint b of e.) If s_j lies in a crossing quadrilateral face f, then, as is easily verified, e and e^+ must cross the same pair of opposite edges of f, completing the induction step. Otherwise, since s_j is shielded, there is a graph edge $e' \in E^*$ and a subsegment $s' \in \Xi(e')$ that connects τ_{j-1} and τ_j on the same side of s_j as e^+. Three cases can arise:

- $e' = e^+$: The induction step is complete.
- e' *crosses* τ_{j-1} *at a point that lies between* u *and the crossing with* e^+: Since G is quasi-planar, e' cannot cross e or e^+. Moreover, e' cannot have an endpoint within the interior of the triangle $\triangle$ bounded by e, e^+, and τ_{j-1}, by the induction hypothesis and the fact that all faces of $\mathcal{A}(E_0)$ are simply connected. Hence, e' must end at a and lie inside $\triangle$ near a. However, this contradicts the choice of e^+ as the closest neighbor of e near a. Thus this case is impossible.
- e^+ *crosses* τ_{j-1} *at a point that lies between* u *and the crossing with* e': In this case, e^+ cannot cross s_j or s' or terminate inside f. Thus, it must meet τ_j. This completes the induction step and hence the proof of the claim.

Note that the same analysis also applies when $j = k$, that is, when τ_j is the endpoint b of e. Therefore, e and e^+ have the same pair of endpoints. Contradiction. □

As there are no shielded edges, the total number of edges of E^*, and thus also of E, is $O(n)$. This completes the proof of Theorem 1.

3 Discussion

In this section we discuss some consequences of the above results.

Theorem 5. *Let* $G(V, E)$ *be a graph with* n *vertices that can be drawn in the plane with no four pairwise crossing edges. Then the number of edges of* G *is* $O(n \log^2 n)$.

Proof. We first estimate the number C of crossings between the edges of G. Let e be an edge of G, and let G_e be the subgraph of G consisting of all edges that cross e. Then G_e is a quasi-planar graph. Thus, by Theorem 1, the number of edges of G_e is $O(n)$, which implies that $C = O(n|E|)$. One can then combine this estimate with the analysis in [6], to conclude that $|E| = O(n \log^2 n)$. □

Corollary 6. *Let $k \geq 4$ be an integer, and let G be a graph with n vertices that can be drawn in the plane with no k pairwise crossing edges. Then the number of edges of G is $O(n \log^{2k-6} n)$.*

Proof. This is an immediate consequence of the analysis in [6], which proceeds by induction on k, based on the improved bound of Theorem 5 for $k = 4$. □

Theorem 5 and Corollary 6 improve the bounds given in [6] by a factor of $\Theta(\log^2 n)$.

There are several interesting problems that are left open in this paper. The first problem is to find the best constant of proportionality in the bound of Theorem 1. A trivial lower bound is roughly $6n$, obtained by overlaying two edge disjoint triangulations of a point set. The constant 6 can be slightly improved.

Another open problem is as follows. For a quasi-planar graph G, let $\chi = \chi(G)$ be the smallest number with the property that the edges of G can be colored with χ colors, so that the edges in each color class form a planar graph. Clearly, if G has n vertices, then the number of edges of G is at most $3\chi(G)n$. Thus, a plausible conjecture is that $\chi(G)$ is bounded from above by a constant. Recall that this conjecture is true with $\chi(G) \leq 5$, if there exists a plane drawing of G in which no three edges are pairwise crossing and the vertices are in convex position (see also [2, 3] for a weaker constant bound and for related results concerning more general classes of graphs). Moreover, if there exists such a drawing of G in which the vertices lie on two parallel lines, then one can easily show that $\chi(G) \leq 2$. Does there exist a constant upper bound for $\chi(G)$ when all edges of G cross a common line? A weaker conjecture is that there exists a subset E' of pairwise noncrossing edges of G such that $|E'| \geq \beta |E|$ for some absolute constant $\beta > 0$. The existence of such a subset E' would imply, by planarity, that $|E'| = O(n)$, which would provide another proof of Theorem 1.

Acknowledgments

We wish to thank Zoltán Füredi, Jiří Matoušek, and Otfried Schwarzkopf for helpful discussions concerning the problem studied in this paper. We also extend our gratitude to Bernd Gärtner and Emo Welzl for bringing this problem to our attention in 1989.

References

1. M. Golumbic, *Algorithmic Graph Theory*, Academic Press, New York, 1980.
2. A. Gyárfás, On the chromatic number of multiple interval graphs and overlap graphs, *Discrete Math.* 55 (1985), 161–166.
3. A. Gyárfás and J. Lehel, Covering and coloring problems for relatives of intervals, *Discrete Math.* 55 (1985), 167–180.
4. A. V. Kostochka, On upper bounds for the chromatic number of graphs, in: *Modeli i metody optimizacii, Tom 10*, Akademiya Nauk SSSR, Sibirskoe Otdelenie, 1988, 204–226.

5. J. Pach and P.K. Agarwal, *Combinatorial Geometry*, Wiley–Interscience, New York, 1995.
6. J. Pach, F. Sharokhi and M. Szegedy, Applications of the crossing number, *Proc. 10th Annual ACM Symp. on Computational Geometry* (1994), pp. 198–202. Also to appear in *Algorithmica*.

Universal 3-Dimensional Visibility Representations for Graphs

Helmut Alt[1] * and Michael Godau[1] ** and Sue Whitesides[2] ***

[1] Freie Universität Berlin Berlin, Germany
[2] McGill University Montreal, Canada

Abstract. This paper studies 3-dimensional visibility representations of graphs in which objects in 3-d correspond to vertices and vertical visibilities between these objects correspond to edges. We ask which classes of simple objects are *universal*, i.e. powerful enough to represent all graphs. In particular, we show that there is no constant k for which the class of all polygons having k or fewer sides is universal. However, we show by construction that every graph on n vertices can be represented by polygons each having at most $2n$ sides. The construction can be carried out by an $O(n^2)$ algorithm. We also study the universality of classes of simple objects (translates of a single, not necessarily polygonal object) relative to cliques K_n and similarly relative to complete bipartite graphs $K_{n,m}$.

1 Introduction

This paper considers 3-dimensional visibility representations for graphs. Vertices are represented by 2-dimensional objects floating in 3-d parallel to the xy-plane (these objects can be swept in the z direction to form thick objects if desired). There is an edge in the graph if, and only if, the objects corresponding to its endpoints can see each other along a thick line of sight parallel to the z-axis. A thick line of sight is a tube of arbitrarily small by positive radius whose ends are contained in the objects. Throughout this paper, we use the term "visibility representation" to refer to this particular model.

The corresponding notion of 2-dimensional visibility has received wide attention due to its applications to such areas as graph drawing, VLSI wire routing, algorithm animation, CASE tools and circuit board layout. See [DETT] for a survey on graph drawing in general; for 2-dimensional visibility representations, see for example [DH], [TT], [KKU], [W].

Exploration of 3-dimensional visibility is still in the early stages. From the point of view of geometric graph theory, it is natural to consider visibility representations of graphs in dimensions higher than 2. From the point of view of

* alt@inf.fu-berlin.de Institut für Informatik, FU Berlin, Takustr. 9 , 14195 Berlin, Germany. This research was supported by the ESPRIT Basic Research Action No. 7141, Project ALCOM II.

** godau@inf.fu-berlin.de

*** sue@cs.mcgill.ca Written while the author was visiting INRIA-Sophia Antipolis and Freie Universität Berlin. Research supported by NSERC and FCAR grants.

visualization of graphs, it is basic to ask whether 3-dimensional representations give useful visualizations. For a 3-dimensional representation to be useful for visualization, it should be powerful enough to represent all graphs, or at least basic kinds of graphs. This motivates us to ask which classes of objects are *universal*, i.e., can give visibility representations for all graphs, or all graphs of a given kind?

The visibility representation considered in this paper has also been studied in [BEF+] (an abstract was presented at GD'92), in [Rom], and in [FHW]. In these papers, the objects representing vertices are axis-aligned rectangles, or disks, and the properties of graphs that can be represented by these objects are studied. By contrast, this paper begins with families of graphs (all graphs, or all graphs of a specific kind), and explores simple ways to represent all graphs in the family.

Section 2 considers which translates of a given, fixed figure are universal for cliques K_n and complete bipartite graphs $K_{m,n}$. Section 3 uses counting arguments based on arrangements to show that no class of polygons having at most some fixed number k of sides is strong enough to represent all graphs. Section 4 shows that every graph on n vertices has a visibility representation by polygons each of which has at most $2n$ sides. These sections also contain additional results not listed here in the introduction.

2 Graphs realizable by translates of a figure

In this section we will investigate which complete and which complete bipartite graphs can be realized as visibility graphs of *translates* of one fixed figure. Here a *figure* is defined as an open bounded set whose boundary is a Jordan-curve. We say that a graph G can be *realized* by a figure F iff G is the visibility graph of translates of F. It will turn out, for example, that with many figures arbitrary complete graphs can be realized whereas each figure can only realize a finite number of stars, i.e. complete bipartite graphs of the form $K_{1,n}$.

2.1 Complete graphs

The realization of complete graphs K_n by translates of special figures like squares and disks has been investigated by Fekete, Houle, and Whitesides [FHW] and by Bose et al. [BEF+]. In [FHW] it was shown that K_7 can be realized by a square, whereas any K_n, $n \geq 8$ cannot. On the other hand, any K_n can be realized by a disk. We will consider more general figures in the following theorem.

First, we need the following definitions:
A curve C is called *strictly convex*, iff for any two points $p, q \in C$ the interior of the line segment $\overline{pq}$ does not intersect C. We say that a figure F has a *local roundness* if there is some open set U such that $U \cap \partial F$ is a strictly convex curve.

Theorem 2.1 a) *Any K_n can be realized by any nonconvex polygon.*
b) *For any convex polygon P there is an $n \in \mathbb{N}$ such that no $K_m, m \geq n$ can be realized by P.*

c) To any K_n there is a convex polygon realizing it.

d) Any figure F with a local roundness can realize any K_n.

Proof. a)We first observe that the figure in Fig.1 can realize any K_n. If P is a nonconvex polygon then it has at least one nonconvex vertex. Arranging copies of P in a neighborhood of this vertex as in Fig. 1 realizes any K_n. b)(Sketch)

Let $P_1, \ldots, P_k$ be a sequence of (projections of) translates of a convex n-gon ordered by increasing z-coordinates, $e_1, \ldots, e_k$ the corresponding translates of one edge, and H_i the halfplane bounded by the straight line through e_i which contains P_i, $i = 1, \ldots, k$. We define a linear order on $e_1, \ldots, e_k$ (more precisely, on the set of lines passing through them) by: $e_i \le e_j \iff H_i \supseteq H_j$. By geometric considerations it can be shown:

Claim: If P_1, P_2, P_3 are translates of a convex polygon realizing K_3, then not all sequences e_1, e_2, e_3 of translates of one edge can be monotone in the above order.

For example in Fig. 2 e_1, e_2, e_3 is monotone increasing, d_1, d_2, d_3 monotone decreasing, but c_1, c_2, c_3 is not monotone.

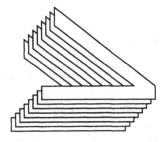

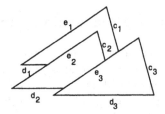

Fig. 1. Realization of an arbitrary K_n with a nonconvex polygon

Fig. 2. Triangles realizing K_3.

Now let $f(k) = (k-1)^2 + 1$ for $k \in \mathbb{N}$ and let for $n \in \mathbb{N}$ $N := f^n(3)$ (n-fold iteration of f; actually $N = 2^{2^n} + 1$). Using an argument from [BEF+] we will show that K_N cannot be realized by any convex n-gon. Suppose otherwise and let $e^1, \ldots, e^n$ be the edges and $P_1, \ldots, P_N$ the translates of the n-gon. Since $N = (f^{n-1}(3) - 1)^2 + 1$ by the Theorem of Erdös-Szekeres [ES] the sequence $e_1^1, \ldots, e_N^1$ of translates of edge e^1 has a monotone subsequence of length $f^{n-1}(3)$. Considering the corresponding subsequence of polygons it must have a subsequence of length $f^{n-2}(3)$ where both the e^1- and e^2-sequences are monotone. Iterating this process we would obtain a subsequence of length $f^0(3) := 3$ where all edge-sequences are monotone in contradiction to the claim above. c) follows

from the fact that any K_n can be realized by disks and any disk can be approximated to arbitrary precision by convex polygons. d) Consider a nondegenerate segment of $F's$ boundary that is strictly convex. We can select a suitable subsegment σ with the following property: If l is the straight line through $\sigma's$ endpoints then no line perpendicular to l intersects σ in more than one point. Assume also w.l.o.g. that l is horizontal, so σ looks as in Fig. 3.

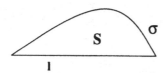

Fig. 3. Curve segment σ

Let S be the convex figure bounded by σ and the line segment between its endpoints. We will show by an inductive construction:

Claim: For any K_n there exists a realization by n translates $S_1, \ldots, S_n$ of S with the following properties:

i) Let $S'_1, \ldots, S'_n$ be the projections of $S_1, \ldots, S_n$ into the xy-plane. There exists a horizontal straight line g such that all the horizontal segments of $S'_1, \ldots, S'_n$ lie strictly below g.

ii) There is a visibility for any pair S_i, S_j, $i \neq j$ strictly above g.

iii) Let s_{ij} be the intersection point of S'_i, S'_j. For $i = 1, \ldots, n-1$ some non-degenerate part c_i of S_i's boundary and some part of its interior are visible from $z = \infty$ in any neighborhood of $s_{i,n}$.

iv) The z-coordinate of S_i is i for $i = 1, \ldots, n$.

The claim is obviously true for $n = 1$.

Suppose now by inductive hypothesis that we positioned $S_1, \ldots, S_n$ satisfying the claim. We choose some point p on the boundary of S_n to the right of all $s_{1,n}, \ldots, s_{n-1,n}$ as intersection point $s_{n+1,n}$ (see Fig. 4). Now we position S_{n+1} in the plane $z = n + 1$ as follows:

First we put it exactly over S_n. Then we move it upwards slightly so that i) is still correct. Then it is moved to the left until it intersects S_n in p (see Fig. 4). The total motion can be made arbitrarily small, in fact, small enough so that iii) is still satisfied. ii) is satisfied by part iii) of the inductive hypothesis since s_{n+1} covers all points $s_{1,n} \ldots s_{n-1,n}$.

2.2 Complete Bipartite Graphs

[BEF+] considers the realization of complete bipartite graphs by unit disks and unit squares. It is shown that $K_{2,3}$ and $K_{3,3}$ can be realized but $K_{j,3}$, $j \geq 4$ cannot. Here we will consider translates of more general convex objects and in particular the realization of stars $K_{1,n}$. In fact, we will show:

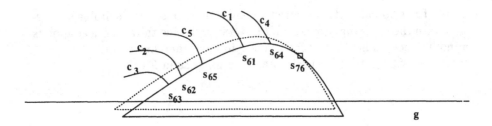

Fig. 4. Construction of S_7.

Theorem 2.2 *a)* $K_{1,5}$ *but no* $K_{1,n}$, $n \geq 6$ *can be realized with parallelograms.*

b) If B is a strictly convex body $K_{1,6}$ but no $K_{1,n}$, $n \geq 7$ can be realized by B.

c) To any figure F there exists an $n \in I\!N$ such that for all $k \in I\!N$, $k \geq n$ $K_{1,k}$ is not realizable by F.

d) To any $K_{n,m}$ there exists a quadrilateral realizing it.

Proof. a) A realization of $K_{1,5}$ by parallelograms is quite straightforward. $K_{1,n}$ $n \geq 6$ is not possible since one parallelogram cannot intersect 5 or more disjoint parallelograms of the same size.b) (Sketch) Here we use some results from convexity theory obtained by Hadwiger [H] and Grünbaum [G]. In fact, they showed that at most 8 translates of a convex body B in two dimensions can touch B without intersecting it or each other. The number 8 is only achieved by parallelograms, otherwise it is 6. For strictly convex bodies we observe that the tangent rays from B separating two neighboring touching translates all point into different directions and their slopes form a strictly monotone sequence (see Figure 5). From this it is possible to conclude that if one of the translates is removed one can distribute the others so that they still touch B but not each other any more. Then they can be pushed slightly inward B and we have a realization of $K_{1,5}$. Placing another copy on the other side exactly over B gives a realization of $K_{1,6}$.

The impossibility of $K_{1,7}$ is derived with similar arguments from the fact that no 6 translates of B can intersect B without intersecting each other (see [G]).c) Consider a realization of $K_{1,n}$ and its projection onto the xy-plane.

Then no point of the plane can be covered by the projections of more than three of the figures. Furthermore the figure representing the center of the star must be intersected by all others, so all projections must lie within in a circle whose diameter is at most three times the diameter of F. These two properties imply that the number of figures is limited.d) The construction is shown in Fig. 6.

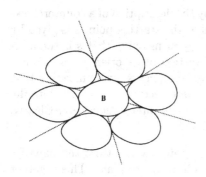

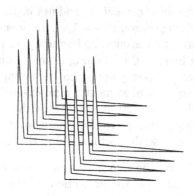

Fig. 5. B touched by 6 of its translates.

Fig. 6. Realization of $K_{4,5}$ by quadrilaterals

3   An upper bound on the number of graphs representable by k-gons

In this section we will show that not each graph has a visibility representation by k-gons for some fixed $k \in I\!N$. In fact, we will even see that in order to represent all graphs with n vertices by polygons, some of those must have more than $\lfloor \frac{\alpha n}{\log n} \rfloor$ vertices for some constant $\alpha > 0$.

Definition 1. A graph is said to be **k-representable** iff there is a visibility representation with (not necessarily convex) simple polygons each having at most k vertices.

The interesting fact that for every k there is a graph which is not k-representable follows from the following theorem.

Theorem 2. There is an $\alpha > 0$ and there are graphs $G_2, G_3, G_4, ..., G_n, ...$ such that G_n has n vertices and is not $\lfloor \frac{\alpha n}{\log n} \rfloor$-representable.

The theorem follows quite easily from the following lemma.

Lemma 3. There is a β such that for all n, k there can be at most $2^{\beta n k \log(nk)}$ many graphs with a fixed vertex set $V = \{v_1, ..., v_n\}$ which are k-representable.

Proof. We consider an arbitrary k-representable graph $G = (V, E)$ with $V = \{v_1, ..., v_n\}$. Obviously, if G is k-representable then there exists a representation by polygons $P_1, ..., P_n$ parallel to the xy-plane with at most k edges each. Without loss of generality we can assume that P_i has z-coordinate i for $i = 1, ..., n$.

Consider the projections of all the polygons into the xy-plane. Extend each edge s of each polygon to a line l_s, obtaining a family $\mathcal{L}$ of at most $m := nk$ not necessarily distinct straight lines. Each edge s and, thus, each line l_s can be oriented by the convention that the polygon lies, say, left of s. Now, G can be uniquely identified by the information in the following items.

1. the *arrangement* of the lines in $\mathcal{L}$.
2. Each polygon $P_i, i = 1, ..., n$ is identified by the description of a counterclockwise tour around its boundary. In particular, the starting point s is given by a line $l \in \mathcal{L}$ containing it and a number $n_0 \leq m$ meaning that s is the n_0th intersection point when traversing l in direction of its orientation. Then a sequence of at most k numbers $n_1, ..., n_r \in \{1, ..., m\}$ is given, meaning that the tour starts at s, goes straight on l for n_1 intersections, then turns into the oriented line crossing there, goes straight for n_2 intersections, etc. Clearly, this describes a tour within the arrangement.

Cleary, the information in the above items uniquely identifies the pairwise intersections of the projections of the polygons into the xy-plane. This together with the convention that P_i has z-coordinate equal to i makes it possible to determine all visibilities, and hence G itself.

It remains to count the number of different possibilities for the data in the above items:

1. As is well known (see [A]) the number of different arrangements of m oriented straight lines is at most $2^{\beta_1 m \log m}$ for some constant $\beta_1 > 0$.
2. For each polygon there are m possibilities for the starting line l, and at most m possibilities for each number $n_0, ..., n_r$, $r \leq k$. So the number of possibilities per polygon is bounded by m^{k+2}. Altogether, the number of possibilities is at most $m^{(k+2)n}$, which is at most $2^{\beta_2 m \log m}$ for some constant $\beta_2 > 0$.

Multiplying the upper bounds in 1 and 2 gives the desired total upper bound of $2^{\beta m \log m}$ where $\beta = \beta_1 + \beta_2$.

Since there are exactly $2^{\binom{n}{2}}$ graphs with vertex set V there are at least $2^{\binom{n}{2}}/n!$ (pairwise nonisomorphic) graphs with n vertices which is more than $2^{\delta n^2}$ for some $\delta > 0$. Theorem 2 follows from this lower bound and Lemma 3.

On the other hand, every graph with n vertices is $(2n + 1)$-representable, which will be shown in the next section.

4 The Construction

This section gives a general construction which produces for any graph $G = (V, E)$ a 3-dimensional visibility representation for G. The construction can be carried out in a straight-forward manner by an algorithm that runs in $O(n^2)$ time, where n is the number of vertices of G. Each vertex is represented by a polygon of $O(n)$ sides (the polygons may differ in shape).

If desired, the basic construction can be modified easily and with the same time complexity to produce convex polygonal (or polyhedral) pieces. Furthermore, these pieces can be made to have all vertex angles of at least $\pi/6$. By using the technique of [CDR], it is also possible to implement the algorithm in $O(n^2)$ time with respect to a Turing machine model of computation.

4.1 The Basic Pieces

Let W denote a regular, convex $2n$-gon centered at the origin O, and let $w_1, w_2, \ldots w_{2n}$ denote the locations of its vertices. We use W to define the basic pieces representing the vertices of G. For this purpose, let X denote a regular, convex n-gon with vertices located at the odd-indexed vertices of W. Imagine adding triangular "tabs" to X to obtain W as follows. Call edge w_{2i-1}, w_{2i+1} of X *tab position* i, and for each i from 1 to n, add a triangle whose vertices are $w_{2i-1}, w_{2i}, w_{2i+1}$ to X at tab position i. W is X together with its tabs (see Fig. 7).

The pieces of our construction are obtained from X in a similar way, except that the tabs may vary in size. The construction may attach to tab position i of X a tab T_i with vertices w_{2i-1}, t_i, w_{2i+1}. Vertex t_i is called the *tab vertex* of T_i. In general, T_i lies inside the corresponding tab on W, with vertex t_i lying on the radial line through O and w_{2i}.

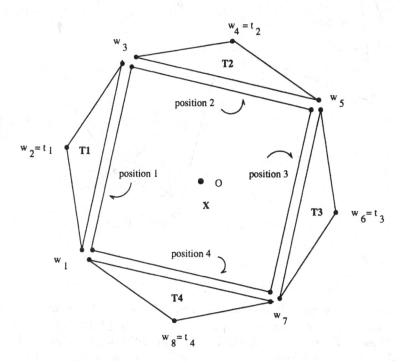

Fig. 7. Regular n-gon X for $n = 4$ tabs.

Definition 4. Let p_{2i} denote the point of intersection of the radial line through O and w_{2i} with the line through w_{2i-1} and w_{2i+1}. The *size* s_i of tab T_i is defined by $s_i = nd(t_i, p_{2i})/d(w_{2i}, p_{2i})$.

A tab of full size n has its tab vertex t_i positioned at w_{2i}.

We depth first search G, assigning to each vertex a number i indicating the order in which the search discovers the vertex. The ith vertex discovered is represented by a polygon P_i consisting of a wedge-shaped portion of X with tabs of various sizes adjoined. See Fig. 8.

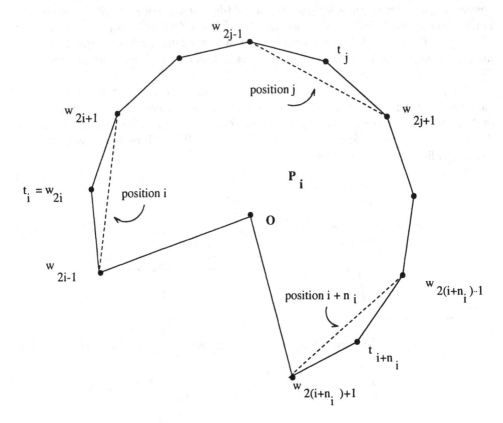

Fig. 8. Piece P_i.

The bounding wedge of P_i is defined by two radial segments emanating from O, one to w_{2i-1} and the other to $w_{2(i+n_i)+1}$, for some $n_i \geq 0$ to be determined. Between these radial segments, X has $1 + n_i$ tab positions. Each piece P_i has a tab of full size n at its lowest indexed tab position, i.e., at position i. Hence P_i has a tab vertex $t_i(P_i) = w_{2i}$. For $i < j \leq i + n_i$, the existence and location of the tab vertex $t_j(P_i)$ of tab $T_j(P_i)$ depends on the size $s_j(P_i)$ assigned to tab $T_j(P_i)$.

The idea behind the construction is as follows. Realize a depth first search tree for G by polygonal pieces floating parallel to the x, y-plane. Arrange these pieces so that the piece $P(v)$ representing a vertex v lies above the pieces representing vertices in the subtree rooted at v, with the x, y-projection of $P(v)$ containing exactly the projections of the pieces $P(w)$ for which w belongs to the subtree rooted at v. Thus each piece has the possibility of seeing its ancestors and descendants, but nothing else.

Unless G itself is a tree, depth first search discovers back edges, i.e., edges of G that do not appear as tree edges in the depth first search tree. A familiar property of depth first search trees for graphs is that each back edge must connect an ancestor, descendant pair in the tree. The purpose of adding tabs of varying sizes is to control which ancestors and descendants see each other.

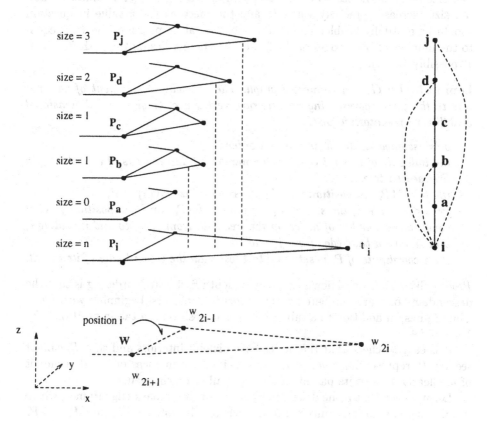

Fig. 9. Back edges from i and their inverted staircase of tabs.

Suppose the depth first search tree has a back edge between i and ancestor j of i. Our construction creates a visibility between the tab T_i of full size n in position i on P_i and a tab in position i on P_j. See Fig. 9.

Of course there may be back edges in the tree joining i to k, where k lies on the path from i to its ancestor j. (Consider $k = b, c, d$ in the figures.) In this case, our construction creates a visibility between the tab in position i on P_k and the full sized tab in position i on P_i. Note that the visibility between the tabs in position i on P_k and P_j must be blocked if the graph G contains no edge between j and k. Hence, for example, the tabs in position i on P_b and P_j must be blocked from seeing each other by intervening tabs.

Blocking inappropriate visibilities between tabs is achieved by creating an inverted staircase of tabs above the tab of full size on P_i and the tab in position i on P_j. The tab on P_i has full size n. The tab in position i on the piece immediately above P_i is assigned size 0, as this piece sees P_i in any case. The tab on the next piece above P_i is also assigned size 0 unless there is a back edge from i to the vertex corresponding to this piece; in this case, the tab size is increased to 1. Tab size remains the same or increases with increasing integer z values. In fact, tab size increases precisely when P_i and the piece at the z value in question should be mutually visible. Thus the size of the tab in position i on P_j is equal to the number of back edges of the form i, k, where k lies on the path from i to j (possibly $k = j$).

Lemma 5. *Let G be a connected graph. The following assignment of parameters to the piece representing an arbitrary vertex v of G gives a 3-dimensional visibility representation for G:*

- *v is assigned its depth first search order i;*
- *the index n_i of v is set equal to the number of descendants of v in the depth first search tree;*
- *the tab $T_i(P_i)$ in position i on P_i is assigned size $s_i(P_i) = n$;*
- *for $i < j \leq i + n_i$ the size $s_j(P_i)$ of the tab $T_j(P_i)$ on P_i at position j is set equal to the number of nodes on the tree path from j, up to and including i, that receive a back edge from j; and*
- *the z coordinate of P_i is set equal to 1 less than the z coordinate of its parent.*

Proof. (Sketch) A well-known property of depth first search ordering is that the descendants of v are numbered with consecutive integers, beginning with $i + 1$. Thus P_i has, in addition to a tab of full size at position i, a tab in position j for $1 < j \leq i + n_i$.

It is easy to check that the pieces have disjoint interiors and that P_i cannot see any P_k representing a vertex w unless w is either an ancestor or a descendant of v. Clearly, P_i sees its parent (if any) and all of its children.

Let us check that if the depth first search tree has a back edge from v, where v is numbered i, to some ancestor u of v, where u is numbered k, then P_i and P_k are mutually visible. P_k has a tab in position i. This tab aligns with the tab of full size in position i on P_i. Furthermore, the tab on P_k has size greater than the intervening tabs in position i, as the number of back edges from i on the path from i to k is at least one greater than the number of back edges on the path from i to k, up to but not including k. Hence P_i and P_k have a line of visibility between their tabs at position i. Thus all back edges are represented.

It can also be checked that no inappropriate visibilities are present.

It is straightforward to design an algorithm that, by computing tab sizes efficiently, carries out the construction in $O(n^2)$ time. Summarizing we obtain

Theorem 6. *Every graph on n vertices is 2n-representable. Furthermore, a representation can be constructed in $O(n^2)$ time.*

Corollary 7. *The construction of Lemma 5 can be modified to produce convex pieces, fat pieces, polyhedral pieces, or pieces having any combination of these properties.*

Proof. To produce convex pieces, use a W with sufficiently many vertices ($12n$) that each piece has a vertex angle at O of at most $\pi/6$. To produce fat pieces, move the vertex at O sufficiently close to the chord through the first and last vertices of P_i shared with W. To produce polyhedral pieces, take the cross product of P_i with a short line segment parallel to the z axis.

References

[A]   N. Alon, "The number of polytopes, configurations and real matroids", *Mathematika 33* (1986), pp. 62–71

[BEF+] P. Bose, H. Everett, S. Fekete, A. Lubiw, H. Meijer, K. Romanik, T. Shermer and S. Whitesides, "On a visibility representation for graphs in three dimensions," *in* Snapshots of Computational and Discrete Geometry, v. 3, *eds.* D. Avis and P. Bose, McGill University School of Computer Science Technical Report SOCS-94.50, July 1994, pp. 2 - 25

[CDR] J. Canny, B. Donald and E. K. Ressler, "A rational rotation method for robust geometric algorithms", Proc. 8th ACM Symp. Comput. Geom., 1992, 251 - 260.

[DETT] G. Di Battista, P. Eades, R. Tamassia and I. Tollis, "Algorithms for drawing graphs: an annotated bibliography", *Computational Geometry Theory and Applications*, v. 4, 1994, 235 - 282. Also available from wilma.cs.brown.edu by ftp.

[DH]   A. Dean and J. Hutchison, "Rectangle-visibility representations of bipartite graphs," *in* Proc. Graph Drawing '94, Princeton, NJ, 1994. Lecture Notes in Computer Science LNCS v. 894, Springer-Verlag, 1995.

[ES]   P. Erdös, Gy. Szekeres, "A Combinatorial Problem in Geometry", *Compositio Math. 2*, 1935, 463-470.

[F]   S. Felsner, personal communication, 1995.

[FHW] S. Fekete, M. Houle and S. Whitesides, "New results on a visibility representation of graphs in 3D," *in* Proc. Symposium on Graph Drawing (GD '95), Passau, Germany, 1995, Springer-Verlag LNCS series (these proceedings).

[G]   B. Grünbaum, "On a Conjecture of H. Hadwiger", *Pacific J. Math.*, 11, 215-219.

[H]   H. Hadwiger, "Über Treffanzahlen bei translationsgleichen Eikörpern", *Arch. Math.*, Vol. VIII, 1957, 212-213.

[KKU] E. Kranakis, D. Krizanc and J. Urrutia, "On the number of directions in visibility representations of graphs," *in* Proc. Graph Drawing '94, Princeton, NJ, 1994. Lecture Notes in Computer Science LNCS v. 894, Springer-Verlag, 1995, 167 - 176.

[Rom] K. Romanik, "Directed VR-representable graphs have unbounded dimension," *in* Proc. Graph Drawing '94, Princeton, NJ, 1994. Lecture Notes in Computer Science LNCS v. 894, Springer-Verlag, 1995, 177 - 181.

[TT]   R. Tamassia and I. Tollis, "A unified approach to visibility representations of planar graphs," *Discrete Comput. Geom.* v. 1, 1986, 321 - 341.

[W]   S. Wismath, "Characterizing bar line-of-sight graphs," *in* Proc. ACM Symp. on Computational Geometry, 1985, 147 - 152.

KGB
A Customizable Graph Browser

Hartmut Benz

University of Stuttgart, Department of Computer Science, Institute of Parallel and
Distributed High Performance Systems, Breitwiesenstr. 20-22, D-70565 Stuttgart,
email: Hartmut.Benz@informatik.uni-stuttgart.de

Overview

This paper presents the architecture of a generic, customizable graph browser **KGB**[1].
The **KGB** has been designed to handle *very large* and *dynamically changing graphs*
which are frequently used as repository management graphs in large applications. Special emphasis has been put on the *flexible presentation* of the information encoded in
the graph and the *reduction of the user's workload* when adapting the presentation to
his or her special needs. The **KGB** was built as a debugging and visualization tool for
document hierarchies of the CASE tool **KOGGE**[2] currently under development at the
"Institute for Software Engineering", University of Koblenz [3].

To automatically structure very large graphs and specifically select the subset of
visible graph elements, the **KGB** implements three methods of abstraction. To achieve
flexibility in the presentation, seperate visualization techniques can be applied to the
attributes of each vertex and edge.

A *View into a graph* is defined by a set of applicable abstractions, presentation description and user interaction descriptions (e.g. the mapping from user generated input
events to graph browser operations). A View is described as a simple rule based language. To handle dynamically changing graphs the View Description is interpreted by
the **KGB** (rather than compiled into it).

Architecture

The architecture of the **KGB** is object oriented. Its primary constituents are a *source
graph interface* which connects the **KGB** with the user's graph implementation [4, 6].
The *abstraction component* selects the elements of the graph to be visualized at a time
and which parts of the graph can be grouped to increase readability. The *layout component* determines drawing positions as well as graphical representation for each visible graph element [5]. Finally, the *user interaction component* deals with the adequate graphical representation on the screen and the interpretation of user actions in
the course of the dialog.

---

[1] German: *Konfektionierbarer Graphen-Browser*. The paper is a result of my diploma thesis at
the University of Koblenz, Germany (cf. [1]).

[2] German: *Koblenzer Generator für Graphische Entwurfsumgebungen*, Koblenz generator for
graphical software engineering systems

Abstraction

Common visualization techniques like windowing, zooming, fish-eye view or graph folding are not suitable to render graphs larger than a few thousand vertices and edges [7, 8, 9, 11]. Beyond this threshold either the required drawing area gets too large or the size of the graph elements approaches the size of a pixel. In any case the amount of the displayed information easily overloads human ability to successfully perceive and utilize it, so that other techniques have to be used.

Abstraction describes a method that hides irrelevant detail and by that enhances larger interrelations. It is therefore an adequate method to filter and structure large amounts of information. The **KGB** currently offers three variations of abstraction which are shortly described in the next three paragraphs.

Abstraction by Global Exclusion. This method permits to selectively exclude certain graph elements from browsing. It is based on the idea that a large and complex graph contains several semantic structures. By globally excluding those graph elements the user is currently not interested in, only the important structures are visualized. Global exclusion rules are defined intaractively as follows:

> *Graph elements of class C are hidden.*

Abstraction by Local Restriction. This method permits to exclude graph elements that are irrelevant to the user's current focus of attention. The method is based on the idea of locality, the observation that semantically closely related graph elements are commonly "only a few incidences apart". Nevertheless, this abstraction method puts a high cognitive load on the user to keep a mental map. Furthermore, a positionally stable layout is necessary since moving the focus of attention introduces problems similar to those of interactive graph editing. Local restriction rules are defined interactively as follows:

> *Graph elements of ...*
> *class C have distance 1.*
> *class $\mathcal{E}$ have distance 0.7.*
> *Maximum distance is 3;*

Abstraction by Subgraph Abstraction. This method permits to select (semantically connected) subgraphs and represent them each by a single abstraction element. The element retains the information: "there is a subgraph". Subgraph abstraction is a common feature in graph visualization and editing tools. Alas, the usual interactive selection with a mouse is inefficient for very large graphs and completely useless for frequently changing graphs.

This problem can be solved by automatic subgraph selection mechanisms. The **KGB** currently provides two rule-driven selection engines. Hard-coded algorithms, although more efficient, have been postponed due to their inflexibility. They can, of course, be used to solve specialized selection tasks. Subgraph abstraction rules are defined intaractively as follows:

> *Abstraction "Dir" from a Directory_Vertice with all inbound adjacencies;*

The graph browser currently uses two subgraph description languages which will not be described any further in this paper. Both languages are computable in polynomial time (ref. [1, 2]).

Subgraph abstraction obviously introduces a *hierarchical structure* into the graph. As opposed to the other methods, subgraph abstraction can be applied to a graph more than once. This has three important consequences:

- Subgraph abstraction has to be defined on hierarchical graphs.
- Repeated application can lead to arbitrarily deep abstractions. The abstraction hierarchy has the structure of a tree.
- The resulting tree depends on the sequence of abstractions.

Information visualization

Generally, a graph contains global and local information. *Global* or structural information consists of the graph elements (their existence) and their incidence structure. The visualization of this information is performed by placing the graph elements and is widely researched as *graph layout*.

Local information consists of the attributes that graph elements are quite regularly annotated with. The visualization of this information is often crucial to the user's ability to understand the semantics of the graph at all.

The **KGB** acknowledges this importance. The user can freely and interactively describe which (subset of the) attributes of a graph element are to be visualized and which graphical representation they should receive. The **KGB** supports a great variety of graphical representations to choose from. For example:

> *edges Class_A: use Name as text label;*
> *edges: label color from Int2Col(Weight);*

View into a Graph

A *View into a graph* consists of a set of rules defining possible abstractions and visual representations of attributes. In addition to these definitions, there are rules to choose one (of many) layout algorithms and to associate user actions to graph browser operations [10]. For example:

> *Layout Sugiyama("2,2,1");*
> *double-click-right calls ShowDetailedInformation;*
> *<RETURN> on Abstraction_vertex calls OpenAbstraction;*

Views can be named and the user can easily create, modify or switch between them while using the **KGB**.

Genericity

The amount of source code required to adapt the tool to a new task has been minimized. The complete adaptation (including source code fragments) is described using a simple abstract language. From this description a *graph browser compiler* automatically generates a dedicated **KGB**. Thus, the **KGB** is a generic tool.

Future Work

Primary objective of the future development is the portation from NeXTSTEP to X-Windows. Future development will concentrate on improving the user interface and extending the abstraction methods. A user interface supporting more direct manipulative techniques is highly desirable. Additional abstraction methods like *fish-eye view* or *concentrator vertices* [11] should be included. Furthermore, the subgraph selection methods used in subgraph abstraction should be combined with the interactive graph layout proposed in [7]. In addition, different approaches to visualize the hierarchies introduced by the subgraph abstraction should be explored [9].

References

1. Benz, H.; *"KGB – Ein konfektionierbarer Graphen-Browser"*; Diplomarbeit an der Universität Koblenz-Landau, Abt. Koblenz, FB Informatik, 1993
2. Capellmann, C.; Franzke, A.; *"GRAL—Eine Sprache für die graphbasierte Modellbildung"*; Diplomarbeit an der Universität Koblenz, FB Informatik, 1991
3. Carstensen, M.; Meissner, A.; Rhein, U.; *"Forschungsschwerpunkt CASE. Dritter Zwischenbericht."*; Universität Koblenz, 1991
4. Dahm, P.; Ebert, J.; Litauer, C.; *"Benutzerhandbuch EMS-Graphenlabor V3.0"*; Koblenz 1994; (Unpublished manuscript) ftp://ftphost.uni-koblenz.de/outgoing/GraLab/
5. Di Battista, G.; Eades, P.; Tamassia, R.; *"Algorithms for Drawing Graphs: An Annotated Bibliography"* ftp://wilma.cs.brown.edu /pub/papers/compgeo/gdbiblio.*
6. Ebert, J.; *"A Versatile Data Structure For Edge-Oriented Graph Algorithms"*; Communications of the ACM (6) 1987
7. Henry, T. R.; Hudson, S. E.; *"Interactive Graph Layout"*; Proceedings of the ACM SIGGRAPH Symposium on User Interface Software, 1991
8. Himsolt M.; *"GraphEd: An Interactive Graph Editor"*; Proc. STACS 89, Lecture Notes in Computer Science, vol. 349, pp. 532-533, Springer-Verlag, 1989. ftp://ftp.forwiss.uni-passau.de/pub/local/graphed
9. Johnson, B.; *"TreeViz: Treemap Visualization of Hierarchically Structured Graphs"*; in [12], pp. 369–370
10. Paff, G. (Editor); *"User Interface Management Systems"*; Proc. of the Workshop on User Interface Management Systems held in Seeheim, FRG, Nov. 1-3, 1983, Springer 1985
11. Paulisch, F. N.; Tichy, W. F.; *"EDGE: An Extendible Graph Editor"*; in Software–Practice and experience, Vol. 20(S1), June 1990; John Wiley & Sons, Ltd.
12. *"Proc. of CHI 1992 (Monterey, California, May 3-7, 1992)"*; ACM, New York, 1992

Adocs: a Drawing System for Generic Combinatorial Structures

François Bertault

INRIA Lorraine, 615 rue du Jardin Botanique, BP 101,
F-54600 Villers-les-Nancy, France

Abstract. Existing graph drawing systems imply the use of specific algorithms for each kind of data structures. This paper provide a description of the Adocs program. The program is based on a generic description of the structures, and thus allows to draw structures of an infinite number of classes. It can be used in order to produce graphical output for Gaïa, an uniform random generator of combinatorial structures. It could be used with other programs and is well suited for drawing compound objects.

1   Introduction

The theory of decomposable structures [1], used in order to describe combinatorial objects, allows to describe large classes of data structures. With only a small set of constructors, we can describe structures frequently used in software, like permutations, trees, partitions, or functionals graphs. The Adocs [1] program produces a graphical representation for such objects.

2   Description of Combinatorial Structures

Definition 1. The objects classes we can deal with are specified by a set of productions of the form `A = <rhs>`, where `A` is the name of the class being defined, and `<rhs>` is an expression involving elementary classes, constructors and other classes specifications.

Elementary classes are :
- Epsilon : object of size 0
- Atom : object of size 1

Constructors available are :
- Union(A,B,...) : disjoint union of the classes A,B,...
- Prod(A,B, ...) : product of the classes A, B,...
- Set(A) : sets (with repetitions) whose elements are in A
- Sequence(A) : sequences of elements of A
- Cycle(A) : directed cycles of elements of A
- Subst(A,B) : B-objects whose atoms are replaced by A-objects

---

[1] available via anonymous ftp from ftp.loria.fr, in pub/loria/eureca/ADOCS/

Example 1. The class N of functional digraphs can be defined as a set of cycles of binary trees. A binary tree (class **Tree**) can be defined as a leaf or the product of a node with two trees :

```
{N=Set(Cycle(Tree)), Tree=Union(Leaf,Prod(Node,Tree,Tree)),
Node=Atom, Leaf=Atom}
```

Definition 2. Inputs of **Adocs** are objects described with the following initial objects and lowercase operators :

Initial objects are :
- **Identifiers** : a word with alpha-numerical characters, or a set of characters between double quotes.

Operators are :
- **prod(a,b,...)** : product of the objects a, b,...
- **set(a,b,..)** : set of objects a, b,...
- **sequence(a,b,...)** : sequence of objects a,b,...
- **cycle(a,b,...)** : directed cycle of objects a,b,...

Example 2. set(cycle(prod(A,a1,a2)),cycle(prod(B,"f(3)","*%$"))) is a valid functional digraph object. Figure 1 shows an object described using all the operators.

3 Representation

The default settings for each operator representation are :

- **prod(o1,o2,..,om)** draws a rooted tree, where o1 is the father of the objects o2 ... om, with non-oriented edges between father and sons.
- **sequence(o1,o2,..,om)** places o1 .. om on a straight line, starting with o1 on the left, with oriented edges between adjacent elements.
- **cycle(o1,o2,..,om)** places o1 .. om along a circle, with oriented edges.
- **set(o1,o2,..,om)** places o1 .. om along a spiral, with no edges.

We can add informations to each operator in order to constraint the representation. Thus, we can have different representations of a same structure (Fig. 2). The instructions are as follows, where <operator> is either **prod, sequence, cycle** or **set** :

- **<operator>[Tree](o1,o2,..om)** to represent a rooted tree (default for **prod**).
- **<operator>[Radial](o1,o2,..om)** to represent a free-tree.
- **<operator>[Line](o1, o2, ... om)** to place elements o1 .. om on an horizontal line (default for **sequence**).
- **<operator>[Circle](o1, o2, ... om)** to place elements o1 .. om on a circle (default for **cycle**).
- **<operator>[Spiral](o1, o2, ... om)** to place elements o1 .. om on a spiral (default for **set**).

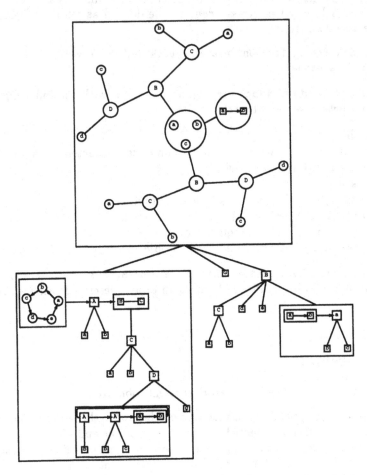

Fig. 1. Rooted tree with free-tree, sequence of trees, cycle and set

4 Implementation

Different layout algorithms are used in the **Adocs** program. The algorithm for drawing rooted trees is based on the Reingold and Tilford algorithm [3]. The algorithm has been extended to m-ary trees with nodes of different sizes. For drawing free-trees, we use an algorithm proposed by P. Eades [2]. The spiral algorithm is heuristic and tries to place a set of circles in the smallest possible enclosing circle. All algorithms used need improvement and other algorithms could be included, but **Adocs** can already draw quite large structures (more than 10000 nodes) in a few seconds.

The **Adocs** interface is very simple. The description of the object we want to draw is written into a file (for example **descr**). The Unix command **Adocs descr** produces the **descr.ps** Postscript file containing the drawing of the structure. Other outputs can also be provided, like LaTeX pictures or a specific simple graphical description (involve **AdocsTex** or **AdocsGr** instead of **Adocs**). **Adocs**

```
prod(prod[Radial](C,H,H,H),        prod[Line](prod(C,H,H,H),
    cycle(a,b,c,d,e),                  cycle[Line](a,b,c,d,e),
    sequence(A,                        sequence[Tree](A,
        prod(B,a,b),                       prod(B,a,b),
        C))                                C))
```

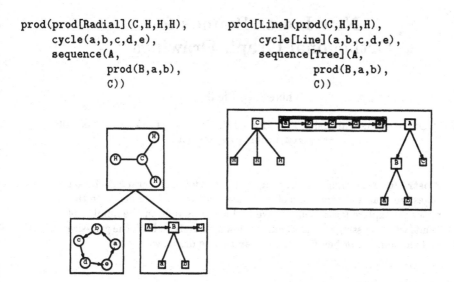

Fig. 2. Two different drawing of a same structure

can then easily be connected to other programs, like structure generators or other drawing programs.

For example Adocs can directly draw the structures calculated by the random structure generator Gaïa [2] [4]. Given a structure class specification and a size of the object we want to obtain, Gaïa generates with uniform probability an object of the class, that we can represent with Adocs.

5 Conclusion

Adocs could also be used with other programs, like data structures browsers. In particular, the generic description proposed for data structures is well suited for compound drawing of data structures.

References

1. Philippe Flajolet, Paul Zimmermann, and Bernard Van Cutsem. A calculus for the random generation of labelled combinatorial structures. *Theoretical Computer Science*, 132(1-2):1–35, 1994.
2. Eades Peter D. Drawing free trees. *Bulletin of the Institute for Combinatorics and its Applications*, 5(2):10–36, 1992.
3. Edward M. Reingold and John S. Tilford. Tidier drawings of trees. *IEEE Transactions on Software Engineering*, SE-7(2):223–228, March 1981.
4. Paul Zimmermann. Gaïa: a package for the random generation of combinatorial structures. *MapleTech*, 1(1):38–46, 1994.

[2] available via anonymous ftp from ftp.inria.fr, in lang/maple/INRIA/combstruct

New Lower Bounds for
Orthogonal Graph Drawings *

Therese C. Biedl

RUTCOR, Rutgers University, P.O. Box 5062, New Brunswick, NJ 08903-5062,
therese@rutcor.rutgers.edu

Abstract. An orthogonal drawing is an embedding of a graph such that edges are drawn as sequences of horizontal and vertical segments. In this paper we explore lower bounds. We find lower bounds on the number of bends when crossings are allowed, and lower bounds on both the grid-size and the number of bends for planar and plane drawings.

1 Introduction

Orthogonal graph drawings are an important tool for graph layout, e.g. for Data Flow Diagrams or Entity Relationships Diagrams. Two important measurements of the quality of a drawing are the grid-size and the number of bends. Every 4-graph has an orthogonal drawing of grid-size $\mathcal{O}(n) \times \mathcal{O}(n)$ with $\mathcal{O}(n)$ bends. Minimizing the number of bends is $\mathcal{NP}$-complete [6], and so is the question whether a graph can be embedded in a grid of prescribed size [8, 5]. Therefore, one tries to find heuristics where the obtained worst-case sizes are a priori known and small. Different algorithms have been developed, depending on the connectivity and whether the graph is planar or not. See Table 1 for an overview.

	Triconnected		Biconnected		Connected	
	Grid-size	Bends	Grid-size	Bends	Grid-size	Bends
	Nonplanar					
Simple	hp $\frac{7}{4}n - 2$ *	$2n + 2$ [1]	hp $\frac{7}{4}n - 2$ *	$2n + 2$ [1]	n [1]	$2n + 2$ [1]
Multigraph	$n + 1$ [1]	$2n + 4$ [1]	$n + 1$ [1]	$2n + 4$ [1]	$\frac{4}{3}n - 1$ [4]	$\frac{8}{3}n + 2$ [4]
With Loops		−		$2n - 1$ [4]	$2n - 1$ [4]	$4n$ [4]	$4n$ [4]
	Plane					
Simple	hp $\frac{4}{3}n + 2$ [4]	$\frac{4}{3}n + 4$ [4]	n [11]	$2n + 2$ [2]	$\frac{6}{5}n + 1$ [4]	$\frac{12}{5}n + 2$ [4]
Multigraph	$n + 1$ [13]	$2n + 4$ [14]	$n + 1$ [13]	$2n + 4$ [14]	$2n - 1$ [4]	$4n - 2$ [4]
With Loops		−		$2n + 1$ [4]	$4n + 4$ [4]	$2n + 1$ [4]	$4n + 4$ [4]

Table 1. Known algorithms. "hp" means that this is a bound on the half-perimeter. We give the (to our knowledge) first citation of each result. The results marked * are by Papakostas and Tollis (private communication).

---

* A full version of this paper can be found in [3]. This paper was written while the author was visiting TU Berlin and working at Tom Sawyer Software.

To measure the goodness of these algorithms, we want to find graphs which need at least a certain grid-size or at least a certain amount of bends. In this paper we deal with these lower bounds. We summarize our results in Table 2.

Nonplanar Drawings		Triconnected	Biconnected	Connected	Non-Connected
Non-planar	Simple	$\frac{10}{7}n$	$\frac{10}{6}n$	$\frac{11}{6}n$	$\frac{12}{5}n$
Planar	Simple	$\frac{6}{5}n$	$\frac{10}{7}n$	$\frac{11}{7}n$	$2n$
	Multigraph	$\frac{10}{7}n$	$2n$	$\frac{7}{3}n$	$4n$
	With Loops	–	$3n$	$3n$	$6n$

Plane Drawings		Triconnected	Biconnected	Connected	Non-Connected
Simple	Grid-size	$\frac{2}{3}n+1$	$n-1$	$\frac{6}{5}(n-1)-1$	?
	Bends	$\frac{4}{3}n+4$	$2n-2$ *	$\frac{12}{5}(n-1)-2$	?
Multigraph	Grid-size	$\frac{2}{3}(n-2)+3$	$n+1$	$2n-3$	$2n-1$
	Bends	$\frac{4}{3}(n-2)+8$	$2n+4$ *	$4n-6$	$4n$
With Loops	Grid-size	–	$n+2$	$2n+1$	$4n-1$
	Bends	–	$3n$	$4n+4$	$6n$

Table 2. Lower bounds for orthogonal drawings. "–" means that this case is impossible. "?" means that we didn't find lower bounds better than for the connected case. The results marked * were already discovered by [15].

2 Definitions

Let G be a graph with n vertices. We always assume that G is a *4-graph*, i.e. it has a maximum degree of 4. G is called *4-regular* if every vertex has degree 4. By *subdivision* of an edge e we understand that we delete e, add a new vertex, and connect it with the two endpoints of e. Edges of the form (v,v) (*loops*) are not necessarily forbidden. Also, two vertices may be connected by more than one edge (*multiple edge*). Graphs without loops and multiple edges are called *simple*, graphs without loops, but possibly with multiple edges, are called *multigraphs*.

G is called *connected* if for any two vertices there is a path between them. It is called *biconnected* if for any vertex v the graph $G - \{v\}$ is connected. It is called *triconnected* if for any two vertices v, w the graph $G - \{v, w\}$ is connected. A triconnected 4-graph with more than three vertices can never have a loop.

A graph is called *planar* if it has a drawing without crossing (*planar drawing*). This defines a circular ordering of the edges incident to a vertex v (*combinatorial embedding*). A planar drawing splits the plane into different components, called *faces*. The unbounded component is called the *outerface*. The combinatorial embedding defines a planar drawing which is unique except for the choice of the outerface. A planar graph is called *plane* if both a combinatorial embedding and the outerface are specified.

An *(orthogonal) drawing* of G is an embedding of G in the plane such that all edges are drawn as sequences of horizontal and vertical line segments. It is called *planar* if no drawings of edges intersect. It is called *plane* if it is planar, if G was a plane graph, and the drawing exactly reflects the given embedding and the outerface. A point where the drawing of an edge changes its direction is called a *bend* of this edge.

A column (row) of the drawing is called *vertex-used* if it contains a vertex, *line-used* if it contains a vertical (horizontal) part of an edge, and *used* if it is vertex-used or line-used. The *width* of the drawing is the number of used columns minus 1, the height is the number of used rows minus 1. A drawing with width n_1 and height n_2 has *grid-size* $n_1 \times n_2$, *half-perimeter* $n_1 + n_2$, and *area* $n_1 \cdot n_2$.

3 Lower Bounds for Non-Planar Drawings

Very little is known about lower bounds for the grid-size. There exist 4-graphs with crossing number $\Omega(n^2)$ which therefore need a grid of the same area [16]. Leighton [9] proved that the planar tree of meshes needs $\Omega(n \log n)$ area in any orthogonal drawing. In both cases the constants are very small.

We deal here with the number of bends, and develop various lower bounds, depending on the connectivity of the graph. The only known results are for simple biconnected graphs: Storer [11] showed a lower bound of $\frac{8}{7}n$ bends, and Papakostas and Tollis improved it to $\frac{8}{5}n$ (private communication).

3.1 Lower Bounds for Small Graphs

We use the following special graphs: The *complete graph* K_5, the *octahedron* O, the *quadruple edge graph* Q, and the *double loop* L, which are shown below. We develop a few easy lemmas to get lower bounds for these graphs.

Fig. 1. K_5, the octahedron, the quadruple edge, and the double-loop.

Lemma 1. *In a drawing of a 4-regular graph, every used column has two bends. Formulated differently: if a drawing of a 4-regular graph has width w, then there are at least $2w + 2$ bends.*

Proof. If a column is vertex-used, let u be the top vertex and w be the bottom vertex in the column. By 4-regularity all connections at u and w are used. So the top connection from u must have a bend, and so must the bottom connection of w. If a column is line-used only, every begin- and endpoint of a line is a bend. Hence we have at least two bends.

In a drawing of width w there are $w + 1$ used columns (by definition of width), and therefore at least $2w + 2$ bends by the above. $\qquad\square$

Lemma 2. *A 4-regular graph has $2\lceil\sqrt{n}\rceil + 4$ bends in any drawing.*

Proof. We only sketch this proof. It is clear that we need at least either $\lceil\sqrt{n}\rceil$ columns or $\lceil\sqrt{n}\rceil$ rows to accomodate the vertices. Since we have a 4-regular graph, we need two more rows and two more columns at the extreme ends. This proves the claim together with Lemma 1. □

Lemma 3. *A simple 4-regular graph has at least 12 bends in any drawing.*

Proof. The proof is an easy (but lengthy) case analysis to show that in any embedding we must have at least 6 rows or 6 columns. We then have at least 12 bends by Lemma 1 (or analogously for rows). We skip this for brevity. □

These two lemmas imply that L needs 6 bends, Q needs 8 bends, and K_5 and O need 12 bends in any drawing.

3.2 Constructing bigger graphs

The lower bounds for non-connected graphs are easy to get by taking many copies of O, K_5, Q, and L, respectively. To obtain bigger connected graphs, we need to study how subdividing an edge changes the lower bound.

Lemma 4. *Subdividing an edge lowers the lower bound on the bends by at most 1.*

Proof. Assume G needs b bends, and after subdividing one edge, we get G_s. Let G_s be drawn with c bends. If we remove the vertex that came from subdividing, this adds at most one bend, so we get a drawing of G with $c + 1$ bends. Consequently $c \geq b - 1$. □

Theorem 5. *There are the following lower bounds for connected graphs:*
1. *planar simple graphs: $\frac{11}{7}n$ bends*
2. *simple graphs: $\frac{11}{6}n$ bends*
3. *planar multigraphs: $\frac{7}{3}n$ bends*
4. *planar graphs with loops: $3n$ bends*

Proof. We demonstrate only case (1) in detail, the other cases are analogous. Take an octahedron and subdivide one edge. We get a graph with 7 vertices that by Lemma 4 needs 11 bends. Take k copies and connect the vertices of degree 2. This graph is connected, has $7k$ vertices and needs $11k = \frac{11}{7}n$ bends. □

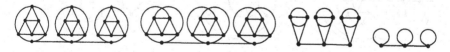

Fig. 2. Connected graphs that need many bends in any drawing.

Theorem 6. *There are the following lower bounds for biconnected graphs:*

1. *planar simple graphs: $\frac{10}{7}n$ bends*
2. *simple graphs: $\frac{10}{6}n$ bends*
3. *planar multigraphs: $\frac{6}{3}n = 2n$ bends*
4. *planar graphs with loops: $3n$ bends*

Proof. Again we demonstrate only case (1). Take an octahedron and subdivide two edges on the outerface. We get a graph with 8 vertices that by Lemma 4 needs 10 bends. Take k copies and identify the vertices of degree 2. This graph is biconnected, has $7k$ vertices and needs $10k = \frac{10}{7}n$ bends. $\qquad\square$

Fig. 3. Biconnected graphs that need many bends in any drawing.

Theorem 7. *There are the following lower bounds triconnected graphs:*

1. *planar simple graphs: $\frac{6}{5}n$ bends.*
2. *simple graphs: $\frac{18}{13}n$ bends.*
3. *planar multigraphs: $\frac{18}{13}$ bends.*

Proof. Again we demonstrate only case (1). Take an octahedron and subdivide the three edges on the outerface. We get a graph with 9 vertices that by Lemma 4 needs 9 bends. Take $2k$ copies and identify the vertices of degree 2 in such a way that the resulting graph is planar and triconnected (see also Fig. 4). The graph then has $(6 + \frac{3}{2})2k = 15k$ vertices and needs $9 \cdot 2k = \frac{18}{15}n = \frac{6}{5}n$ bends. $\qquad\square$

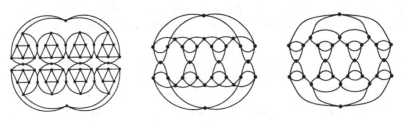

Fig. 4. Triconnected graphs that need many bends in any drawing.

4 Lower Bounds for Planar Drawings

4.1 Plane graphs

Assume after taking away all edges on the outerface of a plane graph H we get the graph G. Then we say that H *contains G on the inside*.

Lemma 8. *Let H contain G on the inside, where G has minimum degree 2. If H can be drawn in a $w \times h$-grid then G can be drawn in a $(w-2) \times (h-2)$-grid.*

Proof. Consider a drawing Γ_H of H in a $w \times h$-grid. Deleting the edges on the outerface of H we get a drawing Γ_G of G. Since G has minimum degree 2, the highest row of Γ_G must be used by a horizontal line. This line belongs to an edge e on the outerface of G. Since G is on the inside of H, e is not on the outerface of H. Γ_H reflects the embedding of H, so we must have a line drawn above e in Γ_H. Therefore Γ_H has at least one more unit in top-direction. The same holds for the other three directions. $\square$

Triconnected graphs Kant showed lower bounds of $\frac{2}{3}(n-2)+2$ on the grid-size and $\frac{4}{3}(n-2)+2$ bends for triconnected plane graphs. We improve this slightly with the following graph class:

Definition 9. Define the graph classes $\{T_i'\}$ and $\{T_i\}$ as follows:

- T_1' is a 3-cycle.
- T_i' is obtained by taking a copy of T_{i-1}', and adding three vertices in the outerface. Then we add a 6-cycle between the three new vertices and the three vertices of degree 2 of T_{i-1}', such that T_i' contains T_{i-1}' on the inside.
- T_i is obtained by taking a copy of T_i' and adding a 3-cycle between the three vertices of degree 2 of T_i' such that T_i contains T_i' on the inside.

See Fig. 5 for an illustration of this graph.

Lemma 10. *T_i needs a width and height of $2i+1$ in any plane drawing.*

Proof. We first show a lower bound for T_i', namely, it needs a width of $2i - 1$. This is shown by induction on i. Since T_1' is a triangle, it needs a 1×1-grid. Now consider T_i', $i \geq 2$. By construction it contains T_{i-1}' on the inside, so if we could embed T_i' with width less than $2i+1$, then by Lemma 8 we could embed T_{i-1}' in a grid of width less than $2i - 1 = 2(i-1)+1$, a contradiction.

Now finally consider T_i. By construction it contains T_i' on the inside, so it needs two more units in width than T_i', which gives a lower bound of $2i+1$ on the width. The proof for the height is similar. $\square$

Lemma 11. *T_i needs $4i+4$ bends in any plane drawing.*

Proof. This is trivial: T_i is 4-regular and has a width of $2i+1$ in any drawing. By Lemma 1 it therefore must have $2(2i+1)+2 = 4i+4$ bends in any drawing. $\square$

Theorem 12. *There are the following lower bounds for plane triconnected graphs:*

- *simple graphs:* $(\frac{2}{3}n + 1) \times (\frac{2}{3}n + 1)$-*grid and* $\frac{4}{3}n + 4$ *bends.*
- *multigraphs:* $(\frac{2}{3}(n - 2) + 3) \times (\frac{2}{3}(n - 2) + 3)$-*grid and* $\frac{4}{3}(n - 2) + 8$ *bends.*

Proof. We are done in the simple case, since T_i has $n = 3i$ vertices. Obtain graph $\hat{T}_i$ as follows: subdivide two different edges on the outerface of T_i, and add a double edge between the two new vertices, such that the double edge encloses T_i. Some calculation shows that $\hat{T}_i$ has $3i + 2$ nodes, needs a width and height of $2i + 3$, and $4i + 8$ bends. This proves the claim for multigraphs. □

Biconnected graphs Storer showed a lower bound of $n - 2$ on the grid-size of biconnected simple graphs [11]. Tamassia, Tollis, and Vitter showed a lower bound of $2n + 4$ bends for multigraphs and $2n - 2$ bends for simple graphs [15]. We use their graph class to show a slightly better bound on the grid-size.

Definition 13. Define the graph classes $\{B_i'\}$ and $\{B_i\}$ as follows:

- B_1' is a double edge.
- B_i' is obtained by taking a copy of B_{i-1}', adding two vertices in the outerface, and adding a 4-cycle alternating between the two new vertices and the two vertices of degree 2 of B_{i-1}', such that B_i' contains B_{i-1}' on the inside.
- B_i is obtained by taking a copy of B_i' and adding a double edge between the two vertices of degree 2 of B_i' such that B_i contains B_i' on the inside.

The following lemma is proved exactly as in Lemma 10 and Lemma 11. The second claim was known before [15], though proved by different means.

Lemma 14. B_i *needs a width and height of* $2i + 1$ *and* $4i + 4$ *bends.*

Fig. 5. T_4 and B_4, both drawn optimally.

Theorem 15. *There are the following lower bounds for plane biconnected graphs:*

- *simple graphs:* $(n - 1) \times (n - 1)$-*grid and* $2n - 2$ *bends*
- *multigraphs:* $(n + 1) \times (n + 1)$-*grid and* $2n + 4$ *bends*
- *graphs with loops:* $(n + 2) \times (n + 2)$-*grid and* $3n$ *bends*

Proof. For multigraphs we are done by Lemma 14 since B_i has $2i$ vertices.

For simple graphs, let $\bar{B}_i$ be the graph obtained from B_i by subdividing one of each of the double edges. Some calculation shows that $\bar{B}_i$ has $2i + 2$ vertices, needs a height and width of $2i + 1$ and $4i + 2$ bends in any drawing.

For graphs with loops, let $\hat{B}_i$ be the graph obtained from B_i by subdividing one edge on the outerface, and adding a loop incident to the new vertex, such that the loop encloses B_i. Some calculation shows that $\hat{B}_i$ has $2i + 1$ vertices, and needs a width and height of $2i + 3$ in any drawing. This proves the lower bound on the grid-size. For the number of bends, the claim was shown in Theorem 6. □

Connected graphs The known lower bounds for connected plane graphs are $\frac{8}{3}(n-2)$ bends for multigraphs, and $4(n-2)$ bends for graphs with loops [15]. We improve these bounds and develop new ones for simple graphs.

Definition 16. Define the graph classes $\{C_i'\}$ and $\{C_i\}$ as follows:
- C_1' is a loop.
- C_i' is obtained by taking a copy of C_{i-1}', adding one vertex in the outerface, and adding a double edge between the new vertex and the vertex of degree 2 of C_{i-1}', such that C_i' contains C_{i-1}' on the inside.
- C_i is obtained by taking a copy of C_i' and adding a loop at the vertex of degree 2 of C_i' such that C_i contains C_i' on the inside.

See Fig. 6 for an illustration of this graph class. The following lemma is proved exactly as in Lemma 10 and Lemma 11.

Lemma 17. C_i *needs a width and height of* $2i + 1$ *and* $4i + 4$ *bends.*

Definition 18. The graph class CM_i is essentially defined as the graph class C_i, with the exception that both loops are replaced by a quadruple edge with one edge subdivided. See also Fig. 6.

The following lemma is proved similar as in Lemma 10 and Lemma 11.

Lemma 19. CM_i *needs a width and height of* $2i + 5$ *and* $4i + 12$ *bends.*

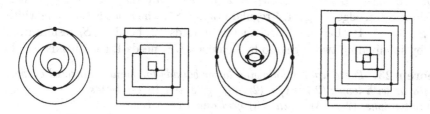

Fig. 6. C_4 and CM_4, both drawn optimally.

Definition 20. Define the graph classes $\{CS_i\}, \{CS_i'\}$ and $\{CS_i''\}$ as follows:
- CS_1'' is a 3-cycle.
- CS_i' ($i \geq 1$) is obtained by taking a copy of CS_i'' and adding two vertices in the outerface. Then we add a 4-cycle alternating between the two new vertices and two vertices of degree 2 of CS_i'', such that CS_i' contains CS_i'' on the inside.
- CS_i ($i \geq 1$) is obtained by taking a copy of CS_i' and adding one vertex in the outerface. Then we add a 3-cycle between this new vertex and the two vertices of degree 2 of CS_i', such that CS_i contains CS_i' on the inside.
- CS_{i+1}'' ($i \geq 1$) is obtained by taking a copy of CS_i and adding two vertices in the outerface. Then we add a 3-cycle between the two new vertices and the vertex of degree 2 of CS_i, such that CS_{i+1}'' contains CS_i on the inside.

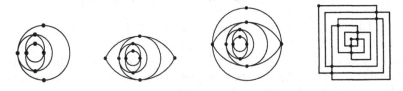

Fig. 7. CS_2'', CS_2', and CS_2; and an optimal drawing.

Lemma 21. CS_i *needs a width and height of* $6i - 1$ *in any plane drawing.*

Proof. We show only the width by induction on i; showing also a lower bound for CS_i'' of $6i - 5$ and a lower bound for CS_i' of $6i - 3$. CS_1'' is a triangle that needs a width of $1 = 6 \cdot 1 - 5$. Assume the claim was shown for CS_i''.

CS_i' contains CS_i'' on the inside and needs two more units in width than CS_i'', which gives a lower bound of $6i - 3$. CS_i contains CS_i' on the inside and needs two more units in width than CS_i', which gives a lower bound of $6i - 1$. Finally, CS_{i+1}'' contains CS_i on the inside and needs two more units in width than CS_i, which gives a lower bound of $6i + 1 = 6(i + 1) - 5$. □

Lemma 22. CS_i *needs* $12i - 2$ *bends in any plane drawing.*

Proof. CS_i has two vertices of degree 2. If we remove those and connect their neighbors by an edge, we get a 4-regular graph CS_i^*, which needs the same width as CS_i, $6i - 1$. By Lemma 1 CS_i^* therefore needs $12i$ bends. CS_i results from CS_i^* by subdividing two edges, so by Lemma 4 CS_i needs $12i - 2$ bends. □

Theorem 23. *There are the following lower bounds for plane connected graphs:*
- *graphs with loops:* $(2n + 1) \times (2n + 1)$*-grid and* $4n + 4$ *bends.*
- *multigraphs:* $(2n - 3) \times (2n - 3)$*-grid and* $4n - 4$ *bends.*
- *simple graphs:* $(\frac{6}{5}(n - 1) - 1) \times (\frac{6}{5}(n - 1) - 1)$*-grid and* $\frac{12}{5}(n - 1) - 2$ *bends.*

Proof. One shows easily that C_i has i vertices, CM_i has $i + 4$ and CS_i has $5i + 1$ vertices. The results follow with Lemma 17, 19, 21 and 22. □

Non-connected graphs For brevity we skip the definition and proofs for non-connected graphs. For simple graphs we did not find graphs with better lower bounds than those for connected graphs.

Fig. 8. Non-connected 4-graphs which need a big grid and many bends.

Theorem 24. *There are the following lower bounds for plane graphs:*
  - *graphs with loops:* $(4n-1) \times (4n-1)$*-grid and* $8n$ *bends.*
  - *multigraphs:* $(2n-1) \times (2n-1)$*-grid and* $4n$ *bends.*

4.2 Combinatorial embedding can be chosen

To the author's knowledge no research has been done into lower bounds for planar drawings. We provide some results here which are close to optimality in the number of bends.

Triconnected Graphs For triconnected planar graphs there exists only one combinatorial embedding. Therefore, if we consider all possible choices of the outerface of T_i, then we get a lower bound for any planar drawing of T_i.

Lemma 25. T_i *needs a width and height of* i *and* $4i-2$ *bends in any planar drawing.*

Proof. Assume that T_1' had the vertices $\{v_1, v_2, v_3\}$ and that we obtained T_i' by adding the vertices $\{v_{3i-2}, v_{3i-1}, v_{3i}\}$. Let Γ be the embedding of T_i defined in Definition 9, this induces an embedding of T_i'. We know that T_i' in this embedding needs a width of $2i-1$, and can also show that it needs $4i-3$ bends. Assume we are given some planar orthogonal drawing of T_i, this induces a planar embedding Γ'. The outerface of Γ' can have degree 3 or 4.

If the degree is 3, then we may assume that the outerface of Γ' is either the outerface of Γ, or one of the three faces adjacent to it. After deletion of the edges (v_{3i}, v_{3i-1}), (v_{3i-1}, v_{3i-2}), (v_{3i-2}, v_{3i}) we have a copy of T_i' embedded as in Γ. So we need a width and height of $2i-1 \geq i$ and $4i-3$ bends of T_i'. The three deleted edges form a triangle and need a bend, so we get a lower bound of $4i-2$ bends.

If the degree is 4, we assume after possible renumbering that the outerface is $\{v_{3j-3}, v_{3j-1}, v_{3j}, v_{3j+1}\}$ for some $1 < j < i$. Splitting the graph at v_{3j-2}, v_{3j-1} and v_{3j} we get a copy of T_j' and a copy of T_{i-j+1}', embedded as in Γ. So the number of bends in this embedding is at least $4j-3+4(i-j+1)-3 = 4i-2$. At the very least the width and height of the drawing must be $\min\{2j-1, 2(i-j+1)-1\}$ which is smallest for $j = \frac{i+1}{2}$ and then equals i. $\qquad\square$

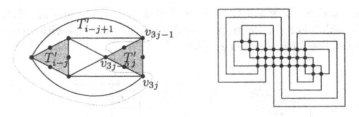

Fig. 9. T_i embedded with outerface-degree 4 leads to a bend-optimal drawing.

Theorem 26. *We have a lower bound of an $\frac{n}{3} \times \frac{n}{3}$-grid and $\frac{4}{3}n - 2$ bends for planar drawings of triconnected graphs.*

Biconnected graphs For biconnected graphs the combinatorial embedding is not unique. However, even though B_i has many different combinatorial embeddings, all give the same planar drawing, except for possible renaming of the vertices and the choice of the outerface. Therefore, by considering different choices of the outerface, we get a lower bound for planar drawings of B_i.

We do not explain the details here, and leave it to the reader to show that B_i needs an $i \times i$-grid and $4i$ bends in any planar drawing. We can then again go over to $\bar{B}_i$ to get the lower bounds for simple graphs.

Theorem 27. *There are the following lower bounds for planar biconnected graphs:*
- *multigraphs: $\frac{n}{2} \times \frac{n}{2}$ and $2n$ bends.*
- *simple graphs: $(\frac{n}{2} - 1) \times (\frac{n}{2} - 1)$ and $2n - 6$ bends.*

5 Remarks and Open Problems

In this paper we have considered lower bounds: for the number of bends in the non-planar case and on both the number of bends and the grid-size in the planar and plane case. Various results have been proved, which either give completely new lower bounds or considerably improved the old ones.

For plane graphs, the results are almost optimal, and the difference is only a small constant, if at all. For planar graphs, the results are fairly good in terms of the number of bends, but improvement should be possible for the grid-size.

Much work remains to be done for non-planar drawings. For the number of bends, there is a small gap in the factor between the lower and the upper bound. No algorithm is known that draws a planar graph with fewer bends if we allow for crossings. We suspect that such an algorithm should be possible.

An even bigger problem are lower bounds on the grid-size for non-planar drawings. The current proofs give only a fairly small constant. It would be also interesting to see more techniques for proving lower bounds on the grid-size of non-planar drawings.

Finally, we would like to pose the open problem of lower bounds for graphs of higher maximum degree. Usually, graphs with higher degree are drawn orthogonally by assigning boxes instead of points to vertices. With such a representation every planar graph can be drawn without bends (see 1D visibility representations [10, 12]). But not all graphs can be drawn without bends: such a drawing is a 2D visibility representation, and can exist only for graphs which are the union of two planar graphs. It would be interesting to see which graphs can be drawn without bends at all, and what are lower bounds for those that can't.

References

1. T. Biedl, Embedding Nonplanar Graphs in the Rectangular Grid, *Rutcor Research Report* 27-93, 1993. Available via anonymous ftp from *rutcor.rutgers.edu*, file */pub/rrr/reports93/27.ps.gz*.

2. T. Biedl, G. Kant, A better heuristic for orthogonal graph drawings, *Proc. of the 2nd European Symp. on Algorithms (ESA 94), Lecture Notes in Comp. Science* 855, Springer-Verlag (1994), pp. 124-135.

3. T. Biedl, New Lower Bounds for Orthogonal Graph Drawings, *Rutcor Research Report* 19-95, 1995. Available via anonymous ftp from *rutcor.rutgers.edu*, file */pub/rrr/reports95/19.ps.gz*.

4. T. Biedl, Orthogonal Graph Drawings: Algorithms and Lower Bounds, Diploma thesis TU Berlin (to appear).

5. M. Formann, F. Wagner, The VLSI layout problem in various embedding models, *Graph-Theoretic Concepts in Comp. Science (16th Workshop WG'90)*, Springer-Verlag, Berlin/Heidelberg, 1992, pp. 130–139.

6. A. Garg, R. Tamassia, On the computational complexity of upward and rectilinear planarity testing, *Proc. Graph Drawing '94, Lecture Notes in Comp. Science* 894, Springer Verlag (1994), pp. 286-297 .

7. G. Kant, Drawing planar graphs using the *lmc*-ordering, *Proc. 33th Ann. IEEE Symp. on Found. of Comp. Science* 1992, pp. 101-110, extended and revised version to appear in *Algorithmica, special issue on Graph Drawing*.

8. M.R. Kramer, J. van Leeuwen, The complexity of wire routing and finding minimum area layouts for arbitrary VLSI circuits. *Advances in Computer Research, Vol. 2: VLSI Theory*, JAI Press, Reading, MA, 1992, pp. 129–146.

9. F.T. Leighton, New lower bounds techniques for VLSI, *Proc. 22nd Ann. IEEE Symp. on Found. of Comp. Science* 1981, pp. 1-12.

10. P. Rosenstiehl, R.E. Tarjan, Rectilinear planar layouts and bipolar orientations of planar graphs, *Discr. and Comp. Geometry* 1 (1986), pp. 343–353.

11. J.A. Storer, On minimal node-cost planar embeddings, *Networks* 14 (1984), pp. 181–212.

12. R. Tamassia, I.G. Tollis, A Unified Approach to Visibility Representations of Planar Graphs, *Disc. Comp. Geom.* 1 (1986), pp. 321-341.

13. R. Tamassia, I.G. Tollis, Efficient embedding of planar graphs in linear time, *Proc. IEEE Int. Symp. on Circuits and Systems* (1987), pp. 495–498.

14. R. Tamassia, I.G. Tollis, Planar grid embedding in linear time, *IEEE Trans. Circ. Syst.* 36 (9), 1989, pp. 1230-1234.

15. R. Tamassia, I.G. Tollis, J.S. Vitter, Lower bounds for planar orthogonal drawings of graphs, *Inf. Proc. Letters* 39 (1991), pp. 35–40.

16. L.G. Valiant, Universality considerations in VLSI circuits, *IEEE Trans. on Comp.* C-30 (2), 1981, pp. 135-140.

The Effect of Graph Layout on Inference from Social Network Data

Jim Blythe[1] and Cathleen McGrath[2] and David Krackhardt[2]

[1] School of Computer Science, Carnegie Mellon University, Pittsburgh, PA 15213
[2] Heinz School of Public Policy and Management, Carnegie Mellon University

Abstract. Social network analysis uses techniques from graph theory to analyze the structure of relationships among social actors such as individuals or groups. We investigate the effect of the layout of a social network on the inferences drawn by observers about the number of social groupings evident and the centrality of various actors in the network. We conducted an experiment in which eighty subjects provided answers about three drawings. The subjects were not told that the drawings were chosen from five different layouts of the same graph. We found that the layout has a significant effect on their inferences and present some initial results about the way certain Euclidean features will affect perceptions of structural features of the network. There is no "best" layout for a social network; when layouts are designed one must take into account the most important features of the network to be presented as well as the network itself.

1 Introduction and Problem Statement

Social network analysis is a fast-growing field of social science that uses graph theory to analyze the structure of relationships among a set of individuals or groups. In this field, the vertices of a graph represent actors in a community and the edges or arcs typically represent patterns of relationships, such as communication, trade or friendship, within the community. In the remainder of this paper, we shall refer to a graph interpreted in this way as a "social network", and to a drawing of a social network as a "sociogram". We have been designing layout and visualization tools specifically aimed at social network analysis that include automatic layout facilities as well as the graphical presentation of some measures used in the field (Krackhardt et al 1994).

In this paper we explore the influence of network layout on individuals' perceptions of common social network measures, in particular "prominence" and "bridging", which are two facets of centrality, as well as the number of groups present in a network. Prominence deals with how visible or involved an actor is based on his or her position in the network. Importance as a bridge deals with an actor's strategic positioning between groups, and most closely relates to betweenness centrality which we define below. Grouping most closely relates to the number of maximal cliques present in the graph.

1.1 Social Networks

Focussing on a particular application area allows for a more concrete metric of a good drawing than the general case, because of the more restricted kinds of information to be

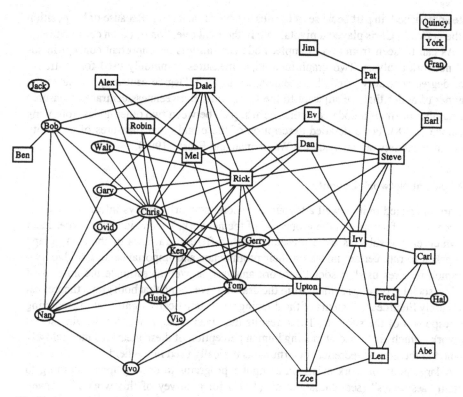

Fig. 1. A network of friendship ties in a small U.S. high-tech firm with 36 employees. This sociogram was layed out by hand.

presented. We present a brief example of how social network data is often collected and analyzed. Krackhardt (1992) studies the role of close friendship or "philos" in organizational change in a small U.S. high-tech firm with 36 employees. The 33 employees who participated in the study were presented with a list of all other employees and asked to mark those they considered to be "personal friends". The question was repeated for the other employees so that each respondent reported who all the employees considered to be personal friends. A "philos" relation is said to exist from i to j if i and j agreed that i considers j a personal friend. Figure 1 shows the sociogram for the philos relationship. Arrows are omitted because most relations are symmetric.

After the study was completed, a union attempted certification at the firm — to be voted an official union by the workers of the firm — and failed. A network analysis can help explain this relatively rare occurrence (Krackhardt 1992). In Fig. 1, the actors who are eligible to vote on the certification are depicted as ovals, and the managers as squares. Hal was the person who originally contacted the union, and he was chosen as their representative leading up to the vote on certification. His position in the network is far from central, and his influence is correspondingly low. Chris, a considerably more central actor, had originally supported the union but was increasingly ambivalent as the vote approached, in part because of his ties to several managers. Because of his position in the network, Chris played a pivotal role in the final rejection of union certification.

As can be seen from this example, node centrality is an important concept in social network analysis. Two graph theoretical measures commonly used for centrality are "degree centrality", and "betweenness centrality". Degree centrality measures the number of nodes that are adjacent to the focal node. Betweenness centrality measures the number of times a node is on the shortest path between two other nodes, following Freeman (1978). For a detailed description of these and other measures of centrality used in social network analysis see Wasserman and Faust. (1994).

1.2 Social Network Layout

We are interested in how well a layout of a social network conveys information about the centrality of actors and the groupings of the network. We conducted an experiment in which eighty subjects were presented with different drawings of the same graph and asked to rate certain nodes on their prominence and importance as a bridge, two common features used in social network analysis, as well as estimate the number of distinct social groupings evident from the drawing. We found that both graph theoretical and purely Euclidean elements of the drawing were statistically significant in predicting the responses of the subjects. These results may lead to aesthetics for drawing social networks which are aimed at making human perception of these measures more closely aligned to the graph theoretical determinants typically used in the field.

A large body of work exists on computer programs to draw graphs according to certain "aesthetics" (see Battista et al. (1994) for a survey of this work). However, almost all of this work considers aesthetics that attempt to improve graph readability from a very general point of view without considering specific applications, and uses general aesthetics such as the regular spacing of the nodes and the minimization of bends and edge crossings. In addition only a few studies explicitly question the aesthetics in use and there has been very little work on analyzing how well such aesthetics actually improve the information conveyed in a graph drawing. Ding and Mateti (1990) consider the subjective factors that go into the drawing of a diagram intended to explain data structures in computer programs. They produce their factors by examining the pictures that appear in a number of text books in that field. Batini et al (1985) make an experimental study of the aesthetics used in Entity Relationship diagrams from the field of software engineering. We are not aware of any studies that consider the information one might want to convey in a social network drawing.

In the rest of this section we describe the features of a drawing that are used to predict the values a human might put on social network measures while looking at a drawing. In the next section we describe our experimental design and the results of the study.

Our experiment was conducted using 5 different drawings of one particular network having 12 actors and 24 ties. This framework allows us to hold structural relationships among nodes constant while varying their spatial relationships. The drawings vary in the proximity of nodes to each other and the positioning of nodes in the center or periphery of the graph. We investigate the influence of two different kinds of factors on perceptions from sociograms: those based purely on the structure of the graph and those based on the spatial properties of the layout. After the structural factors are taken into account, we explore how spatial factors are found to influence respondents' perceptions of prominence, bridging, and grouping. While controlling for structural characteristics we study the influence of:

1. The proximity of a node to the center of the layout on the perception of its prominence.
2. The positioning of a node between pairs of other nodes on the perception of its importance as a bridge.
3. The spatial clustering of groups of nodes on the perception of grouping.

2 Experimental Design

80 graduate students who had just completed a course in organizational theory (a course that emphasized the importance of understanding networks in organizations) volunteered to be subjects in the experiment. The subjects were given a questionnaire containing three of the five graph drawings and told the following:

> The following three graphs are modelled after networks of communications observed in three different merger and acquisition teams of an investment banking firm. A connection between two team members means that they discuss work related matters with each other. If no line exists between two team members then they never discuss work with each other.

All nodes were labelled with first names in the drawings presented to the subjects. By providing a context, the investment banking firm and group member names, we attempt to focus the subjects' attention on the social aspect of communication networks. In every drawing, each node was mapped to a new name. For analysis and discussion purposes, we have renamed the nodes of interest with letters from A to E. The five drawings are shown in Fig. 2.

We used an incomplete balanced latin square design to control for order in the presentation of the sociograms by displaying each drawing first, second, and third exactly once. Each drawing is preceded by every other drawing except one exactly once.

We asked respondents three questions about the same five focal nodes in each sociogram: 1) how many subgroups were in the sociogram; 2) how "prominent" was each player in the sociogram; and 3) how important a "bridging" role did each player occupy in the sociogram. The subject rated the prominence and importance of the same five nodes (with different first name labels) for each drawing by circling a number from 1 to 7 on a Likert scale that went from not prominent (or not important as a bridge) to most prominent (or most important as a bridge). The question about prominence was worded as follows: *"Some individuals have a more prominent role in their team than other individuals. Please rate the following people according to how prominent within their team they appear to you by circling the appropriate number next to each name."* The question about importance as a bridge was worded in the following way: *"Some individuals are important because they form a bridge between subgroups. Please rate the following people according to how important they appear to be as bridges between*

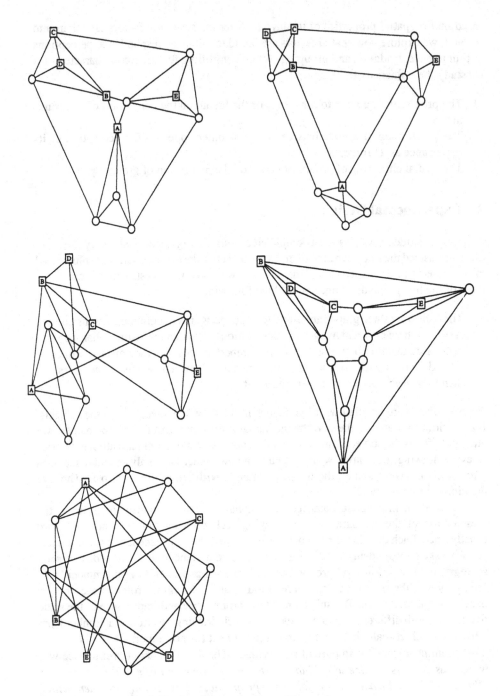

Fig. 2. The five different drawings of the graph.

subgroups by circling the appropriate number next to each name." Finally, we created the questionnaire using a format that allowed one drawing and questions about that drawing to be viewed at the same time, without viewing any other drawing.

2.1 The Network

Table 1. Centrality Measures

	Degree Centrality	Freeman Betweenness Centrality
A	5	8.67
B	5	8.67
C	4	4.67
D	3	0.00
E	3	0.00

The network itself is symmetric, with 12 actors and 24 ties. We used a small graph so that the subjects would not be overwhelmed by the amount of information presented to them. The overall measure of network density is 36 percent. Table 1 shows the values for the two measures of centrality defined in section 1.1 for nodes A through E. Nodes A and B are automorphically equivalent as are nodes D and E. The network has four cliques.

We adhered to the general standards of aesthetics discussed earlier when laying out the graphs, that is, we tried to avoid line crossings, and we tried to avoid nodes landing on top of lines. Drawing 5, the circle, did not adhere to the standards. It is included because of its general acceptance as a means for presenting social network data.

3 Results

The results of our analysis of individuals' reports of each node's prominence and importance as a bridge and of the overall number of groups in the network support our hypotheses that the spatial relationships influence perception in the graph layouts. We use analysis of variance to test whether the average prominence or importance as a bridge for each node changes with each layout. While the ordering of prominence and importance as a bridge does not change across drawings, the relative values assigned to prominence and importance as a bridge do change.

We also use regression analysis on all nodes combined to test the influence of our spatial measures when structural measures are included. We find that both structural and spatial characteristics of the nodes predict the respondents' reports of nodes' prominence and nodes' importance as a bridge. For grouping, we compare the distribution of groups across all five layouts and find some differences in the shapes of the distributions across the five layouts.

The following sections give details of the analysis for prominence, importance as a bridge, and grouping. In the tables that follow, one star (*) denotes significance at the 0.05 level, two stars (**) at the 0.01 level and three stars (***) at the 0.001 level.

Table 2. Individual mean centered scores for prominence

	Drawing 1	Drawing 2	Drawing 3	Drawing 4	Drawing 5	F Value
A	0.65	1.50	1.07	1.10	0.36	3.63 **
	(1.79)	(1.16)	(1.25)	(1.60)	(1.91)	
B	0.60	1.35	0.95	1.08	0.27	3.07 *
	(1.86)	(1.38)	(1.46)	(1.60)	(1.77)	
C	-0.31	-0.12	0.23	0.26	-0.39	2.42 *
	(1.31)	(0.93)	(1.02)	(1.44)	(1.61)	
D	-0.07	-0.98	-1.18	-0.25	-0.94	5.63 ***
	(1.36)	(1.30)	(1.56)	(1.49)	(1.34)	
E	-0.11	-1.60	-1.16	-0.41	-1.10	6.99 ***
	(1.65)	(1.31)	(1.57)	(1.50)	(1.66)	

3.1 Prominence

Before we analyzed the prominence data, we converted the prominence that was reported on a 7-point Likert scale to a mean centered prominence score. We did this by subtracting each respondent's average prominence score across all three layouts from each node's prominence score assigned by the respondent. This controls for individuals' tendencies to rate high or low in general. If a node has a negative mean centered prominence score, the respondent rated that node less prominent than the average.

We use analysis of variance to test the null hypothesis that each node's average mean centered prominence score is the same across all five layouts. The average mean centered prominence scores and results of the analysis of variance are reported in Table 2. When we compare the prominence scores for all five drawings, the null hypothesis that the nodes' average mean scores are all the same can be rejected. Thus, respondents' evaluation of nodes' prominence is not constant across different drawings, rather the degree of prominence assigned by the respondents changes as the drawing changes.

We further estimate a linear equation using Ordinary Least Squares (OLS) to predict prominence using degree centrality (the number of edges adjacent to each node) and the Euclidean distance from the center of each node (normalized across drawings). We estimate this equation on one data set made up of all 5 nodes A through E in all 5 drawings combined, with a total of 1127 observations after accounting for missing responses. Each node's degree centrality remains constant across drawings because it is a structural characteristic of the node, while the node's normalized Euclidean distance from the center of the graph varies across drawings. We find that (1) degree centrality has a positive and significant relationship to reported prominence and (2) normalized Euclidean distance from the center of the graph has a negative and significant relationship to reported prominence. Both effects are significant at the 0.001 level. These two results support our initial hypothesis that both the structural characteristics of the network as well as the spatial representation of nodes in a sociogram influence individuals' perceptions of prominence.

We included dummy variables for individual respondents to control for individual effects, so we did not use mean centered prominence scores, but rather the reported prominence scores. More details about the regression analysis for prominence and importance as a bridge are found in a longer version of this paper (McGrath et al. 1995).

3.2 Importance as a Bridge

Table 3. Individual mean centered scores for importance as bridge

	Drawing 1	Drawing 2	Drawing 3	Drawing 4	Drawing 5	F Value
A	1.70	1.99	1.45	1.31	0.65	5.73 ***
	(1.44)	(0.90)	(1.28)	(1.84)	(1.66)	
B	1.73	1.85	1.47	1.48	0.75	4.15 **
	(1.41)	(1.25)	(1.14)	(1.66)	(1.58)	
C	0.26	0.24	0.58	0.83	-0.42	5.77 ***
	(1.26)	(1.03)	(1.24)	(1.38)	(1.48)	
D	-1.72	-2.05	-1.73	-1.31	-0.86	5.31 ***
	(1.34)	(0.99)	(1.18)	(1.67)	(1.46)	
E	-1.61	-2.09	-1.68	-1.56	-1.04	3.09 *
	(1.49)	(0.97)	(1.41)	(1.42)	(1.79)	

We perform the same analysis of variance on measures of importance as a bridge. Results are shown in Table 3. Again, we calculate a mean centered bridging score for each node. The null hypothesis that the average mean centered bridging score is equal across drawings can be rejected.

We also estimate an equation to predict the reported importance as a bridge of each node. We estimate it as a linear function of the node's betweenness centrality, using Freeman's betweenness centrality measure (1978), and the average angle value between the node and every other pair of nodes in the drawing. The betweenness angle for a node F with respect to a pair of other nodes X and Y is the angle of deviation between the line from X to F and the line from F to Y. It is independent of the ordering for X and Y. If F lies directly between X and Y, this angle is zero, and as F move further from X and Y the angle increases to a maximum value of 180 degrees, as shown in Fig. 3. We sum the betweenness angle for each node between all other pairs of nodes, and normalize this value.

From linear regression, we find that the coefficient of betweenness centrality is positive and statistically significant suggesting (as expected) that betweenness centrality has a positive relationship with reported importance as a bridge. Also, the coefficient of normalized angle between pairs of nodes is negative and statistically significant.

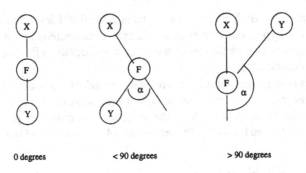

Fig. 3. Angle of deviation as a spatial measure of betweenness.

This supports our hypothesis that spatially positioning a node between pairs of other nodes will increase viewers' perception of that node's importance as a bridge. Again, the effects are significant at the 0.001 level.

3.3 Grouping

Table 4. Number of groups reported in each drawing

Graph Drawing	Mean Number of Groups	
	Mean (Standard Deviation)	
1	5.09	(3.44)
2	4.29	(3.73)
3	3.58	(1.62)
4	4.79	(3.94)
5	3.49	(2.92)

We investigate the effect of the spatial clustering of nodes on the subjects' perceptions of the number of groups in the network. Table 4 shows the mean number of groups reported for each drawing.

We consider the shape of the distributions of the number of groups reported. Figure 4 shows the distribution of responses for the number of groups for all 5 drawings. It can be seen that drawing 5, the circle, has a distribution that is much flatter than the rest, suggesting that respondents could not determine groupings with consistency in the circle drawing. Drawing 3, which presented the nodes clumped in two groups, one large and one small was not interpreted as two groups by many people. We used the Kolmogorov-Smirnov test to determine if each pair of distributions of reported number of groups could be considered statistically significantly different. The Kolmogorov-Smirnov Test for two samples tests the null hypothesis that two random samples are taken from the same distribution. Table 5 shows that the distribution of number of groups for Drawing 5 differs from the distribution of number of groups for every other drawing. In addition, Drawing 1 differs from Drawings 2 and 3.

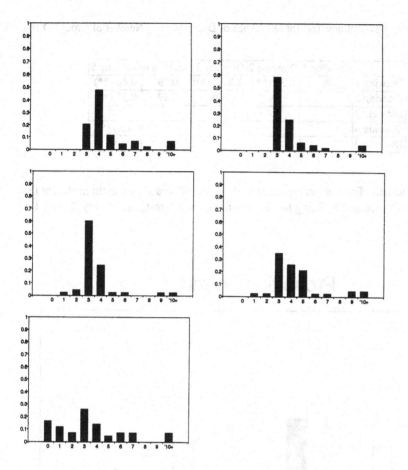

Fig. 4. Distributions of the number of groups reported for each drawing

4 Discussion

We have shown that both structural and spatial factors influence individual's perceptions of prominence, importance as a bridge and grouping. We conducted the first empirical study of the influence of these factors, and while it is dangerous to draw strong conclusions from this small experiment, it would appear that in many cases the influence of spatial layout on the drawing can be predicted to some degree. In particular we found that (1) the perceived prominence of an actor decreases as the corresponding node moves away from the center of the social network, (2) the number of groups perceived by an individual can be altered by the relative proximity of nodes in the sociogram and (3) an actor's perceived importance as a bridge between two groups can be decreased by moving the node further from the center of the bisector of the two groups.

In general it is not possible to say that one drawing is the "best" drawing of a particular social network. Often the best drawing is the one that highlights the characteristic of the network that is being discussed, and there may be no single drawing that best highlights

Table 5. Kolmogorov-Smirnov Test for Difference of Distribution of Number of Groups D stat (p-value)

	Drawing 1	Drawing 2	Drawing 3	Drawing 4	Drawing 5
Drawing 1	X	0.38 (**)	0.46 (***)	0.19	0.40 (**)
Drawing 2		X	0.08	0.19	0.35 (**)
Drawing 3			X	.27	.28
Drawing 4				X	0.30 (*)
Drawing 5					X

every characteristic. For our example network, a "good" drawing should highlight the order of prominence and bridging for the five nodes of interest, and clearly display the group structure.

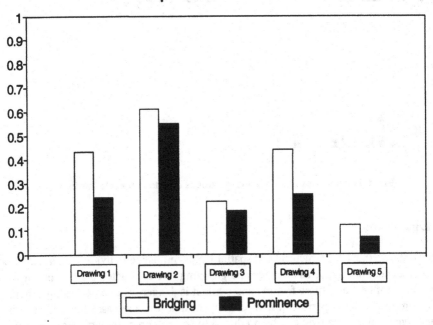

Fig. 5. The proportion of responses with correct ordering of nodes for each layout.

We compare the number of times respondents report the correct order of nodes for each drawing. For both prominence and bridging, the correct order is $A = B > C > D = E$. Figure 5 shows that Drawing 2 has the highest proportion of correctly ordered responses for both prominence and bridging. This suggests that Drawing 2 is the "best" drawing to convey prominence and bridging. On the other hand, Fig. 4 shows that Drawing 1 has the highest proportion of correct responses for the number of groups, 4, in the network.

The results of our study suggest objective measures that can be applied to a sociogram to estimate how well it conveys important structural information about the network. The measures allow us to compare two drawings of the same network, given a set of structural characteristics they are desired to convey, and also suggest a new set of drawing aesthetics that can be used to help produce good drawings of sociograms automatically. However, this is a preliminary study and there is clearly much more work to be done in this area. While we have focussed on social networks in this paper, we believe the same approach can be used to provide more concrete metrics in many application areas of graph drawing.

References

C. Batini, L. Furlani, and E. Nardelli. What is a good diagram? a pragmatic approach. In *4th International Conference on the Entity Relationship Approach*, Chicago, 1985.

G. Di Battista, P. Eades, R. Tamassia, and I. Tollis. Algorithms for drawing graphs: an annotated bibliography. *Computational Geometry: Theory and Applications*, 1994.

C. Ding and P. Mateti. A framework for the automated drawing of data structure diagrams. *IEE Transactions on Software Engineering*, 16(5):543–557, 1990.

L.C. Freeman. Centrality in social networks conceptual clarification. *Social Networks*, 1:215–239, 1978.

David Krackhardt. *The Strength of Strong Ties: the Importance of Philos in Organizations*, in Nohria and Eccles (eds) "Networks and Organizations: Theory and Practice", chapter 3. Harvard Business School Press, Boston, 1992.

David Krackhardt, Jim Blythe, and Cathleen McGrath. Krackplot 3.0: An improved network drawing program. *Connections*, 17(2):53–55, 1994.

Cathleen McGrath, Jim Blythe, and David Krackhardt. The effect of spatial arrangement on inference from graphs. Working note, Heinz School of Public Policy and Management, Carnegie Mellon University, 1995.

Stanley Wasserman and Katherine Faust. *Social Network Analysis: Methods and Applications*. Cambridge University Press, Cambridge, 1994.

Drawing Nice Projections of Objects in Space*

Prosenjit Bose
University of British Columbia
Vancouver, British Columbia, Canada

Francisco Gomez
Universidad Politecnica de Madrid
Madrid, Spain

Pedro Ramos
Universidad Politecnica de Madrid
Madrid, Spain

Godfried Toussaint
McGill University
Montreal, Quebec, Canada

1. Introduction

In our world we are frequently concerned with describing and analyzing three-dimensional (3-D) rigid objects in 3-D space. However, we often have at our disposal only a 2-D medium, such as paper or a computer-graphics screen, on which to display a necessarily incomplete representation or picture of the objects we are interested in. Therefore it is desirable to obtain 2-D representations of our objects that approximate the real objects as faithfully as possible in some sense [KK93], [Ga95]. A sub-field of *visualization* closely related to the class of problems considered here is *graph-drawing* [DETT]. One of the archetypal problems in graph-drawing consists of asking, for a given graph, a "nice" drawing of it. A graph in this context is not a rigid object in 3-D space but a more abstract topological structure which permits the shortening, lengthening and bending of its edges to achieve the desired goal. By contrast, we are concerned with rigid *metrical* objects in 3-D space which are composed of points (vertices) and line segments (edges) and we would like to obtain "nice" *projections* of these objects on some plane that will afford them.

We are concerned here with *parallel* or orthogonal projections [FDFH] rather than perspective projections. Parallel projections may be considered as perspective projections in the limit as the view point approaches a location infinitely far away from the object being viewed. Intuitively, we may think of our object as a wire-frame sitting in 3-D space above the horizontal (*xy*) plane, and the parallel projection of the object on the *xy*-plane as the shadow cast by the wire frame when a light source shines from a point infinitely high along the positive *z*-axis. Obtaining "nice" parallel projections of an object then reduces to the problem of finding a suitable 3-D *rotation* for the object such that its shadow on the *xy*-plane contains the desired properties.

To date such problems have received scant attention in the computational geometry literature. When the objects are convex polyhedra (solid bodies) several questions have been explored. For example, a problem of interest in robotics concerns the determination of whether a convex polyhedron may be translated through a "door" that has the shape of a convex polygon. Geometrically this problem reduces to determining if the polyhedron has a shadow that fits in the door [St82], [To85]. Algorithms have also been found for determining the projections of a convex polyhedron that minimize or maximize the *area* of the shadow that the polyhedron makes on a plane when placing a light source at infinity [MS85], [BGK95]. In computer graphics, good projections for radiosity computation are those that yield the most number of facets visible from the viewpoint [Co90]. On the other

---
* Research of the first author was supported by an NSERC & Killam Fellowship. Research of the second and third authors was carried out during their visit to McGill University in 1995 and was self-supported. The fourth author was supported by NSERC Grant no. OGP0009293 and FCAR Grant no. 93-ER-0291.

hand, when the objects are 3-D polygonal objects (skeletons or wire-frames) very little is known. Hirata et al., [HMTT] give bounds on the worst-case combinatorial complexity of the simplest projections of the skeletons of 3-D convex subdivisions onto a plane. Such simple projections have application to the design of efficient 3-D point location query algorithms [PT92]. Closer in spirit to the work presented here, Kamada and Kawai [KK88] present an $O(n^6 \log n)$ time algorithm for computing the projection of a wire-frame, that in a sense maximizes the projected minimum distance between parallel segments. Finally, Bhattacharya and Rosenfeld [BR94] have studied a special class of orthographic projections called Wirtinger projections for 3-D polygons.

In the work presented here the objects considered are polygonal structures in 3-D. Such objects include sets of disjoint line segments, 3-D simple polygons, knots, trees, and more generally, sets of segments in which the segments may touch each other at their end points, such as skeletons of 3-D Voronoi diagrams or other subdivisions such as those in [HMTT]. There are many specific geometrical characteristics of the vague notion of the "niceness" of a projective drawing of an object. Some of these are more desirable than others depending on the application in mind. One requirement of "nice" is that all the significant features of the 3-D object should be visible in the projection. In other words, no vertex should lie behind another, no edge should look like a vertex and no edge should hide another edge. Furthermore, no three edges may have an interior point in common. This type of projection, closely related to Wirtinger projections [BR94], is useful in visualizing knots, and in knot theory is called a *regular* projection [Re83], [Li93]. Another requirement for effective visualization is *simplicity*. One measure of simplicity is the number of crossings of edges in the projection. It is desirable to obtain the projection that minimizes the number of crossings. We will refer to such projections as *minimum-crossing* projections. If the minimum number of crossings is zero we call such projections *crossing-free*. In some applications we may have a 3-D directed tree as an object of interest. Such a tree may represent a system of veins in the human brain for example, where the direction of an edge represents the direction of blood flow in the corresponding vein segment. Here it is of interest to determine if there exists a projection such that all the directions of the edges of the tree are monotonically increasing in a specified direction on the projection plane. In general we call such projections *monotonic* projections. More specifically, a projection is monotonic if the projected image on the projection plane is monotonic. A planar polygonal chain is monotonic if there exists a direction such that every line orthogonal to this direction, that intersects the chain, yields a point as the intersection. A planar polygon is monotonic if it can be partitioned into two chains each of which is monotonic with respect to the same direction. A tree is monotonic if it contains a root and a direction such that all paths from the root to the leaves are monotonic with respect to that direction. In this paper we investigate the above four types of projections for objects which are sets of disjoint line segments, simple polygons, polygonal chains and trees.

We should add here that the notions of minimum crossing drawings and monotonic drawings are classic visualization problems that have been well studied in the context of graph drawing [DETT]. The general question of given a graph, can one find an embedding in the plane that minimizes the number of crossing edges, is NP-complete [GJ83]. In fact this problem is also NP-complete for a variety of special cases [SSV94]. A lot of work has been done for drawing graphs in a monotonic way in the plane. These drawings are known in the graph-drawing literature as *upward planar* drawings. The general problem of determining for a given directed graph, whether it can be drawn in the plane such that every edge is monotonically increasing in the vertical direction and no two edges cross is NP-complete, as is the problem of deciding if an undirected graph can be drawn in the plane such that every edge is a horizontal or vertical segment and no two edges cross [GT95].

In this paper we consider the following problems. Given a polygonal object (geometric graph, wire-frame or skeleton) in three dimensional euclidean space (such as a simple polygon, knot, skeleton of a Voronoi diagram or solid model mesh), we consider the problem of computing "nice" parallel (orthographic) projections of the object. If we imagine the viewer to be positioned at infinity above the xy-plane we are asked for a rotation of the object in space such that its projection on the xy-plane has the desired "niceness" properties. We consider a variety of definitions of "nice." One such definition, well known in the graph-drawing literature, is a projection with *few crossings*. We consider the most general polygonal object, i.e., a set of disjoint line segments. We show that given a set of n line segments in space, deciding whether it admits a *crossing-free* projection can be done in $O(n^2 \log n + k)$ time and $O(n^2)$ space, where k is the number of such intersections and $k = O(n^4)$. This implies for example that given a simple polygon in 3-space we can determine if there exists a plane on which the projection is a simple polygon, within the same complexity. Furthermore, if such a projection does not exist, a *minimum-crossing* projection can be found in $O(n^4)$ time and $O(n^2)$ space. Another definition of "nice" is that of a *regular* projection (of interest to knot theorists) where the projection has the property that no point of the projected image corresponds to more than two points of the original object in space. We show that a set of line segments in space (which includes polygonal objects as special cases) always admits a regular projection, and that such a projection can be obtained in $O(n^3)$ time. A description of the set of all directions which yield regular projections can be computed in $O(n^3 \log n + k)$ time, where k is the number of intersections of a set of quadratic arcs on the direction sphere and $k = O(n^6)$. Finally, when the objects are polygons and trees in space, we consider *monotonic* projections, i.e., projections such that every path from the root of the tree to every leaf is monotonic in some direction on the projection plane. We solve a variety of such problems. For example, given a polygonal chain P, we can determine in $O(n)$ time if P is monotonic on the projection plane, and in $O(n \log n)$ time we can find *all* the viewing directions with respect to which P is monotonic. In addition, in $O(n^2)$ time, we can determine all directions for which a given tree or a given simple polygon is monotonic.

2. Regular and Wirtinger projections

Let S be a set of n distinct and disjoint line segments in E^3 specified by the cartesian coordinates of their end-points (vertices of S) and let H be a plane. Let S_H be the parallel projection of S onto H. A parallel projection of S is said to be *regular* if no three points of S project to the same point on H and no vertex of S projects to the same point on H as any other point on S [Li93]. This definition implies that for disjoint line segments (1) no point of S_H corresponds to more than one vertex of S, (2) no point of S_H corresponds to a vertex of S and an interior point of an edge of S, and (3) no point of S_H corresponds to more than two interior points of edges of S. Therefore the only crossing points (intersections) allowed in a regular projection are those points that belong to the interiors of precisely two edges of S. This condition is crucial for the successful visualization and manipulation of knots [Li93]. Knots are defined as polygons in 3-D and are special cases of sets of line segments where not all segments are disjoint. Note that a vertex where two edges are joined together in the case when the line segments form a 3-D polygon counts as (not two) but one vertex. Regular projections of 3-D polygons were first studied by the knot theorist K. Reidemeister in 1932 [Re32] who showed that all 3-D polygons admit a regular projection and in fact almost all projections of polygons are regular. This result was re-discovered by Bhattacharya and Rosenfeld [BR94] for a restricted class of regular projections known as *Wirtinger* projections. Regular projections allow two consecutive edges of a 3-D polygon to project to two colinear consecutive edges on H. Therefore some shape features of the polygon are lost in regular projections. For visualization applications this may not be desirable. Those regular projections in which it is also required that no two consecutive edges of the 3-D polygon have colinear projections, are known as *Wirtinger* projections. The above authors did not address the algorithmic complexity of actually finding regular or Wirtinger projections. In this section we study the complexity of computing a single regular or Wirtinger projection as well as constructing a description of all such projections for the more general input consisting of disjoint line segments. These results include therefore results for 3-D chains, polygons, trees and geometric graphs in general. The description of all projections allows us to obtain regular or Wirtinger projections that optimize additional properties. For example, one may be interested in obtaining the most tolerant projection in the sense that it maximizes the deviation of the view-point required to violate the regularity property.

Given three line segments (edges of S) in E^3, all the directions d that result in a non-regular projection of S in which we have a point of S_H that corresponds to three interior points of edges of S, are specified by the family of line transversals of the three edges in question. Using results by Avis and Wenger [AW87], [AW88] on transversals of three skew lines in space it can be shown that each triple of segments of S yields an arc, on the unit sphere of directions, which corresponds to those directions that do not admit a regular projection. These arcs are produced by the intersections of a conic and the sphere of directions. Furthermore, these arcs have measure zero on the sphere and therefore cannot cover it. Using a similar argument it follows that the other cases also lead to measure-zero forbidden directions thus establishing the following lemma.

Lemma 2.1: *A set of line segments in space always admits a regular projection.*

To compute a regular projection of a set of line segments, or a description of all the directions that admit a regular projection, one may in theory compute the arrangement of the $O(n^3)$ arcs on the sphere. However, the intersection of two quadratic surfaces yields arcs on the sphere that are space curves of degree four and computing the arrangement of such curves is difficult in practice. A much better approach is to project these arcs from the sphere to the plane $z=1$ since then we only need to compute the arrangement of a set of quadratic arcs on the plane. There exist several optimal segment-intersection algorithms for computing the arrangement of a set of arcs on the plane. The algorithms of Chazelle & Edelsbrunner [CE92] or Amato, Goodrich and Ramos [AGR95] do not appear to be able to be modified to handle quadratic curve segments. However, recently Balaban [Ba95] discovered an optimal algorithm that computes all intersections of quite general curves, including quadratics, that has time and space complexities $O(n \log n + k)$ and $O(n)$, respectively, where k is the number of intersections among the curves. Thus we obtain the following results.

Theorem 2.2: *Given a set of line segments in space, a regular projection can be obtained in $O(n^3)$ time. A description of the set of all directions which yield regular projections can be computed in $O(n^3 \log n + k)$ time, where k is the number of intersections of the arcs on the direction sphere and $k = O(n^6)$.*

One may wonder if it is worth using the optimal quadratic curve segment intersection algorithm of Balaban in practice given that there is a suboptimal but very simple algorithm due to Bentley and Ottman [BO79] that also handles quadratic curve segments and has time and space complexities $O(n \log n + k \log n)$ and $O(n)$, respectively, where k is the number of intersections among the curves. Balaban has conducted experiments comparing his optimal algorithm to the Bentley-Ottman algorithm for as many as 4,000 segments and the latter algorithm was twice as fast. In fact, Balaban suggests that in practice the suboptimal algorithm should be used unless the number of segments is at least 200,000.

Recall that a Wirtinger projection [BR94] of a 3-D polygon is a special type of regular projection in which no two adjacent edges project to a pair of colinear edges. We can use the above approach to compute Wirtinger projections of polygons also. For Wirtinger projections we have, in addition to the $O(n^3)$ forbidden curve segments on the direction sphere, a set of n additional forbidden great circles. Each pair of adjacent edges of the 3-D polygon yields a plane that contains them. Translate this plane to the origin and intersect it with the sphere of directions. This intersection is a forbidden great circle of directions since for each view point on this circle the two adjacent edges appear to be colinear. In total we still have $O(n^3)$ forbidden curve segments and great circles. We therefore conclude the following.

Theorem 2.3: *Given a polygon P in space, a Wirtinger projection of P can be obtained in $O(n^3)$ time. A description of the set of all directions which yield a Wirtinger projection of P can be computed in $O(n^3 \log n + k)$ time, where k is the number of intersections of arcs and great circles on the direction sphere and $k = O(n^6)$.*

3. Minimum-crossing projections

Whereas a regular projection of a set of line segments always exists, this is not true of crossing-free projections. To establish this it suffices to construct a counter example with three line segments very close to each other and parallel to the three orthogonal axes of the cartesian coordinate system. Here we are interested in computing a description of all the directions (if any exist) that admit crossing-free projections. Furthermore, if no crossing-free projections exist we are interested in finding projections that minimize the number of crossings. Recall that for graph-drawing problems, obtaining a minimum-crossing drawing is NP-complete [GJ83], [SSV94]. By contrast, for the projective drawing versions of these problems we provide polynomial time solutions.

Given two line segments (edges of S) in E^3, all directions d that result in a non-crossing-free projection of S in which we have a point of S_H that corresponds to two points of different edges of S, are specified by the family of line transversals of the two edges in question. In E^3 two edges of S yield a tetrahedron as a description of this family of transversals. This tetrahedron in turn determines four great-circle arcs on the unit sphere of directions that define a convex spherical quadrilateral. Thus each pair of segments of S yields a spherical quadrilateral on the direction sphere that corresponds to a set of directions which results in a crossing occurring between these two line segments. Such quadrilaterals are termed *forbidden*. This leads to the following lemma.

Lemma 3.1: *A set of disjoint line segments in space admits a crossing-free projection iff there exists a point on the sphere of directions that it is not covered by a forbidden quadrilateral.*

The set of $O(n^2)$ forbidden spherical quadrilaterals determined by all pairs of segments in E^3 determines a spherical arrangement on the sphere of directions. We may convert this arrangement to another arrangement of straight-line (possibly unbounded) quadrilaterals on a plane by projecting the forbidden quadrilaterals to the plane $z=1$. To determine if S admits a crossing-free projection then reduces to the problem of determining if the transformed straight-line quadrilaterals cover the plane. We can do this by computing the *contour of the union* of these quadrilaterals. If the contour of the union is empty, then there is no direction that yields a projection without crossings.

Several algorithms have been developed for computing the contour of the union of a set of polygons. Some of these [SB92], [CN89], [NP82] are customized versions of the Bentley-Ottman line-segment intersection algorithm [BO79]. All of them compute the entire arrangement induced by the quadrilaterals and assign to each face in the arrangement, the number of quadrilaterals that intersect it. Faces numbered with zero form the contour of the union. Nievergelt & Preparata [NP82] present a version of the algorithm tailored specifically for convex polygons whose time and space complexities are $O(n \log n + k)$ and $O(n)$, respectively, where n is the number of segments in the polygons and k is the number of intersections of the segments. Souvaine & Bjorling-Sachs [SB92] propose another algorithm that computes the contour of the union from the *vertical map* by using topological sweep in time linear in the size of the map. This algorithm achieves the same time and space bounds as the algorithm of Nievergelt & Preparata [NP82]. However, in order to apply either of these algorithms to compute the contour of the union in our context, they require minor modifications to handle the unboundedness of some of our quadrilaterals. Thus we have the following theorem.

Theorem 3.2: *Given a set of n line segments in space, deciding whether it admits a crossing-free projection can be done in $O(n^2 \log n + k)$ time and $O(n^2)$ space, where k is the number of edge intersections and $k = O(n^4)$.*

If a set of line segments does not admit a crossing-free projection it is of interest to compute the projection that minimizes the number of crossings. To solve this problem we can proceed in a similar manner to that described above but this time search the entire arrangement to find the region covered with the minimum number of quadrilaterals. Therefore we have the following result.

Theorem 3.3: *Given a set of n line segments in space, a minimum-crossing projection can be found in $O(n^4)$ time and $O(n^2)$ space.*

Besides the obvious application of minimum-crossing projections to visualization, we mention here that they also have applications to point location problems in 3-D. Consider a 3-D convex subdivision of space. Recall that the point location algorithm of Preparata & Tamassia [PT92] projects the skeleton of the subdivision onto the xy-plane to obtain a new planar subdivision with additional vertices at all intersection points. This planar subdivision is then pre-processed for planar point location before doing binary search on the z direction. We can apply our algorithm to the original subdivision to minimize the memory required by the planar point location portion of their algorithm.

4. Monotonic projections

The general notion of monotonicity is another characteristic of polygonal objects that aids in their visualization. A simple polygonal chain in 3-D may not admit a crossing-free projection but it may admit a projection which is monotonic in some direction. Here we are interested in determining questions such as: does a given structure admit a monotonic projection in some unspecified direction? Such problems closely resemble the NP-complete problem of determining for a given directed graph, whether it can be drawn in the plane such that every edge is monotonically increasing in the vertical direction and no two edges cross [GT95]. Again, by contrast we provide polynomial time solutions to a variety of similar orthographic projective versions of these drawing problems. First we consider the monotonicity of polygonal chains in E^3. Specifically, we address three questions. Given a polygonal chain P and a direction d, is P monotonic with respect to direction d? Recall that a polygonal chain $P = v_1, v_2,..., v_n$ is *monotonic* in direction d provided that the intersection of P with every plane with normal d is empty, or a point. We show how to answer this question in $O(n)$ time, where n is the number of vertices of P. Next, given a polygonal chain P, we ask if P is monotonic in some direction? We present an algorithm that determines whether a polygonal chain is monotonic in $O(n)$ time. Finally, given a polygonal chain P, it is of interest to determine *all* directions of monotonicity of P. We show how to compute all the directions for which P is monotonic in $O(n \log n)$ time.

Given two points a and b, let ab denote the vector directed from a to b and ba the vector directed from b to a. A plane can be defined by a point p contained in that plane and the normal vector n of the plane. Given a point p and a vector n, the plane defined by them is denoted by $H(p,n)$. Given a plane $h=H(p,n)$, we define the two half-spaces determined by this plane as follows. The open and closed half-spaces h^+ are defined as $\{x / px \cdot n > 0\}$ and $\{x / px \cdot n \geq 0\}$, respectively. Similarly, the open and closed half-spaces h^- are defined as $\{x / px \cdot n < 0\}$ and $\{x / px \cdot n \leq 0\}$, respectively. Henceforth, all half-spaces are open unless explicitly stated otherwise. To avoid ambiguity and simplify the discussion, we adopt the convention that if P is monotonic in direction d, then v_1 is a minimum for P with respect to d. We first address the question of deciding whether a polygonal chain is monotonic in a given direction. A key property of chains monotonic with respect to direction d is that their sub-chains are also monotonic with respect to d.

The above implies that it suffices to determine all directions for which a line segment is monotonic in order to compute all directions for which a polygonal chain is monotonic. Now a line segment is monotonic in every un-oriented direction except those perpendicular to the line segment. By our convention, we are interested in the oriented directions where line segment $[ab]$ is monotonic and a is minimum with respect to the given direction. The point a is a minimum with respect to all directions $D = \{d / d \cdot ab > 0\}$. Let $h = H(O, ab)$ (where O is the origin). It follows that all directions for which $[ab]$ is monotonic can be represented by the intersection of the half-space h^+ with the unit sphere S^2 that represents all directions in space (the sphere of directions). Given a polygonal chain $P = v_1, v_2,..., v_n$ and a direction d, we would like to determine if P is monotonic with respect to d. We simply verify that each of the line segments of P: $[v_1,v_2], [v_2,v_3],..., [v_{n-1},v_n]$ is monotonic with respect to d, i.e., that $v_j v_{j+1} \cdot d > 0$. We conclude with the following.

Theorem 4.1: *Given a polygonal chain P and a direction d, in $O(n)$ time, one can determine if P is monotonic with respect to d.*

We can determine if a polygonal chain $P = v_1, v_2, ..., v_n$ is monotonic for some direction in the following way. Let h_i^+ represent the half-space determined by the plane $H(O, v_iv_{i+1})$. Let D be the intersection of the h_i^+ over all i. Then the set of all directions for which P is monotonic is described by $D \cap S^2$. Determining if D, the intersection of a set of half-spaces is non-empty can be accomplished in linear time using linear programming [Me83]. Therefore we conclude with the following.

Theorem 4.2: *Given a polygonal chain P, one can determine if P is monotonic in $O(n)$ time.*

As noted above, $D \cap S^2$ describes the set of all the directions from which P is monotonic. Since the intersection of a set of half-spaces can be computed in $O(n \log n)$ time [PS85], we conclude with the following.

Theorem 4.3: *Given a polygonal chain P, one can determine in $O(n \log n)$ time all the directions with respect to which P is monotonic.*

Now we turn to the monotonicity of simple polygons and trees in E^3. The polygonal chains, simple polygons and trees in E^3 are all graphs embedded in E^3. In order to continue the discussion in this more general setting, we define a *geometric graph*. A *geometric graph* is a two-tuple (V, E), where V is a finite set of distinct points in general position in E^3, and E is a family of closed straight-line segments with end-points in V. The elements of V and E are called *vertices* and *edges*, respectively. For more definitions and terminology concerning graphs, the reader is referred to [BM76]. In the previous section, the geometric graphs that we considered were paths. In this section, we concentrate on trees and cycles (polygons). We begin by describing some properties of geometric graphs. Given a vertex v of a geometric graph G, we denote the set of *edges* adjacent to v by $EA(v)$.

Lemma 4.4: *Vertex v is a minimum with respect to d for $EA(v)$ if and only if $\forall e \in EA(v)$, v is a minimum with respect to d for e.*

Given a vertex v of a geometric graph G, we denote by $MD(v)$ the set of directions for which v is a minimum for the set $EA(v)$. Let $e = [vv_i]$ be an edge in $EA(v)$. By e we denote the vector $v\,v_i$. Let $h(e) = H(O,e)$. We see that $MD(v)$ is the intersection of $h^+(e)$ over all e contained in $EA(v)$. A vertex v of a geometric graph is a *proper local minimum* with respect to direction d provided that v is a minimum for the set $EA(v)$ in direction d. A vertex v is a *local minimum* with respect to direction d if $\forall e \in EA(v)$, the edge e is contained in the closure of $H^+(v,d)$.

We now address several questions concerning the monotonicity of trees. Suppose we are given a rooted tree T, and a direction d. The first question we address is to determine if T is monotonic in direction d. Notice that two things are specified in this question, the *root* of the tree and the proposed *direction* of monotonicity. The next four questions we address are the following: (1) Given a rooted tree T, does there exist a direction d for which T is monotonic? (2) Given an unrooted tree T and direction d, does there exist a root such that T is monotonic with respect to d? (3) Given an unrooted tree T, does there exist a direction d and a root of T such that T is monotonic with respect d? (4) Given an unrooted tree T, find all roots and directions for which T is monotonic.

Recall that a tree T is called a *rooted tree* if a unique vertex v of T is specified to be the root, otherwise the tree is *unrooted* or *free*. A rooted tree T is monotonic in direction d provided that the path from the root to every vertex is monotonic in direction d. The key behind the efficient solution of all above-mentioned problems depends on the following characterization of the monotonicity of rooted trees.

Lemma 4.5: *A rooted tree T is monotonic in direction d if and only if the root r of T is a proper local minimum and no other vertex is a local minimum with respect to direction d.*

Proof: ($\Rightarrow$) Assume T is monotonic with respect to d. If r is not a proper local minimum then at least one root to leaf path in T is not monotonic. Suppose there exists a vertex v of T that is not the root, such that v is a local minimum. Let $h = H(v,d)$. We see that $EA(v)$ must be in the closure of h^+. Let P be the unique path from r to v in T. P must be monotonic since T is monotonic. Also, the root r is a minimum and v is a maximum for P with respect to direction d, by the convention of monotonic paths. Therefore, $P\backslash v$ must be contained in h^-. Let v_j be the vertex preceding v in P. Since v_j is adjacent to v, we see that $[vv_j] \in EA(v)$. But this implies that $[vv_j]$ is contained in the closure of h^+ which contradicts the monotonicity of P.

($\Leftarrow$) Given a vertex v, we will represent the plane $H(v,d)$ by h_v. Assume that the root r of T is the only proper local minimum and no other vertex of T is a local minimum. Let v be an arbitrary vertex of T. We will show that the path P from r to v must be monotonic in direction d. Let the path $P = v_1 (=r), v_2,..., v_{k-1}, v_k (=v)$. Suppose P is not monotonic with respect to d. Let v_i be the first vertex of P such that $[v_{i-1}v_i]$ is not monotonic with respect to d. Since $[v_{i-1} v_i]$ is not monotonic with respect to d, we conclude that v_i must be contained in the closure of $h_{v_i}^-$. Since v_i is not a local minimum, and T is acyclic, there must exist a vertex v_j in $EA(v_i)$ different from v_{i-1}, such that v_j is contained in $h_{v_i}^-$. Similarly, since v_j is not a local minimum, there must exist a v_m in $EA(v_j)$ different from v_j, such that v_m is contained in $h_{v_j}^-$. By continuing this argument, it follows that since there are only a finite number of vertices in T, there must be a vertex v_t such that $EA(v_t)$ is contained in the closure of $h_{v_t}^+$, contradicting the fact that no vertex of T is a local minimum.

Suppose we are given a tree T, a root r of T and a direction d and want to determine whether T is monotonic with respect to d. By Lemma 4.5, since r is the only proper local minimum and no other vertex is a local minimum with respect to d, the direction d cannot be contained in the closure of $MD(v)$ for all vertices of T other than the root. Since $MD(v)$ is an intersection of half-spaces, determining whether or not a direction d is contained in $MD(v)$ can be done in $O(|M(v)|)$ time. But $O(|M(v)|)$ is $O(d(v))$ where $d(v)$ is the degree of v in the tree T. Therefore, since the sum of the degrees of the vertices of a tree is linear in the number of vertices of the tree, we conclude that in $O(n)$ time where n is the number of vertices of T, we can determine if a rooted tree is monotonic in a given direction d.

Theorem 4.6: *Given a rooted tree T, and a direction d, one can decide in $O(n)$ time if T is monotonic with respect to d.*

Suppose next that the root of the tree is no longer specified. Then, to determine if T is monotonic in direction d, we must first find a root. By Lemma 4.5, the root must be the only proper local minimum and no other vertex is a local minimum. Therefore, there must exist exactly one vertex r of T such that d is contained in $MD(r)$, which becomes the root. All other vertices v of T must have the property that d is not in the closure of $MD(v)$. Again this can be determined in linear time. Therefore, we conclude with the following.

Theorem 4.7: *Given an unrooted tree T, and a direction d, one can decide in $O(n)$ time where n is the number of vertices of T, if there exists a root r of T such that T is monotonic with respect to d.*

Before continuing, we need a few more preliminaries. Instead of representing all directions in 3-space by the sphere of directions S^2, we will represent all directions in E^3 by points on the surface of the axis-parallel cube AC centered at the origin O and with edge length 2. A point p on AC represents the direction Op. Although this representation is not standard, it will simplify many of the algorithms to follow.

Now, suppose that the root r of T is specified but the direction is not. For T to be monotonic in some direction d, we see that d must be in $MD(r)$ but outside $MD(v)$ for all vertices v of T. This problem is in fact as difficult as the general problem where given an unrooted tree T, find all possible roots and directions for which T is monotonic. As such we will present the solution to the general problem below.

The intersection of $MD(v)$ and AC represents the set of directions for which v is a proper local minimum. Since $MD(v)$ is the intersection of a set of half-spaces, the intersection of $MD(v)$ with a facet F of AC is either empty, the facet itself, or a convex polygon. For each facet F_i of AC $(1 \leq i \leq 6)$ and each vertex v_j of T $(1 \leq j \leq n)$, we compute the intersection $I(i,j) = MD(v_j) \cap F_i$. On each facet F_i, notice that the set $I_i = \{I(i,j) / 1 \leq j \leq n\}$ is simply a collection of convex polygons. This collection of polygons has the following property. If a point $p \in F_i$ is contained in the interior of k polygons of the set I_i, then there are k vertices of T that are local minima with respect to the direction Op and each of the vertices that are proper local minima is identified by the polygon which contains p. That is, if p is contained in polygon $I(i,3)$ then vertex v_3 is a proper local minimum with respect to direction Op. Therefore, to determine if there are any directions with respect to which T is monotonic, we want to determine if there are any regions in each facet F_i $(1 \leq i \leq 6)$ that are covered by only one polygon of the set I_i $(1 \leq i \leq 6)$. In fact, we want to find all regions that are covered by only one polygon. This set of regions represents the set of all the directions and roots from which T is monotonic.

Let A_i represent the subdivision induced on facet F_i by the set of polygons I_i. This subdivision can be computed deterministically in $O(n \log n + t)$ time where t is the total number of intersection points of all polygons in I_i. The complexity of A_i is $O(n^2)$. Consider the graph G_i which has a node for every cell of A_i, and an edge between two nodes if the corresponding cells are incident to the same edge of A_i. The graph G_i has $O(n^2)$ nodes and edges. Start at any node a_1 of G_i and compute in $O(n)$ time how many polygons of I_i cover it. Store this number with a_1. Start from a_1 with a depth first search. Every edge (a_1, a_m) of G_i we traverse corresponds to going inside or outside a polygon of I_i, in which case we take the number of a_1 and add or subtract one from it and assign this number to a_m. Thus the whole process of assigning values to nodes of G_i can be done in $O(n^2)$ time. Let M_i represent the set of nodes with the minimum number assigned to it. If this number is one, then each of the cells represented by a node in M_i represents a set of directions and a root from which T is monotonic. The root of T is specified by the vertex generating the convex polygon covering the cell.

Theorem 4.8: *In $O(n^2)$ time, we can determine all directions for which T is monotonic.*

Consider now the problem of determining the monotonicity of a simple polygon in E^3. We begin with a few definitions. A simple polygon in E^3 is a geometric graph that is a cycle. A simple polygon P is monotonic in direction d provided there exist two vertices u,v of P such that both paths from u to v are monotonic in direction d.

The characterization of monotonicity of simple polygons in E^3 is similar to that of trees and therefore the solution for trees is applicable in this case. Therefore, we conclude with the following.

Theorem 4.9: *Given a simple polygon P in E^3 and a direction d, in $O(n)$ time it can be determined if P is monotonic with respect to d.*

Theorem 4.10: *Given a simple polygon P in E^3, in $O(n^2)$ time, we can determine all the directions with respect to which P is monotonic.*

5. Conclusion

Our results on regular and minimum-crossing projections of line segments have immediate corollaries for polygonal chains, polygons, trees and more general geometric graphs in 3-D since these are all special cases of sets of line segments. Our results also have application to graph drawing for knot-theorists. Let K be a knot with n vertices. To study the knot's combinatorial properties, knot theorists obtain a planar graph G called the diagram of K by a regular projection of K. Many of their algorithms are applied to G and therefore their time complexity depends on the space complexity of G. By combining our algorithms we can obtain regular projections with the minimum number of crossings thereby minimizing the time complexity of their algorithms.

6. Acknowledgments

The authors thank Suneeta Ramaswami, Fady Habra, Ivan Balaban, Bernard Chazelle and Mike Goodrich for correspondence. The second author thanks Lola for her support.

7. References

[AGR95] Amato, N. M., Goodrich, M. T. and Ramos, E. A., "Computing faces in segment and simplex arrangements," *Proc. Symp. on the Theory of Computing*, 1995.

[AW88] Avis, D., and Wenger, R., "Polyhedral line transversals in space," *Discrete and Computational Geometry*, vol. 3, 1988, pp. 257-265.

[AW87] Avis, D., and Wenger, R., "Algorithms for line stabbers in space," *Proc. Third ACM Symp. on Computational Geometry*, 1987, pp. 300-307.

[Ba95] Balaban I. J., "An optimal algorithm for finding segments intersections," *Proc. ACM Symp. on Comp. Geom.*, Vancouver, Canada, June 1995, pp. 211-219.

[BO79] Bentley, J.L. and Ottmann, T.A., "Algorithms for reporting and counting geometric intersections", *IEEE Trans. Comput.* vol. 8, pp. 643-647, 1979.

[BR94] Bhattacharya, P. and Rosenfeld, A., "Polygons in three dimensions," *J. of Visual Communication and Image Representation*, vol. 5, June 1994, pp. 139-147.

[BM76] Bondy, J. and Murty, U. S. R., Graph Theory with Applications, Elsevier Science, New York, 1976.

[BGK95] Burger, T., Gritzmann, P. and Klee, V., "Polytope projection and projection polytopes," TR No. 95-14, Dept. Mathematics, Trier University.

[CE92]   Chazelle, B. and Edelsbrunner, H., "An optimal algorithm for intersecting line segments in the plane," *J. ACM*, vol. 39, 1992, pp. 1-54.

[CN89]   Chiang, K., Nahar, S., and Lo, C., "Time-efficient VLSI artwork analysis algorithms in GOALIE2", *IEE Trans. CAD*, vol. 39, pp. 640-647, 1989.

[Co90]   Colin, C., "Automatic computation of a scene's good views," *MICAD'90*, Paris, February, 1990.

[DETT]   Di Battista, G., Eades, P., Tamassia, R. and Tollis, I. G., "Algorithms for drawing graphs: an annotated bibliography," *Computational Geometry: Theory and Applications*, vol. 4, 1994, pp. 235-282.

[FDFH]   Foley, J. D., van Dam, A., Feiner, S. K. and Hughes, J. F., *Computer Graphics: Principles and Practice*, Addison-Wesley, 1990.

[Ga95]   Gallagher, R. S., Ed., *Computer Visualization: Graphic Techniques for Engineering and Scientific Analysis*, IEEE Computer Society Press, 1995.

[GJ83]   Garey, M.R., and Johnson, D.S., "Crossing number is NP-complete," *SIAM J. Alg. Discrete Methods*, vol. 4, 1983, pp. 312-316.

[GT95]   Garg, A. and Tamassia, R., "On the computational complexity of upward and rectilinear planarity testing," eds., R. Tamassia and I. G. Tollis, *Proc. Graph Drawing'94*, *LNCS* 894, Springer-Verlag, 1995, 286-297.

[HMTT]   Hirata, T., Matousek, J., Tan, X.-H. and Tokuyama, T., "Complexity of projected images of convex subdivisions," *Comp. Geom.*, vol. 4., 1994, pp. 293-308.

[KK88]   Kamada, T. and Kawai, S., "A simple method for computing general position in displaying three-dimensional objects," *Computer Vision, Graphics and Image Processing*, vol. 41, 1988, pp. 43-56.

[KK93]   Keller, P. R. & Keller, M. M., *Visual Cues: Practical Data Visualization*, IEEE Computer Society Press, 1993.

[Li93]   Livingston, C., *Knot Theory*, The Carus Mathematical Monographs, vol. 24, The Mathematical Association of America, 1993.

[Me83]   Megiddo, N., "Linear-time algorithms for linear programming in R^3 and related problems," *SIAM Journal of Computing*, vol. 12, 1983, pp. 759-776.

[MS85]   McKenna, M. & Seidel, R., "Finding the optimal shadows of a convex polytope," *Proc. ACM Symp. on Comp. Geom.*, June 1985, pp. 24-28.

[NP82]   Nievergelt, J., Preparata, F., "Plane-sweep algorithms for intersecting geometric figures", *Communications of ACM* vol.25, pp. 739-747, 1982.

[PS85]   Preparata, F., and Shamos, M., *Computational Geometry: An introduction*, Springer-Verlag, New York, 1985.

[PT92]   Preparata, F. and Tamassia, R., "Efficient point location in a convex spatial cell-complex," *SIAM Journal of Computing*, vol. 21, 1992, pp. 267-280.

[Re83]   Reidemeister, R., *Knotentheorie*, Ergebnisse der Mathematic, Vol. 1, Springer-Verlag, Berlin, 1932; L. F. Boron, C. O. Christenson and B. A. Smith (English translation) *Knot Theory*, BSC Associates, Moscow, Idaho, USA, 1983.

[SB92]   Souvaine, D. and Bjorling-Sachs, I., "The contour problem for restricted-orientation polygons", *Proc. of the IEEE*, vol. 80, pp. 1449-1470, 1992.

[SSV94]   Shahrokhi, F., Szekely, L., and Vrt'o, I., "Crossing number of graphs, lower bound techniques and algorithms: A survey," *Lecture Notes in Comp. Science*, vol. 894, Princeton, New Jersey, 1994, pp. 131-142.

[St82]   Strang, G., "The width of a chair," *The American Mathematical Monthly*, vol. 89, No. 8, October 1982, pp. 529-534.

[To85]   Toussaint, G. T., "Movable separability of sets," in *Computational Geometry*, G. T. Toussaint, Ed., Elsevier Science Publishers, 1985, pp. 335-375.

Optimal Algorithms to Embed Trees in a Point Set

Prosenjit Bose*   Michael McAllister**   Jack Snoeyink***

Department of Computer Science, University of British Columbia,
Vancouver, BC, V6T 1Z2 Canada

Abstract. We present optimal $\Theta(n \log n)$ time algorithms to solve two tree embedding problems whose solution previously took quadratic time or more: rooted-tree embeddings and degree-constrained embeddings. In the rooted-tree embedding problem we are given a rooted-tree T with n nodes and a set of n points P with one designated point p and are asked to find a straight-line embedding of T into P with the root at point p. In the degree-constrained embedding problem we are given a set of n points P where each point is assigned a positive degree and the degrees sum to $2n - 2$ and are asked to embed a tree in P using straight lines that respects the degrees assigned to each point of P. In both problems, the points of P must be in general position and the embeddings have no crossing edges.

1 Introduction

The problem of deciding whether a set of points admits a certain combinatorial structure, as well as computing an embedding of that structure on the point set, has been a recurrent theme in many fields. The list of problems falling into this category is virtually endless. We mention a few of the structures that are current topics of research.

The triangulation of a point set is a structure that has spurred much research because of its many applications in areas such as finite element methods, graphics, medical imaging, Geographic Information Systems (GIS), statistics, scattered data interpolation, and pattern recognition, to name a few [14, 15].

The combinatorial structure of interest in this paper is the tree, which is well-studied in the literature. For example, the study of spanning trees of a set of points has a long history. From a graph drawing perspective (see [5] for a survey of graph drawing), the traditional questions ask whether a (rooted or free) tree $T = (V, E)$ can be embedded in the plane such that some criterion is satisfied: e.g., that the area of the resulting embedding is small [4, 9], that the symmetry present in the tree is revealed in the embedding [11], or that T is isomorphic to

---

* Partially supported by an NSERC and a Killam Postdoctoral Fellowship
** Partially supported by an NSERC Postgraduate Fellowship
*** Partially supported by an NSERC Research Grant and a B.C. Advanced Systems Institute Fellowship.

the minimum-weight spanning tree [6, 13] or proximity graph [1, 2] of the points in which the vertices are embedded. In essence, the tree is given as input; one needs to construct a set of points in which to embed the tree such that it satisfies the criterion.

The two tree embedding problems that we study have a slightly different perspective: the points are given as input, and the tree may or may not be. We say that an n-node tree $T = (V, E)$ can be *straight-line embedded* onto a set of n points P, if there exists a one-to-one mapping $\phi: V \to P$ from the nodes of T to the points of P such that edges of T intersect only at nodes. That is, edges $(\phi(u_1), \phi(v_1)) \cap (\phi(u_2), \phi(v_2)) = \emptyset$, for all $u_1v_1 \neq u_2v_2 \in E$. We show, in the final section, that to obtain a straight-line embedding of any tree in a set of n points requires $\Omega(n \log n)$ time.

The first problem, called the rooted-tree embedding problem, was originally posed by Perles at the 1990 DIMACS workshop on arrangements: Given n points P in general position and an n-node tree T rooted at node ν, can T be straight-line embedded in P with ν at a specified point $p \in P$? Perles showed that this was possible if p was on the *convex hull of P*, which is the smallest convex set containing the points P. Pach and Törőcsik [17] showed that it could if p was not the *deepest point* of P, obtained by repeatedly discarding points on the convex hull. Finally, Ikebe et al. [8] showed that there was always such an embedding. All three algorithms use quadratic time. We show that one can use a deletion-only convex hull structure [3, 7] to obtain $O(n \log^2 n)$ time and then improve this to $\Theta(n \log n)$ time. If p is the point with greatest y-coordinate then the $O(n \log^2 n)$ algorithm can embed the tree such that all paths from the root to a leaf are vertically monotone.

The second problem, degree-constrained embedding or dc-embedding, is similar, although the tree T is not specified. Consider a point set $P = \{p_1, p_2, \ldots, p_n\}$ in general position in the plane where each p_i is assigned a positive integral value d_i as its *degree*; the degrees satisfy $\sum_{i=1}^{n} d_i = 2n - 2$. Can some tree T be straight-line embedded on the set of points P such that a tree node of degree d_i maps to point p_i, for all i? Tamura and Tamura (now Ikebe) [20] showed that such a tree always exists and presented an $O(n^2 \log n)$ time algorithm to compute one. We present an optimal $\Theta(n \log n)$ time algorithm for this problem.

Similar embedding problems can be posed for planar graph embeddings in a set of points; these problems remain open. The problem of embedding a tree into a set of points with vertically monotone paths from the root to all leaves in $\Theta(n \log n)$ time is also open.

2 Hull trees

Our algorithms for embedding trees use segments from the convex hull of the unassigned points as tree edges; the convex hull property prevents our tree edges from intersecting. Consequently, we need efficient access to the convex hull of points. Moreover, we need the ability to delete points from the convex hull as we embed tree vertices at them. Overmars and van Leeuwen [16] describe a data structure to store the convex hull that permits arbitrary insertion and deletion of points into the set of points. In our algorithms, we do not need to insert points into the convex hull but simply delete them; we opt for hull trees [3, 7], which provide better amortized time complexities for point deletions. This section provides a brief introduction to hull trees.

A hull tree of a set of n points P stores the upper or lower convex hull of P; the entire convex hull of P can be represented by two hull trees. A hull tree for P's upper hull is a binary tree in which each leaf is a point of P and internal nodes represent an edge in the upper hull of the node's leaves (figure 1). Each internal node in the tree stores

- the upper hull edge between the leaves in the node's left subtree and the node's right subtree.
- the number of leaves in its subtree.
- in the dc-embedding problem, each leaf has a degree value given in the problem's input. An internal node to the hull tree stores the sum of the degrees of the leaves in its subtree.

If the hull tree is initially balanced then it maintains an $O(\log n)$ height.

We use two of the hull tree operations described by Hershberger and Suri [7]: point deletion and set partition. The point deletion operation removes a point from the hull tree. Each of the hull edges at internal nodes from the point's leaf node to the root of the hull tree may need to be recomputed as a result of the deletion; a bottom-up merge of hulls along the path to the root accomplishes this task.

In the set partition operation, we are given a vertical line and we want to split, or partition, the hull tree into two parts: one hull tree for the points left of the vertical line and one hull tree for the remaining points. Assume that the vertical line goes through a point q. The path from the root to the leaf for q in the hull tree contains all the hull edges that cross the vertical line. Split the hull tree along this

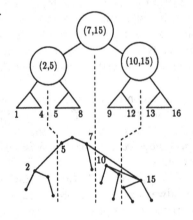

Fig. 1. Top level of a hull tree (above) and upper hull edges (below).

path, duplicating the path in each part to maintain connectivity. As with the deletion operation, recompute the hull edges that appear along the path in each part to get two hull trees. Finally, use the point deletion operation to remove the duplicated point from one of the hull trees as necessary.

Each of the hull tree operations use $O(\log n)$ amortized time. The set partition operation takes $O(\log n)$ time to divide the hull tree into two parts and to duplicate the path. The remaining time in the set operation is the same as in point deletion; it is the time required to recompute the hull edges along one path. Create a potential function for the hull tree by assigning each internal node a value equal to the number of hull edges that appear above the node's edge. The initial potential of the tree is $O(n \log n)$. When recomputing the hull edges, we either keep the same edge or the replacement hull edge has fewer hull edges below it thus lowering the overall cost of the tree. If we find the replacement hull edge at a node v by walking along the two hulls of the left and right subtrees of v then updating the edges along the path takes $O(\log n + k)$ time where k is the number of hull edges over which we walked and the amount by which the potential function decreases. Consequently, the path update takes $O(\log n)$ amortized time.

3 Embedding a rooted-tree with the root on the convex hull

Before embedding a rooted-tree in a set of points where any of the points can be designated to embed the root, we consider the simpler problem where the root of the tree is designated to appear at a specified point of the convex hull. In this restricted case, we embed the tree and preserve the order of the children about each tree node. Ikebe et al. [8] and Pach and Törőcsic [17] each provide a quadratic time algorithm whose time can be reduced to $O(n \log^2 n)$. After briefly sketching this lower complexity algorithm, we present an $O(n \log n)$ algorithm.

Ikebe et al. embed a tree T into a set of points P with the root at a point p on the convex hull of P by locating rays $\ell_0, \ell_1, \ldots, \ell_m$ from p such that there are exactly $|T_i|$ points between ℓ_{i-1} and ℓ_i where $T_1, \ldots, T_m$ are the subtrees of the root of T. The lines $\ell_1, \ldots, \ell_m$ are found by linear time median search. The subtrees $T_1, \ldots, T_m$ are then recursively embedded in the points between adjacent ℓ_i. This leads to an algorithm with a $\Theta(n^2)$ worst-case time.

If the points of P are placed in a convex hull maintenance structure that supports deletions in $O(\log n)$ amortized time [3, 7] then we can find the lines ℓ_i without resorting to a full median search. Let T_L be the leftmost subtree of T and let T_R be $T \setminus T_L$. Assume that $|T_L| \leq |T_R|$. Delete, one at a time, $|T_L|$ points from the convex hull that appear as the left neighbour of p. These are precisely the points between ℓ_0 and ℓ_1. Rebuild the convex hull maintenance structure for the deleted points; recursively embed T_L in the new convex hull structure and T_R in the convex hull structure left after the deletions. The revised complexity of the algorithm is $O(n \log^2 n)$ from the recurrence $T(n) = T(n - k) + T(k) + O(\min(k, n - k) \log n)$ where $1 \leq k \leq n - 1$.

This $O(n \log^2 n)$ embedding algorithm can guarantee that all paths in the embedding from the root ν to each leaf is vertically monotone if p is the highest point of P. When the algorithm recursively embeds T_L, it selects the deleted points with greatest y-coordinate as the root for T_L. Similarly, when the root ν has degree 1, the algorithm chooses the point of P with second greatest y-coordinate as the root for the single subtree of ν. The the deleted points with greatest y-coordinate is found in $O(|T_L|)$ time; the point of P with second greatest y-coordinate is found in $O(\log n)$ time by keeping the points of P sorted by

y-coordinate in a balanced tree and updating the tree along with the convex hull maintenance structure. The time recurrence and time complexity for the algorithm remain unchanged.

The cost of recomputing convex hulls in the variant of Ikebe et al.'s algorithm remains an expensive operation. Our algorithm uses the same notion of isolating the points for one subtree but only uses vertical separation lines and the upper hull of the points, assuming that p lies on the upper hull. When the root of the tree is not in, or immediately adjacent to, the isolated set of points, the algorithm embeds the tree along a path on the upper hull to reach the subset. By handling the leftmost or rightmost subtree of the root and then deleting the points used along the upper hull, the algorithm prevents embedded tree edges from crossing Since all division lines are vertical, the set partition operation on hull trees in section 2 divides the hull tree in $O(\log n)$ amortized time and provides a better time complexity.

Theorem 1 *Suppose that we are given an n-node tree T with root node ν and a set of n points P in general position with the point p on the upper hull $UH(P)$. There is an algorithm that takes $O(n \log n)$ time to straight-line embed T in P with ν at point p.*

Proof: Create a balanced hull tree for the upper hull of P and label each point of P with a value of *any*; the label can take the values *left*, *right*, or *any*. When we solve subproblems recursively, the hull trees and labels are not recomputed or reset.

Let T_L be the leftmost subtree of ν, let ν' be the root of T_L (then ν' is the child of ν in T), and let T_R be $T \backslash T_L$. Let ℓ be a vertical line that has $|T_R| - 1$ points of $P \backslash \{p\}$ to its right. Finally, let q be the left neighbour of p on $UH(P)$, if it exists, as in figure 2. If ν has degree one then T_R is only ν. Embed ν at p. To complete the embedding, let r be the neighbour of p on $UH(P)$ as dictated by p's label, where a label of *any* dictates either neighbour along $UH(P)$. We prove later that there is such a neighbour on $UH(P)$ in the label's direction. Delete p from $UH(P)$ and recursively embed T_L rooted at ν' into the point set $P \backslash \{p\}$ with ν' going to r.

Otherwise ν has degree at least 2 and there are three possibilities:

- line ℓ intersects edge (p, q) as shown in figure 2
- line ℓ is to the left of point q
- line ℓ is to the right of point p

If ℓ intersects edge (p, q) then partition $UH(P)$ along ℓ into two upper hulls, $UH(P_L)$ to the left of ℓ and $UH(P_R)$ to the right of ℓ. Recursively embed T_L rooted at ν' into the set P_L at point q and embed T_R rooted at ν into the set P_R at point p.

If ℓ lies to the left of q then we shift the root of T and the embedding point in P leftward until we obtain the previous case. The shift assigns point q a label of *right* and recursively embeds T, now rooted at ν', into P at the point q.

We must ensure that ν is eventually embedded at p after the shift. In the recursive call, T_R is the rightmost child of ν' in T and the dividing line ℓ' lies to the left of ℓ. Consequently, either $T_L \setminus \nu'$ is embedded completely left of the vertical line through q with ν' still to be embedded at q, or the root of T is shifted left during the embedding of T_L. Assume the former case since the latter case eventually leads to it as the root of T continues to shift left. Then the remaining upper hull to the right of q is identical to $UH(P)$ right of q since all deletions in the embedding of T_L occur to the left q. In particular, p is still the right neighbour of q on the upper hull. Once $T_L \setminus \nu'$ is embedded, ν' is a leaf with T_R as its only child. The algorithm embeds ν' at q and the label at q, still *right* since the root shifts in T were leftward, generates a recursive call to embed T_R into the remaining points with ν going to point p.

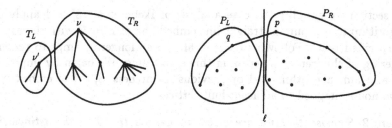

Fig. 2. Divided tree T and point set P.

In the final case, ℓ lies to the right of p. If R is the rightmost subtree of ν in T then the vertical line that has exactly $|R|$ nodes to its right lies to the right of ℓ so we can embed T rooted at ν into P at point p by descending R rather than T_L as above with left and right interchanged.

Each of the above steps is accomplished in $O(\log n)$ amortized time. When the root is a leaf of the tree, a point gets deleted from the hull tree in amortized $O(\log n)$ time. For the remaining steps, we find the separating line ℓ with a binary search down the hull tree in $O(\log n)$ time. Once we have the separating line, we either partition the point set for T_L and T_R in $O(\log n)$ time to get two hull trees for the recursive subproblems or the root of T is shifted.

Each step either embeds a point of T into P or shifts the root of T in one direction and does not reverse that direction until the root that started the shifting is embedded into P. Each of these actions occurs $O(n)$ times for an overall $O(n \log n)$ time complexity.

Finally, the initial hull tree for $UH(P)$ is computed in $O(n \log n)$ time and the points of P receive their initial labels in $O(n)$ time. ∎

A sample embedding appears in figure 3.

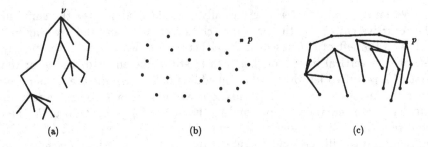

Fig. 3. A rooted-tree (a), a point set with distinguished point p (b), and the embedding of the tree (c).

4 Embedding a rooted-tree

In this section we simplify the case analysis of Ikebe et al. [8] and apply our new algorithm to compute a straight-line embedding of T in P with the root of T at a specified point. Following Ikebe et al., we no longer attempt to preserve the ordering of children at a vertex in this embedding. We begin by improving a quadratic-time algorithm used by previous researchers [8, 17] to embed trees with two nodes mapped to adjacent hull vertices.

Theorem 2 *Suppose that we are given an n-node tree T with distinguished nodes ν and η ($\nu \neq \eta$), and a set of n points P in general position having edge (p, q) on the convex hull $CH(P)$. There is an algorithm that, in $O(n \log n)$ time, embeds T in P with ν at p and η at q.*

Proof: Assume that we have a convex hull maintenance structure for P that supports deletions of hull vertices in amortized $O(\log n)$ time; such a structure can be built initially in $O(n \log n)$ time [3, 7]. Let T_η be the induced subtree of $T \setminus \nu$ that contains η and let T_ν be the complement, $T \setminus T_\eta$.

We can find a line through p with $|T_\nu| - 1$ points on one side and q on the other side by repeatedly deleting the point of the hull adjacent to p that is different from q. When done, delete p and apply theorem 1 to embed T_ν in the deleted points with ν at p. This takes $O(|T_\nu| \log n)$ time.

Point q is on the hull of the points that remain. Let p' be the hull vertex adjacent to q where the open segment $\overline{pp'}$ does not intersect the hull. Let ν', the child of ν in subtree T_η, be the root of T_η. Recursively embed T_η with ν' at p' and η at q. The total time required for data structure building, point deletion, and tree embedding is $O(n \log n)$. ∎

Now we can embed a rooted-tree T in a point set P with the root at a chosen point p. The basic idea is illustrated in figure 4: Use a centroid node to partition T into a subtree T_m and two forests T_α and T_β such that we can find in P the vertices of an empty triangle $\triangle pqr$ with rays from p, q, and r that divide P into convex sets in which T_m, T_α and T_β can be embedded according to theorems 1 or 2. The partition of the tree influences the partition of the point set, and

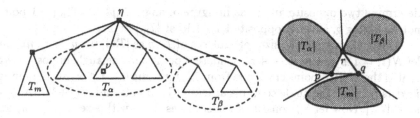

Fig. 4. Partitioning T at a centroid η and embedding in P

vice versa. Some special cases (when the point p is on the convex hull, the root is a centroid, or a forest is empty) are handled along the way.

Theorem 3 *Given an n-node tree T with root node ν and a set of n points P in general position, we can embed T in P with ν at a chosen point $p \in P$ in $O(n \log n)$ time.*

Proof: If p is on the convex hull $CH(P)$, then we can use the algorithm of theorem 1. Otherwise, sort the points of P radially around p.

Let η be a centroid node of T— that is, if we remove η and its incident edges from T, then we are left with connected subtrees with at most $|T|/2$ nodes [10]. For our purposes, the size of a subtree is the number of nodes other than ν that it contains. Let subtree T_m be a maximum-size subtree of $T \setminus \eta$. We now determine forests T_α and T_β and three rays from p that form angles $\leq \pi$ whose interiors contain $|T_\alpha|$, $|T_m| - 1$, and $|T_\beta|$ points, as in figure 5.

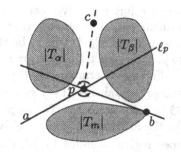

Fig. 5. Find bisector ℓ_p, then $\overrightarrow{pb}$ and $\overrightarrow{pc}$

First, find a line ℓ_p through p that bisects the points—each open halfspace contains $(n - 1)/2$ points. To see that such a line exists, consider the integer function $D(\theta)$ whose value is the difference between the number of points in the left and right halfspaces of a line through p at angle θ. By our general position assumption, the value of D changes by ± 1 when the line hits or passes a point. Since $D(\theta) = -D(\theta + \pi)$, the function $D(\theta)$ has a zero.

Second, choose a point $a \notin P$ on ℓ_p and a point $b \in P$ left of $\overrightarrow{pa}$ so that the interior of angle $\angle apb$ contains $|T_m| - 1$ points and is as large as possible. Recall that $|T_m|$ does not count ν if $\nu \in T_m$. There are essentially two choices—a can be chosen on either side of p, and then b is determined as the the $|T_m|$th point counterclockwise around p from $\overrightarrow{pa}$. If there is a point of P on $\overrightarrow{pa}$ then perturb a into $\angle apb$. When done, the lines $\overleftrightarrow{pa}$ and $\overleftrightarrow{pb}$

determine two opposite angles as in figure 5: angle $\angle apb$ has $|T_m| - 1$ points not including b, and the opposite has at least $|T_m|$ points.

Third, enumerate the sizes of subtrees of $T \setminus \eta$ as $|T_m|, n_1, n_2, \ldots, n_k$, and let $N(i) = 1 + n_1 + n_2 + \cdots + n_i$. Choose a point c in the angle opposite $\angle apb$ that is the $N(i)$th point clockwise from $\overrightarrow{pa}$, for some $0 \le i \le k$. Such an index i exists because the angle contains at least $|T_m|$ points and $N(j+1) - N(j) = n_j \le |T_m|$. Now let T_α consist of the subtrees $T \setminus \eta$ with sizes $n_1, \ldots, n_i$ and let T_β be the rest of the subtrees.

In two special cases we can finish the embedding easily: If $\eta = \nu$, we embed ν into p and embed T_m, T_α, and T_β by the algorithm of theorem 1. If η has degree 2, then T_β is empty—in this case, we embed η into c and embed T_α and T_m into their appropriate angles with (c, b) connecting η to T_m. Whether ν goes with T_α or with T_m, it can be embedded at p according to theorem 2.

Otherwise, determine the convex hull of points inside $\angle bpc$, including b and c but not p. We can assume, without loss of generality, that node

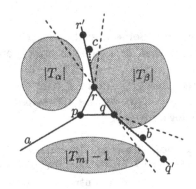

Fig. 6. Finding (q, r)

ν is in T_m or T_α. Let (q, r) be the hull edge that intersects $\overrightarrow{ap}$ and is closer to p. Note that triangle $\triangle pqr$ is empty of points of P, as in figure 6.

Finally, determine $q' \in P$ such that the open region bounded by $\overline{pa}$, $\overline{pq}$, and $\overrightarrow{qq'}$ contains $|T_m| - 1$ points; this can be done by sorting points right of $\overrightarrow{ap}$ radially around q. The slope of $\overrightarrow{qq'}$ lies between the slopes of $\overrightarrow{rq}$ and $\overrightarrow{pb}$ for the following reasons: If q' is not left of $\overrightarrow{rq}$ (dotted in figure 6) then the open region bounded by $\overline{pa}$, $\overline{pq}$, and $\overrightarrow{qq'}$ does not include any point of P from $\angle cpb$. If $\overrightarrow{qq'}$ does not intersect $\overrightarrow{pb}$, then the open region bounded by $\overline{pa}$, $\overline{pq}$, and $\overrightarrow{qq'}$ includes all points inside $\angle apb$.

Similarly, determine $r' \in P$ such that the interior of the open region bounded by $\overline{pa}$, $\overline{pr}$, and $\overrightarrow{rr'}$ contains $|T_\alpha|$ points. The slope of $\overrightarrow{rr'}$ lies between the slopes of $\overrightarrow{pc}$ and $\overrightarrow{qr}$. Thus, the three unbounded regions defined by $\triangle pqr$, $\overrightarrow{pa}$, $\overrightarrow{qq'}$, and $\overrightarrow{rr'}$ are convex.

We use theorems 1 or 2 to embed T_α, T_m, and T_β in the appropriate regions with ν at p, the root of T_m at q, and η at r. Sorting, computing convex hull structures, and embedding each take $O(n \log n)$ steps. ∎

5 Finding degree-constrained embeddings

A problem similar to the rooted-tree embedding of section 3 is to find a tree with non-crossing straight line edges in a set of points when the vertex degree for each point in the plane is given but the tree itself is not provided. Tamura and Tamura [20] called this a degree-constrained embedding (dc-embedding), proved that such an embedding exists if and only if the sum of the degrees for n points is $2n - 2$, and provided an algorithm to find a dc-embedding in $O(n^2 \log n)$ time. Using hull trees and partitions, we obtain an $O(n \log n)$ time algorithm for the same problem:

Theorem 4 *If we are given n points $P = \{p_1, p_2, \ldots, p_n\}$ in general position where each point is labeled with a positive integer d_i such that $\sum_{i=0}^{n} d_i = 2n - 2$ then there is an algorithm that takes $O(n \log n)$ time to find a dc-embedding on P.*

Proof: Create a deletion only upper hull maintenance structure for the points of P as described in section 2. For the convenience of the proof, assume that the names of the points in P are sorted by x-coordinate: $p_i < p_j$ for $i < j$. Assume that $n > 1$. Finally, let $S(j) = 2j - 1 - \sum_{i=1}^{j} d_i$.

The upper hull of P falls into one of three categories:

1. there is a point of degree 1 and a point of degree greater than 1 on the hull
2. all points on the hull have degree 1
3. all points on the hull have degree greater than 1

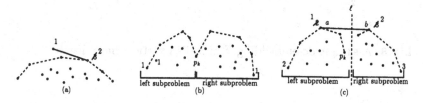

Fig. 7. Three cases of recursion for the dc-embedding algorithm.

In the first case, there must exist two such points that are adjacent along the hull. Join these points by an edge, delete the point of degree 1 from the hull, and decrease the degree of the other point by 1 (see figure 7a). The rest of the dc-embedding is then built recursively.

If all the hull points have degree 1 then either $n = 2$, or $d_2 \geq 2$, or there exists an index k such that $S(k - 1) > 0$ and $S(k) \leq 0$. If $n = 2$ then join the two vertices and stop. If $d_2 \geq 2$ then join p_1 to p_2 by a tree edge, delete p_1 from the upper hull $UH(P)$, decrease d_2 by 1, and recurse. If $d_2 = 1$ then $S(2) = 1 > 0$ and $S(n - 1) = 0$ so the third condition holds for some $1 < k < n$. By definition, $S(k) = S(k-1) + 2 - d_k$ which implies that $d_k \geq 3$. Partition P and $UH(P)$ at p_k with p_k belonging to both smaller sets (see figure 7b). In the left subset, assign p_k a degree of $S(k - 1) + 1$ which is at least 1 and at most $d_k - 1$. In the right subset, assign p_k the remaining degree from d_k. Compute a dc-embedding for each subset recursively; the resulting trees will be connected through p_k.

Finally, if all the hull points have degree greater than 1 then there exists an index k such that $S(k) = 0$ since $S(1) < 0$ and $S(n-1) \geq 0$ and the difference $S(j) - S(j-1)$ increases by at most 1 whenever p_j has degree 1. Let ℓ be a vertical line between p_k and p_{k+1} and let a and b be the left and right endpoints of the upper hull edge of P that intersects ℓ (figure 7c). Partition P and its upper hull along ℓ, join points a and b by a tree edge, decrease the degrees of points a and b by 1 each, and recursively find the dc-embedding for the subsets of P left and right of ℓ.

The time complexity of each step is $O(\log n)$ amortized time. The index k that satisfies the given conditions is found through a binary search in the hull tree in $O(\log n)$ time. The deletions and hull partitions of the steps are each done in $O(\log n)$ amortized time.

Each of the steps occurs $O(n)$ times since it either embeds a tree edge or partitions the convex hull where the partition vertex becomes a leftmost or rightmost hull vertex and is ineligible for a later partition. ∎

A sample dc-embedding appears in figure 8.

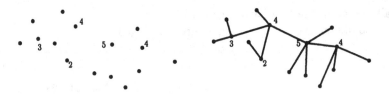

Fig. 8. A sample dc-embedding. Unlabelled vertices have degree 1.

6 Lower bounds

In this section, we provide $\Omega(n \log n)$ lower bounds on quadratic algebraic decision trees for computing a straight-line embedding of a tree onto a set of n points and for computing a dc-embedding on n points. Theorem 5 establishes the optimality of our algorithms for solving these problems. The same bound can be established with a reduction to the unit cost RAM model used by Paul and Simon [18] for their lower bound on sorting.

Theorem 5 *Finding a straight-line embedding of any tree T with n nodes into a set P of n points requires $\Omega(n \log n)$ time.*

Proof: An Euler tour of a tree T embedded into a set of n points gives a chain on $2n$ points in which no segments cross. A careful implementation of Melkman's algorithm [12] will then compute the convex hull of P in $O(n)$ time. The $\Omega(n \log n)$ lower bound for computing the convex hull of P [19] implies the same lower bound for the tree embedding problem. ∎

References

1. P. Bose, G. Di Battista, W. Lenhart, and G. Liotta. Proximity constraints and representable trees. In R. Tamassia and I. G. Tollis, editors, *Graph Drawing (Proc. GD '94)*, volume 894 of *Lecture Notes in Computer Science*, pages 340–351. Springer-Verlag, 1995.

2. P. Bose, W. Lenhart, and G. Liotta. Characterizing proximity trees. Report TR-SOCS-93.9, School of Comp. Sci., McGill Univ., Montreal, Quebec, Canada, 1993.

3. B. Chazelle. On the convex layers of a planar set. *IEEE Transactions on Information Theory*, IT-31:509–517, 1985.

4. P. Crescenzi and A. Piperno. Optimal-area upward drawings of AVL trees. In R. Tamassia and I. G. Tollis, editors, *Graph Drawing (Proc. GD '94)*, volume 894 of *Lecture Notes in Computer Science*, pages 307–317. Springer-Verlag, 1995.

5. G. Di Battista, P. Eades, R. Tamassia, and I. G. Tollis. Algorithms for drawing graphs: an annotated bibliography. *Comput. Geom. Theory Appl.*, 4:235–282, 1994.

6. P. Eades and S. Whitesides. The realization problem for Euclidean minimum spanning trees is NP-hard. In *Proc. 10th Annu. ACM Sympos. Comput. Geom.*, pages 49–56, 1994.

7. J. Hershberger and S. Suri. Applications of a semi-dynamic convex hull algorithm. *BIT*, 32:249–267, 1992.

8. Y. Ikebe, M. Perles, A. Tamura, and S. Tokunaga. The rooted tree embedding problem into points in the plane. *Discrete & Computational Geometry*, 11:51–63, 1994.

9. G. Kant, G. Liotta, R. Tamassia, and I. Tollis. Area requirement of visibility representations of trees. In *Proc. 5th Canad. Conf. Comput. Geom.*, pages 192–197, Waterloo, Canada, 1993.

10. D. E. Knuth. *Fundamental Algorithms*, volume 1 of *The Art of Computer Programming*. Addison-Wesley, second edition, 1973.

11. J. Manning and M. J. Atallah. Fast detection and display of symmetry in trees. *Congressus Numerantium*, 64:159–169, 1988.

12. A. Melkman. On-line construction of the convex hull of a simple polyline. *Information Processing Letters*, 25:11–12, 1987.

13. C. Monma and S. Suri. Transitions in geometric minimum spanning trees. In *Proc. 7th Annu. ACM Sympos. Comput. Geom.*, pages 239–249, 1991.

14. A. Okabe, B. Boots, and K. Sugihara. *Spatial Tessellations: Concepts and Applications of Voronoi Diagrams*. John Wiley & Sons, 1992.

15. J. O'Rourke. *Computational Geometry in C*. Cambridge University Press, 1994.

16. M. Overmars and J. van Leeuwen. Maintenance of configurations in the plane. *Journal of Computer and System Sciences*, 23:166–204, 1981.

17. J. Pach and J. Törőcsik. Layout of rooted trees. In W. T. Trotter, editor, *Planar Graphs*, volume 9 of *DIMACS Series*, pages 131–137. American Mathematical Society, 1993.

18. W. Paul and J. Simon. Decision trees and random access machines. Logic and Algorithmics, Monograph 30, L'Enseignement Mathématique, 1987.

19. F. P. Preparata and M. I. Shamos. *Computational Geometry: an Introduction*. Springer-Verlag, New York, NY, 1985.

20. A. Tamura and Y. Tamura. Degree constrained tree embedding into points in the plane. *Information Processing Letters*, 44:211–1214, 1992.

An Experimental Comparison of Force-Directed and Randomized Graph Drawing Algorithms

Franz J. Brandenburg*, Michael Himsolt* and Christoph Rohrer

University of Passau, 94030 Passau, Germany
{brandenb, himsolt}@informatik.uni-passau.de

Abstract. We report on our experiments with five graph drawing algorithms for general undirected graphs. These are the algorithms FR introduced by Fruchterman and Reingold [5], KK by Kamada and Kawai [11], DH by Davidson and Harel [1], Tu by Tunkelang [13] and GEM by Frick, Ludwig and Mehldau [6]. Implementations of these algorithms have been integrated into our GraphEd system [9]. We have tested these algorithms on a wide collection of examples and with different settings of parameters. Our examples are from original papers and by our own. The obtained drawings are evaluated both empirically and by GraphEd's evaluation toolkit. As a conclusion we can confirm the reported good behaviour of the algorithms. Combining time and quality we recommend to use GEM or KK first, then FR and Tu and finally DH.

1 Introduction

Graph drawing has become an important area of research in Computer Science. There is a wide range of applications including data structures, data bases, software engineering, VLSI technology, electrical engineering, production planning, chemistry and biology. Simply speaking, graph drawing is concerned with the problem of obtaining aesthetically pleasing drawings of graphs. However, what means aesthetically pleasing? A list of criteria has been laid down, including uniformity of the edge length and the node distribution, crossings and the display of symmetries. Recent developments have brought a number of powerful and sophisticated algorithms, which attempt to cope with these aesthetic criteria. An excellent survey and classification can be found in [2].

In this study we consider five well-known algorithms for straight-line drawings of general undirected graphs. Our implementations are called FR, KK, DH, Tu and GEM, respectively. They are based on the algorithms introduced by Fruchterman and Reingold [5], Kamada and Kawai [11], Davidson and Harel [1], Tunkelang [13] and by Frick, Ludwig and Meldau [6]. The algorithms take arbitrary undirected graphs as input and produce straight-line drawings. They modify a given drawing iteratively and attempt to minimize some cost function or

* This research is partially supported by the Deutsche Forschungsgemeinschaft, Grant Br 835/6-1, Forschungsschwerpunkt "effiziente Algorithmen für diskrete Probleme und ihre Anwendungen"

the energy of the drawing. Their strategies are different, using the spring embedder or cost functions and randomization and the simulated annealing paradigms. Each algorithm has been described in detail in an original research paper, where examples of its performance and some comparisons can be found.

However, it is difficult to recall and check these results. This is partially due to the random character of the algorithms. Also, it is the concrete implementation that is in use. The implementation has a big influence particularly on the run time. The drawings depend on the settings of parameters and to a less extend on the initial drawings.

We have integrated FR, KK, DH, Tu and GEM into the GraphEd system, which is a platform for our experiments and evaluations, and we have run the programs on a wide collection of examples. This report includes only some graphs which have been named elsewhere. Further data can be found in [12]. The drawings are evaluated both empirically and by GraphEd's evaluation toolkit [10]. Our experiments confirm the good behaviour of the algorithms reported in the literature. They reach the intended goals, in particular uniform distributions of the edge lengths and the nodes. For many graphs their symmetries are well displayed.

Related experimental work on graph drawing has been presented by Davidson and Harel [1] and Fruchterman and Reingold [5], which mutually compare their drawings, by Frick et al. [6] comparing their GEM with GraphEd's FR and an earlier implementation of KK, Tunkelang [13], Himsolt [10] and di Battista et al. [3]. Himsolt considers a broad collection of graph drawing algorithms with different graph drawing standards and di Battista et al. test and compare three algorithms producing rectlinear drawings. Our spirit is similar with a focus on algorithms for straight line drawings of general graphs.

2 Algorithms

In this section we give an outline of the algorithms and explain particularities of our implementations. We followed the original descriptions unless otherwise stated.

2.1 FR

FR is the GraphEd implementation of the algorithm by Fruchterman and Reingold [5]. Based on the force-directed method, FR computes attractive and repulsive forces and simultaneously moves all nodes according to these forces, where the moved distance is bounded by a temperature t. This process is iterated for some rounds. The free parameters of the implemented algorithm are the optimal node distance, the maximal number of iterations and the temperature t. These parameters can be chosen by the user. The algorithm terminates if either the maximal force acting at a node falls below a user defined threshold or if the maximal number of iterations is exceeded.

2.2 KK

KK is the algorithm by Kamada and Kawai [11]. It computes the total energy of the drawing from the actual and the graph theoretic distances between nodes. Solving a system of partial differential equations by the Newton-Raphson method, the locally best node is moved to reduce the total energy of the drawing. This is repeated until the energy falls below a preset threshold.

C. Rohrer has tuned KK for the GraphEd system by a more efficient implementation of the Newton-Raphson method. This leads to a significant speed-up over an earlier version of KK, which was under study by Frick et al. [6] and Himsolt [10]. In the original description of the algorithm the number of iterations depends on the threshold and the structure of the drawing and thus is hard to predict. Instead, our implementation uses 10*n iterations, where n is the size of the graph.

2.3 DH

DH is the simulated annealing approach by Davidson and Harel [1]. The goal is to minimize a cost function $f(G) = \sum \lambda_i f_i(G)$, which sums over (1) the node distribution, (2) the edge length, (3) the edge crossings, (4) the node-edge distances and (5) the borderlines. Each of these components has its individual weight λ_i, which can be set by the user. Our implementation has a few individual features. They have been introduced to increase the flexibility of our DH and to save some running time, a critical factor for DH, as already stated in [1]. First, we drop the borderlines. Our concern are graphs that are (bi-) connected. Secondly, we normalize the weight parameters λ_i. For a given drawing, the related costs vary by orders of magnitude. E.g., the edge length sums over the quadratic distance between its nodes, the node distribution sums over the inverse quadratic distances and crossings are counted as integers. Therefore these costs are first normalized to 10, and then the user can set relative weights $(\lambda_1, \lambda_2, \lambda_3, \lambda_4)$ with $1 \leq \lambda_1, \lambda_2 \leq 10, 0 \leq \lambda_3 \leq 1000$ and $0 \leq \lambda_4 \leq 10$. The larger range for λ_3 is chosen to force planarity. If $\lambda_3 = 0$ or $\lambda_4 = 0$, then the computations for the detection of possible edge crossings and for the node-edge distances are suppressed.

For the motion of a vertex DH has two options. The first one follows [1] and attempts to move a randomly chosen node to some randomly chosen point on a circle around v with radius r. Starting with an initially high value the radius r decreases geometrically in each round. However, with a decreasing radius, the neighbourhood shrinks and the validity of the underlying theory is in question. Note that DH is an approximation to the simulated annealing process. In the second option, for each round, the radius r is randomly generated and for the trials of node movements the offsets are chosen randomly within a circle of radius r. Furthermore, we have added a re-enforcement heuristic, which again can be switched on/off. This concept is adopted from GEM [6]. If the last movement of a node has lead to an improvement of the cost function, its next update is restricted to almost the same direction, i.e. a point in the sector of width $\pi/2$. As in [1] the DH algorithm finishes with a fine-tuning phase, doing a linear number of moves where only improvements of the cost functions are allowed.

2.4 Tu

Tu is our implementation of Tunkelang's incremental algorithm [13]. Tunkelang uses a template of 16 locations. These are the 8 local neighbour positions and 8 positions at distance d resp. $d\sqrt{2}$, where d is the so-called quality parameter set by the user. $d = 4$ is the default value. Tu inserts the nodes one after another in some precomputed order, here breadth first from the graph theoretic center. For a new node, Tu checks the template positions of each of its neigbours and of the corners of the screen as a candidate position and chooses the locally best. After each insertion, there is a recursive finetuning. All neighbours of the current node are checked for an improved position, using the template from above for the candidate positons. The cost function is that of DH.

2.5 GEM

The graph embedder GEM has been introduced by Frick et al. [6]. It is a tuned and randomized version of a spring embedder, and combines ideas from FR and DH. GEM has been explained in detail in [6], where it is compared with FR and an earlier version of KK. A. Ludwig has made the implementation of GEM available to us.

2.6 Comparison and Measurements

We can present only an excerpt of our tests and examples. There are too many graphs and varieties of parameter settings.

For our measurements we consider the run time on a Sparc 10, the ratio of the length of the longest and the shortest edges, the normalized standard deviation of the edge length and the number of edge crossings. Furthermore, the distribution of the nodes, the ratio of the farthest and nearest pair of nodes, the number of edge crossings and the area have been computed. But this data gives a less significant picture.

In our tests we have made several runs, starting from a randomly generated input drawing, and collected the data of the best, the worst and the average run. The so obtained data is not significantly different, except for the number of crossings and when the number of iterations is too low. There it happened that the best drawings of some planar graphs such as trees, grids and triangular nets came out planar and others had many crossings.

Figures 1, 2 and 3 show test data on complete graphs up to K_{24} and on a set of mixed graphs, which consists of 59 graphs from the papers by Davidson and Harel [1] and Fruchterman and Reingold [5], ordered by the sum of the nodes and the edges. Further examples have been tested in [12]. The drawings are computed with our default settings as mentioned before. For DH they are $(1, 2, 0, 0)$ for the relative weights and $900n$ iterations split into 30 rounds of $30n$ trials and followed by $80n$ fine-tuning iterations, where n is the size of the graph. Since edge crossings and node edge distances are ignored, the inner loops of all five algorithms are of the same asymptotic complexity.

Figure 4 shows a comparison of DH without crossing costs $(1, 2, 0, 0)$ and with high crossing costs $(1, 2, 100, 0)$.

2.7 Evaluations

1. Overall, we can confirm the behaviour of the algorithms reported in the original papers. The algorithms reach the intended goals and produce drawings with uniform edge length and uniform node distribution. In this sense, the obtained drawings are aesthetically pleasing. This can also be said empirically for the visual impressions of the pictures.

2. The algorithms are stable against random input graphs. They converge towards one of the usually few stable drawings, which are of comparable quality, both empirically and by the collected data on the edge length etc.

3. FR, KK, GEM and DH without crossing optimization often produce drawings with a similar appearance. They display symmetry and perfom particularly well on (almost) complete graphs and on graphs from regular polytops and with a 3-D appearance, such as hypercubes, dodecahedron, icosahedron etc. or tori. Distorted drawings occurred for loosely connected graphs.

4. Tu often yields drawings that are different from those of the other programs. Thus, Tu is worth a trial, if the others fail. Tu does not display symmetry, which may be useful, if symmetries are not important. It performs well on grids and net structures, where the other algorithms sometimes failed to untangle distored inputs. However, its behaviour is hard to predict. Its quality parameter has an estimated exponential impact on the run time.

5. DH is the most flexible, but also the most time consuming algorithm. From our experiments we recommend to use more iterations than proposed by Davidson and Harel. This gives better and more stable results. However, DH is difficult to steer. It is difficult to find the proper mix for the relative weights. High penalties for crossings and close node edge distances often destroy the balance and the symmetry of the drawings and the uniformity of the edge length and the node distribution. And due to our implementation they cost time. This is underpinned by the data shown in Fig. 4. Some drawings produced by DH are illustrated in Fig. 5.

6. GEM and KK are very competitive in speed; the difference in speed in Fig. 1 is neglectible. They outperfom the others.

7. FR is fast on small graphs, but slows down on larger graphs with more than 60 nodes and edges.

8. KK produces smooth drawings with a low ratio of the longest and shortest edges and a small deviation of the edge length.

3 Conclusion

Each of the tested algorithms is a good tool for straight-line drawings of general undirected graphs. But there is no universal winner. For each algorithm we have found examples, where it produces the most pleasing drawings.

If you use these algorithms with the GraphEd system, we recommmend to try GEM or KK first, or FR if the graph is small with up to 60 nodes and edges. Next, try FR or Tu and finally DH. If time doesn't count, then play with the parameters of DH or Tu until you get a pleasing drawing. This recommendation takes quality and time into account.

If you have some knowledge of your input graph, e.g. planarity, and you insist on having a planar drawing, then do heavy-duty preprocessing and apply modified and adapted versions of these force-directed or randomized algorithms as a beautification step, as proposed in [7]. DH with high weights on crossings does not perform well, and it is difficult to find the proper mix of the parameters for reasonable or good drawings.

Acknowledgements. We wish to thank A. Ludwig for making the GEM available to us, and F. Dichtl for tests with the algorithms.

References

[1] Davidson, R., Harel, D.: Drawing graphs nicely using simulated annealing. Department of Applied Mathematics and Computer Science (1991)

[2] Di Battista, G., Eades, P., Tamassia, R., Tollis, I.G.: Algorithms for drawing graphs: an annotated bibliography. Comput. Geom. Theory Appl. **4** (1991) 235–282

[3] Di Battista, G., Garg, A., Liotta, G., Tassinari, E., Tamassia, R., Vargiu, F.: An experimental comparison of three graph drawing algorithms. Proc. 11th AMC Sympos. Comput. Geom. (1995)

[4] Eades, P.: A heuristic for graph drawing. Congressus Numeratium **42** (1984) 149–160

[5] Fruchtermann, T.M.J., Reingold, E.M.: Graph drawing by force-directed placement. Software, Practice and Experience **21** (1991) 1129–1164

[6] Frick, A., Ludwig, A., Mehldau, H.: A fast adaptive layout algorithm for undirected graphs. Proc. Workshop on Graph Drawing 94. LNCS **894** (1994) 389–403

[7] Harel, D., Sardas, M.: Randomized graph drawing with heavy-duty preprocessing. Department of Applied Mathematics and Computer Science Weizmann Institute of Science, Rehovot/Israel, Technical Report **CS93-16** (1993)

[8] Himsolt, M.: Konzeption und Implementierung von Grapheneditoren. Dissertation, Universität Passau, Shaker Verlag Aachen (1993)

[9] Himsolt, M.: GraphEd: A graphical platform for the implementation of graph algorithms. Proc. Workshop on Graph Drawing 94, LNCS **894** (1994) 182–193

[10] Himsolt, M.: Comparing and evaluating layout algorithms within GraphEd. J. Visual Languages and Computing **6** (1995)

[11] Kamada, T., Kawai, S.: An algorithm for drawing general undirected graphs. Inf. Proc. Letters **31** (1989) 7–15

[12] Rohrer, C.: Layout von Graphen unter besonderer Berücksichtigung von probabilistischen Algorithmen. Diplomarbeit, Universität Passau (1995)

[13] Tunkelang, D.: A practical approach to drawing undirected graphs. Carnegie Mellon University (1994)

Time (Mixed Graphs)

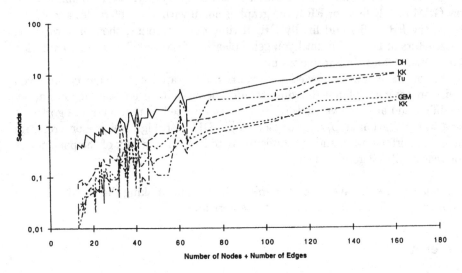

Time (Complete Graphs)

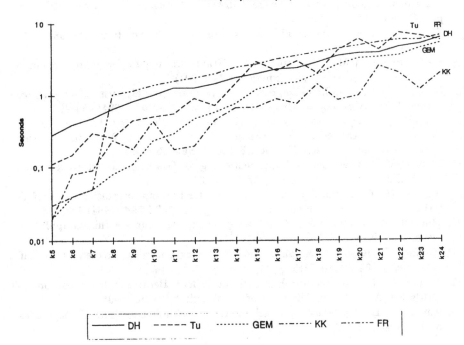

Fig. 1. Runtimes

Longest Edge / Shortest Edge (Mixed Graphs)

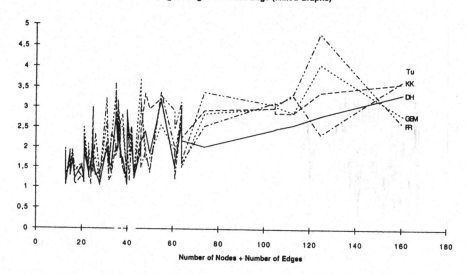

Longest Edge / Shortest Edge (Complete Graphs)

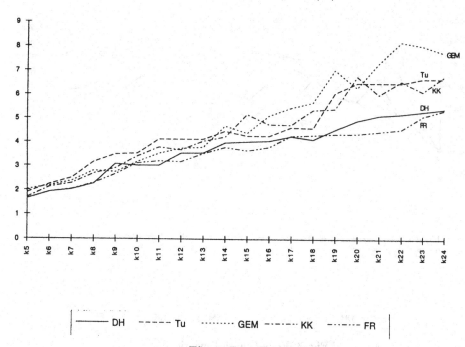

Fig. 2. Edge Length Ratios

Deviation of Edge Length, Normalized (Mixed Graphs)

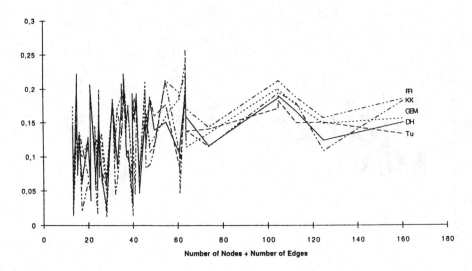

Deviation of Edge Length, Normalized (Complete Graphs)

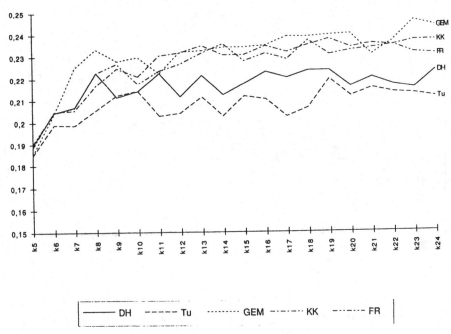

Fig. 3. Edge Length Deviation

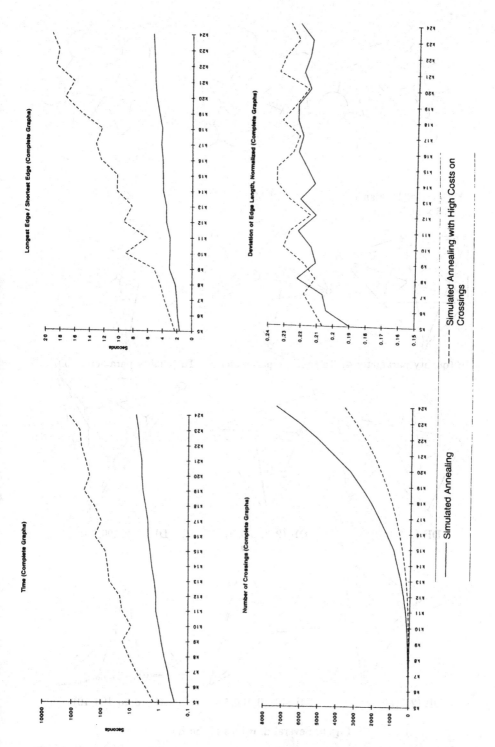

Fig. 4. Comparison of the DH algorithm with and without crossing optimization

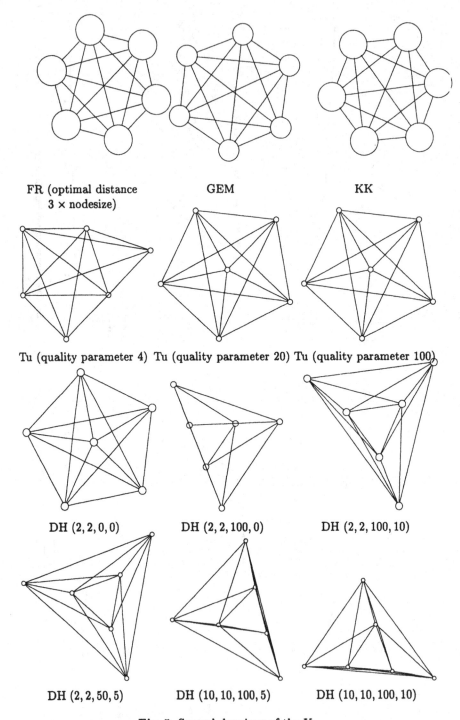

FR (optimal distance   GEM   KK
3 × nodesize)

Tu (quality parameter 4) Tu (quality parameter 20) Tu (quality parameter 100)

DH (2, 2, 0, 0)   DH (2, 2, 100, 0)   DH (2, 2, 100, 10)

DH (2, 2, 50, 5)   DH (10, 10, 100, 5)   DH (10, 10, 100, 10)

Fig. 5. Several drawings of the K_6

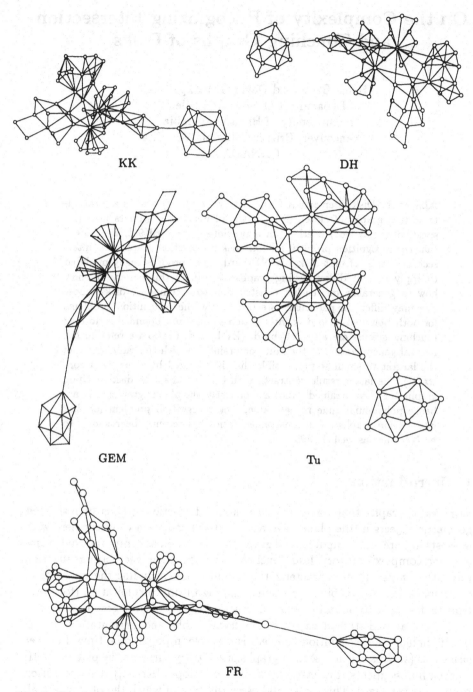

Fig. 6. Graphs from the graph drawing competition drawn with several algorithms

On the Complexity of Recognizing Intersection and Touching Graphs of Disks

Heinz Breu[*] and David G. Kirkpatrick[**]
Department of Computer Science
University of British Columbia
Vancouver, British Columbia V6T 1Z4
CANADA

Abstract. Disk intersection (respectively, touching) graphs are the intersection graphs of closed disks in the plane whose interiors may (respectively, may not) overlap. In a previous paper [BK93], we showed that the recognition problem for *unit* disk intersection graphs (i.e. intersection graphs of unit disks) is **NP**-hard. That proof is easily modified to apply to unit disk touching graphs as well. In this paper, we show how to generalize our earlier construction to accomodate disks whose size may differ. In particular, we prove that the recognition problems for both bounded-ratio disk intersection graphs and bounded-ratio disk touching graphs are also **NP**-hard. (By bounded-ratio we refer to the natural generalization of the unit constraint in which the radius ratio of the largest to smallest permissible disk is bounded by some fixed constant.) The latter result contrasts with the fact that the disk touching graphs (of unconstrained ratio) are precisely the planar graphs, and are hence polynomial time recognizable. The recognition problem for disk intersection graphs (of unconstrained ratio) has recently been shown to be **NP**-hard as well [Kra95].

1 Introduction

Families of graphs that have realizations as intersection graphs of restricted geometric objects in the plane have attracted the attention of researchers with interests in pure and computational graph theory as well as computational geometry and complexity theory. Issues include the recognition, geometric realization (layout), non-geometric characterization, application (including the modeling of communication and visibility problems) and algorithmic exploitation (for problems that appear to be intractable for general graphs) of such graphs.

In several well-studied cases (eg. interval graphs and permutation graphs [BL76, Spi85]) the recognition problem is solvable in polynomial time. In other situations (such as the intersection graphs of arbitrary curves in the plane [Kra91] or even line segments restricted to two or more slopes [Kra94]) the recognition problem is **NP**-hard. This paper addresses the case in which the objects are all closed disks.

---

[*] On leave from Hewlett-Packard Laboratories.
[**] Supported in part by NSERC grant A3583.

A *disk intersection graph* G is the intersection graph of a set of closed disks in the plane. That is, each vertex of G corresponds to a disk in the plane, and two vertices are adjacent in G if and only if the corresponding disks intersect. The set of disks is said to *realize* the graph.

Note that two disks intersect if and only if the distance between their centers is at most the sum of their radii. Therefore, disk graphs can be realized equally well as a set of weighted points in the plane; two vertices are adjacent in the graph exactly when the Euclidean distance between their associated points is at most the sum of the associated weights.

A *disk touching graph* G is the intersection graph of a set of closed disks in the plane whose interiors are constrained to be disjoint. Thus, G is *realized* by a set of interior-disjoint disks, where two disks touch (have a common boundary point) if and only if the associated vertices are adjacent in G.

If all disks have the same size they are said to realize a *unit* disk intersection graph or *unit* disk touching graph. Clearly, not every disk intersection graph or disk touching graph has a unit realization. For example, every star $K_{1,t}$ is a disk touching graph, but only those with $t < 6$ have realizations as disk intersection graphs. (Note that the actual unit of size is not critical, since a set of disks realize the same graph even if the coordinate system is scaled by any convenient amount.)

In addition to their intrinsic interest as geometric graphs, there are several motivations for studying disk intersection and touching graphs. The former provide a natural two-dimensional generalization of interval graphs (cf. [FG65, Rob68]), for which a great deal is known (for example, polynomial time recognition, efficient algorithms for problems that are **NP**-hard in general, and efficient approximation algorithms). In addition, disk intersection graphs (or their unit restriction) have been used to model several physical problems, for example radio frequency assignment [Hal80] and ship-to-ship communications (attributed to Marc Lipman by [Rob91]). They have also been used as test cases for heuristic algorithms designed for arbitrary graphs [JAMS91]. More applications are described by [CCJ90] and [MHR92]. Disk touching graphs (also known as disk packing graphs) play an important role in the construction of high resolution embeddings of planar graphs [MP92]. Unit disk touching graphs can also be seen as a natural generalization of grid graphs [BC87] (in which the disks are constrained to be centred at integer grid points).

In a previous paper [BK93] the problem of determining if a given graph is a unit disk intersection graph (equivalently, the problem of determining if the sphericity (cf. [Hav82a, Fis83]) of a graph is less than or equal to two) was shown to be **NP**-hard. This answered an open question mentioned in [CCJ90] and [MHR92]. The reduction can be modified without difficulty to prove that the recognition of unit touching graphs is also **NP**-hard.

In this paper, we show how to generalize our construction presented in [BK93] to accomodate disks whose size may differ. In particular, we prove that the recognition problems for both bounded-ratio disk intersection graphs and bounded-ratio disk touching graphs are also **NP**-hard. More formally, for every $\rho \geq 1$, we define the following:

ρ-BOUNDED DISK INTERSECTION GRAPH RECOGNITION
INSTANCE: Graph $G = (V, E)$.
QUESTION: Does G have a realization as the intersection graph of a set of
disks, whose radii fall in the range $[1, \rho]$.

ρ-BOUNDED DISK TOUCHING GRAPH RECOGNITION
INSTANCE: Graph $G = (V, E)$.
QUESTION: Does G have a realization as the touching graph of a set of disks,
whose radii fall in the range $[1, \rho]$.

Our main results are the following:

Theorem 1. *ρ-BOUNDED DISK INTERSECTION GRAPH RECOGNITION
is **NP**-hard, for every fixed $\rho \geq 1$.*

Theorem 2. *ρ-BOUNDED DISK TOUCHING GRAPH RECOGNITION is **NP**-
hard, for every fixed $\rho \geq 1$.*

It is interesting to note that the unconstrained DISK TOUCHING GRAPH
RECOGNITION problem (equivalently, the ∞-BOUNDED DISK TOUCHING
GRAPH RECOGNITION problem) has a familiar polynomial time solution by
virtue of the fact[3] that a graph is an (unconstrained) disk touching graph if and
only if it is planar. On the other hand, the unconstrained DISK INTERSEC-
TION GRAPH RECOGNITION problem remains **NP**-hard [Kra95].

As indicated previously, our proofs are similar in form to that presented in
the unit case. The next section recalls the essential structure of our **NP**-hardness
reduction. Section 3 sets out some properties of disk packings (specifically, re-
lating the number of disks in a given packing to the size of its boundary). These
properties are used in Section 4 to describe the components of a generic reduction
from a variant of SATISFIABILITY to both of the problems described above.
Section 5 offers some concluding remarks.

2   Overview of reduction for unit disk graphs

Our proof (cf. [BK93]) of the **NP**-hardness of the UNIT DISK INTERSEC-
TION GRAPH RECOGNITION problem is a reduction from a variant of CNF
SATISFIABILITY. Specifically, we show that every conjunctive normal form
Boolean formula $\mathcal{F}$, in which every clause contains at most three literals and
every variable appears in at most three clauses, can be transformed into a graph
$G_{\mathcal{F}}$ with the property that $G_{\mathcal{F}}$ is a unit disk intersection graph if and only if $\mathcal{F}$
is satisfiable.

---

[3] This result, frequently attributed to W.P. Thurston, was evidently first discovered
in 1935 by P. Koebe (cf. [Sac94])

This transformation proceeds in three stages. We begin by defining the bipartite graph $G_{\mathcal{F}}^{SAT}$ determined by the literal-clause incidence relation in $\mathcal{F}$. It is straightforward to see that $\mathcal{F}$ is satisfiable if and only if the edges of $G_{\mathcal{F}}^{SAT}$ can be oriented so as to satisfy certain out- and in-degree constraints at the clause and literal vertices (essentially an edge is oriented from a clause c to a literal u in c if u is "chosen" to satisfy c). Next it is shown that the graph $G_{\mathcal{F}}^{SAT}$ can be embedded on a grid without overlapping edges (here we exploit the restricted form of the input formula; in fact, even edge crossings can be avoided by starting with a more restrictive—yet still NP-hard—variant of SATISFIABILITY [Kra95].) Edges of the embedded $G_{\mathcal{F}}^{SAT}$ can be viewed as a sequence of unit-length grid segments. Hence the entire embedded graph $G_{\mathcal{F}}^{SAT}$ can be described as a conglomerate of fixed sized modules (including edge segments and clause and literal junctions). The orientability of $G_{\mathcal{F}}^{SAT}$ is easily recast in terms of orientations of these modules. Finally, we show how to realize each of these modules by a small unit disk intersection graph whose several feasible realizations reflect the several permissible orientations of the corresponding module.

The component unit disk intersection graphs are themselves composed of simpler pieces called *cages*. A cage is simply a chordless cycle. Since a cage is realized as a ring of connected disks, every cage has a fixed *capacity* (informally, the maximum number of disjoint disks that can be packed into the interior of some realization). If two cages share a sequence of three of more vertices on their boundary and are realized with neither cage embedded inside the other (as will always be the case), then any connected subgraph attached to one or more of the interior vertices of this sequence must be entirely embedded in one or other of the two cages. (We refer to such a subgraph in the context of its associated cages as a *flipper*.) This is the mechanism used both to express binary "choices" in the modules and, in the event that a cage does not have the residual capacity to host one of its incident flippers due to earlier "choices", to propagate "choices".

Most of the technical details of the proof concern the construction of networks of cages and flippers that permit disk realizations of all and only the desired forms. Since the desired realizations are all achieved with adjacent cage vertices at unit distance (i.e with the corresponding disks merely touching), and the impossibility of undesired realizations is based on capacity arguments, the proof is easily adapted to apply to unit disk touching graph recognition as well. An extension of the proof to the bounded-ratio problems requires a more careful and general treatment of our lowest level building blocks, cages and flippers. This is taken up in the next section.

3 Hexagon packings

Our reduction from CNF-SATISFIABILITY to the recognition problems for bounded-ratio disk intersection and touching graphs is, in fact, slightly simpler than its precurser in that all of the constituent cages are the same size (some sufficiently large multiple of six, dependent in part on the value ρ) and all are realizable (when the given formula is satisfiable) as hexagonal chains of touching disks whose centres lie on a regular hexagon. (We refer to the latter as a *hexagonal realization* of the cage.)

This uniformity implies that our entire construction, at least to the level of detail of cages, respects an underlying hexagonal grid. This substantially simplifies the issues of fabricating and inter-connecting modules. Of course, as before, we must argue that the realizations that we wish to consider are, up to topological equivalence, the only possible realizations. Here we rely on arguments concerning the capacity of cages which, unlike those in our earlier proof, can be arbitrarily large. The *hexagonal capacity* of a cage is the maximum number of interior-disjoint unit disks that can be packed into the interior of a hexagonal realization of the cage. The *unconstrained capacity* of a cage is the maximum number of interior-disjoint disks that can be packed into the interior of *any* realization of the cage. (Note that in both cases the maximum is achieved by a realization of the cage in which adjacent vertices touch only.) Our main tool for arguing about the unrealizability of certain embeddings is the following:

Lemma 3. *As the size of a cage C increases, the ratio of its unconstrained capacity to its hexagonal capacity approaches $2\sqrt{3}/\pi$ (≈ 1.103).*

Proof. The realization of a cage C by a set of maximum-sized touching disks all of whose centres are co-circular clearly has maximum internal area, among all disk intersection realizations of C. It follows that as the size of C increases, the ratio of the unconstrained capacity to the hexagonal capacity of C approaches the ratio of the area of a circle to the area of a regular hexagon with the same circumference.

Corollary 4. *For all sufficiently large cages C, embeddings of C containing more interior-disjoint disks than $2\sqrt{3}/\pi$ times the hexagonal capacity of C, are unrealizable.*

In our constructions cages are joined at their corners; specifically, any corner vertex and its two adjacent vertices may be shared by two (otherwise disjoint) cages. In such a situation the shared corner vertex cannot respect the hexagonal realization of both cages. In general, this corner vertex is the attachment point for a subgraph (which we refer to as a *flipper*). In any realization the flipper is embedded entirely inside one or other of the two cages. We refer to the number of vertices in the flipper (in the case of disk touching graphs) or the number of independent vertices in the flipper (in the case of disk intersection graphs) expressed as a fraction[4] of the hexagonal capacity of the cage, as its *size*. Figure 1 illustrates[5] a pair of joined cages that share a flipper of size 1. This shared flipper is shown embedded in the right cage.

Typically, a cage may have two or more neighbouring cages, and hence several incident flippers. By Corollary 4, any subset of flippers whose total size exceeds

---

[4] More precisely, we mean the fraction achieved asymptotically as the size of cages increase.

[5] This and subsequent illustrations apply to disk touching graphs. The modifications for disk intersection graphs are straightforward (cf. [Bre95]).

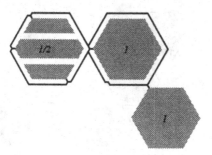

Fig. 1. A pair of joined cages with one shared flipper (and four other flippers).

$2\sqrt{3}/\pi$ cannot be simultaneously embedded inside any realization of their common cage. Thus, for example, it is impossible for the left cage in Figure 1 to house the shared flipper (of size 1) in addition to *any* of its resident flippers (of sizes 1/4 and 1/2). Similarly, the right cage cannot contain both of its incident flippers (of size 1); one of these is shown displaced below the cage.

In general, flippers are constructed from a single connected portion of the hexagonal grid. (They are depicted as such in all of our figures.) The shapes of specific flippers are chosen to permit the simultaneous internal realization (in a hexagonal realization of the cage) of *any* subset of flippers whose total size does not exceed 1. For example, the left cage in Figure 1 has three internally embedded flippers, one of size 1/2 and two of size 1/4.

4 Modules

We now have in place the tools with which we can specify and verify our general construction. We begin by describing the basic building blocks (modules). Each module has up to four designated terminals (a cage corner together with its incident flipper). Modules are connected by identifying some pairs of terminals. We conclude by describing how modules are combined to mimic the bipartite graph orientability problem described in Section 2.

4.1 Clause modules

A clause module is designed to model a clause vertex in the graph $G_{\mathcal{F}}^{SAT}$ with three incident edges at least one of which must be oriented away from the vertex (towards a "chosen" literal). The module, in one of its feasible hexagonal realizations, is illustrated in Figure 2 below. It is easy to verify that all seven embeddings, in which one or more of the flippers incident on vertices T, B or R are embedded externally, have hexagonal realizations. By Corollary 4, the remaining case, in which all flippers are embedded internally, is unrealizable.

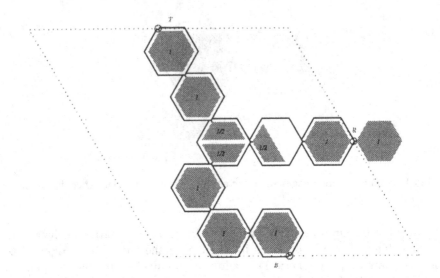

Fig. 2. Clause module with terminals T, B and R.

4.2 Consistency checking modules

In a satisfying truth assignment for $\mathcal{F}$ a variable or its negation (but not both) can be chosen to satisfy up to three clauses. This is modelled, in the graph $G_{\mathcal{F}}^{SAT}$, by constraining the edges incident with the vertex associated with each variable or its negation to all be oriented away from the vertex. Thus, with each variable v we associate a module formed from two submodules, one for each of the two literals v and $\bar{v}$. The submodule for the positive literal is depicted in one of its feasible realizations in Figure 3. Construct the submodule for the negative literal by rotating Figure 3 by 180 degrees. We can then construct the module for the variable by identifying terminal R with the rotated copy of terminal R. The flipper on terminal R in Figure 3 may be embedded outside the submodule, as illustrated (in which case it must be embedded inside the submodule corresponding to the negative literal). Alternatively, the flipper may be embedded inside the submodule, in which case it easy to see (by Corollary 4) that the flippers on terminals T, B, and L must be embedded outside the submodule.

4.3 Connector modules

The "choices" associated with clause modules need to be propagated to the appropriate consistency checking modules. This propagation can be achieved through the use of connector modules, built by chaining together cages where each successive pair has a shared (unit) flipper. Figure 4 illustrates one realization of a typical connector module. It is easy to see that any embedding of the module, in which at least one of the flippers incident on terminals L and R is

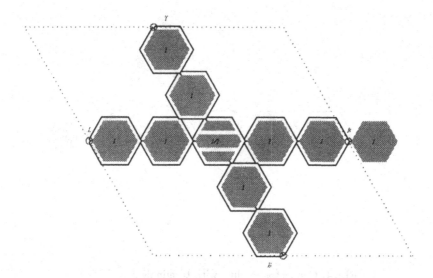

Fig. 3. Positive literal submodule with terminals L, R, T, and B.

embedded externally, has a hexagonal realization. Corollary 4 guarantees that the embedding with both of these flippers internal is unrealizable.

4.4 Crossover modules

As we have described things, connector modules alone are not sufficient to model the interconnection of clause and consistency checking modules. Since the graph $G_{\mathcal{F}}^{SAT}$, under our assumptions about $\mathcal{F}$, is not necessarily planar, the interconnections may be forced to cross. One way to avoid this is to restrict attention to formulas $\mathcal{F}$ for which the associated graph $G_{\mathcal{F}}^{SAT}$ is planar. (SATISFIABILITY restricted to this class of formulas is known to remain **NP**-hard [Kra94].) Alternatively, we can formulate a crossover module. One realization of such a module is described Figure 5. It is a simple exercise to confirm, using Corollary 4, that any embedding of this module, in which the flippers incident on both L and R or those on both T and B are embedded internally, is unrealizable. Similarly, all other embeddings of these flippers have hexagonal realizations.

5 Conclusions

We showed, in the preceeding section, how to construct modules suitable for implementing the reduction of the orientability problem for graphs $G_{\mathcal{F}}^{SAT}$ to the realizability problem for bounded ratio disk intersection or touching graphs. (One of the interesting features of this proof is the fact that it depends very little—at least at the level of abstraction that we have been able to describe it here—on the notion of contact.)

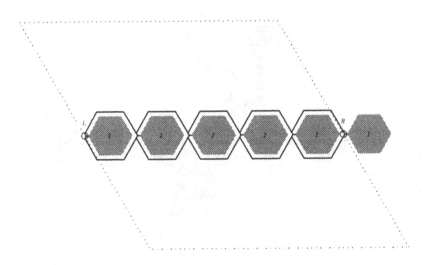

Fig. 4. Connector module with terminals L and R.

Although our proof follows the same general approach as that used in establishing the **NP**-hardness of the unit disk intersection (and touching) graph recognition problems, it differs in several important respects. Two of these, specifically the uniformity of our cages, and the reliance on asymptotic packing capacity bounds rather than properties of particular small configurations of disks, serve to simplify as well as generalize the reduction. This generalization can be exploited to prove analogous results for higher dimensions, for objects others than disks (eg. squares), and for certain grid-constrained versions of our problems [Bre95]. Although grid-constrained versions of our recognition problems (like grid graph recognition [BC87, Grä95]) are in fact **NP**-complete, it is not clear that the unconstrained versions are in **NP**. (Membership in **PSPACE** follows directly from results of Canny [Can88].)

In the case of disk touching graphs, the result in this paper depends critically on the specified bound on disk ratios. Indeed, as noted previously, in the absence of such a bound the recognition problem is straightforward. Our result says not only that the use of arbitrarily large disks is essential to the realization of planar graphs as touching graphs of disks (this is clear from degree considerations alone; in fact it follows from a result of [MP92] that an exponential—in the size of the graph—sized ratio may be required to realize some graphs), but that the need for large disks cannot be determined by efficiently checked conditions (unless **P** = **NP**). Our techniques do not allow us to address the case of non-constant but sub-exponential ratio bounds.

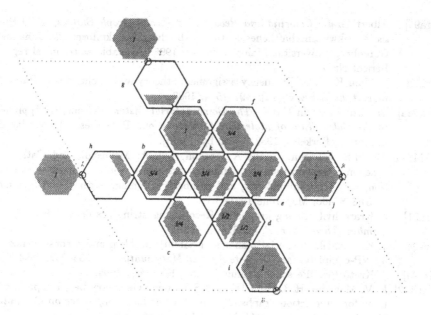

Fig. 5. Crossover module with terminals L, R, T, and B.

References

[Bre95]   Heinz Breu. *Algorithmic Aspects of Constrained Unit Disk Graphs.* Ph.D. thesis, University of British Columbia, Vancouver, Canada. In preparation.

[BC87]   Sandeep N. Bhatt and Stavros S. Cosmadakis. The complexity of minimizing wire lengths in VLSI layouts. *Information Processing Letters*, 25(4):263–267, 1987.

[BK93]   Heinz Breu and David G. Kirkpatrick. Unit Disk Graph Recognition is NP-Hard. Technical Report 93–27, Department of Computer Science, University of British Columbia, August, 1993. (To appear in *Computational Geometry: Theory and Applications.*)

[BL76]   K.S. Booth and G.S. Luecker. Testing for the consecutive ones property, interval graphs and graph planarity using PQ-tree algorithms. *J. Comput. System Sci.*, 13:335–379, 1976.

[Can88]   John Canny. Some algebraic and geometric computations in PSPACE. In *Proceedings of the 20th Annual Symposium on the Theory of Computing*, pages 460–467, Chicago, Illinois, 2–4 May 1988.

[CCJ90]   Brent N. Clark, Charles J. Colbourn, and David S. Johnson. Unit disk graphs. *Discrete Mathematics*, 86(1/3):165–177, 1990.

[FG65]   D.R. Fulkerson and O.A. Gross. Incidence matrices and interval graphs. *Pacific Journal of Mathematics*, 15(3):835–355, 1965.

[Fis83]   Peter C. Fishburn. On the sphericity and cubicity of graphs. *Journal of Combinatorial Theory, Series B*, 35:309–318, 1983.

[GJ79]   Michael R. Garey and David S. Johnson. *Computers and Intractability: a Guide to the Theory of NP-completeness.* W.H. Freeman and Company, 1979.

[Grä95]   Albert Gräf. *Coloring and Recognizing Special Graph Classes*. PhD thesis, Musikwissenschaftliches Institut, Abteilung Musikinformatik, Johannes Gutenberg-Universität Mainz, February 1995. Available as technical report Bericht Nr. 20.

[Hal80]   William K. Hale. Frequency assignment: theory and applications. *Proceedings of the IEEE*, 68(12):1497–1514, 1980.

[Hav82a]   Timothy Franklin Havel. *The Combinatorial Distance Geometry Approach to the Calculation of Molecular Conformation*. PhD thesis, University of California, Berkeley, 1982. Cited by [Fis83].

[JAMS91]   David S. Johnson, Cecilia A. Aragon, Lyle.A. McGeogh, and Catherine Schevon. Optimization by simulated annealing: an experimental evaluation; part II, graph coloring and number partioning. *Operations Research*, 39(3):378–406, May-June 1991.

[Kra91]   J. Kratochvíl. String graphs II. Recognizing string graphs is NP-hard. *J. Combin. Theory Ser. B*, 52: 67–78, 1991.

[Kra94]   J. Kratochvíl. A special planar satisfiability problem and a consequence of its NP-completeness. *Discrete Applied Mathematics*, 52: 233–252, 1994.

[Kra95]   J. Kratochvíl. Personal communication, February, 1995.

[MHR92]   M.V. Marathe, H.B. Hunt III, and S.S. Ravi. Geometry based approximations for intersection graphs. In *Fourth Canadian Conference on Computational Geometry*, pages 244–249, 10–14 August 1992.

[MP92]   S. Malitz and A. Papakostas. On the angular resolution of planar graphs. In *Proceedings of the 24th Annual Symposium on the Theory of Computing*, pages 527–538, May 1992.

[Rob68]   Fred S. Roberts. Indifference graphs. In *Proof Techniques in Graph Theory*, pages 139–146, Academic Press, New York and London, February 1968. Proceedings of the Second Ann Arbor Graph Theory Conference.

[Rob91]   Fred S. Roberts. *Quo Vadis, Graph Theory*. Technical Report 91–33, DIMACS, May 1991.

[Sac94]   Horst Sachs. Coin graphs, polyhedra, and conformal mapping. *Discrete Mathematics*, 134: 133–138, 1994.

[Spi85]   Jeremy Spinrad. On comparability and permutation graphs. *SIAM Journal on Computing*, 14(3):658–670, August 1985.

Fast Interactive 3-D Graph Visualization

Ingo Bruß*, Arne Frick**

Universität Karlsruhe, Fakultät für Informatik, D-76128 Karlsruhe, Germany

Abstract. We present a 3-D version of GEM [6], a randomized adaptive layout algorithm for nicely drawing undirected graphs, based on the spring-embedder paradigm [4]. The new version, GEM-3D, contains several improvements besides the adaptation to 3-D geometry.

The main result of this work is that for the first time, 3-D layout and presentation techniques are combined available at interactive speed. Even large real-life graphs with hundreds of vertices can be meaningfully displayed by enhancing the presentation with additional visual clues (color, perspective and light) and the possibility of interactive user navigation.

In the demonstration, we interactively visualize many graphs (artificial and real-world) of different size and complexity to support our claims. We show that GEM-3D is capable of producing a textbook-like drawing of the PETERSEN graph, a notoriously hard case for automatic drawing tools. To the best of our knowledge, this has not been achieved before by automatic layout algorithms purely based on heuristics.

1 Introduction

Visualizations of large discrete data structures are becoming increasingly more important in the literature [11, 16, 17] due to their practical relevance. Many discrete data structures from the real world can be modeled as (large) graphs, thus creating a need for automatic layout strategies to display them.

This paper introduces a new 3-D graph layout algorithm based on the well-known *spring-embedder* approach [4, 7, 9]. Spring-embedder algorithms have the ability to produce graph embeddings in the plane that look like projections of 3-D layouts onto a drawing area. It is therefore natural to extend a spring-embedder algorithm to 3-D and explore the effects. We chose the GEM (*graph embedder*) algorithm [6] for this study due to its excellent performance in terms of both layout time and the quality of the resulting drawings. In addition, GEM scales well to graphs with several hundred vertices, thus making it quite successful in practice. Despite being randomized, GEM turned out to deliver very stable results. The GEM algorithm combines the spring-embedder approach with ideas from simulated annealing by assigning each vertex a *local temperature*.

The remainder of this paper is organized as follows. As a general approach, we distinguish between the layout and presentation aspects of the discussed topics. Section 2 reviews related work on both aspects. In Sect. 3 we present the GEM-3D layout

---

* EMail: `bruss@wbkst10.mach.uni-karlsruhe.de`
** EMail: `frick@informatik.uni-karlsruhe.de`

algorithm and the presentation and interaction techniques used by the GEM3DDRAW system. Section 4 discusses the results. We evaluated both the algorithmic performance and the quality of the resulting drawings on the layout side. Regarding the presentation, we state observations and experiences made with the presentation techniques employed. The summary lists possible applications and directions for further research in this area.

2 Related work

Even more than in 2-D, the display of graphs in 3-D is actually a two-fold task, consisting of a *layout* phase, drawing the graph æsthetically, and a *presentation* phase, that applies viewing strategies, techniques and tools to present a meaningful view on the graph to the observer.

2.1 Layout

Previous research on 3-D graph drawing algorithms has focused on restricted kinds of drawings, e.g. tree and orthogonal drawings [2]. Recently, it has been extended to the layout of hierarchical information [15] in 3-D. The idea of drawing arbitrary undirected graphs in 3-D seems to have first appeared in [7]. This paper and the accompanying system demonstration extend those ideas and show that it is now possible to draw even large undirected graphs with several hundred vertices and possibly thousands of edges in 3-D with interactive speed.[1] It extends a similar result for computing drawings of undirected graphs in 2-D, that confirmed an earlier conjecture in [7] by using an improved adaptive cooling schedule[6]. In the remainder of this section, we briefly recall the basic working of the underlying class of algorithms, called *spring-embedder* algorithms, before we review GEM, the predecessor to the algorithm presented in this paper.

Spring-embedder algorithms use a physical model based forces that are exerted on the vertices in order to improve their positions according to several æsthetics [4]. Once the vertices are placed, the edges are drawn as straight lines between the vertices. The model states that vertices repel each other, while adjacent vertices are attracted to each other. These simple rules define a dynamic system that can be driven into a local energy minimum. The easiest strategy to do so is to use the *gradient descent* method, according to which only downhill moves are allowed, until no further improvements are possible [9]. Other strategies of achieving convergence are the use of simple cooling schedules that restrict the allowed moves over time [7] or to apply *simulated annealing* [3, 13].

The resulting drawings satisfy a surprising number of æsthetic criteria commonly used to evaluate the quality of drawings. In particular, vertices are almost evenly spaced in the drawing area, and the deviation of the edge lengths is low. Further important criteria believed to determine the æsthetic appearance of a drawing include *edge crossings* and *symmetries*. Statistical evidence to support the former claim has recently been

---

[1] In this context, "interactive speed" means that the initial layout should not take less than a few seconds, perhaps more if there is some sort of progress report such as animation. In the navigation phase, the user must receive immediate response on his commands, which are mostly mouse-based.

given in [14], while the data on the latter is not yet conclusive. Two important observation on spring-embedder algorithms can be made. Although the minimization of edge crossings and the maximization is not part of their energy functions to minimize, global symmetries are often found in the resulting drawings, if they exist in the graph. Secondly, graphs are often embedded with few edge crossings, or appearing as a 2-D projection of a 3-D layout. The latter property inspired us to construct 3-D layouts using the spring-embedder model. The objective has been to achieve æsthetic layouts of general undirected graphs and their presentation with interactive speed.

As the performance of annealing-based approaches is not very promising in terms of interactive speed [3, 13], we chose to base our research on the GEM algorithm presented in [6], a randomized adaptive algorithm. Based on the spring-embedder paradigm, several key factors contribute to its outstanding performance, among them a gravitational force and several heuristics to speed up convergence. Surprisingly, the resulting algorithm is very robust under several types of randomization. For example, vertex positions are updated one at a time according to a random permutation chosen initially.

Vertex updates are grouped into *rounds*, in which each vertex is updated exactly once. The direction of a position update is computed from the force resulting from the attractive, repulsive and gravitational force, while the length of the update move depends on the last update and an adaptive component called the *local temperature* of the vertex. Local temperatures are high initially and adapt to the movements of the respective vertex. Once GEM detects an *oscillation* of a vertex (see Fig. 1a) or a *rotation* of a subgraph (see Fig. 1b) during a vertex update step, the corresponding local temperature is lowered, thus further approaching the desired energy minimum. On the other hand, if a vertex keeps going in the same direction (cf. Fig. 1c), this is interpreted as a signal to accelerate it for further moves by increasing its local temperature.

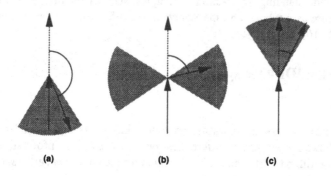

Fig. 1. Examples of vertex oscillation (a), rotation (b), and acceleration (c).

2.2 Presentation

There exist many techniques to display large graphs, e.g.

1. All information associated with the vertices and edges of the graph is displayed. The obvious drawback is that the details quickly become too small.
2. All information is drawn into a virtual drawing area, only part of which is visible at any given time. The user may browse the drawing by scrolling and arc traversal. This approach, however, tends to obscure the global structure of the underlying graph [8].
3. Several views are displayed at once: a global overview map preserving the overall structure, and one or more zoom-in maps displaying local information with details not present in the overview.
4. Distorting views such as the *fish-eye lens* approach [17] are useful to preserve the overall structure. The area within a *focal region* will be visible in detail, while other areas will be displayed distorted, i.e. smaller or with less detail. The rate of distortion rises with the distance from the focal region.

Another kind of distorting view is defined by arranging the objects under display in 3-D space, and to use a perspective projection onto the view plane. This leaves the details closest to the view plane unchanged, while objects further away are displayed smaller according to their distance to the view plane. We believe that 3-D views are more natural and therefore more intuitive than fisheye views, because they are closer to the human visual perception, although statistical evidence to support this belief is still missing.

There are several techniques to display 3-D pictures on a 2-D computer screen, among which *animation* is probably the simplest. Based on a single picture, the illusion of depth is created by having pictures differ slightly over time. *Stereopsis* requires two different pictures at any time, one for each eye. It may be realized by e.g. polarizing filters or holography. For simplicity reasons, we based the presentation of our layout results only on animation, although nothing prohibits the use of other techniques. Depth perception may be supported by graphical depth cues such as *color* and *light* to improve user understanding. Interaction techniques, such as *rotation*, *zoom*, *translation* of the structure as a whole, and the *selection* of artifacts within the structure further contribute [5, 16] to this objective.

3   The GEM3DDRAW system

3.1   Layout

Dependent parameters In contrast to previously known spring-embedder algorithms, GEM-3D allows for vertices of different *shape* and *extent*. This information is used in the layout algorithm to determine a lower bound for the *minimum vertex separation*

$$V_{\min} = v_{\text{avg\_size}},$$

which is given by the average vertex size $v_{\text{avg\_size}}$. The minimum vertex separation also influences another parameter, the so-called *optimal* or *desired* edge length, previously defined as a constant that would usually depend on the size of the drawing area [9]. In GEM-3D, the desired edge length is computed as

$$E_{\text{des}} = v_{\text{avg\_size}} \cdot \overline{v.deg},$$

where $\overline{v.deg}$ is the average edge degree of the vertex set. This serves the purpose of avoiding tightly packed layouts for highly connected graphs.

We shall now describe the GEM-3D algorithm in more detail. In doing so, we shall especially mention the differences regarding the 2-D version. The transformation of GEM from 2-D to 3-D was mostly straight forward, as the algorithm contains nothing inherently two-dimensional. Of course, the computation of Euclidean distances had to be adapted. The notion of opening angles was extended to opening cones. The most difficult part was the adaptation of the convergence-speedup heuristics. Figure 2 shows the algorithm at an abstract level. As convergence to a local minimum has not yet been formally proven, we ensure termination by using the constant R_{max} as an "emergency exit".

-- *Input:*		1
--	$G = (V, E)$ graph where	2
--	$V = $ **set of record**	3
--	s -- *shape type and size information*	4
--	ξ -- *current position*	5
--	p -- *last impulse*	6
--	t -- *local temperature*	7
--	d -- *skewness gauge (see text)*	8
--	$E = $ **set of record**	9
--	v_1 -- *adjacent vertex 1*	10
--	v_2 -- *adjacent vertex 2*	11
--	d -- *[optional:] directedness flag*	12
--	R_{max} *maximal number of rounds*	13
-- *Output:* for each $v \in V$, a position is computed		14
		15
compute T_{min}, *the desired minimal temperature*		16
compute E_{des}, *the optimal edge length*		17
compute E_{min}, *the minimal edge length*		18
forall $v \in V$ **do**		19
initialize v;		20
nrounds := 0;		21
while $\sum_i$ v.$t > T_{min}$ **and** *nrounds*< R_{max} **do**		22
forall $v \in V$ **do**		23
compute v.p;		24
update v.ξ *and* v.t;		25
nrounds := *nrounds*+1;		26

Fig. 2. Main loop of the GEM-3D algorithm.

Initialization and vertex choice remain the same as in the 2-D version. Vertices may be positioned randomly initially, and vertices are chosen for updates according to a random permutation.

Impulse computation We consider the computation of a new impulse for a given vertex u. The attractive, repulsive and gravitational forces $\mathbf{F_a}, \mathbf{F_r}$ and $\mathbf{F_g}$ are added up and smeared a little by adding a Brownian motion component consisting of a small random vector ρ with expectation $\mathbf{0}$:[2]

$$p = \sum_{v \in V} \mathbf{F_r}(u, v) + \sum_{(u,v) \in E} \mathbf{F_a}(u, v) + \mathbf{F_g}(u) + \rho, \tag{1}$$

resulting in the *current impulse* of the vertex, that is further scaled by the local temperature $\mathbf{v}.t$ and then applied to $\mathbf{v}$ as movement relative to its current position.[3]

Position and temperature update Figure 3 shows the abstract algorithm for updating the position and temperature of a vertex v. The position of v is updated by the current impulse, which is scaled with the current temperature of v. Afterwards, the temperature of a vertex is updated according to the rotation and oscillation detection heuristics. As shown in Fig. 1, a vertex is accelerated (decelerated) if it moves into the same (opposite) direction twice in a row, which can be detected by storing the impulse computed at the last update. As in the 2-D case, the definition of "same" and "opposite" depend on the opening angle α_o. The influence of oscillation detection is determined by a constant $\sigma_o < 1$.

Unfortunately, the definition of "rotation" cannot be naturally extended from 2-D to 3-D. In GEM, there is a skewness gauge indicating trends towards rotations by adding 1 for each approximate right angle movement to the left and -1 for each approximate right angle movement to the right. If the absolute value of the resulting counter becomes larger than a threshold τ, then the local temperature is reduced. The objective is to dampen vertex movements in situations such as rotating subgraphs.

The extension of this mechanism would obviously require an infinite number of skewness gauges, as there are infinitely many planes of rotation. Therefore, three alternatives were considered.

1. three counters, one for each of the coordinate planes $x - y$, $x - z$, and $y - z$. The counters would be used to detect rotations in the projections of the last and current impulse vector onto the coordinate planes.
2. a single counter to count the number of approximate 90° angles.
3. a global cooling schedule instead of rotation detection. In fact, a rotation detection scheme with large opening angle σ_r and small sensitivity σ_r is very similar to a cooling schedule, because it will almost always fire.

---

[2] The gravitational force draws vertices towards the barycenter of the current layout in order to keep disconnected components from drifting apart. However, we found that the constants should be lower than in the 2-D case. Otherwise, the graph will be embedded too tight into the drawing space. In addition, a low gravitational force effects smoother embeddings than a higher.

[3] In equation 1, the summation goes over all vertices for the repulsive force and all adjacent vertices for the attractive force.

```
-- Input:                                                    1
--   v          vertex to be updated                         2
--   p          current impulse of v                         3
-- Output:                                                   4
--   v          with updated ξ, t, d, p                      5
                                                             6
scale p by v.t                                               7
v.ξ := v.ξ + p;                                              8
β:=∠p, v.p;                                                  9
case β of                                                    10
    Acceleration then Increase v.t                           11
    Oscillation then Decrease v.t                            12
    Rotation then small decrease of v.t                      13
esac                                                         14
v.p:=p;                                                      15
```

Fig. 3. Position and temperature update algorithm.

3.2 Presentation

User support for *navigation* is available by user-controlled rotation, translation and zooming. In addition, a vertex may be selected by its name, using a pop-up menu or an entry widget in the user interface. The selected vertex will subsequently be rotated to the front of the display to also show its context.

In order to support *depth perception* of a 3-D image from a 2-D picture, we used several of the techniques mentioned in Sect 2.2. *Perspective* contributes to the perception of depth in the drawing by changing the size of vertices depending on their distance from the view plane. *Colors* can effectively help to reveal structure. In addition, a new technique called *artificial fog* was introduced. It modulates the color of objects depending on their distance from the view plane, e.g. by adding more and more of a dark color to make distant objects not only smaller, but also darker.

4 Results

In this section we summarize the results achieved so far with our prototype of GEM-3D. Again, algorithmic results are distinguished from results based on the presentation techniques. For obvious reasons, the pictures shown below cannot convey the same information as a self-guided, animated exploration of the graphs. Therefore, we have made our implementation publicly available for major platforms.[4]

---

[4] URL: http://i44www.info.uni-karlsruhe.de/~frick.

4.1 Layout

In this section, we consider both the quality and the performance of the layout algorithm.

Quality As for the quality of the resulting layouts, we observed that

- Like all spring-embedder algorithms, GEM-3D is good at separating vertices as much as possible in the drawing space (see Fig. 4 and 6).
- Vertices do not touch each other in general. No overlaps have been encountered yet.
- Although the focus of the algorithm is on fast convergence, symmetries are still often found (see Fig. 4 and 7).
- The 3-D topology of 3-D graphs is displayed (see Fig. 4 and Fig. 5).
- A textbook-like embedding of the PETERSEN graph can be produced (see Fig. 8 and Fig. 9) with an observed chance of 1 in 3–4 tries.

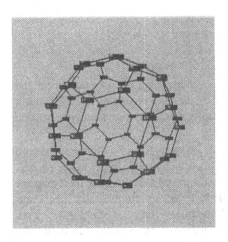

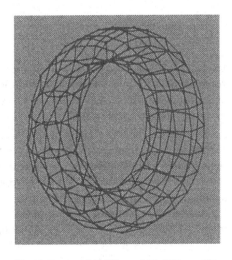

Fig. 4. The buckminsterfullerene molecule C_{60}. These examples shows the 3-D display, symmetry detection and vertex separation properties of GEM-3D.

Fig. 5. Torus with $|V| = 200, |E| = 400$. This figure took 6 seconds to be generated.

Performance All measurements were done on an SGI Indy workstation with a MIPS R4600 processor in normal daily use. Whenever times are given, we shall also provide the number of rounds necessary to compute the respective layout, as the inner loop computing the forces depends only on the graph size, while the outer loop appears to convey the complexity of the graph in terms of the number of rounds necessary to converge.

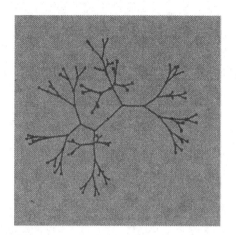

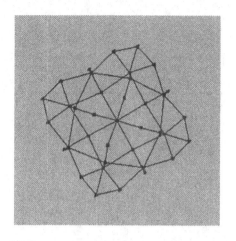

Fig. 6. Complete binary tree of height 7. This is another example of the vertex separation property.

Fig. 7. The highly symmetric graph in Fig. 26 of [7]. Although appearing to be a 2-D drawing, the third coordinate dimension is used to separate vertices, as the color differences indicate.

It turns out that the larger the graph gets, the faster is GEM-3D over its predecessor: The layout of the graph in Fig. 10 took only 65 rounds on the average, using 5 seconds of user time, an 80% increase over GEM, which required more than 100 rounds on the average. The 20x10 torus from Fig. 5 required a total user time of 6 seconds for the layout and then another 2 seconds to bring up the view. We determined empirically that the number of rounds required for convergence is usually sublinear in the number of vertices. For the worst case, the choice of $R_{max} = 5$ guarantees a linear number of rounds, resulting in a total complexity of $O(|V|^3)$ vertex updates. Even in this rarely occuring worst case, the resulting drawings are usually quite acceptable.

Rotation and oscillation detection both continue to contribute to the overall performance. The continuously good performance over a wide range of parameters and heavy use of randomization renders GEM-3D a very good candidate as a 3-D layout algorithm for undirected graphs.

4.2 Presentation

Experiments with GEM-3D support the initial hypothesis that depth cues help to convey much information in little screen space. On the other hand, graphs like the $K_{3,3}$ cannot be recognized by non-expert users without a coloring of the vertices hinting at the partition (see the right side of Fig. 11). This is an example of a more general problem with displaying information as graphs: Edges only convey neighborhood information, but sometimes the data structure contains information not being displayed by edges.

The selection mechanism proved to be useful to instantly focus on selected vertices, which may be hidden inside a large conglomerate of vertices and edges.

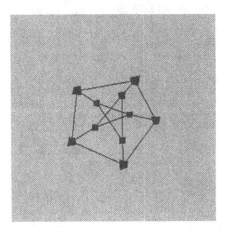

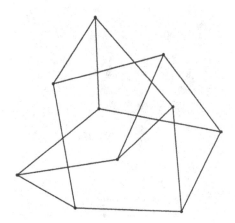

Fig. 8. Text-book drawing of the PETERSEN graph produced by GEM-3D. There is a fairly good chance to find such a drawing.

Fig. 9. Typical drawing produced by 2-D spring-embedder algorithms.

5 Conclusions

In this paper, we have shown how to extend the GEM algorithm to 3-D coordinate dimensions while significantly improving the layout speed and introducing new presentation techniques. The overall objective of drawing and presenting large graphs with interactive speed has been achieved.

We applied GEM-3D to real-world graphs, as well as to artificial ones. The results show that the algorithm draws 3-D structures as such, providing the viewer with helpful insights into their topology. For large, not too dense graphs, the resulting drawings are of high quality. In this case, the algorithm positions highly interconnected connected vertices into well-separated clusters. This is especially useful if combined with user interaction and depth cues (color and size of vertices).

3-D drawings may help to reveal structure in graphs unaccessible to known automatic 2-D graph layout algorithms. A good example, that may also be of some theoretical interest, is the fact that GEM-3D is able to produce a textbook-like drawing of the PETERSEN graph. Until now, it was impossible to achieve this result using spring-embedder algorithms, probably due to the fact that the textbook drawing violates the æsthetic criteria coded into the forces (spacing of vertices, minimization of edge length deviation).

Due to its potential to visualize even large structures in a very intuitive way, GEM-3D is now used in CAKETOOL, a large AI software system, to visualize Bayesian networks [1]. This provides evidence that the time has come for interactive 3-D graph visualization combined with automatically generated layouts. Other possible applications include the display of social networks [12] and the use as component in multimedia applications for displaying large hypertext-like structures.

Depth cues will be even more helpful if supported by light sources placed outside

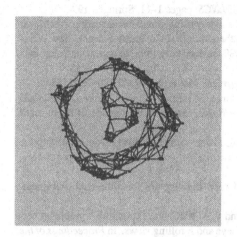

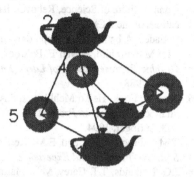

Fig. 10. Example of a large real-world graph ($|V| = 128, |E| = 508$), defined in [10] as *miles(128,0,0,0,0,10,4711)*.

Fig. 11. Example of how color, vertex shapes and labels help to reveal the structure of a graph, in this case a $K_{3,3}$. Vertices are colored, shaped and labeled according to their layer.

the drawing space. Experiments to this end are promising. Another promising approach is to substitute the rotation detection part by a simple global cooling schedule, as outlined in Sect. 3.1.

The current version of GEM-3D can draw directed graphs by displaying appropriate arrows on the edges. However, this does not imply that the forces between the connected vertices are treated as unidirectional. Therefore, the layout presently remains the same for the directed and undirected version of the graph. In the future, we plan to study the effect of one-way forces exerted by directed edges. Of course, this will lead us further away from the original physical interpretation of the state equations involved.

GEM3DDRAW was designed to be extensible. This allowed for the addition of further node shapes and visualization techniques (fog, perspective). Further research will explore the use of constraints to accommodate other æsthetic criteria and user knowledge about the data structure.

In conclusion, the GEM family of algorithms remains an attractive choice for the practitioner, while at the same time posing a challenge to theoreticians to actually *prove* the run-time and convergence behavior we observe so consistently, even under heavy randomization.

References

1. Ingo W. Bruß. Konzeption und Realisierung einer Visualisierungskomponente für komplexe Datenstrukturen unter dem Werkzeug CAKETool. Master's thesis, Universität Karlsruhe, 1995.

2. R. F. Cohen, P. Eades, T. Lin, and F. Ruskey. Three–dimensional graph drawing. In *Proceedings of Graph Drawing'94*, volume 894 of *LNCS*, pages 1–11. Springer, 1994.

3. R. Davidson and David Harel. Drawing graphs nicely using simulated annealing. Technical Report CS89-13, Department of Applied Mathematics and Computer Science, The Weizmann Institute of Science, Rehovot, Israel, 1989. revised July 1993, to appear in Communications of the ACM.

4. P. Eades. A heuristic for graph drawing. *Congressus Numerantium*, 42:149–160, 1984.

5. Kim M. Fairchild, Steven E. Poltrock, and George W. Furnas. *SemNet: Three-Dimensional Graphic Representations of Large Knowledge Bases*, chapter 5. Lawrence Erlbaum associates, 1988.

6. Arne K. Frick, Heiko Mehldau, and Andreas Ludwig. A fast adaptive layout algorithm for undirected graphs. In *Proceedings of Graph Drawing'94*, volume 894 of *LNCS*, pages 388–403. Springer, 1994.

7. T.M.J. Fruchterman and E.M. Reingold. Graph drawing by force-directed placement. *Software–Practice and Experience*, 21, 1991.

8. J.G. Hollands, T.T. Carey, M.L. Matthews, and C.A. McCann. Presenting a graphical network: A comparison of performance using fisheye and scrolling views. In *Proceedings of the 3rd International Conference on Human-Computer Interaction*, pages 313–320, September 1989.

9. T. Kamada and S. Kawai. An algorithm for drawing general undirected graphs. *Information Processing Letters*, 31, 1989.

10. Donald E. Knuth. *The Stanford GraphBase: A Platform for Combinatorial Computing*. ACM Press, New York, 1993.

11. J.D. Mackinlay, George G. Robertson, and S.K. Card. The perspective wall: Detail and context smoothly integrated. In *Proceedings of the ACM SIGCHI Conference on Human Factors in Computing Systems*, pages 173–179. ACM, 1991.

12. Cathleen McGrath, Jim Blythe, and David Krackhardt. The effect of graph layout on inference from social network data. In *Proceedings of GD'95*, 1995.

13. Burkhard Monien, Friedhelm Ramme, and Helmut Salmen. A parallel simulated annealing algorithm for generating 3D layouts of undirected graphs. In *Proceedings of GD'95*, 1995.

14. Helen C. Purchase, Robert F. Cohen, and Murray I. James. Validating graph drawing æsthetics. In *Proceedings of GD'95*, 1995.

15. S.P. Reiss. 3-D Visualization of Program Information. In R. Tamassia and I. Tollis, editors, *Graph Drawing DIMACS International Workshop GD '94*, number 894 in LNCS, pages 12–24. Springer Verlag, 1994.

16. George G. Robertson, J.D. Mackinlay, and S.K. Card. Cone trees: Animated 3-D visualizations of hierarchical information. In *Proceedings of the ACM SIGCHI Conference on Human Factors in Computing Systems*. ACM, 1991.

17. Manojit Sarkar and Marc H. Brown. Graphical fisheye views of graphs. *Comm. of the ACM*, 37(12):73–84, December 1994.

GD-Workbench: A System for Prototyping and Testing Graph Drawing Algorithms*

Luciano Buti[1], Giuseppe Di Battista[2], Giuseppe Liotta[3], Emanuele Tassinari[1], Francesco Vargiu[1], and Luca Vismara[4]

[1] Dipartimento di Informatica e Sistemistica, Università di Roma "La Sapienza"
Via Salaria 113, 00198 Roma, Italy
{buti,tassinar,vargiu}@dis.uniroma1.it
[2] Dipartimento di Discipline Scientifiche, Sezione Informatica
Terza Università di Roma
Via Segre 2, 00146 Roma, Italy
dibattista@iasi.rm.cnr.it
[3] Department of Computer Science, Brown University
115 Waterman Street, Providence, Rhode Island 02912–1910
gl@cs.brown.edu
[4] Istituto di Analisi dei Sistemi ed Informatica, Consiglio Nazionale delle Ricerche
Viale Manzoni 30, 00185 Roma, Italy
vismara@iasi.rm.cnr.it

Abstract. We present a tool for quick prototyping and testing graph drawing algorithms. The user interacts with the system through a diagrammatic interface. Algorithms are visually displayed as directed paths in a graph. The user can specify an algorithm by suitably combining the edges of a path. The implementation exploits the powerful functionalities of Diagram Server and has been experimented both as a research support tool and as a back-end of an industrial application.

1 Overview

We present GD-Workbench (GDW), a system for quick prototyping and testing graph drawing algorithms. The user interacts with GDW through a diagrammatic interface. Graph drawing algorithms are visually displayed as directed paths in a graph. The potential users of GDW are:

- Graph drawing *researchers* that aim at *experimenting* existing algorithms against real-life or randomly generated graphs; an experimental work performed with an early version of GDW is described in [2].
- Graph drawing *researchers* that aim at *implementing* new algorithms and at quick understanding how such algorithms can exploit existing algorithms as subroutines.

* Work supported in part by Progetto Finalizzato Sistemi Informatici e Calcolo Parallelo of the Consiglio Nazionale delle Ricerche, by ESPRIT Basic Research Action No. 7141 (ALCOM II), by the National Science Foundation under grant CCR-9423847, and by N.A.T.O.- C.N.R. Advanced Fellowships Programme.

- Graduate course *teachers* that aim at easily demonstrating in their classes the behavior of graph drawing algorithms.
- *Professionals*, already skilled with graph drawing, that want to select the best algorithm for a given application; as an example, GDW has been used in cooperation with the *Italian Authority for Computer Engineering in the Public Administration*: about 2, 500 Entity-Relationship diagrams have been automatically drawn and the whole work (algorithm-selection, graphic features setting, fine-tuning, large-scale drawing, and printing) has been performed with GDW.

GDW has the following main functionalities:

- *Path-management.* It allows to construct a new algorithm by composing steps of existing algorithms. The new *algorithm* is visually represented as a *path in a graph.* It also allows to visualize several algorithms at the same time, to discard previously visualized algorithms, and to show info on algorithms.
- *Schema-management.* It allows to load graphs (we will use the terms schema and graph as synonyms) and to create them either with a graph editor or with random graph generators.
- *Test-management.* It allows to draw the current graphs with the currently visualized algorithms and to generate reports on the aesthetic features of the drawings.

Although several interesting and powerful tools have been recently devised in the graph drawing field (to give only a few examples we mention [5, 4, 7, 6, 3]), we believe that GDW has several innovative characteristics. In particular, the user-interaction paradigm of GDW has flexibility and friendliness features that, to our knowledge, have no counterparts in existing tools. The flexibility and friendliness of GDW are both in providing an easy interaction with the algorithms and in showing diagrams. Existing tools usually focus on just one of these two aspects.

Rather than presenting all the details of the architecture and of the implementation, we prefer to start the description of the system with an introductory example (Section 2). However, the main architectural issues are outlined in Section 3. Section 4 describes further examples of usage of GDW. Future research directions are sketched in Section 5.

The paper is supplied with several figures, most of them snapshots of the screen.

2 An Introductory Example

We show how a user can simply construct his/her own graph drawing algorithm with GDW, by combining pieces of existing algorithms.

2.1 The Taxonomy

GDW presents the algorithms to the user through a *taxonomy* of classes of graphs. The most general class of graphs of the taxonomy is *Multigraph*; a multigraph is a graph that has both directed and non-directed edges. All the other classes of the taxonomy are subclasses of *Multigraph*. Each class is provided with a set of *methods* that map an object of a class into an object of another class. A method is a layout functional step, taken from an existing algorithm. A drawing algorithm *A* is a sequence of methods that is visually represented on the taxonomy as a path (*algorithmic path*); the edges of the algorithmic path describing *A* are the methods that compose *A* and the vertices are the classes of the taxonomy the methods are associated to.

The taxonomy is a very general structure to classify graph drawing algorithms and has been already exploited for the internal structure of the algorithms database of Diagram Server [1, 3].

2.2 Constructing a New Algorithm

Suppose the user wants to draw graphs with a polygonal graphic standard (i.e. all edges are polygonal lines) and to this aim wants to construct a "new" algorithmic path. He/she opens a window displaying the taxonomy and executes the following steps:

- All the classes of the taxonomy are displayed on the screen. The dashed edges of the taxonomy show containment between classes. Class *Multigraph* is white colored. White classes (in this case only *Multigraph*) are the already selected classes for the algorithmic path (Fig. 1).

  The classes *Connected*, *Planar*, and *Digraph* are red colored. Red classes are the ones that can be reached by applying an available method of the last selected class of the currently constructed algorithmic path (in this case class *Multigraph*).

- The user clicks on one of these three classes, say *Planar*. The system displays the set of available methods that transform an element of *Multigraph* into an element of *Planar*. The user can now select one method of the set. In this case the set consists of just one method, namely `MakePlanar` (Fig. 2).

- Now class *Multigraph* is white, class *Planar* is colored white and red, and the classes *Connected*, *Digraph*, *ConnectedPlanar*, and *FourPlanar* are red. *Planar* is white and red because (i) it is already selected for the algorithmic path and (ii) there is method `MakePlanar` (inherited from the class *Multigraph*) that transforms an element of *Planar* into an element of the class itself. The user can now click on any reachable class, i.e. either a red or a white and red class, say *ConnectedPlanar* (Fig. 3).

  The system displays the set of methods that transform an element of *Planar* into an element of *ConnectedPlanar*. The user can now select one method of the set. In this case the set consists of two methods, namely `MakeConnected` and `IsConnected`. In our example, the user selects `MakeConnected`.

- The user, by performing other similar operations, constructs an algorithmic path whose final class is *Polygonal*. The complete path is now on the screen and consists of classes *Multigraph*, *Planar*, *ConnectedPlanar*, *Biconnected-Planar*, *PlanarSTDigraph*, *ReducedPlanar*, *Straightline*, and *Polygonal*, plus the methods connecting them (Fig. 4).

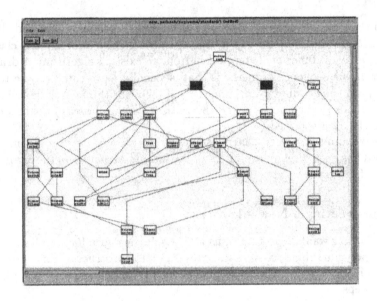

Fig. 1.

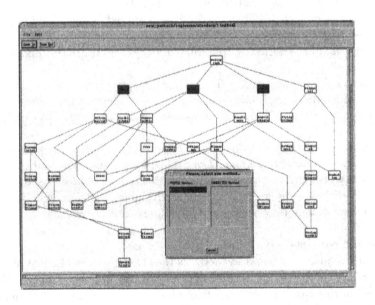

Fig. 2.

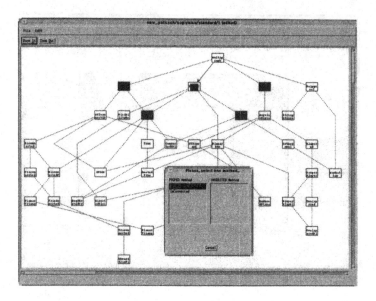

Fig. 3.

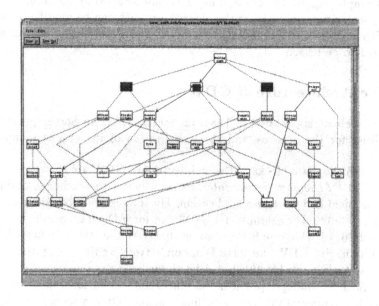

Fig. 4.

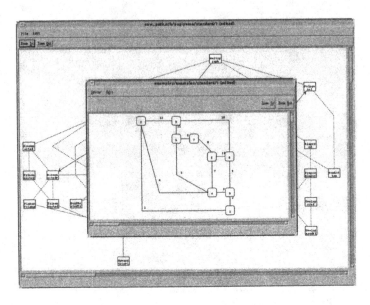

Fig. 5.

– The user can now execute the algorithmic path to obtain a drawing of an input graph (Fig. 5). Of course, if he/she is not satisfied by the resulting drawing, with similar operations the algorithmic path can be modified, choosing different classes and/or different methods. The algorithmic path can also be stored to be reused.

3 The Architecture of GDW

GDW is a client application of Diagram Server. Diagram Server provides the capabilities for the drawing and the visualization of the graphs managed by GDW.

In Fig. 6 the main blocks of the architecture of GDW are shown.

The *GDW / Diagram Server Interface* coordinates the exchange of data between the client and the server application. The interaction is based on a message passing technique. For example: (i) GDW can force Diagram Server to wait for a user action; (ii) Diagram Server can notify GDW that the user has clicked on a menu item; (iii) GDW can force Diagram Server to enter the status in which vertices and edges can be added or deleted; etc.

The *Schema Manager* provides the functionalities for loading, creating (automatically or manually), and discarding schemas. Once a schema is loaded, it is active and it can be used during the testing of the algorithmic paths.

A schema can be randomly generated by using the *Random Graph Generator* of GDW. It is possible to generate different types of graphs, including connected graphs, biconnected planar graphs, and trees. The component uses two different strategies for the generation of the graphs: (i) they are generated from scratch,

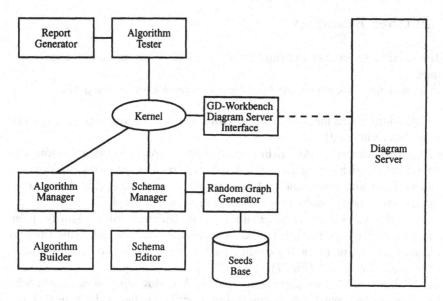

Fig. 6. The Architecture of GDW.

by using, first, randomly insertion of edges and, second, local adjustments that force them to belong to the chosen class; (ii) they are generated starting from a core set of existing real-life based graphs (stored in the *Seeds Base*), by means of a list of operations that preserve the similarity with the starting graph.

Alternatively, a schema can be constructed by using the *Schema Editor*, a powerful interactive editor that allows to add, delete, move, reshape vertices and edges.

The *Algorithm Manager* manages the algorithmic paths. They can be loaded, created, visualized, and discarded. As for the schemas, once an algorithmic path is loaded, it becomes active.

The *Algorithm Builder* is the component devoted to the creation of new algorithmic paths. Algorithmic paths can be created also starting from existing ones. The Algorithm Builder, through the GDW / Diagram Server Interface, asks Diagram Server to display the taxonomy on a window. The actions performed by the user on the taxonomy are captured by Diagram Server and notified to the tool, that executes the proper operations (e.g. opening of a dialog window, sending a message for highlighting a class of the taxonomy, sending a message for the insertion of an edge).

The *Algorithm Tester* and the *Report Generator* are the components for the testing and the evaluation of the active algorithmic paths on the active schemas. Partial and full reports on the tests can be generated as well as diagrams with disparate graphic features.

4 Further Examples

In this section we exploit the functionalities of GDW by means of a set of examples.

The first example shows the facilities to manage algorithmic paths.

- AlgorithmicPaths menu allows to activate, discard, create, highlight, get info on algorithmic paths.
- ActivatePath item of AlgorithmicPaths menu displays a dialog window with a set of algorithmic paths that have been stored in previous working sessions. Once an algorithmic path is selected, it is activated, it is shown in the taxonomy, and it can be executed on a given set of graphs.
- Displaying several active algorithmic paths. Different colors identify different algorithmic paths. White classes and edges describe subpaths that are shared by two or more algorithmic paths. For example, the yellow path is the algorithm bend2 (Fig. 7).
- Info about the active algorithmic paths. A dialog window is shown with the correspondence between colors and algorithmic paths. By selecting one color, the list of classes and methods of the corresponding algorithmic path is displayed in a text window (Fig. 8).
- Highlighting an algorithmic path. When several algorithms are shown at the same time, the screen may become difficult to read. Each active algorithmic path can be highlighted by selecting it in a dialog window.
- By clicking on a class (say *PlanarTriangulated*) the system displays the available methods for that class and for each method suitable bibliographic references (Fig. 9).

The following example illustrates the capabilities of GDW in managing graphs.

- Schemas menu allows to activate, discard, and get info either on single schemas or on directories of schemas.
- ActivateSchema item of Schemas menu displays a dialog window with the stored schemas is shown. Once a schema is selected it can be drawn using the active algorithmic paths.
- InfoSchemas. The identifiers of the active schemas are shown on a dialog window.
- Random graph generation. GDW is provided with a random graph generator The user, by means of a dialog window, can select the class of graphs to be generated, their number and size. In this case two biconnected graphs with ten vertices are generated (Fig. 10).
- Alternatively, the user may construct graphs by using a powerful interactive graph editor.

The functionalities of GDW in testing algorithms are shown by the following example.

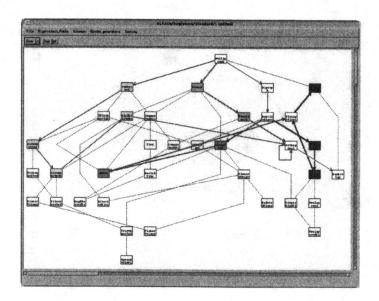

Fig. 7.

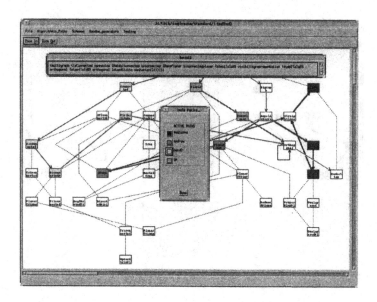

Fig. 8.

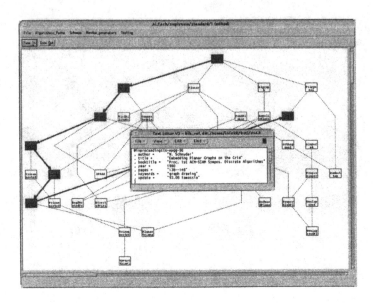

Fig. 9.

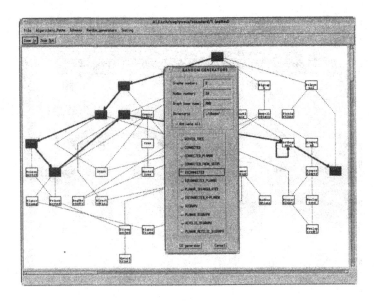

Fig. 10.

- Testing menu allows to apply the active algorithmic paths on the active schemas (randomly generated or manually constructed) and to setup several output options.
- Output Options item of Testing menu displays a dialog window in which the user can select the diagrams to be displayed on the screen and choose to generate the reports.
- The diagrams obtained by applying the active algorithmic path (**bend2**) to two randomly generated graphs are shown in Fig. 11.
- In the left window of Fig. 12 the reports on the previous application are displayed, while in the right window the average results of the reports are displayed.

5   Future Work

We will improve and expand our tool in the following directions.

- We plan to interconnect GDW with an object-oriented software development platform, in order to support the whole development cycle of a graph drawing algorithm.
- In order to better show experimental reports, we aim at integrating into GDW a system for visualizing graphics.
- We aim at extending our experiments by further interacting with the Italian Public Administration, which is a promising source of case studies.

References

1. P. Bertolazzi, G. Di Battista, and G. Liotta. Parametric graph drawing. *IEEE Trans. Softw. Eng.*, 21(8):662–673, 1995.
2. G. Di Battista, A. Garg, G. Liotta, R. Tamassia, E. Tassinari, and F. Vargiu. An experimental comparison of three graph drawing algorithms. In *Proc. 11th Annu. ACM Sympos. Comput. Geom.*, 1995.
3. G. Di Battista, G. Liotta, and F. Vargiu. Diagram Server. *J. Visual Languages and Computing* (special issue on Graph Visualization, I. F. Cruz and P. Eades, editors), 6(3), 1995.
4. C. Ding and P. Mateti. A framework for the automated drawing of data structure diagrams. *IEEE Trans. Softw. Eng.*, SE-16(5):543–557, 1990.
5. P. Eades, I. Fogg, and D. Kelly. SPREMB: a system for developing graph algorithms. *Congressus Numerantium*, 66:123–140, 1988.
6. M. Himsolt. GraphEd: A graphical platform for the implementation of graph algorithms. In R. Tamassia and I. G. Tollis, editors, *Graph Drawing (Proc. GD '94)*, volume 894 of *Lecture Notes in Computer Science*, pages 182–193. Springer-Verlag, 1995.
7. F. N. Paulish and W. F. Tichy. EDGE: An extendible graph editor. *Softw. - Pract. Exp.*, 20(S1):63–88, 1990.

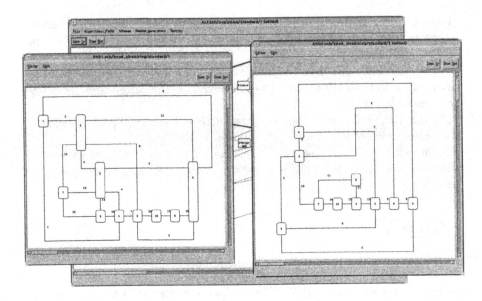

Fig. 11.

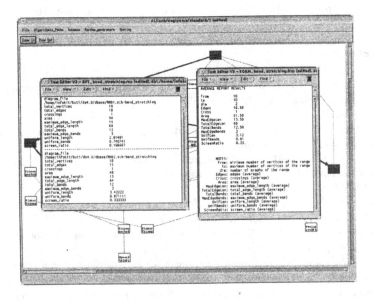

Fig. 12.

CABRI-Graph: A Tool for Research and Teaching in Graph Theory

Yves CARBONNEAUX, Jean-Marie LABORDE and Rafaï Mourad MADANI

IMAG-LSD2, BP 53 - 38000 Grenoble Cedex 9, France
e-mail : Yves.Carbonneaux, Jean-marie.Laborde, Mourad.Madani@imag.fr

Abstract. We present the graph visualization and compuational system CABRI-Graph, an interactive tool that can be used by student, teacher or researcher. The user may construct graphs interactively, select algorithms or graph transformations from a menu, and view the results directly on the screen

1. Introduction

When studying a practical problem one have often to draw a symbolic representation of the problem which may allow a better visualization and then a better understanding of the situation. A natural way is to represent the interactive objects of the situation by points and lines expressing relationships. That is the graph representation. As practical problems have always complicated representations, the use of some graph software becomes unavoidable.

There exist three categories of graph software : systems like GraphBase [6] which consist of a library of graph algorithms, systems like EDGE[7] or DaVinci [4] concentrate on graph visualization and editing, and system like GraphEd [5] which is a combination of an editor and a library of algorithms, and also set of graph layout algorithms.

The goal of the CABRI-Graph[1] software is to create interactively a graph. CABRI-Graph [1, 2] allows you to handle graphs as you would do it on paper, with all the facilities and power a microcomputer can bring. The main interest of CABRI-Graph ("CAhier de BRouillon Interactif", i.e. "computerized sketchpad") lies in its highly interactive behaviour.

2. CABRI-Graph

CABRI-Graph consists essentially of a graph editor, associated with a toolkit allowing different computations, such as evaluation of graph invariants (chromatic number, stability number, chromatic polynomial, ...) or the performing of some classical transformations. Transformations upon graph drawing are also available (planar drawing of planar graphs, graph embedding, ...).

---

[1] CABRI-Graph and related documents are available by anonymous ftp from the site ftp.imag.fr, /pub/Cabri/ .

CABRI-Graph is structured as two parts, namely the interactive interface and the toolkit, which can be subdivided in three subparts: random generator of graphs, graph transformations and graph computations.

2.1 The User Interface

CABRI-Graph conforms strictly to the interface standards recommended by Apple Macintosh. CABRI-Graph manages a so called graph-clipboard of its own which can be edited as any other graph window. The graph-clipboard allows to exchange graphs in picture format with other text processing or drawing sofware (MacDraw, MacPaint, MacWrite, Word, ...).

All operations are menu, mouse or key and mouse driver. Working window has a menu bar, and there are keyboard shortcuts for frequently used commands. We have tried to make the user interface as easy as possible; it takes a few minutes to learn the basic commands of CABRI-Graph. The graphs may have an unlimited number of vertices and edges. Only memory capacity of hardware machin can stop the size of the graph.

2.2 The Toolkit

A number of actions, transformations or computations may be applied to a graph and are obtained through the menus; each one concerns the graph itself or the subgraph induced by the selected vertices. Meanwhile it is already possible to perform interesting experiments with the random generator of graph.

2.2.1 Random Graph

The random generator of graphs [2], satisfying a number of parametrizable properties (fig. 1) has been specially developped to allow the user (student, teacher, researcher in graph Theory, ...) to experiment with theorems: one may also invalidate a hasty conjecture, or on the contrary, bring some support to a more elaborate one.

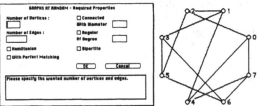

Fig. 1. Random generator of graph.

2.2.2 Transformation and Computation

The product of graphs is often difficult to visualize. CABRI-Graph propose a tool for computing three types of product: cartesian product, cross-product and complete product of the graph on the screen and the graph on the clipboard.

It can be interesting to use CABRI-Graph to draw nice figure for a publication. This work can be made extremely easy by taking advantage of different features of the editor CABRI-Graph, such as the random generator and the operations product and identify.

For instance CABRI-Graph includes the construction of cliques, of line-graphs, the computation of connected components and the other basic properties and invariants.

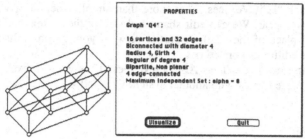

Fig. 2. Q4 graph and properties.

CABRI-Graph also propose different graphical representations designed to emphazise some properties of the graph (fig. 2). For instance, the 4-cube in figure 2 have been drawn by "Product by K2" (the "Transform" menu) with an edge as basis. The "Functions" menu makes you able to compute and visualize the properties of the drawn graph.

3. The Editor or Graphical Representation of Graph

The objects manipulated by CABRI-Graph are sets of points together with edges inter-connecting them and the edges. The incidence relation between vertices and edges is not the only mathematical structure involved, but also the geometrical aspects (the edges can be lines or curves).

CABRI-Graph offers various options to customize the appearance of vertices and edges :

Vertex Vertex can have arbitrary shape (square, circle or triangle), arbitrary size and arbitrary color - fundamental eight color without black; the black color is to represent a selected vertex.

Edge[2] Edge can have also arbitrary shape (line or curve), arbitrary size and arbitrary color - fundamental eight color without white; the white color is to represent a selected edge.

Fig. 3. A graph with a curved edge.

---

[2] We have introduced in CABRI-Graph the possibility to draw curved edges and thus multiple edges. To allow all different forms of edges (loops, laces, circle arcs, ...), we use cubic Bézier curves which permit appropriate forms [3].

Label CABRI-Graph supports arbitrary labels for vertices. There is no restriction on the contents or the lenght of the labels. Label is placed automatically, but the user can place it elsewhere. Furthermore, labels may have arbitrary fonts, size, and style.

The command "Edit Vertices ..." is more than simple labeling of vertices. The dialog window appear. We can edit shape and size of the vertex, and also string, font, size and place of the label. A partial view of your graph is displayed. The vertex you are editing is pointed out.

CABRI-graph saves graphs as ASCII text files. Thus you can exchange graphs written in the same format with another system like UNIX.

4. Conclusion

CABRI-Graph has many usefull facilities as to manipulate, to visualize.graphs and properties, to compute graph invariants or graph products.

It has been extensively used as a support for teaching, and much more for research. CABRI-Graphes allow to elaborate conjectures or to elaborate proofs and part of proofs for them. The lay point is to check possible intermidiate step in the whole proof process.CABRI is an acronym for "CAhier de BRouillon Interactif" (interactive notebook). The project is developped at the Laboratoire de Structures Discrètes et de Didactique", IMAG (Grenoble Institute of Computer Science and Epllied Mathematics). CABRI-Graph is written in C and runs on AppleMacintosh.

References

1. Baudon Olivier (1990), CABRI-Graphes, un CAhier de BRouillon Interactif pour la Théorie des Graphes, These de 3-ième cycle, Grenoble , France.
2. Bordier J. & Laborde J-M. (1991), An Interactive Tool for Graph Theory, 7th Annual Apple EUC, Conference proceeding, Paris, France.
3. Carbonneaux Y., Madani R. M. & Laborde J-M. (1995), Characterization of Bezier Curves based on Fixed Points, submitted.
4. Fröhlich M. & Werner M. (1994) Demonstration of the Interactive Graph Vizualisation System *da Vinci,* Lecture Notes of Computer Science, Springer-Verlag, n° 894, pp. 266-269.
5. Himsolt M. (1994) GraphEd: A Graphical Platform for the Implementation of Graph Algorithms, Lecture Notes of Computer Science, Springer-Verlag, n° 894, pp. 182-193.
6. Knuth Donald E. (1994) The Stanford GraphBase, A Platform for Combinatorial Computing, ACM Press (Addison-Wesley Publishing Company).
7. Newberry Paulisch (1993) F: The Design of an Extensible Graph Editor, Lecture Notes of Computer Science, Springer-Verlag, n° 704.

Graph Folding:
Extending Detail and Context Viewing into a
Tool for Subgraph Comparisons

M. Sheelagh T. Carpendale, David J. Cowperthwaite,
F. David Fracchia and Thomas Shermer

School of Computing Science, Simon Fraser University
Burnaby, B.C. V5A 1S6 Canada
carpenda@cs.sfu.ca

Abstract. It is a difficult problem to display large, complex graphs in a manner which furthers comprehension. A useful approach is to expand selected sections (foci) of the graph revealing details of subgraphs. If this expansion is maintained within the context of the entire graph, information is provided about how subgraphs are embedded in the overall structure. Often it is also desirable to realign these foci in order to facilitate the visual comparison of subgraphs. We have introduced a distortion-based viewing tool, three-dimensional pliable surface (3DPS) [1], which allows for multiple arbitrarily-shaped foci on a surface that can be manipulated by the viewer to control the level of detail contained within each region. This paper extends 3DPS to include the repositioning of foci so as to bring together spatially separated regions for the purpose of comparison while retaining the effect of detail in context viewing. The significance of this approach is that it utilizes precognitive perceptual cues about the three-dimensional surface to make the distortions comprehensible, and allows the user to interactively control the location, shape, and extent of the distortion in very large graphs.

Keywords: distortion viewing, graph layout, 3D interactions, information visualization, interface design

1 Introduction

An increasing number of large and complex graphs are being generated in a great variety of fields; in computing alone they are used to express such things as visual languages, software, hypertext, natural language parsing, and databases. Part of the appeal of graphs is that they are both a mathematical and a visual formalism. While there are several available visual interpretations of a graph, creating a display that actually aids in their comprehension is far from trivial. For instance, simply spreading a complex graph across the screen usually results in dense and confusing visual clutter. However, as Tufte [19] reminds us it is not in the nature of information to be confusing, rather it is the display that needs consideration.

Ideally, one would like to take advantage of our natural visual pattern recognition abilities by being able to see the entire graph. Also, the details of subgraphs and how these subgraphs are embedded in the overall structure are of interest. However, with existing solutions such as panning and zooming, the desire to examine detail often conflicts with the ability to maintain global context. Zooming out or compressing the data to fit within the space of the screen can result in its becoming too dense to discern detail. Zooming in or magnifying to reveal sufficient detail results in the loss of context.

Multiple views allow for the simultaneous display of detail and global structure in separate images, however the integration of these distinct views must be performed consciously by the user. Evidence as to how we combine information from multiple sensory channels has arisen from a number of studies in experimental psychology [2, 10, 11]. Information perceived as a single event is integrated automatically, while that perceived as distinct events requires a more strenuous re-integration. Even though the user may be cognitively aware that views in multiple windows pertain to a single information space, perceptually they remain distinct. For example, the effort of maintaining which subgraph belongs where and of its exact embedding has to be performed consciously by the user. If the desired detail view can be provided in a manner that smoothly integrates it into the global context then it preserves the possibility of visual gestalt.

Several viewing methods have been presented that allow the user to access detail within context through the use of various distortion techniques. This is a growing body of work pioneered by Furnas's [3] paper on generalized fisheye views. These techniques are usually based on the fisheye lens metaphor, creating a magnified focus for chosen sections and displaying the rest of the graph in decreasing scale as distance from the focus increases. Some of the main themes are: finding a balance between current interest and relative importance of the information [12, 13, 17, 18], using a mathematical curve to achieve magnification (arctan [6], hyperbola [8]), and using perspective projection to create the detail in context views [9, 16]. For a more detailed survey see [14]. These approaches provide various integrated views displaying both the required details while maintaining global context.

However, one advantage of separate views in multiple windows which has previously been lost with a detail in context viewing tool is the ability to move and reposition individual views. This is often used to align images of separate subgraphs so that visual comparisons are facilitated. We extend our distortion viewing technique, 3DPS [1], to allow this freedom while maintaining integration advantages that come with the perception of the graph as a single event.

The brief overview of 3DPS in Section 2 is followed by an explanation of how folding extends 3DPS to provide repositioning of foci (Section 3). Methods used to enable comprehension of the resulting form are presented in Section 4. Section 5 contains examples of folding. Finally, Section 6 summarizes the results.

2 3D Pliable Surface

The intention behind distortion viewing is to magnify a chosen section or focus until the desired level of detail is revealed. To compensate for the extra screen space used by this magnification, the rest of the image is distorted and/or compressed. Ideally this compression is gradual enough to provide good visual integration between the focus and its context.

For the basic distortion function we chose the three-dimensional Gaussian curve as its bell shape curves away from the focus at its apex and inflects to curve back into the surface (Figure 1). These Gaussian curves transform the

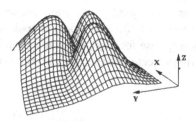

Fig. 1. 3D surface of blended Gaussian curves

two-dimensional flat surface into a three-dimensional curved surface. The three-dimensional nature of this distortion approach offers several advantages. For instance, using single-point perspective to view the surface from above provides magnification of detail to scale. Second, it provides a useful metaphor for the actions performed to create the distortions; pulling a section towards oneself to see it better, or in this case magnify it, appears to be a natural response.

Magnification of Single Focus to Scale: Magnification is provided by raising a 3D Gaussian curve perpendicular to the surface. The center of the gaussian

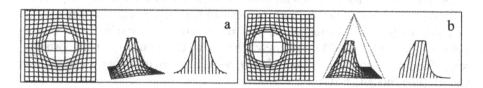

Fig. 2. (a) Single focus with flattened top, leftmost image shows the top view, center image shows the 3D view, and rightmost image shows the profile with ToEye vectors along which it is raised. (b) This single focus is to one side of center. Dotted lines denote the viewing frustum. The ToEye vectors directed at the viewpoint are shown

curve is projected up to the height h_c. To provide a flat region where only scaling occurs, the curve may be truncated; limited to a fraction f of this maximum height (Figure 2a). The points of the graph in the central magnified region are all projected up to the same height, $h_c f$. The height h_p of all other points on the curve is a simple relationship of the distance d_p to the center of the region, the height h_c, and its standard deviation s_c:

$$h_p = h_c \exp^{-\frac{s_c}{d_p}}$$

To keep foci from any point on the surface inside the field of view (Figure 2b) we use vectors directed from the plane to the view-point, *ToEye*, rather than move the view-point to align with the focus as in [16]. The ToEye vector from the center of a focus is used throughout the curve. This provides the desired magnification and ensures magnification to scale across the tops of the foci. Normalizing the z-component of these vectors instead of the length provides equivalent magnification response for any point on the surface. Within the focus region, where all points are projected to the same height, scaling is still preserved.

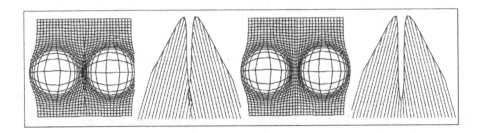

Fig. 3. Examples of two foci using ToEye vectors. When viewed from above the visual effect is as if the foci are being pulled up perpendicular to the plane On the left the inter-focal region is unblended; On the right blended

Multiple Foci: Using ToEye vectors for each focus allows for multiple foci within the field of view and magnification to scale for each focus region (Figure 3). However, a point under multiple curves will have a projection vector associated with each curve. A blending is performed using the curve's ToEye vector and a weight contributed by the curve's height at this point [1] This allows for larger foci to be positioned more closely, while still maintaining a continuous smooth (unwrinkled) surface.

Foci with Arbitrary Shapes: The single-point foci can be extended to other arbitrary shapes as well, for example lines or polygons (Figure 4). The exact shape and location of a focus can be drawn on the screen by the user. Now the

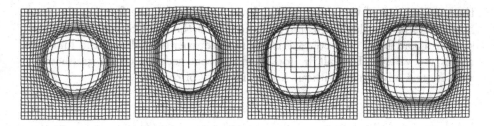

Fig. 4. Various foci types: from left to right: point, line, convex and concave polygons

height h_p of a point outside of a focus but within a region is determined not by its distance from the center of the region but by its distance to the edge of the defined focus. If the point is either on the line or within the polygonal focus it is projected to the full height h_c of the curve. The center of the arbitrary region is still used to determine the vector to the eye.

Distortion Control: In any distortion viewing tool compromises are made between the magnification in each foci and compression in the rest of the image. Our model offers the user considerable control not only over how much compression there is but where minimum and maximum compression occurs. The pattern of compression is a direct result of the slope of the curve. Therefore giving user control of the parameters (height h_c and standard deviation s_c) that affect the slope and providing an auxiliary function (half sine wave) that can be subtracted to actually adjust curvature allows to user to set preferences such as increased magnification in the region immediately adjacent to the focus or a more gradual integration into the context.

3   Surface Folding

In many cases it is desirable to provide the ability to bring detail views of spatially separated regions of an image together in order to facilitate visual comparisons. Traditionally this has meant the use of magnified views in subwindows which are moved independently of the original image and hence have no direct visual connection to the rest of the image. Folding allows for this freedom to reposition magnified sections without detaching them from the rest of the image.

Folding: Extending 3DPS to include the repositioning of multiple foci while retaining the effect of a detail in context viewing tool, is possible in part because of the three-dimensional nature of the distortions. A focus, or magnified section, is the top of a 'hill'. The steepness of the sides of this 'hill focus' can be adjusted to minimize interference with other foci. The top, or focus can then be moved without changing the location of the base of the hill. The stable base maintains the same section of the graph within focus, and keeping the focus at the same height retains the degree of magnification. It is the sides of the hill that are stretched and bent. The context is maintained over the sides of the hills and across the valleys. This allows foci from different sections of the graph to be repositioned adjacently, without losing the sense of context.

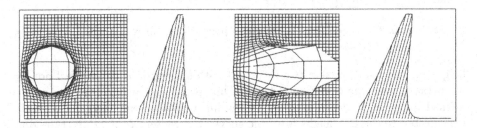

Fig. 5. Single off-center focus. On the left, pulled towards viewpoint; on the right, folded across the image

As the surface appears to be a solid object, we as humans, will assume it to be complete. This is an asset, because if the surface is perceived as complete then it can be stretched, folded, and warped without sections of it perceptually disappearing. Unfolding or viewing it from a different angle will expose temporarily obscured sections. The result is a tool that can be used analogously to folding a printed map to expose the areas of interest. This allows for the repositioning of foci without loss of the perception of the image as a single event.

Surface folding is achieved by shearing the projection (ToEye) vectors. To create a focus that is simply magnified a section is pulled up towards the viewpoint. When viewed from the top it appears that the focus is rising straight up from its base. However, when viewed from the side one can see that off-center foci are slightly tipped so that they point towards the viewpoint (Figure 5). This is accomplished by pointing their ToEye vectors at the viewpoint (Figure 2). In

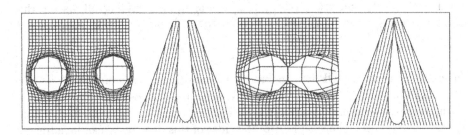

Fig. 6. A pair of foci, repositioned to be adjacent

this manner the whole projection of the focus can be readily shifted by pointing the ToEye vector elsewhere. If the ToEye vector is directed at any point on the plane parallel to the original surface which also contains the viewpoint, all the properties of height, magnification and scaling for the foci remain constant while its position in x and y change. Figure 5 shows side and profile views of the same single focus, on the left it is pulled towards to viewpoint which appears to be straight up, on the right it is folded or pulled across the viewing frustrum. Notice how a small change in profile view translates to a considerable visual difference when viewed from above. Figure 6 shows the same views of the two foci. This time the folding is used to bring two separated foci together.

Foci Motion: There is an important distinction between moving a focal point and folding a focus. Moving a focus through the graph allows for a sequential roving search, the image in the focus changing as the focus moves over different areas of the graph. Folding the focal point maintains the same view within the focus; this view is repositioned over other sections of the graph or aligned with other focal points. This allows spatially separate areas to be positioned adjacently while maintaining a continuous surface between them. At any moment the graph on this surface can be viewed by rotating and adjusting the three-dimensional image or by unfolding the graph.

Elision: While the principle intention in introducing surface folding was to allow for subgraph comparisons, it is possible to think of this as a method for hiding sections of the graph that are not of immediate concern. The result is a tool that can be used analogously to folding a printed map to hide and/or expose the areas of interest. Since folding is directly reversible, sections that have temporarily been hidden are readily retrievable. In fact, one can unfold to allow closer examination of the connections between the expanded detail in the focus and the rest of the graph.

4 Comprehension Factors

A primary goal in the creation of 3DPS is that the distortions remain comprehensible, allowing the user to understand the relative magnification or compression of the various sections of the resulting image. Other distortion methods can be quite readable when applied to regularly spaced information, particularly grids or text; unfortunately not all information is laid out so regularly. We make a distinction between the graph as a 2D image encoding the information and the surface on which the graph is displayed. As visual cues are provided about the surface, distortions will still be readable even when there are gaps in the image. The separation of the image and distortions of the surface means that the original topology of the image is maintained across the surface. Once the surface is manipulated the image is placed upon it. Displaying a surface in such a manner

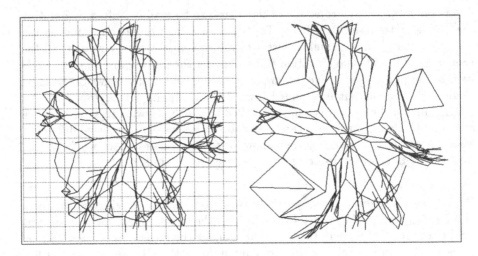

Fig. 7. This series uses the same graph and same distortion throughout. On the left is the undistorted graph; on the right is the graph with 3 foci and no visual cues

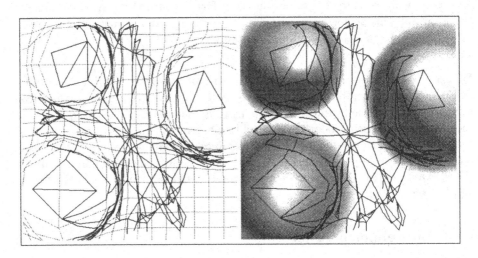

Fig. 8. On the left same 3 foci with shading. On the right, the same graph with 3 foci and both grid and shading

as to reveal its three-dimensional form provides the perceptual information that describes the distortion.

Using Nodes and Edges: Nodes are currently displayed either as single points or as squares. Edges are segmented small enough so that they will lie snuggly against the surface. This provides additional information about the surface. Information spaces that contain long lines now aid in the description of the surface.

Using a Grid: Perspective can be used to provide distortion information. However, understanding three-dimensions from perspective appears to be a learned skill and culturally tied [4]. Perspective has been indicated with the outlines of a three-dimensional shape [9] or by the visual pattern of the data [16]. The choice of smooth curves for distortion and allowing for data with irregular layouts means neither outlines nor patterns in the data will reveal the nature of the distortions. However, a regular grid can be displayed over the entire surface, providing both curve and perspective information. Grid lines indicate relative magnification as well as serving as a texture gradient.

Using Shading: Another choice for revealing form is to employ shading. It has been well established that humans can discern three-dimensional shape from shading alone [15, 18], and there is considerable evidence to support the fact that this is a low-level precognitive skill [7]. Such a low-level visual routine will interfere less with conscious processing and may even provide an aspect of the interface that requires no learning [20].

The series in Figures 7 and 8 shows the progressive addition of each visual cue. The first image is of the undistorted graph laid out with the spring embedder algorithm from GraphEd [5]. The second image shows the graph with three foci and no additional visual cues, third the grid is added and the fourth shows the use of shading. All of these visual cues are optional and are displayed in shades of grey so that while they are readily visible apart or in unison they do not dominate the image.

5 Examples

This section contains some examples of graph folding in action. The first set of examples is of a random graph; in Figure 9 on the left is a simplistic layout, placing each vertex on a grid,

resulting in an image that is a confusion of lines. On the right the same graph has been laid out using the spring embedder algorithm in GraphEd [5]. This reveals several small fans of pendant vertices. Figure 10 magnifies three sections of the denser center part to reveal similar subgraph structure.

The second set of examples(Figures 12 and 10) is of an iterated K12 graph. The outer subgraphs are so densely packed that one can not tell for sure what they are. Two of these have been magnified and compared to verify that they are indeed K12's.

6 Conclusions

While detail in context distortion viewing tools hold much promise as a method for interacting with representation of large graph, the approaches of other systems had made the repositioning of focal regions difficult if not impossible. This

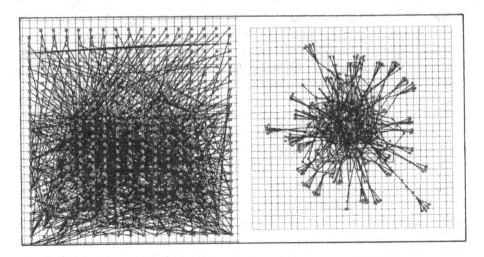

Fig. 9. On the left, a random graph with vertices placed in the grid. On the right, the same graph laid out with spring algorithm

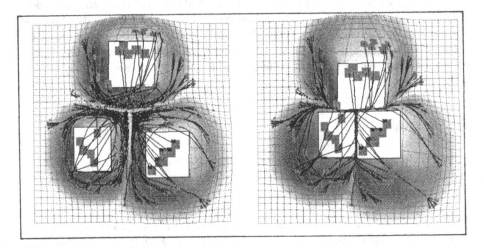

Fig. 10. A random graph with spring layout and three foci folded

paper has introducing folding, a novel concept which combines the advantages of detail in context viewing with the freedom of movement provided by magnified views in separate windows.

The distortion viewing tool 3DPS that we have extended to include folding in itself has several advantages over existing tools; briefly these include:

1. Arbitrarily-shaped focal regions. Sarkar et. al. [18] approximate this advantage with convex polygons. We allow for interactive user specification of a chosen outline for a focal region.

Fig. 11. On the left; iterated K12. On the right; two foci magnified to reveal subgraphs of K12

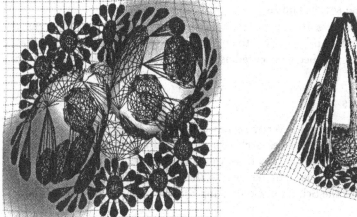

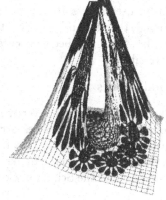

Fig. 12. On the left; iterated K12 folded. On the right; iterated K12 folded side view

2. Distortion control. By adjusting the slope parameters the user can determine the distribution of compression. For instance, distortion can be contained within a relatively small section of the surrounding graph leaving most of the context undistorted. In other techniques the pattern of distortion is controlled by the system, user choices being limited to such things as a global choice between Cartesian or polar transformation [6, 17].

3. Multiple foci. Sarkar et. al [18] introduced multiple foci with two different techniques. However, both approaches have limitations For instance, their orthogonal approach cause strips of uni-direction magnification that create extra unrequested foci at their intersections. Their polygonal approach which

is more similar to ours in appearance, required limitation on focal size and positioning as two large foci that were too close together would cause position reversals in the inter-focal regions. Also this polygonal technique can require iteration of the algorithm to produce an acceptable final image. Our approach requires no iteration and our blending function allows large foci to interact and in fact move through each other. Furthermore, this is the first provision of multiple foci in a three-dimensional distortion approach. This is significant in that it is partly the three-dimensional nature of the distortion that makes folding possible.

4. Precognitive perceptual cues. These are used to reveal the nature of the distortions. Being able to understand the distortion provides knowledge about the degree of compression, information about the original undistorted topology of the graph, and the cumulative result of the history of the user's actions.

While the use of shading provides instant recognition of the patterns of distortion, it causes some problems. Finding the right balance between light and dark intensities is difficult to achieve, especially if one wants to have convincing shading both on the screen and in print.

Presently, user access to the parameters that affect distortion patterns is unconstrained, therefore it is possible to create curves that obscure some context. However, just what has been obscured is always evident and the actions are readily reversible.

Acknowledgments

Thanks to Brian Fisher for verifying our cognitive references, and Art Liestman for ongoing support. This research was supported by graduate scholarships and research and equipment grants from the Natural Sciences and Engineering Research Council of Canada. Thanks also to the Algorithms Lab, Graphics and Multimedia Lab and the School of Computing Science, Simon Fraser University.

References

1. M. S. T. Carpendale, D. J. Cowperthwaite, and F. D. Fracchia. 3-dimensional pliable surfaces: For effective presentation of visual information. In *toappear: UIST: Proceedings of the ACM Symposium on User Interface Software and Technology*, 1995.

2. B. D. Fisher and Z. W. Pylyshyn. The cognitive architecture of bimodal event perception: A commentary and addendum to Radeau. *Cahiers de Psychologie Cognitive/Current Psychology of Cognition*, 13(1):92–96, Feb. 1994.

3. G. W. Furnas. Generalized fisheye views. In *Human Factors in Computing Systems: CHI'86 Conference Proceedings*, pages 16–23, 1986.

4. N. Goodman. *Languages of Art; An Approach to a Theory of Symbols*. Indianapolis: Bobbs-Merrill, 1968.

5. M. Himsolt. GraphEd: A graphical platform for the implementation of graph algorithms. In *Graph Drawing, DIMACS International Workshop, Proceedings*, pages 182–193, 1994.

6. K. Kaugers, J. Reinfelds, and A. Brazma. A simple algorithm for drawing large graphs on small screens. In *Lecture Notes in Computer Science: Graph Drawing*, pages 278 – 282, 1995.

7. D. A. Kleffner and V. S. Ramachandran. On the perception of shape from shading. *In Perception and Psychophysics*, 52(1):18–36, 1992.

8. J. Lamping and R. Rao. Laying out and visualizing large trees using a hyperbolic space. In *UIST: Proceedings of the ACM Symposium on User Interface Software and Technology*, pages 13 – 14, 1994.

9. J. D. Mackinlay, G. G. Robertson, and S. K. Card. The perspective wall: Detail and context smoothly integrated. In *CHI'91 Conference Proceedings*, pages 173 – 180, 1991.

10. D. W. Massaro. *Speech perception by ear and eye: a paradigm for psychological inquiry*. Hillsdale, N.J., Erlbaum Associates, 1987.

11. D. W. Massaro. Attention and perception: An information integration perspective. *Special Issue: Action, attention and automaticity. / (In Acta Psychologica)*, 60(2-3):211–243, Dec. 1985.

12. K. Misue and K. Sugiyama. Multi-viewpoint perspective display methods: Formulation and application to compound digraphs. In *Human Aspects in Computing: design and Use of Interactive Systems and Information Management*, pages 834–838. Elsevier Science Publishers, 1991.

13. E. G. Noik. Layout-independent fisheye views of nested graphs. In *Proceedings of the 1993 IEEE Symposium on Visual Languages*, pages 336 – 341, 1993.

14. E. G. Noik. A space of presentation emphasis techniques for visualizing graphs. In *Graphics Interface'94*, pages 225–233, 1994.

15. V. S. Ramachandran. Perception of shape from shading. *Nature*, 331(14):163–166, 1988.

16. G. Robertson and J. D. Mackinlay. The document lens. In *UIST: Proceedings of the ACM Symposium on User Interface Software and Technology*, pages 101 – 108, 1993.

17. M. Sarkar and M. H. Brown. Graphical fisheye views. *Communications of the ACM*, 37(12):73–84, 1994.

18. M. Sarkar, S. Snibbe, O. J. Tversky, and S. P. Reiss. Stretching the rubber sheet: A metaphor for viewing large layouts on small screens. In *UIST: Proceedings of the ACM Symposium on User Interface Software and Technology*, pages 81 – 91, 1993.

19. E. Tufte. *Envisioning Information*. Cheshire, Connecticut: Graphics Press, 1990.

20. C. Ware. The foundations of experimental semiotics: a theory of sensory and conventional representation. *Journal of Visual Languages and Computing*, 4:91–100, 1993.

Upward Numbering Testing for Triconnected Graphs

M. Chandramouli,   A. A. Diwan

Dept. of Computer Science and Engineering,
Indian Institute of Technology, Powai,
Bombay 400 076,
INDIA

Email: {mouli, aad}@cse.iitb.ernet.in

Abstract. In this paper, we look at the problem of upward planar drawings of planar graphs whose vertices have preassigned y-coordinates. We give a linear time algorithm for testing whether such an embedding is feasible for triconnected labelled graphs.

1   Introduction

For directed graphs, a notion similar to planarity (for undirected graphs) is that of an *upward planar* drawing, that is a planar drawing of the graph such that all the edges are directed upward (monotonic curves from a lower y-coordinate to a higher one). There has been a lot of work on upward planarity testing of various classes of graphs [8, 3]. Recently, Garg and Tamassia [6] have proved that the problem of upward planarity testing for arbitrary graphs is NP-complete.

In this paper, we consider the problem of *upward numbering testing* for planar graphs (see Section 3 for the definition). The main contribution of this paper is a linear time algorithm for *upward numbering* testing for triconnected labelled graphs. Battista and Nardelli [1] have given a linear time algorithm for recognising upward numberings for single source labelled digraphs. Lin [11], gave an algorithm for a subclass of digraphs having their sources and sinks on the outerface, called proper s-t boundary hierarchical graphs. Recently Heath and Pemmaraju [7] have given a linear time algorithm for graphs with adjacent vertices having labels which differ by unity. This algorithm leads to a quadratic time algorithm for upward numbering testing for arbitrary graphs.

The rest of the paper is organised as follows. Section 2 looks at earlier work on upward planarity testing. Section 3 contains the algorithm for upward numbering testing. Section 4 gives an interesting connection between the upward drawings of a graph with distinct labels and an intersection graph realization problem.

2   Upward planar drawings of digraphs

In this section, we look at the prior work on upward planarity testing for digraphs. DiBattista and Tamassia [2] have shown that the problem of upward planarity testing is equivalent to the problem of augmenting a given digraph to obtain a planar s-t digraph. Kelly [9], and DiBattista and Tamassia [2], have proved the following theorem.

Theorem 1. *A digraph admits an upward planar drawing iff it is a subgraph of a planar s-t digraph.* □

DiBattista and Tamassia [2] have given a linear time algorithm for producing an upward polyline grid drawing of a s-t digraph.

3   Upward drawings of planar graphs with labels

We now consider the problem of upward drawings of graphs with the additional constraint that the vertices have labels attached to them, which denote the y-coordinate (level) at which they are to be placed.

3.1   Preliminaries

We consider only planar graphs and henceforth refer to them only as graphs. A map $f : V \rightarrow \mathcal{Z}^+$ (set of positive integers) is called a labelling of the graph.

Definition 1 *A generalized s-t numbering of a graph $G = (V, E)$ is a map $f : V \rightarrow \{1, 2, \ldots, N\}$, such that $| \{v : f(v) = 1\} |$ and $| \{v : f(v) = N\} |$ are equal to one, and such that each vertex $v \neq s, t$, where $f(s) = 1$ and $f(t) = N$, has two adjacent vertices u, w for which $f(u) < f(v) < f(w)$ and the vertices s and t are adjacent.*

We call the value $f(v)$ attached to vertex v the *label* associated with v. We denote a labelled graph as a 3-tuple $G = (V, E, l)$ where l is the labelling function.

Definition 2 *A labelling of the vertices of a planar graph such that no two adjacent vertices have equal labels is said to be an* **upward numbering** *if there exists a planar embedding of the graph such that all the vertices labelled i have y-coordinate i and all the edges are strictly monotonic curves.*

Without loss of generality, we can assume that the labels are positive integers between 1 and N. Let 1 and N be the smallest and largest labels in a labelling of a graph. We can make the following assumption about any labelling of a graph.

- We can assume that there is a unique vertex labelled 1 and also a unique vertex labelled N. We can also assume that these two vertices are adjacent.
- We can assume that in any face that there are no three consecutive vertices with increasing labels. Note that as we are considering triconnected graphs, the faces are uniquely defined.

For all labelled graphs, we denote the unique vertex with the smallest label by s and the unique vertex with the largest label by t. Following the terminology of digraphs, we call a vertex a *source (sink)* if it has no neighbouring vertices with labels *smaller (larger)* than it. In [2] it has been shown that every planar *st-digraph* admits an *upward drawing*. We prove a similar result for labelled digraphs. The following theorem follows easily from the algorithms given in [12] to construct polyline representations(planar embeddings such that the edges are polygonal segments) of a planar graph from its s-t numbering.

Theorem 2. *A labelling of a planar graph G is an upward numbering if and only if the graph can be augmented to a planar graph through the addition of edges such that the given labelling is a generalised s-t numbering of the augmented graph, with no two adjacent vertices having the same label.*

$\square$

The following lemma establishes a property of graphs with a generalised s-t numbering.

Lemma 3. *Given a graph $G = (V, E, l)$. Let u and v be two vertices of G. Then l is a generalised s-t numbering only if there exists a path between u and v such that all the labels are less than $\max(l(u), l(v))$ and there exists a path such that all the labels are greater than $\min(l(u), l(v))$.*

$\square$

3.2 Resolving sources and sinks

Our strategy for testing for *upward numbering* is the following. We try to add edges to a given embedding of a graph to *resolve* all sources and sinks other than s and t, that is to add outgoing edges to sinks and incoming edges to sources to get a planar graph, with a generalised s-t numbering. We say that such an embedding is a *feasible embedding*.

As we are considering triconnected graphs we have to examine only a polynomial number of embeddings as the embedding is decided by the choice of outerface. Moreover we want the edge (s, t) to lie on the outerface. As an edge can be shared by exactly two faces we have to consider only two embeddings. If the graph were to have an upward planar drawing then there are at least two embeddings which are feasible namely the embedding having the *s-t edge* as the leftmost edge and the one with it as the rightmost edge. Hence we have to test exactly one of the embeddings with the edge from the vertex labelled s to the vertex labelled t on the outermost face.

3.3 Reducing the face size

When we are testing a given embedding of a triconnected graph, we can perform certain transformations to make the faces smaller in size. The following lemmas describe ways of adding edges to faces in a given embedding of a labelled triconnected graph to decrease the face size.

Let $G = (V, E)$ be a graph and let e be an edge which does not belong to E. Then by $G + e$ we denote the graph G along with the additional edge e. We represent the fact that l is an upward numbering of G iff it is an upward numbering of H by $(G, l) \equiv (H, l)$.

We omit the proof of the following lemmas, which show that testing for upward numbering of a given labelled graph is equivalent to testing on the same graph with certain additional edges. The first lemma shows that if in any face, there exist vertices u and v such that all the vertices between them on one of the face paths joining them have labels lying strictly between those of u and v, then the edge (u, v) can be added to the face. The second lemma shows that an edge, joining a vertex with the largest label in a face and a vertex with the smallest label in a face, can be added to the face.

Lemma 4. *Given a labelled graph $G = (V, E, l)$. Let f be a face such that there are consecutive vertices x_0, x_1, x_2, ...,x_k such that $l(x_0) = i$ and $l(x_k) = j$ and $i < l(x_r) < j$, $1 \leq r < k$. Let $H = (V, E \cup \{(x_0, x_k)\})$. Then $(G, l) \equiv (H, l)$.*

$\square$

Lemma 5. *Given a labelled graph $G=(V,E,l)$. Let f be a face and let x_i and x_j be vertices with $l(x_i) = 1$ and $l(x_j) = N$, where 1 and N are the smallest and largest labels on the face. Let H be the graph obtained by adding the edge (x_i, x_j) to the face f. Then $(G, l) \equiv (H, l)$.*

$\square$

From the previous lemmas, we note the following facts.

1. If there are two vertices labelled n on the face, they can both be joined to a vertex labelled 1.
2. If there are two vertices labelled 1 on the face, they can both be joined to a vertex labelled n.

We can add edges to the faces using Lemma 4 and 5 till no more edges can be added. We now look at the labels on the faces which are remaining. We can show by simple arguments that the face labels must be one of the four cases depicted in Figure 1.

Lemma 6. *In each of the four basic face configurations shown in Figure 1, the only possible resolution of sources and sinks is one in which, either all the sources or all the sinks are resolved in any face.*

Proof: Case 1: There is a unique vertex with the smallest label. In this case all the sources must be resolved in the face. Otherwise let v be a vertex (source) not resolved in the face. Clearly x_1, x_2, x_3 are not resolved in the face. Hence there exist paths from x_1 to x_3 and from x_2 to v, with labellings as in Lemma 3. But these two paths cannot exist at the same time without violating planarity. Hence it follows that all the sources in the face must be resolved in the face and all of them are resolved by connecting them to the source with the smallest label.

Case 2: There is a unique vertex with the largest label. As in Case 1, all the sinks must be resolved in the face by connecting them to the sink with the largest label.

Case 3: Consider the arc S between x_1 and x_4. If there are no vertices labelled 2 or $n - 1$ on this arc then it follows that we can add the edge (x_1, x_4) to the face f. But we have assumed that all such edges have already been added to the face. Hence there must be a vertex labelled 2 or $n - 1$ in S. Also, one of the vertices x_1 or x_4 must be resolved in the face, otherwise by Lemma 3 we have

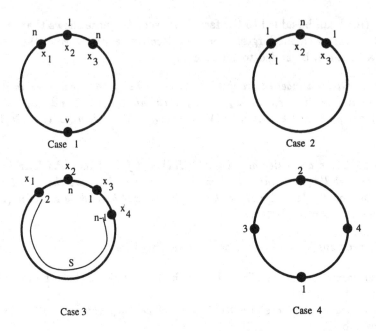

Fig. 1. Basic face configurations

two vertex disjoint paths between x_1 and x_3, and x_2 and x_4 which lie totally outside the face. These paths must cross, violating the planarity of the resolved graph.

If one of the vertices labelled 2 is resolved in the face it follows that all the vertices labelled 2 are resolved in the face. Let v be a vertex labelled 2, which is not resolved in the face. Otherwise, by Lemma 3, there are paths between x_2 and x_4, and x_3 and v which do not share a vertex and which lie completely outside the face in any resolution of the face. Hence these paths must cross, implying that there is no planar resolution. Similarly, we can show that if one of the vertices labelled $n-1$ is resolved in the face, then all the vertices labelled $n-1$ must be resolved in the face.

Subcase A: There is a vertex labelled 2 in S.

x_1 is resolved in the face: Now assuming that all the vertices labelled 2 are resolved in the face by connecting them to x_3 as shown in Figure 2(a). In face f_1, there is no vertex labelled 2 in the arc T. Hence the vertices x_5 and x_6 must have labels which are the largest in the face. If some other vertex has a larger label in arc T, then we can add an edge connecting this vertex and the vertex labelled 2. But we have assumed that all such edges have already been added. We can add the edges (x_3, x_5) and (x_3, x_6) to the face. Hence from Case 1 we get that all the sources must be resolved in the face. The same argument can be applied to each of the faces. For the last face, that is the face containing x_4, either x_7 is adjacent to x_4 or there is a vertex labelled 2 or $n-1$ on the path between them. If there is a vertex labelled $n-1$, then the edge between x_7 and

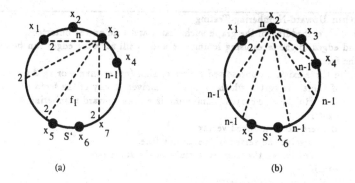

Fig. 2. Resolving all sources or all sinks

the vertex can be added. As we have assumed that all such edges have already been added, there must be a vertex labelled 2 adjacent to x_4. This face is now exactly like the other faces. Hence all the sources must be resolved in f.

Similarly if a single vertex labelled $n - 1$ is resolved in the face, we can show that all the sinks are resolved in the face as shown in Figure 2(b).

Subcase B: In this case there is a vertex labelled $n - 1$ in arc S. The arguments are exactly similar to the above case and it follows that either all the sources or all the sinks are resolved in the face.

Case 4: This case is trivial. $\qquad\qquad\qquad\qquad\qquad\qquad\qquad\qquad\qquad\qquad\qquad\square$

3.4 Algorithm for upward numbering testing

The algorithm for upward numbering testing is shown in Figure 3. We now prove the correctness of the algorithm. We show that at each step of the algorithm there is either a forced vertex, that is a vertex which can be resolved in exactly one face or a vertex which cannot be resolved in any of the faces. If neither of these two conditions hold, then for each unresolved source or sink there are exactly two faces in which they can be resolved. Both these faces are of size four. In this case, we show that the choice of face for resolution is not critical. We omit the proof of the following lemma.

Lemma 7. *Let $G = (V, E, l)$ be a labelled triconnected graph. Let f be a face such that two of the sources on the face are connected to each other by a path lying outside the face, such that the vertices have strictly decreasing labels. Then there exists a sink, which has an unique choice of face for resolution or there exists a sink, which cannot be resolved in any face.*

$\qquad\qquad\qquad\qquad\qquad\qquad\qquad\qquad\qquad\qquad\qquad\qquad\qquad\qquad\qquad\qquad\square$

We use the above property, while proving the next lemma.

Lemma 8. *At every step of the algorithm, there is a forced vertex, or an unresolvable vertex or a vertex for which the choice of face for resolution is not critical.*

```
Algorithm Upward-Numbering-Testing
1.        Find an embedding of the graph such that s and t lie on the outerface.
2.        Add edges to the faces using lemmas 4 and 5 till no more edges can be added to
any of the faces.
3.        while there exists an unresolved source or sink (other than s or t), do
4.            if there is a vertex which cannot be resolved in any of the faces
                    stop. { The current numbering is not an upward numbering }.
5.            endif
6.            if there exists a forced vertex
                    resolve the vertex in the unique face.
                    resolve all the sources or sinks in the same face.
7.            endif
8.            if there is no forced vertex
                    pick any arbitrary vertex.
                    resolve the vertex in any of the faces in which it can be resolved.
                    resolve all the sources or sinks in the same face.
9.            endif
10.       endwhile
```

Fig. 3. Upward numbering testing

Proof: We assume that there are no vertices which are forced or have no choice of face for resolution. We now show that there must exist a vertex for which the choice of face for resolution is not critical. Let s be a vertex such that all vertices, with labels smaller than $l(s)$ have been resolved. As there are no forced vertices or unresolvable vertices, it follows that s can be resolved in at least two faces.

Case 1: s can be resolved in faces f_1 and f_2, which do not share an edge, as shown in the top of Figure 4. Let s_1 and s_2 be the sources with the smallest labels in f_1 and f_2 respectively. By our assumption, it follows that s_1 and s_2 have been resolved. Hence there exists a path P from s_1 and s_2, which have labels less than or equal to $max(s_1, s_2)$. By Lemma 7, we can always assume that the paths do not use any vertex in f_1 or f_2. Consider the largest sink v in the region A marked in Figure 4. If it lies in the interior of the region, then clearly it cannot be resolved in any face which is a contradiction. If it lies on the boundary of A it must lie either on P_1 or on P_2. In either case v is a forced vertex, which contradicts the fact that there are no forced vertices.

Case 2: s can be resolved in two faces f_1 and f_2, which share an edge, as shown in the lower half of Figure 4. Let s_1 and s_2 be the smallest sources in f_1 and f_2 respectively. There are two subcases in this case and they are shown in the lower half of Figure 4.

Subcase A: We again consider the largest sink in region A and argue as in the previous case.

Subcase B: We consider the largest sink in the region A. If it lies on P_1 or P_2 (other than v), we have a forced vertex. We are left now with the case when the

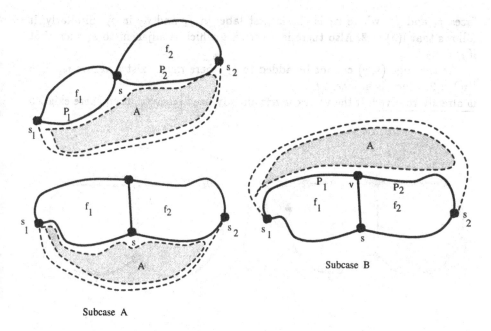

Subcase A

Subcase B

Fig. 4. Faces share a vertex or edge

largest sink is v. If v itself is a forced vertex, we are done. We are now left with the case when v can be resolved in both f_1 and f_2.

If both the face cycles are of size four, then s and v can be resolved in either

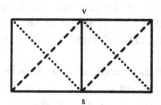

Fig. 5. Face cycles of size four

of the ways shown in Figure 5. Note that only these vertices can be resolved in these faces. Also the manner in which they are resolved within the faces does not affect the choices for other faces.

The other case is that at least one of the face sizes must be larger than four. Let f_2 be the face whose size is larger than four. All the vertices on P_1 have labels between s_1 and v. Similarly, all the vertices on P_2 have labels between s_2 and v. Hence we can assume that these vertices are adjacent, as no such edge can be added to any of the faces. From the previous discussion on the types of labellings on the smallest faces, we get that $l(v)$ is the second largest label in each of the faces f_1 and f_2, where n_1 is the largest label in f_1 and n_2 in f_2. Similarly, it follows that $l(s) = 2$. Also there is a vertex y which is adjacent to s_2 such that $l(y) = n_2$.

As the edge (s, y) cannot be added to f_2, there must exist a vertex w, with $l(w) = 2$, which is adjacent to y.

<u>w already resolved:</u> If the vertex w has already been resolved, then there exists a

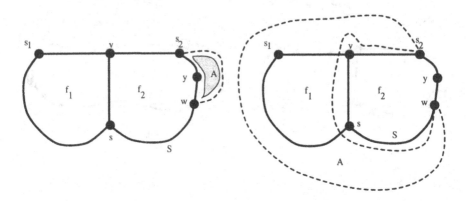

Fig. 6. Vertex w already resolved

path between w and s_2. The path can lie in either of the ways depicted in Figure 6. Consider the largest sink in the enclosed region in the first configuration in Figure 6. If it is in the interior it cannot be resolved. If it lies on the boundary it must be the vertex y which again cannot be resolved. Consider the second configuration shown in Figure 6. In this case consider the largest sink in the region bounded by the dotted line. If it lies in the interior then it cannot be resolved. If it lies on the boundary it must be v, but v cannot the largest sink as it can be resolved in f_1.

<u>w is not already resolved:</u> If there exists another face in which w can be resolved and they do not share an edge, we can argue as previously.

If the two faces share an edge, let s_3 be the smallest source vertex in f_3 (the other face in which w can be resolved). As s_2 and s_3 have already been resolved, there exists a path P between s_2 and s_3. The path P can lie in either of the ways shown in Figure 7.

Figure 7(a): Look at the largest sink in region A. If it lies on the boundary, then it must lie between y and s_3. Hence there is a unique choice of face for resolution. If it lies in the interior it cannot be resolved.

Figure 7(b): Consider the largest sink in the region shown. As $l(v) = n - 1$ it cannot lie between s and w. If it lies between w and s_3 there is a unique face for resolution.

Another possible configuration, is when f_3 and f_2 share the edge (w, z). In this case, it can be argued using exactly similar arguments that there is either a forced vertex or an unresolvable vertex. This completes the proof of the lemma.

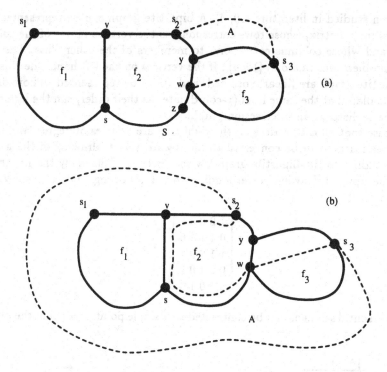

Fig. 7. Vertex w not already resolved

□

Theorem 9. *Algorithm Upward-Numbering-Testing produces a graph with a generalised s-t numbering, of which G is a subgraph, if l is an upward numbering in linear time.*

Proof: The correctness follows from the previous lemmas. We just give a brief sketch of the time complexity arguments. The first phase in which edges are to be added to the various faces can be carried out in time linear in face size for each face. We maintain a list of sources and sinks from which we can add edges and as we traverse the face, we add edges depending on the label assocaited with the new source or sink. The second phase in which the sources and sinks are to be resolved can be carried out by maintaining a graph of sources and sinks and faces in which they can be resolved. Hence the algorithm produces a graph with a generalised s-t numbering in linear time.   □

4   Grid intersection graphs and upward drawings with labels

A bipartite graph is a *grid intersection graph*, if it can be realized as the intersection graph of horizontal and vertical line segments in the plane. Such graphs have been studied in literature [10, 4]. A bipartite graph can be represented by the adjacency matrix, whose rows correspond to the vertices in one of the colour classes and whose columns correspond to members of the other class. The following problem was raised in [10, 4] : If the vertices on the left hand side class of the bipartite graph are linearly ordered, and are to be represented as horizontal segments placed at the same level (y-coordinate) as their order, can the bipartite graph be realized as an intersection graph.

In the case that all the vertices on the right side are vertices of degree two, then such a realization can be converted to an upward planar drawing of the graph (corresponding to the bipartite graph) with labels (as defined by the linear order). The upward drawing corresponding to the following matrix is shown in Figure 8.

$$\begin{bmatrix} 1 & 0 & 0 & 0 & 0 \\ 0 & 1 & 1 & 1 & 0 \\ 1 & 1 & 0 & 0 & 1 \\ 0 & 0 & 1 & 0 & 1 \\ 0 & 0 & 0 & 1 & 0 \end{bmatrix}$$

Each horizontal segment can be contracted to a single point producing the above

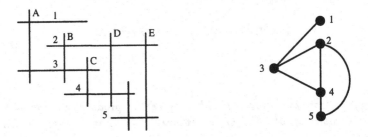

Fig. 8. An upward drawing of an IHV graph with labels

drawing. The process is clearly reversible, that is, such a drawing can be converted to a realization of the corresponding bipartite graph using horizontal and vertical segments alone (see [12] for details). Hence testing for the realization is easy for bipartite graphs arising from triconnected graphs.

5   Discussion

We have given an algorithm to test for *upward numbering* for triconnected graphs. The problem of testing *upward numbering* for arbitrary graphs has some interesting connections with the problem of characterising grid intersection graphs [10, 5].

References

1. G. Di Battista and E. Nardelli. *An Algorithm for Planarity Testing of Hierarchical Graphs*, volume 246 of *Lecture Notes in Computer Science*, pages 277–289. Springer-Verlag, 1987.

2. G. Di Battista and R. Tamassia. Algorithms for plane representations of acyclic digraphs. *Theoretical Computer Science*, 61:175–198, 1988.

3. P. Bertolazzi, G. Di Battista, G. Liotta, and C. Mannino. Upward drawings of triconnected digraphs. *Algorithmica*, to appear.

4. M. Chandramouli. *Upward Planar Graph Drawings*. PhD thesis, IIT Bombay, 1994.

5. M. Chandramouli and A. A. Diwan. Intersection graphs of horizontal and vertical line segments in the plane, 1992. Unpublished manuscript.

6. A. Garg and R. Tamassia. On the computational complexity of upward and rectilinear planarity testing. In *Graph Drawing 94, DIMACS*, 1994.

7. L. S. Heath and S. Pemmaraju. Recognizing leveled-planar dags in linear time. In *Proceedings of Graph Drawing '95*, 1995.

8. M. D. Hutton and A. Lubiw. Upward planar drawing of single source acyclic digraphs. In *Proc. ACM-SIAM Symposium on Discrete Algorithms*, pages 203–211, 1991.

9. D. R. Kelly. Fundamentals of planar ordered sets. *Discrete Mathematics*, 63:197–216, 1987.

10. J. Kratochvil. A special planar satisfiability problem and some consequences of its np-completeness. *Discrete Appl. Math. (to appear)*.

11. X. Lin. *Analysis of Algorithms for Drawing Graphs*. PhD thesis, Department of Computer Science, University of Queensland, 1992.

12. R. Tamassia and I. G. Tollis. A unified approach to visibility representations of planar graphs. *Disc. and Comp. Geometry*, 1(4):321–341, 1986.

On a Visibility Representation of Graphs

F.J. Cobos[1], J.C. Dana[1], F. Hurtado[2], A. Márquez[1] and F. Mateos[1]

[1] Universidad de Sevilla, Facultad de Informática y Estadística, Depto. de
Matemática Aplicada I, Sevilla, Spain
[2] Universidad Politécnica de Cataluña, Depto. de Matemática Aplicada II,
Barcelona, Spain

Abstract. We give a visibility representation of graphs which extends
some very well-known representations considered extensively in the literature. Concretely, the vertices are represented by a collection of parallel
hyper-rectangles in $\mathbf{R}^n$ and the visibility is orthogonal to those hyper-
rectangles. With this generalization, we can prove that each graph admits a visibility representation. But, it arises the problem of determining
the minimum Euclidean space where such representation is possible. We
consider this problem for concrete well-known families of graphs such as
planar graphs, complete graphs and complete bipartite graphs.

1 Introduction

The problem of determining a visibility representation of a graph has been studied extensively in the literature due to the large number of applications (as in
VLSI design, CASE tools, hidden-surface elimination problem, etc., [7, 8, 11, 13,
15]) and, also, by the combinatorial properties of those graphs.

In a visibility representation of a graph, the vertices map to objects in Euclidean space and the edges are determined by certain visibility relations.

Of course, both, the objects and the visibility used play an important role
in characterizing the types of graphs that admit visibility representations. But,
in any case, given a certain class of objects and a concrete visibility, there exist
always graphs that are not representable, in this way, Tamassia & Tollis [14] and
Wismath [16] proved that a graph is a *bar visibility graph* (where the vertices
represent horizontal line segments in the plane and two nodes are connected
by an edge if their two horizontal rectangles can see each other vertically and
non-degenerately) if and only if it admits a planar embedding with all cutpoints
in the exterior face. And Bose et al. [3] proved that K_n is not VR-representable
for $n \geq 103$ (a graph is said to be *VR-representable* if each vertex of the graph
maps to a closed rectangle in $\mathbf{R}^3$ such that the rectangles are disjoint, the planes
determined by the rectangles are perpendicular to the z-axis, and the sides are
parallel to the x or y axes. And, again, two nodes are connected by an edge if
their two horizontal bars can see each other vertically and non-degenerately).

On the other hand, from a more theoretical point of view and since Kuratowski's Theorem [10], several measures of the planarity and/or dimension of a
graph have been considered. But, few of these measures, notably, Boxicity, Grid

intersection graphs [1, 9, 2], are related with visibility representations in the line of the approachs mentioned above.

In this paper, we prove that the representations studied by Tamassia & Tollis [14] and Wismath [16] in $\mathbf{R}^2$ and by Bose et al. [3] in $\mathbf{R}^3$ can be easily generalized to any dimension, and, if we consider for each graph the minimum n where it is possible such a representation we obtain, in this way, a new measure of the complexity of the graph. Thus, we say that a graph is *representable* in $\mathbf{R}^n$ if it can be represented in such a way that each vertex maps to a hyper-rectangle in $\mathbf{R}^n$ (where the hyper-rectangles that we consider are a cartesian product of $n - 1$ closed intervals in $\mathbf{R}$ and a number in the last coordinate; i.e., $[a_1, b_1] \times [a_2, b_2] \times \cdots \times [a_{n-1}, b_{n-1}] \times \{a_n\}$) and two nodes are connected by an edge if there exists a closed cylinder in $\mathbf{R}^n$, orthogonal to the rectangles, of non-zero length and radius such that the ends of the cylinder are contained in each of the hyper-rectangles and it does not intersect any of the other hyper-rectangles. We say that a graph G has *representation index* equal to n (or $RI(G) = n$ for short) if $\mathbf{R}^n$ is the minimum where such a representation is possible.

This paper is organized in the following way, in Sect.2 we prove that any graph has a finite RI, and we see that it is convenient to extend that index. Section 3 is devoted to the study of the RI of planar graphs. And in Sect.4 and 5 we deal with the RI of complete graphs and bipartite graphs respectively.

2 Representation Index

In this section we are going to prove that each graph is representable in some $\mathbf{R}^n$. First, we need the following lemma.

Lemma 1. *Every graph representable in $\mathbf{R}^n$ is representable in $\mathbf{R}^{n+1}$.*

Proof. It is easy to check that of the configuration

$$R_i = [a_1^i, b_1^i] \times \cdots \times [a_{n-1}^i, b_{n-1}^i] \times \{a_i\} \quad i = 1, 2, \ldots p$$

represent G, then the configuration

$$R_i = [a_1^i, b_1^i] \times \cdots \times [a_{n-1}^i, b_{n-1}^i] \times [0, 1] \times \{a_i\} \quad i = 1, 2, \ldots p$$

also represents G. □

Theorem 2. *Given a graph G, there exists $n \in \mathbf{N}$ such that G is representable in $\mathbf{R}^n$.*

Proof. We will prove the theorem by induction on k the number of vertices of G. Obviously the statement is true for small values of k. We now assume that G has $k + 1$ vertices and that the statement is true for graphs with k vertices. Given a vertex p of G, we split a representation $\mathcal{R}$ of $G - p$ into two subset

$$\mathcal{R} = \{R_{t_1}, R_{t_2}, \ldots, R_{t_{n_p}}\} \cup \{Q_{u_1}, Q_{u_2}, \ldots, Q_{u_{(k-n_p)}}\}$$

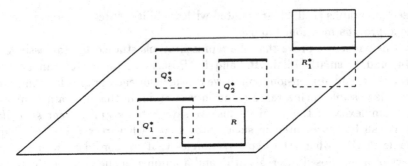

Fig. 1. G is representable in $\mathbf{R}^n$.

In such a way that the hyper-rectangles $R_{t_i} = I_{t_i\,1} \times \cdots \times I_{t_i,\,n-1} \times \{t_i\}$ ($1 \le i \le n_p$) correspond with the vertices p_{t_i} adjacent to p in G. Now, it is easy to see that the configuration $\mathcal{R}^* = \{R_{t_1}^*, R_{t_2}^*, \ldots, R_{t_{n_p}}^*\} \cup \{Q_{u_1}^*, Q_{u_2}^*, \ldots, Q_{u_{(k-n_p)}}^*\} \cup \{R\}$ represents G (being p represented by R), where

$$Q_{u_j}^* = \mathcal{I}_{u_j\,1} \times \cdots \times \mathcal{I}_{u_j\,n-1} \overbrace{\times [-1,0] \times \cdots \times [-1,0]}^{n_p} \times \{u_j\}$$

$$R_{t_1}^* = I_{t_1\,1} \times \cdots \times I_{t_1\,n-1} \overbrace{\times [-1,1] \times \cdots \times [-1,0] \times \cdots \times [-1,0]}^{n_p} \times \{t_1\}$$

$$\vdots$$

$$R_{t_s}^* = I_{t_s\,1} \times \cdots \times I_{t_s\,n-1} \overbrace{\times [-1,0] \times \cdots \times [-1,1] \times \cdots \times [-1,0]}^{n_p} \times \{t_s\}$$

$$\vdots$$

$$R_{t_{n_p}}^* = I_{t_{n_p}\,1} \times \cdots \times I_{t_{n_p}\,n-1} \overbrace{\times [-1,0] \times \cdots \times [-1,0] \times \cdots \times [-1,1]}^{n_p} \times \{t_{n_p}\}$$

and $R = I_1 \times \cdots \times I_{n-1} \overbrace{\times [0,1] \times \cdots \times [0,1] \times \cdots \times [0,1]}^{n_p} \times \{t_{k+1}\}$ with $I_i = [\min_j\{x : x \in I_{j\,i}\}, \max_j\{x : x \in I_{j\,i}\}]$. $\qquad\square$

Observe, that in Lemma 1 the last interval in all hyper-rectangles in the configuration is always $[0,1]$, that means that we are not using the whole $\mathbf{R}^n$ for our representation but only a half ($x_{n-1} \ge 0$) and that all hyper-rectangles are lying on the hyper-plane of equation $x_{n-1} = 0$. Thus, we can give a finer definition of RI saying that a graph G of (old) representation index equal to n has actually $RI(G) = (n-1) + 1/2$ if it admits a repreprestation in $\mathbf{R}^n$ such that any hyper-rectangle is of the form $[a_1, b_1] \times [a_2, b_2] \times \cdots \times [0, b_{n-1}] \times \{a_n\}$.

This new definition will allow us to get a better view of the problem, and we will get our main results using it. Moreover, there are some other additional reasons to consider this more general concept of representation index. For

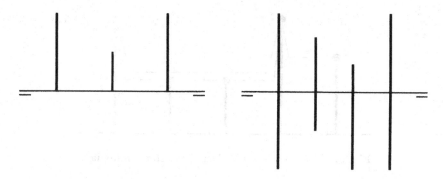

Fig. 2. $RI(K_3) = 1 + 1/2$ and $RI(K_4) = 2$.

instance, observe that if a graph has index $1 + 1/2$, then we can associate it an n-tuple of integer numbers in such a way that each vertex maps to one of the number of the n-tuple and two vertices are joined by an edge if all numbers between them in the n-tuple are smaller. Note that graphs with index $1 + 1/2$ are not the same of those of index 2 (K_4 has index 2 and K_3 has index $1 + 1/2$).

3 Planar Graphs

Bose et al. [3] proved that if G is planar then $RI(G) \le 3$. On the other hand, Tamassia & Tollis [14] and Wismath [16] gave the following theorem characterizing those graphs with representation index smaller or equal to 2.

Theorem 3. *[14, 16] A graph G has representation index smaller or equal to 2 if and only if there is a planar embedding of G with all cutpoints on the exterior face.*

We complete now this theorem, characterizing those graphs with representation index $1 + 1/2$. For that characterization we say that a graph G is *outerhamiltonian* if it has a path (possibly open) containing all vertices of G such that there exists a planar embedding of G with all edges of that path on the exterior face. Observe that Mitchell's algorithm to determine if a graph is outerplanar [12] with some modifications allows to get a linear-time algorithm to determine if a given graph is outerhamiltonian.

Theorem 4. *A graph has representation index $1 + 1/2$ if and only if it is outerhamiltonian.*

Proof. It is obvious that if $RI(G) = 1 + 1/2$ then G is outerhamiltonian.
For the sufficiency there are two cases to consider.
In the first case, we suppose that G is 2-connected. In this case, G is a polygon of vertices $\{v_1, v_2, \ldots, v_n\}$ (where that ordering if one of the two possible orderings of the vertices of the polygon) with some of its non-intersecting diagonals.

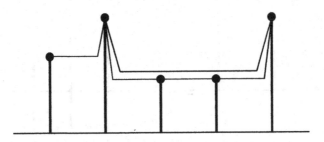

Fig. 3. If $RI(G) = 1 + 1/2$ then G is outerhamiltonian.

We map the vertex v_i to the bar $[0, d_i] \times \{a_i\}$, where $d_i - 1$ for $2 \leq i \leq n - 1$ is the total number of diagonals minus the number of diagonals $\{v_l, v_k\}$ with $l < i < k$, and $d_1 = d_n$ is the total number of diagonals plus 2.

In the second case, if G is not 2-connected, in each block there exist, at most, two cutpoints, place them the first and the last in that block, and we sort the blocks, obtaining, in that way, an ordering of all vertices of the graph. Now, we represent each block as in the first case, but giving the same length (the biggest one) to all bars representing cut-points. □

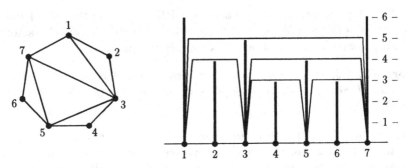

Fig. 4. Construction of a bar-representation in $\mathbf{R}^{1+1/2}$

Finally Bose et al.[3] proved that if G is a planar graph then $RI(G) \leq 3$. Thus, it remains to determine which planar graphs have representation index $2 + 1/2$, this question is still open.

4 Complete Graphs

As it was said before $RI(K_3) = 1 + 1/2$, $RI(K_4) = 2$ and Bose et al. [3] proved that if $n \leq 20$ then $RI(K_n) \leq 3$. In this section, we are going to prove that $RI(K_{10}) = 2 + 1/2$ and that $RI(K_{2n}) \leq RI(K_n) + 1/2$. Observe that this implies Bose et al. [3] result as a particular case. Moreover, we give lowerbounds to the representation index of complete graphs.

Lemma 5. $RI(K_{10}) = 2 + 1/2$.

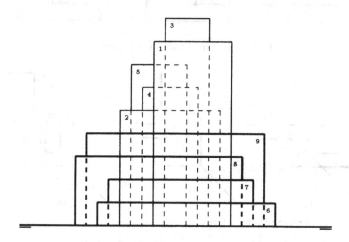

Fig. 5. K_{10} in $\mathbf{R}^{2+1/2}$

Proof. We can give the following configuration (Fig. 5) of 9 rectangles, such that all them are seen from the front. □

Theorem 6. $RI(K_{2n}) \leq RI(K_n) + 1/2$.

Proof. (Outline) We consider two cases. In the first one, we assume that $RI(K_n)$ is not an integer number. If we have a collection of n hyper-rectangles in a half of $\mathbf{R}^m$ representing K_n and such that the last interval of each one is $[0, a_i]$, then we consider a new configuration sustituing those intervals by $[-i, a_i]$, we get a representation of K_{2n} enfrenting to that configuration a new copy of that configuration but flipper over and rotated $90°$ (see Fig.6).

In the second case, we suppose that $RI(K_n)$ is an integer number. If we have a collection of n hyper-rectangles in $\mathbf{R}^m$ representing K_n, then it must exist a hyper-plane ortogonal to all hyper-rectangles and intersecting all them. Thus, we can suppose that $x_{m-1} = 0$ is such a hyperplane. That means that the i-th hyper-rectangle ends with the product $[a_i, b_i] \times \{a_i\}$ with $a_i < 0 < b_i$, then we sustitute those products by $[-i, b_i] \times [0, -a_i] \times \{a_i\}$. In this way, we get n hyper-rectangles in a half of $\mathbf{R}^{m+1}$ representing K_n such that the intervals before the last intervals in the product defining those hyper-rectangle constitute a sort of staircase, we can face in front of them the same configuration, but now changing $[a_i, b_i] \times \{a_i\}$ by $[a_i, b_i] \times [0, i] \times \{-i\}$ in order to get a representation of K_{2n} in a half of $\mathbf{R}^{m+1}$ (see Fig.7). □

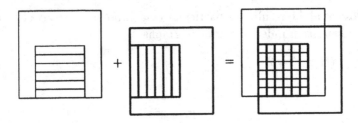

Fig. 6. The first case of Theorem 6.

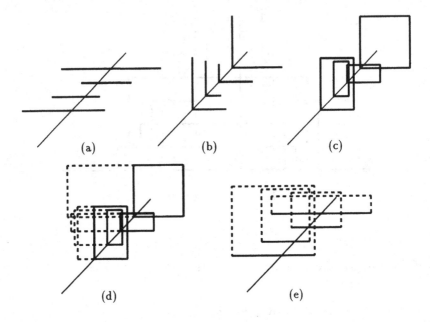

Fig. 7. The second case of Theorem 6.

As an inmediate consequence of Lemma 5 and Theorem 6 we get.

Corollary 7. *For all $n > 2$ if $m \leq 5 \cdot 4^{n-2}$ then $RI(K_m) \leq n$.*

Corollary 7 provides an upperbound to the representation index of the complete graphs. Now, we are going to try to get a lowerbound. As Fekete, Houle and Whitesides do in their paper [6], we will use the following lemma.

Lemma 8. [Attributed by F.R.K. Chung [4] to V. Chvátal and J.M. Steele, among others.] *For all $m > 1$, in every sequence of $\binom{m}{2} + 1$ distinct integers, there exists at last one strongly unimaximal subsequence (with only one local maximum) of length m). On the other hand, there exists a sequence of $\binom{m}{2}$ distinct integers that has no strongly unimaximal subsequence of length m.*

We will use the following easy-to-prove lemma.

Lemma 9. *If the hyper-rectangles $R_i = [a_1^i, b_1^i] \times [a_2^i, b_2^i] \times \cdots \times [a_{t-1}^i, b_{t-1}^i] \times \{a_i\}$ with $1 \leq i \leq n$ ($b_{t-1}^i \geq 0$) represent K_n, and $\{(a_{t-1}^i)\ 1 \leq i \leq n\}$ is a strongly unimaximal sequence, then the hyper-rectangles $R_i' = [a_1^i, b_1^i] \times [a_2^i, b_2^i] \times \cdots \times [0, b_{t-1}^i] \times \{a_i\}$ represent also K_n.*

Bose el al. prove in their paper [3] that $RI(K_{103}) \geq 3$ and Fekete, Houle & Whitesides prove in [6] by using of Lemma 8 that $RI(K_{56}) \geq 3$. It is possible to get this same lowerbound as a consequence of Lemma 9 and Lemma 8, in fact, a more general result can be achieved.

Theorem 10. $RI(K_{\binom{m}{2}+1}) > RI(K_m)$.

Proof. (Outline) If the hyper-rectangles $R_i = [a_1^i, b_1^i] \times [a_2^i, b_2^i] \times \cdots \times [a_{t-1}^i, b_{t-1}^i] \times \{a_i\}$ ($b_{t-1}^i \geq 0$) with $1 \leq i \leq \binom{m}{2} + 1$ represent $K_{\binom{m}{2}+1}$, by Lemma 8, there exists a strongly unimaximal subsequence of (a_{t-1}^i) of length m. Then Lemma 9 assures that K_n has representation index strictly lower than $K_{\binom{m}{2}+1}$. $\square$

5 Complete Bipartite Graphs

In this section we characterize the representation index of all complete bipartite graphs. Firstly, it is easy to observe that $RI(K_{n,m}) \leq 3$. But from the results in Sect. 3, we get that $RI(K_{1,2}) = RI(K_{2,2}) = 1 + 1/2$, and that $RI(K_{2,n}) = RI(K_{2,n}) = 2$ for all $n > 2$.

Now, we are going to prove that $RI(K_{3,n}) = 2 + 1/2$, for $n \geq 3$, and that $RI(K_{n,m}) = 3$ when $n, m \geq 4$.

Lemma 11. $RI(K_{3,n}) = 2 + 1/2$, for $n \geq 3$.

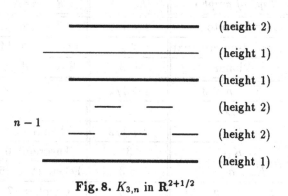

Fig. 8. $K_{3,n}$ in $\mathbf{R}^{2+1/2}$

Proof. As $K_{3,n}$ is not planar, we get that $RI(K_{3,n}) \geq 2 + 1/2$, and we can give the configuration of Fig. 8 to prove that $RI(K_{3,n}) = 2 + 1/2$. □

And studying exhaustly all cases it is possible to prove the following lemma.

Lemma 12. $RI(K_{4,4}) = 3$.

Finally, it is trivial to check that $RI(K_{m,n}) = 3$ for all $m, n \in \mathbf{N}$ such that $4 \leq m \leq n$.

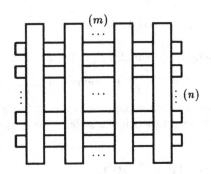

Fig. 9. $K_{n,m}$ in $\mathbf{R}^3$

6 Conclusions

We can sumarize our results in the following table.

Type of graph G	$RI(G)$	Reference
Outerhamiltonian	$1 + 1/2$	Theorem 4
Cutpoint-outerplanar	≤ 2	Theorem 3 [14]
Planar	≤ 3	Bose et al. [3]
$K_n \quad n \leq 3$	$1 + 1/2$	
K_4	2	
K_{10}	$2 + 1/2$	Lemma 5
K_{11}	3	Theorem 10
$K_n \quad n \geq 56$	> 3	Fekete et al. [6]
$K_{5.4^{n-2}} \quad n > 2$	$\leq n$	Corollary 7
$K_n \quad n \geq \binom{m}{2} + 1$	$> RI(K_m)$	Theorem 10

$K_{1,2}$ or $K_{2,2}$	$1 + 1/2$	
$K_{1,n}$ or $K_{2,n}$ $n \geq 3$	2	
$K_{3,n}$ $n \geq 3$	$2 + 1/2$	Lemma 11
$K_{4,4}$	3	Lemma 12
$K_{m,n}$ $4 \leq m \leq n$	3	Bose et al. [3]

References

1. H. Alt, M. Godau and S. Whitesides. Universal 3-Dimensional Visibility Representations for Graphs. Abstract presented at *Graph Drawing'95*, Passau (Germany). 1995.

2. S. Bellantoni, I. Ben-Arroyo Hartman, T. Przytycka and S. Whitesides. Grid intersection graphs and boxicity. North-Holland. Discrete Mathematics, 114:41-49, 1993.

3. P. Bose, H. Everett, S. Fekete, A. Lubiw, H. Meijer, K. Romanik, T. Shermer and S. Whitesides. On a Visibility Representation for Graphs in Three Dimensions. In David Avis and Prosenjit Bose, editors, *Snapshots in Computational and Discrete Geometry*, volume III, McGill University, July 1994. Technical Report SOCS-94.50.

4. F.R.K. Chung. On unimodal subsequences. *J. Combinatorial Theory*, Series A, v. 29:267-279, 1980.

5. P. Erdös and A. Szekeres. A combinatorial problem in geometry. *Compositio Mathematica*, v. 2:463-470, 1935.

6. S.P. Fekete, M.E. Houle and S. Whitesides. New Results on a Visibility Representation of Graphs in 3D. Abstract presented at *Graph Drawing'95*, Passau (Germany). 1995.

7. M. Garey, D. Jhonson and H. So. An application of graph coloring to printed circuit testing. *IEEE Trans. Circuits and Systems.*, CAS-23:591-598, 1976.

8. M.Y. Hsueh and D.O. Pederson. Computer-aided layout of LSI circuit bulding bloks. *Proc. IEEE Int. Symp. on Circuits and Systems.*, pages 474-477, 1979.

9. J. Kratochvil and T. Przytycka. Grid intersection graphs and boxicity on surfaces. Abstract presented at *Graph Drawing'95*, Passau (Germany). 1995.

10. K. Kuratowski. Sur le problème des courbes gauches en topologie. *Fund. Math.*, 15:271-283, 1930.

11. E. Lodi and L. Pagli. A VLSI algorithm for a visibility problem. In P. Bertolazzi and F. Luccio, editors, *VLSI: Algorithms and Architectures*, pages 125-134. Nort-Holland, Amsterdam, 1985.

12. S.L. Mitchell. Linear algorithms to recognize outerplanar and maximal outerplanar graphs. *Information Processing Letters*, 9(5):229-232, 1979.

13. M. Schlag, F. Luccio, P. Maestrini, D.T. Lee and C.K. Wong. A visibility problem in VLSI layout compaction. In F. P. Preparata, editor, *Advances in Computing Research*, volume 2, pages 259-282. JAI Press Inc, Greenwich. CT., 1985.

14. R. Tamassia and I.G. Tollis. A unified approach to visibility representations of planar graphs. *Discrete and Computational Geometry*, 1:321-341, 1986.

15. S. Wimer, I. Koren and I. Cederbaum. Floorplans, planar graphs and layouts. *IEEE Trans. Circuits ans Systems*, 35:267-278, 1988.

16. S.K. Wismath. Characterizing bar line-of-sight graphs. Proc. ACM Symp. on Computational Geometry. Baltimore. MD. 1985.

3D Graph Drawing with Simulated Annealing*

Isabel F. Cruz[1] and Joseph P. Twarog[2]

Department of Electrical Engineering and Computer Science
Tufts University
Medford, MA 02155, USA

Abstract. A recent trend in graph drawing is directed to the visualization of graphs in 3D [1, 5, 6]. A promising research direction concerns the extension of proven 2D techniques to 3D. We present a system extending the simulated annealing algorithm of Davidson and Harel [2] for straight-line two-dimensional drawings of general undirected graphs to three dimensions. This system features an advanced 3D user interface that assists the user in choosing and modifying the cost function and the optimization components on-line.

1 Introduction

The 2D simulated annealing algorithm of [2] starts with an initial configuration obtained by assigning random x and y coordinates to each vertex of the graph. The algorithm proceeds iteratively by computing the cost $c(\sigma)$ of the current configuration σ and randomly selecting a new configuration σ' from the *neighborhood* of σ. The new configuration is chosen if $c(\sigma') \leq c(\sigma)$ or if a random value between 0 and 1 is smaller than $e^{(c(\sigma)-c(\sigma'))/T}$ where T is the current temperature. This iterative process continues until a termination condition is satisfied. The higher T is, the greater the probability that an "uphill" move (i.e., a move that will make the temporary solution worse) is taken. As time goes by T decreases and the configuration stabilizes.

In [2], the *neighborhood* of a configuration contains all configurations that differ from that configuration by one vertex only, and the cost function is the weighted sum of five terms. Each term is a function of an optimization component. The five different components relate to *vertex distribution, borderline proximity, edge length, number of edge crossings*, and *distance between vertices and edges*. The cost function penalizes non-uniform distributions of vertices, vertices that are close to the borderline of the drawing area, long edges, large number of edge crossings, and small distances between vertices and edges. The normalizing factors and the term functions are chosen so that the cost function decreases as the perceived quality of the drawing improves.

2 3D Simulated Annealing

When designing a three-dimensional simulated annealing system, some extensions are straightforward, e.g., perturbing a point within a sphere instead of a

---

* Email addresses of the authors: {isabel,jtwarog}@cs.tufts.edu.

circle. The choice of components for the cost function is however not as immediate. First, what constitutes a good 3D drawing of a graph is not well understood. Secondly, the possibility of changing the viewpoint that is facilitated by the interface will diminish the relevance of edge crossing (this is reinforced by the fact that we are not considering a grid as in [1], thus decreasing the likelihood of two edges crossing).

While distances between vertices and edges are still relevant, we believe the corresponding component (together with the component that relates to the number of edge crossings) is subsumed by a new optimization component that we propose, which measures the distances between pairs of (non-adjacent) edges. We call this component *edge distribution*. Further experimentation is needed to substantiate this choice as in [3, 4] for 2D graph drawing. So that this study can be made, it is paramount to determine what are the most suitable aesthetic criteria in 3D graph drawing. Because the drawings start from a random configuration and are therefore not influenced by preconceived aesthetic criteria, effective but unconventional drawings may be obtained that will contribute to the understanding of the relevant components in three-dimensional drawing.

3 The Interface

In visualizing 3D graphs, an important interaction aspect resides on the ability to dynamically change the view of the graph. The interface to our system (depicted in Figure 1) provides two primary windows. The UVN synthetic camera display allows one to change the orientation of the graph throughout the progressive development of the drawing, lending insight to the graph structure. The other window allows parameters of the annealing process as well as the cost function normalizing factors to be adjusted dynamically. The interface allows for the specification of on-the-fly constraints via the selection window, which incorporates basic spreadsheet functionality (see Figure 2). Specifically, the user can modify the temperature of each individual vertex: by appropriately lowering the temperature of a vertex, its movement will be constrained. Conversely, by raising the temperature those vertices perceived to be inadequately positioned may be "re-annealed".

The application has been implemented according to the object-oriented paradigm with C++ and Motif; it is built upon the framework established by Young [7]. As a result, it is quite extensible and is able to serve as a proper test-bed. For example, adding a new criterion requires only the instantiation of a new subclass consisting of a single cost member function. Parameter controls for such a new cost will be automatically added to the interface.

Future work includes experimentation with alternative cost functions on an extensive suite of graphs.

References

1. R. F. Cohen, P. Eades, T. Lin, and F. Ruskey. Three-dimensional graph drawing. In R. Tamassia and I. G. Tollis, editors, *Graph Drawing (Proc. GD '94)*, volume

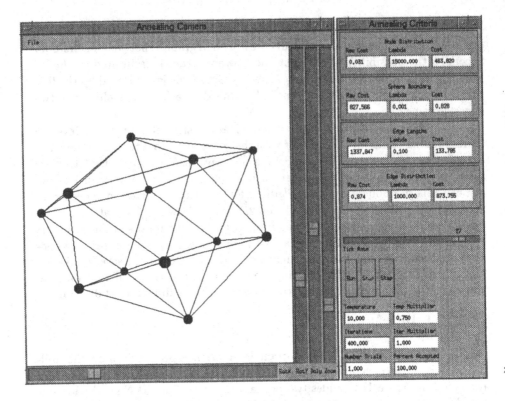

Figure 1: The graphic user interface.

894 of *Lecture Notes in Computer Science*, pages 1–11. Springer-Verlag, 1995.

2. R. Davidson and D. Harel. Drawing graphs nicely using simulated annealing. *Commun. ACM.* To appear.

3. G. Di Battista, A. Garg, G. Liotta, R. Tamassia, E. Tassinari, and F. Vargiu. An experimental comparison of three graph drawing algorithms. In *Proc. 11th Annu. ACM Sympos. Comput. Geom.*, pages 306–315, 1995.

4. M. Himsolt. Comparing and evaluating layout algorithms within GraphEd. *J. Visual Languages and Computing* (special issue on Graph Visualization, edited by I. F. Cruz and P. Eades), 6(3), 1995.

5. T. Jéron and C. Jard. 3D layout of reachability graphs of communicating processes. In R. Tamassia and I. G. Tollis, editors, *Graph Drawing (Proc. GD '94)*, volume 894 of *Lecture Notes in Computer Science*, pages 25–32. Springer-Verlag, 1995.

6. S. P. Reiss. An engine for the 3D visualization of program information. *J. Visual Languages and Computing* (special issue on Graph Visualization, edited by I. F. Cruz and P. Eades), 6(3), 1995.

7. D. A. Young. *Object-Oriented Programming with C++ and OSF/Motif.* Prentice Hall, Englewood Cliffs, N.J., 1992.

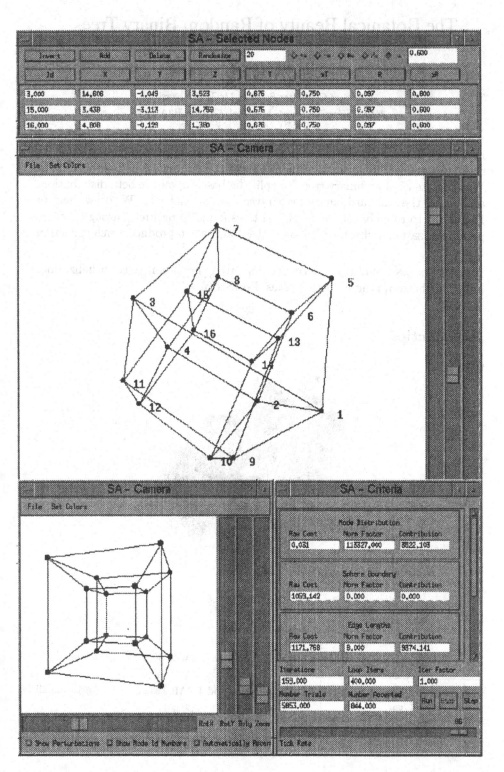

Figure 2: A running simulation.

The Botanical Beauty of Random Binary Trees

Luc Devroye[†] and Paul Kruszewski[‡]

School of Computer Science, McGill University

3480 University Street, Montreal, Canada H3A 2A7

ABSTRACT. We present a simple mechanism for quickly rendering computer images of botanical trees based on random binary trees commonly found in computer science. That is, we visualize abstract binary trees as botanical ones. We generate random binary trees by splitting based upon the beta distribution, and obtain the standard binary search trees as a special case. We draw them in PostScript to resemble actual botanical trees found in nature. Through flexible parameterization and extensive randomization, we can produce a rich collection of images.

KEYWORDS AND PHRASES. Tree drawing, tree simulation, tree visualization, beta distribution, random binary trees, PostScript.

Introduction.

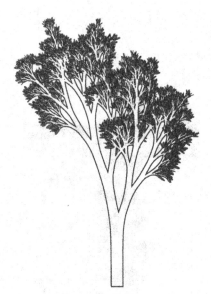

Figure 1. A visualization of a random binary tree with 5000 internal nodes.

[†] Research supported by NSERC Grant A3456 and FCAR Grant 90-ER-0291. Email: luc@cs.mcgill.ca.

[‡] Research supported by a 1967 NSERC Postgraduate Scholarship. Email: kruz@cs.mcgill.ca.

The computer imagery of realistic-looking trees has many applications rang-
ing from the verification of botanical models to computerized landscaping and
animation. In their book, *The Algorithmic Beauty of Plants*, PRUSINKIEWICZ
AND LINDENMAYER (1990) provide an excellent overview of this emerging field.
Through beautiful pictures, they and others have shown the mathematical ele-
gance underlying simple biological systems. In this note, we hope to outline how
computer data structures such as binary trees may be visualized in a similarly
elegant way as botanical trees.

Suffix trees and tries are commonly used for storing text files for string
searching (STEPHEN (1994)). When shown as a drawing in a window, a lot of
information is revealed about the authorship, language, and nature of the text.
Drawings can be used as simple, elegant signatures of files.

For example, the drawing in Figure 1 does not originate in nature. Rather
this image has been created by visualizing a random binary tree in such a way
as to resemble a botanical tree found in nature, i.e., each internal node of the
random binary tree is drawn as a branch in the botanical tree and each external
node is drawn as a leaf.

Our algorithm.

Our algorithm builds on the approach taken in KRUSZEWSKI (1994) which in
turn is inspired by VIENNOT, EYROLLES, JANEY, AND ARQUÈS (1989) (see
also VIENNOT (1990) or ALONSO AND SCHOTT (1995)). Indeed, we are heavily
indebted to these authors for the idea of using combinatorial trees as a basis for
drawing botanical ones. Logically speaking, we first generate a random binary
tree by random splits and then we draw a corresponding botanical tree according
to the resulting structure of the binary tree and to various controlling parameters
($\pi_n \rightarrow \Psi$). In practice, we generate the binary tree and draw a corresponding
botanical one "on the fly", branch by branch, one after the other, in a preorder
traversal. That is, for each subtree rooted at node u with children v and w,
we draw the branch corresponding to u and then recursively draw the branches
corresponding to nodes v and w. Our algorithm is implemented in PostScript
and as such the algorithm runs *entirely inside* the printer.

Overall structure.

We generate the tree by random splits. It is well-known (e.g., DEVROYE (1994))
that many binary tree data structures such as binary search trees, tries, and
PATRICIAS can be simulated by recursive random splits. That is, starting at
the root with n nodes, let X be a $[0, 1]$-valued random variable. Assign the left
and right subtrees $\lfloor nX \rfloor$ and $n - 1 - \lfloor nX \rfloor$ nodes respectively. This splitting
continues with independent identically distributed copies of X on the left and
right subtrees until they each have only one node. Such a tree is called a random
split tree. In our simulations, X is a beta random variable, and we call the
resulting trees random beta trees.

After each node is created, its corresponding branch is drawn as a deformed
rectangle. For each node u with children v and w, we determine for its corresponding branch, its length, width and branching angle. Implicit to our drawing

style is the idea of sap flow through the tree. That is, for each branch, the number of leaves in its subtree is supposed to be the key influence on its growth and there is some relation between size of the logical subtree and layout of the physical branch. Typically for example, the more leaves a branch has above it, the longer and wider that branch is.

Obviously, we are not the first to make such observations. The earliest reference which we could find are from about 1513 by Leonardo Da Vinci (RICHTER (1970)). In his book, *Botany for Painters*, Da Vinci sets up rules to guide artists in representing trees. Although Da Vinci attemps to give scientific explanations why things look as they do, his observations are first and foremost concerned with how things *should* look. We re-iterate that this is also our approach. That is, we are concerned with developing a model which produces convincing synthetic images rather than actually articulating how nature works.

Both VIENNOT ET AL. (1989) and KRUSZEWSKI (1994) use the HORTON-STRAHLER number of a node as the basis for functions of length, width and branching angles. At present, we prefer using subtree sizes. Typically, the width and length of a branch is a nondecreasing function f of $|u|$, the size of the subtree rooted at node u. Often, the æsthetically most pleasing results for the length and the width functions occur when the functions are of the form $c \ln |u|$ or $c\sqrt{|u|}$ where $c > 0$ is a constant (e.g., see Figure 2).

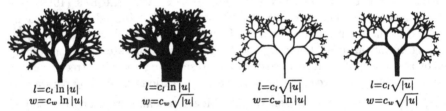

$$l = c_l \ln |u|$$
$$w = c_w \ln |u|$$

$$l = c_l \ln |u|$$
$$w = c_w \sqrt{|u|}$$

$$l = c_l \sqrt{|u|}$$
$$w = c_w \ln |u|$$

$$l = c_l \sqrt{|u|}$$
$$w = c_w \sqrt{|u|}$$

Figure 2. Various length and width functions for the same tree.

However, as Figure 3 shows, many other length functions may be used, such as $\frac{c'}{c+d}$, $\frac{c'}{c+d^2}$ or $\frac{c'}{c+\ln d}$, where d is the depth of the corresponding node and both c and c' are constants.

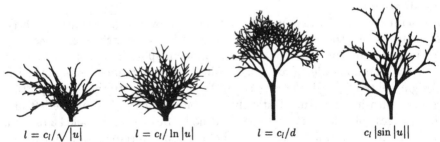

$$l = c_l/\sqrt{|u|}$$

$$l = c_l/\ln |u|$$

$$l = c_l/d$$

$$c_l |\sin |u||$$

Figure 3. Examples of trees with atypical length functions.

Given two sibling branches v and w, it is often the case in nature that the larger branch deviates less in angle from the parent branch u than its sibling.

VIENNOT ET AL. (1989) determine three cases and corresponding angles: a branch is the main branch (typically 10°), secondary branch (typically 25°), or a fork branch where it and its sibling are about equal in value so that both angles are also about equal (typically 30°). Classification of branches is based the HORTON-STRAHLER number. However, since many families of random binary trees have logarithmic HORTON-STRAHLER numbers in the number of nodes (see e.g., DEVROYE AND KRUSZEWSKI (1994,1995)), similar comparisons such as

$$\theta_v = \begin{cases} 10°, & \text{if } \lfloor \log |v| \rfloor > \lfloor \log |w| \rfloor, \\ 25°, & \text{if } \lfloor \log |v| \rfloor < \lfloor \log |w| \rfloor, \\ 30°, & \text{if } \lfloor \log |v| \rfloor = \lfloor \log |w| \rfloor, \end{cases}$$

are equally acceptable (e.g., see the first drawing in Figure 4). Nonetheless, the possibilities for branching angles are endless. In this figure, each tree has the same number of nodes and is drawn from the same probability distribution. The second drawing uses sibling sizes directly to determine angle, i.e., $\theta_v = \frac{|w|}{|v|+|w|} \times 30°$. Finally, the last two drawings rely on depth d of the branch in the tree, i.e., $\theta = \frac{27°}{d+1}$ and $\theta = \frac{23°}{\ln d + 1}$.

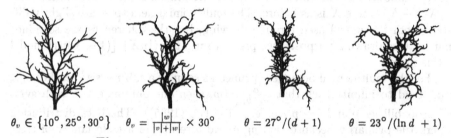

$$\theta_v \in \{10°, 25°, 30°\} \qquad \theta_v = \frac{|w|}{|v|+|w|} \times 30° \qquad \theta = 27°/(d+1) \qquad \theta = 23°/(\ln d + 1)$$

Figure 4. Various angle functions for the same tree.

For greater realism, each angle θ may be multiplied with $\cos(2\pi U)$ where U is uniform[0, 1] to simulate projection in the plane of a random 3-d rotation (e.g., Figure 5 shows this effect on the second drawing in the previous figure). In all cases, to avoid any absurdly asymmetric drawings, at each split, we flip a coin and place θ by $-\theta$ with probability $1/2$.

Figure 5. Simulated random 3-d angles.

Drawing polygonal branches typically results in rough-looking notches where the branches meet. VIENNOT ET AL. (1989) fill in these joints with small triangles. We avoid this problem by drawing smooth, rounded forks rather than individual branches (cf. BLOOMENTHAL (1985)).

That is, the "buds" of the left and right child branches are also drawn with the parent as one smooth unit. We then overlay the buds with the corresponding child forks for a smooth fit. This means that in practice, the size and orientation of a branch is worked out when its parent branch is drawn. Thus, the child forks are laid over the parent fork Ψ. Note that all of the curves are created by the PostScript command `curveto` which implements Bézier curves (see p. 140 in ADOBE (1985)). This layering continues throughout the preorder traversal.

Beta distribution.
We split according to the beta distribution as it yields a rich family of branching patterns. The beta(a, b) has density

$$f(x) = \frac{\Gamma(a+b)}{\Gamma(a)\Gamma(b)}\, x^{a-1}(1-x)^{b-1}, \qquad 0 < x < 1,$$

where $a, b > 0$ are parameters and Γ is the gamma function. For example, beta(1,1) produces random binary search trees. Beta trees are defined in DE-VROYE (1986) as trees in which the sizes of right and left subtrees are multinomial $(n, X, 1 - X)$ where X is as before. The multinomial beta trees are slightly different from the model used for tree drawing, but the differences are so minor that for tree drawing purposes, we prefer to use the $(\lfloor nX \rfloor, \lfloor (1 - X)n \rfloor)$ model of this paper.

For the multinomial beta trees pruned as soon as a subtree size reaches one, if a,b tend to infinity such that $\frac{a}{(a+b)} = p$, then one obtains a trie with symbol probabilities p and $1 - p$ (see e.g., PITTEL (1985)). The beta distribution is versatile primarily because varying a and b results in a wide family of trees with logarithmic average depth and height. That is, the bushiness and elongation of the trees can be controlled by varying the parameters. More formally DEVROYE (1995) shows the following theorems for random split trees in general.

THEOREM 1. *Let D_n be the depth of the last node in a random split tree with n nodes. Then*
$$\frac{D_n}{\log n} \to \frac{1}{\mu} \qquad \text{in probability as } n \to \infty,$$

and $\mathbf{E}\{D_n\}/\log n$ tends to the same limit, where $\mu = 2\mathbf{E}\{Y\log(1/Y)\}$, $Y \in [0, 1]$ is X and $1 - X$ with equal probability, and X is the branch-splitting random variable introduced earlier.

THEOREM 2. *Let H_n be the height of a random split tree with n nodes. Then*

$$\frac{H_n}{\log n} \to \gamma \qquad \text{in probability as } n \to \infty,$$

where $\gamma = \inf\left\{c : e^{t^*}(2m(t^*))^c < 1\right\}$, $m(t) = \mathbf{E}\{Y^t\}$, $t \geq 0$, t^* is the unique solution of $m'(t)/m(t) = -1/c$, and Y is as in Theorem 1.

With random beta trees, one can choose the desired expected depth and height and solve the above formulas to determine explicit values for a and b. Note, for example, that $1/\mu$ can take any value between $1/\log 2$ and ∞. We implement this distribution by the PostScript uniform random number generator **rand** and Cheng's method for beta variates (CHENG (1978) as explained on p. 438 of DEVROYE (1986)).

Figures 6 and 7 show the flexible nature of the beta distribution. Each tree consists of 500 nodes and is drawn using the same rules (for length, width, and angle) but for different beta parameters. As Figure 6 shows, as $a \to \infty$, and $b = a$, the splitting is even and deterministic, and as $a \to 0$, $b = a$, the splitting is asymmetric and unstable.

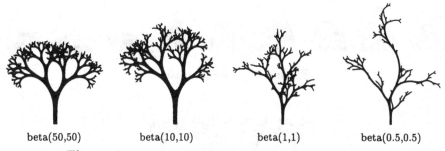

| beta(50,50) | beta(10,10) | beta(1,1) | beta(0.5,0.5) |

Figure 6. Examples of random beta(a,a) trees with 500 nodes.

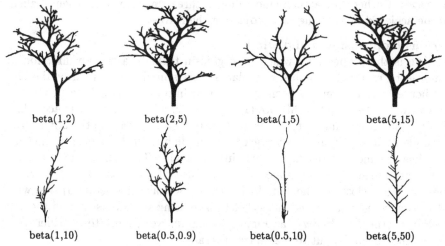

Figure 7. Examples of random beta(a,b) trees with 500 nodes.

Randomization.

Our algorithm is heavily randomized. The underlying tree structure is generated by random splits. Furthermore, all functions such as length, width and branching angle can be perturbed using randomness. Angles can be randomized based upon subtree sizes, depths and split ratios, for example.

Leaves.

Realistic-looking leaves are an important component for any tree drawing program. We use a very simple rectangular shape based on the examples found in SUGDEN (1984). A leaf consists of an apex (top) and a base. Shape is controlled by varying the apex height, base depth and leaf width. As Figures 8 and 9 show, we have nine different apices and four different bases, each constructed with simple Bézier curves.

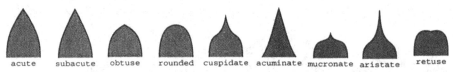

| acute | subacute | obtuse | rounded | cuspidate | acuminate | mucronate | aristate | retuse |

Figure 8. Various leaf apices.

| cordate | cuneate | rounded | truncate |

Figure 9. Various leaf bases.

Added realism can be achieved by drawing two-dimensional projections of the leaves. Rather than actually modelling in three dimensions, sufficient realism can be achieved by rotating and projecting the leaves.

Various tropisms: sun and wind.

Tropism is the property by which an organism turns in a certain direction in response to external stimulus. In plants, this stimulus is primarily the sun and hence heliotropism has been incorporated into many models (e.g., CHIBA, OHKAWA, MURAOKA, AND MIURA (1994)). We simulate heliotropism according to sun position and intensity. With respect to intensity, we use the admittedly naïve idea that the larger the branch the more light it receives over its lifetime and thus the more it reacts by changing its angle. That is, for node u after θ_u is determined, θ_u is multiplied by an intensity factor (based on $|u|$) which pulls branch u closer to the sun. In Figure 10, we take the beta(1,5) tree with 500 nodes from Figure 7 and subject it to increasing sun intensity with the sun directly overhead. However as Figure 10 shows, we neglect to consider that leaves tend to spread out to maximize coverage.

Wind is also an important environmental factor. Both VIENNOT ET AL. (1989) and KRUSZEWSKI (1994) simulate wind by changing the underlying structure; the former always flips larger branch to one side while the later

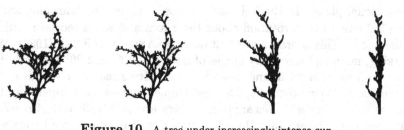

Figure 10. A tree under increasingly intense sun.

Figure 11. A tree under increasingly intense wind.

uses asymmetric tries. As Figure 11 shows, by placing the sun perpendicular to the ground and inverting the intensity function (i.e., larger branches should bend less than smaller ones), reasonable wind can be simulated.

Finally, if we set the wind to blow from above, we can simulate the effect of droughts or flexible branches such as those found in weeping willows.

Figure 12. A weeping willow.

Three-dimensional drawing.

Botanical trees are three-dimensional objects. Therefore, added realism is obtained by drawing the trees in three dimensions and projecting them onto the two-dimensional plane. In the 3-d case, we now consider the branching angle from the 2-d case to be a rotation about the z-axis and add a second rotation about the z-axis. This approach was first taken by ANON AND KUNII (1984). We have forking, main and secondary angles of sizes 30°, 70°, and 20° respectively. Branches are now cylindrical and smooth Bézier curves require a serious computational effort. Currently, we opt for the simpler solution of representing the branches as solid cylinders. This approach is very acceptable when the branches are very thin (e.g., Figure 13). However, drawings of thick-branched trees are rather unappealing. We are currently working on a new 3-d model which will produce the same smooth forks as in the 2-d model.

Figure 13. A three-dimensional tree.

Conclusions.

All of the images were generated in PostScript on a 600 dpi Apple Laser-Writer Pro with 8 megabytes of memory. All files[†] are completely self-contained and are about 27 kilobytes long, of which more than half is documentation. For example, Figure 1 has 5000 branches and takes approximately 33 minutes to print. Image rendering by `ghostview` is about eight times faster.

Many extensions and enhancements can be imagined. Probably, the most desired would be to wrap the program in a graphical user-interface. Currently, different trees are generated by modifying the PostScript code by hand and then re-viewing with the PostScript previewer `ghostview`. A graphical interface would allow the user to freely change parameters and then instantly view resulting changes. We do not grow the tree dynamically to model physical growth. However, if we re-draw the tree after each successive node is added, we could have a reasonable animation of tree growth. Finally, not all trees are binary, we hope to extend our model to arbitrary k-ary trees.

[†] Our programs are available by anonymous `ftp` at `ftp.cs.mcgill.ca` in the directory `pub/tech-reports/library/code/botan-ical.trees/`.

Acknowledgements.

We thank Sue Whitesides for her simulating conversations, useful advice and ongoing encouragement.

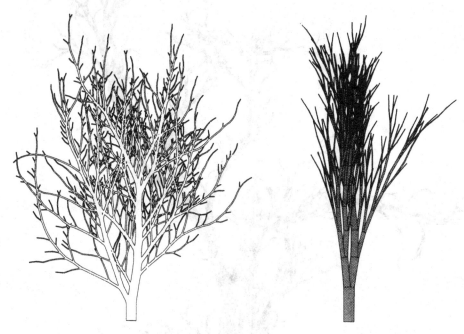

A birch tree from Quebec's Laurentians.   A yucca tree.

Figure 14. Sundry trees.

References.

ADOBE SYSTEMS INC. (1985). *PostScript Language Reference Manual.* Reading, MA: Addison-Wesley.

ALONSO, L. AND R. SCHOTT (1995). *Random Generation of Trees.* Dordrecht, The Netherlands: Kluwer Academic Publishers.

ANON, M. AND T. KUNII (1984). Botanical tree image generation. *IEEE Computer Graphics Applications 4*, 10–34.

BLOOMENTHAL, J. (1985). Modeling the Mighty Maple. In *Proceedings of SIGGRAPH'85, Computer Graphics*, Volume 19, pp. 305–311.

CHENG, R. C. H. (1978). Generating beta variates with nonintegral shape parameters. *Communications of the ACM 21*, 317–322.

CHIBA, N., S. OHKAWA, K. MURAOKA, AND M. MIURA (1994). Visual simulation of botanical trees based on virtual heliotropism and dormancy break. *Journal of Visualization and Computer Animation 5*(1), 3–15.

Figure 15. Passau im Herbst.

DEVROYE, L. (1986). *Non-Uniform Random Variate Generation.* New York: Springer-Verlag.

DEVROYE, L. (1994). *Lectures Notes for Computer Science 690A—Probabilistic Analysis of Algorithms and Data Structures.* Montreal: School of Computer Science, McGill University.

DEVROYE, L. (1995). Universal limit laws for depths in random trees. (Submitted).

DEVROYE, L. AND P. KRUSZEWSKI (1994). A note on the HORTON-STRAHLER number for random trees. *Information Processing Letters 52*, 155–159.

DEVROYE, L. AND P. KRUSZEWSKI (1995). On the HORTON-STRAHLER number for random tries. (Submitted).

KRUSZEWSKI, P. (1994). Using the HORTON-STRAHLER number to draw trees. Technical Report SOCS-94.1, School of Computer Science, McGill University.

PITTEL, B. (1985). Asymptotical growth of a class of random trees. *Annals of Probability 13*, 414–427.

PRUSINKIEWICZ, P. AND A. LINDENMAYER (1990). *The Algorithmic Beauty of Plants*. New York: Springer-Verlag.

RICHTER, J. P. (1970). *The literary works of Leonardo Da Vinci* (3 ed.). New York: Phaidon Publishers Inc.

STEPHEN, G. A. (1994). *String Searching Algorithms*. Lecture Notes Series on Computing. New York: Springer-Verlag.

SUGDEN, A. (1984). *Longman Illustrated Dictionary of Botany*. Essex, UK: Longman Group Limited.

VIENNOT, X. G. (1990). Trees everywhere. In A. Arnold (Ed.), *Proceedings of the 15th Colloquium on Trees in Algebra and Programming, Copenhagen, Denmark, May 15-18, 1990, Lecture Notes in Computer Science*, Volume 431, Berlin, pp. 18–41. Springer-Verlag.

VIENNOT, X. G., G. EYROLLES, N. JANEY, AND D. ARQUÈS (1989). Combinatorial analysis of ramified patterns and computer imagery of trees. In *Proceedings of SIGGRAPH'89, Computer Graphics*, Volume 23, pp. 31–40.

The Strength of Weak Proximity
(Extended Abstract)

Giuseppe Di Battista[1], Giuseppe Liotta[2], Sue Whitesides[3]

[1] Dipartimento di Discipline Scientifiche, Sezione Informatica Terza Universita' di Roma via Segre 2 00146 Roma
dibattista@iasi.rm.cnr.it
[2] Department of Computer Science, Brown University, 115 Waterman Street, Providence, RI 02912.
gl@cs.brown.edu
[3] School of Computer Science, McGill University, 3480 University St. # 318, Montréal, Québec, H3A 2A7 Canada.
sue@cs.mcgill.ca

Abstract. This paper initiates the study of *weak* proximity drawings of graphs and demonstrates their advantages over *strong* proximity drawings in certain cases. Weak proximity drawings are straight line drawings such that if the *proximity region* of two points p and q representing vertices is devoid of other points representing vertices, then segment (p, q) is allowed, but not forced, to appear in the drawing. This differs from the usual, strong, notion of proximity drawing in which such segments must appear in the drawing.

Most previously studied proximity regions are associated with a parameter β, $0 \leq \beta \leq \infty$. For fixed β, weak β-drawability is at least as expressive as strong β-drawability, as a strong β-drawing is also a weak one. We give examples of graph families and β values where the two notions coincide, and a situation in which it is NP-hard to determine weak β-drawability. On the other hand, we give situations where weak proximity significantly increases the expressive power of β-drawability: we show that every graph has, for all sufficiently small β, a weak β-proximity drawing that is computable in linear time, and we show that every tree has, for every β less than 2, a weak β-drawing that is computable in linear time.

1 Introduction and Overview

Given two points u and v of the plane, a *proximity region* of u and v is a portion of the plane, determined by u and v, that contains points relatively close to both of them. A *proximity drawing* of a graph G has been defined in the literature as

* Research supported in part by the National Science Foundation under grant CCR-9423847, by the U.S. Army Research Office under grant 34990–MA–MUR, by FCAR, by Progetto Finalizzato Sistemi Informatici e Calcolo Parallelo of the Italian National Research Council (CNR), and by N.A.T.O.- CNR Advanced Fellowships Programme.

a *straight-line drawing* (vertices of G are mapped to distinct points of the plane, and edges to straight-line segments) such that: (i) for each edge (u, v) of G, the proximity region of the points representing u and v does not contain any other vertex; and (ii) for each pair of non-adjacent vertices u, v of G, the proximity region of the points representing u and v contains at least one other vertex.

Most of the results on proximity drawings take as proximity regions the so-called β-*regions* [11]. Such regions form an infinite family, each element of the family being identified by a value of the parameter β ($0 \leq \beta \leq \infty$). For example, when $\beta = 1$ the proximity region of u and v is the disk with u and v as antipodal points; when $\beta = 2$ the proximity region is the intersection of two disks with centers at u and v and radius the distance $d(u, v)$ between u and v; when $\beta = \infty$ the proximity region is the infinite strip perpendicular to the line segment between u and v. A β-drawing is a proximity drawing such that the proximity regions are β-regions. A graph is β-drawable if it has a β-drawing. A brief survey on proximity drawability is in [4]. Besides their theoretical interest, proximity drawings have been studied for their practical characteristics: neighboring graph vertices are clustered in the drawing, and adjacent edges tend to have large angles. Furthermore, proximity drawings are related to minimum spanning tree drawings, minimum weight drawings of triangulations, and Delaunay drawings (e.g., see [6, 10, 2]).

The purpose of this paper is to initiate a study of *weak* proximity drawings, in particular, weak β-drawings. A weak proximity drawing of a graph G is one that ignores requirement (ii) for traditional, or *strong*, drawings. In other words, if (u, v) is *not* an edge of G, then no requirement is placed on the proximity region of u and v in a weak drawing. For example, Fig. 1 shows a weak proximity drawing of a tree. Here, the proximity region of any two points p and q is the disk having p and q as antipodal points. Note that the drawing is not a strong drawing, as no edges between neighbors of the degree six vertex are included.

There are several motivations for studying weak proximity drawings and in particular, weak β-drawings.

- Strong proximity drawability is very restrictive, perhaps too much so. By relaxing (ii), a graph G can no longer be reconstructed from the locations of its vertices in a weak drawing; however, many graphs that do not admit strong drawings can be drawn weakly. For example, a tree that has a vertex of degree greater than five has no strong β-drawing for any β [3]. Thus the drawing in Fig. 1 illustrates a graph that is *weak* but not *strong* drawable for the circular disk proximity region defined by antipodal points. Also, characterizations and algorithms for strong β-drawability have been devised only for a few classes of graphs.
- Weak and strong visibility drawings (e.g., see [15]) can be considered as a particular class of proximity drawings. In the field of visibility drawing, the coordinated study of both strong and weak types of drawings led to deep and practical results.
- Weak proximity can be considered as an "edge-vertex resolution rule" in the sense that a vertex cannot enter the region of influence of an edge. Thus, the

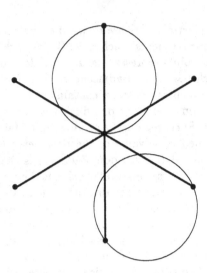

Fig. 1. A weak proximity drawing of a tree.

study of weak proximity can contribute to the body of drawing strategies
that adopt a resolution rule (e.g., see [5, 9]).

– The weak proximity model may well be sufficient for many drawing applications, particularly ones that do not require recovery of the graph solely from
the positions of its vertices.

The main results presented in this paper are as follows.

General graphs. We show that any graph G is weak β-drawable for all β in
the range 0 to some upper bound that is a function either of the number of
vertices or of the maximum vertex degree of G. (Section 3.)

Planar graphs. First, we show how to extend existing strong proximity drawability results on outerplanar graphs to weak drawability results. Second, we
show that, in a certain interval for β, strong and weak β-drawings of triangulated planar graphs coincide. Third, we give new insights on the interplay between angular resolution and proximity drawability. Namely, we show how to
interpret any straight-line drawing algorithm for planar triangulated graphs
as an algorithm for constructing weak proximity drawings. (Section 4.)

Trees. We provide an algorithm to draw any tree as a weak β-drawing for any
value of β less than two. Then we show that for $2 \leq \beta < \infty$, the weak and
the strong proximity models give rise approximately to the same class of
β-drawable trees. Finally, we show the NP-hardness of deciding whether a
tree has a weak proximity drawing for $\beta = \infty$, where the region of influence
is an open strip. (Section 5.)

All our algorithms admit a linear time implementation in real RAM. The above results represent, in many cases, substantial improvements over the known algorithms and characterizations for strong proximity drawability.

2 Preliminaries

We recall some basic definitions concerning proximity drawings.

Given a pair x, y of points in the plane, the *open β-region of influence of x and y*, and the *closed β-region of influence of x and y*, denoted by $R(x, y, \beta)$ and $R[x, y, \beta]$ respectively, are defined as follows:

1. For $0 < \beta < 1$, $R(x, y, \beta)$ is the intersection of the two open disks of radius $d(x, y)/(2\beta)$ passing through both x and y. $R[x, y, \beta]$ is the intersection of the two corresponding closed disks.
2. For $1 \le \beta < \infty$, $R(x, y, \beta)$ is the intersection of the two open disks of radius $\beta d(x, y)/2$, centered at the points $(1-\beta/2)x+(\beta/2)y$ and $(\beta/2)x+(1-\beta/2)y$. $R[x, y, \beta]$ is the intersection of the two corresponding closed disks.
3. $R(x, y, \infty)$ is the open infinite strip perpendicular to the line segment $\overline{xy}$ and $R[x, y, \infty]$ is the closed infinite strip perpendicular to the line segment $\overline{xy}$.
4. $R(x, y, 0)$ is the empty set; $R[x, y, 0]$ is the line segment connecting x and y.

Let G be a graph. A *weak (strong) (β)-drawing* of G is a weak (strong) proximity drawing of G such that for each pair of points x, y the proximity region is $R(x, y, \beta)$. Weak and strong (β)-drawings are called w-(β)-drawings and s-(β)-drawings, respectively. Analogously, a *weak (strong) [β]-drawing* of G is a weak (strong) proximity drawing of G such that for each pair of points x, y the proximity region is $R[x, y, \beta]$. Weak and strong [β]-drawings are called w-[β]-drawings and s-[β]-drawings, respectively.

A graph is *w-(β)-drawable* (*s-(β)-drawable*) if it has a w-(β)-drawing (s-(β)-drawing). Analogously, a graph is *w-[β]-drawable* (*s-[β]-drawable*) if it has a w-[β]-drawing (s-[β]-drawing). When it is clear from the context or when it is not necessary to distinguish between open and closed proximity regions, we simplify the notation by talking about *β-drawings* and *β-drawable* graphs. A class of graphs is w-β-drawable (resp. s-β-drawable) if all its graphs are w-β-drawable (resp. s-β-drawable). A class of graphs is not w-β-drawable (resp. s-β-drawable) if it contains at least one graph that is not w-β-drawable (resp. s-β-drawable).

The following properties extend many results on strong drawability to weak drawability.

Property 1. An s-β-drawable graph is also w-β-drawable.

Property 2. Let Γ be a w-β-drawing of a graph G and let P be the set of points of Γ representing the vertices of G. Let Γ' be an s-β-drawing such that Γ' uses the same set of points P to represent the vertices of G'. Then $G \subseteq G'$.

Attractive drawability inclusion properties follow from weak proximity.

Property 3. If G is a w-β-drawable graph, then $G' \subset G$ is a w-β-drawable graph.

Property 4. If G is a w-$\overline{\beta}$-drawable graph, then G is w-β-drawable for any β such that $0 \le \beta < \overline{\beta}$.

To analyse w-β-drawings we will frequently use two angles $\alpha(\beta)$ and $\gamma(\beta)$, defined as follows.

1. $\alpha(\beta) = \inf\{\angle zxy \parallel z \in R[x, y, \beta]\}$.
2. $\gamma(\beta)$ is only defined for $\beta \ge 2$, and $\gamma(2) = \frac{\pi}{3}$. For $\beta > 2$, let $z \ne y$ be a point on the boundary of $R[x, y, \beta]$ such that $d(x, y) = d(x, z)$. Then $\gamma(\beta) = \angle zxy$.

Property 5. [3]

1. $\beta = \sin \alpha$ for $0 \le \beta < 1$.
2. $\beta = \frac{1}{1 - \cos \alpha}$ for $1 \le \beta \le \infty$.
3. $\beta = \frac{1}{\cos \gamma}$ for $2 \le \beta \le \infty$.

3 General Graphs

In this section, we give a simple, fast method for producing w-(β)- and w-[β]-drawings of arbitrary graphs on n vertices for certain β in the range $0 \le \beta < 1$.

Theorem 6. *Any graph on n vertices is w-[β]-drawable for all values of β such that $0 \le \beta < \sin(2\pi/n)$; any graph on n vertices is w-($\sin(2\pi/n)$)-drawable.*

Sketch of proof. For the trivial case $\beta = 0$, it suffices to place the points on a circular arc of measure $< \pi$. For $\beta > 0$, place n points equally spaced around a circle C of arbitrary radius R. Recall that for $0 < \beta < 1$, the radius r for the circular arcs bounding a proximity region $R(x, y, \beta)$ is given by $d(x, y)/(2\beta)$. For β sufficiently close to 0, the region of influence is a slight widening of the line segment joining x and y. It lies entirely within C and hence contains none of the n points distinct from x and y. Radius r decreases with increasing β. When β increases to the extent that $r \le R$, then points that are not consecutive on the circle cannot be joined by an edge in a w-[β]-drawing (similarly for $r < R$ and w-(β)-drawings). This critical value of β is thus determined by $R = r = d(p, q)/2\beta$, i.e., $\beta = d(p, q)/2R$, where p, q is the closest pair of points that are *non-consecutive* on C. For n equally spaced points, $sin(2\pi/n) = (d(p, q)/2)/R$, so $d(p, q) = 2R sin(2\pi/n)$. Hence, independent of R, the critical value of β is $sin(2\pi/n)$ (approximately $2\pi/n$ for large n).

The statement of Theorem 6 can be strengthtened in certain cases by using a method of [16]. Consider a coloring of G by χ colors, where χ is the chromatic number of G. Divide a circle C of arbitary radius R into χ arcs of equal length.

Cluster points receiving color i about the center of arc i. The closest pair p, q of points that are non-consecutive on C but that are joined by an edge of G must lie in different arcs. Hence given any $t < 2R\sin(\pi/\chi)$, by placing points sufficiently close to the centers of their arcs, we can be sure that $d(p, q) \geq t$. Hence, for $0 \leq \beta < \sin(\pi/\chi)$, w-$\beta$-drawings exist for G. Of course, the chromatic number χ is in general hard to compute. Instead of breaking C into χ arcs, one can instead break C into $d + 1$ arcs, where d is the maximum degree of G. A coloring for G by $d + 1$ or fewer colors can be obtained in linear time by a well-known greedy coloring algorithm. Hence for β between 0 and $\sin(2\pi/(d+1))$, a w-(β)- or w-$[\beta]$-drawing for G can be obtained in linear time. Hence we have:

Theorem 7. *A graph with chromatic number χ is w-β-drawable for all β such that $0 \leq \beta < \sin(\pi/\chi)$. A w-$\beta$-drawing for G can be obtained in linear time for any β such that $0 \leq \beta < \sin(\pi/(d+1))$, where d is the maximum degree of G.*

4   Planar Graphs

Several interesting results on planar graphs can be "imported" from strong proximity results by using Property 1. Other results can be obtained with the same property and little more work. For example, from Lubiw and Sleumer [12] we obtain:

Theorem 8. *Biconnected outerplanar graphs are w-β-drawable for all values of β such that $0 \leq \beta < 2$. Furthermore, such graphs are w-(2)-drawable.*

Sketch of proof. In [12] it is shown that any biconnected outerplanar graph is s-(2)-drawable. The conclusion follows immediately by Properties 1 and 4.

Another example concerns *planar triangulated* graphs, which admit planar drawings with all faces triangular, including the external face.

Theorem 9. *Let G be a planar triangulated graph. For each value of β such that $1 < \beta \leq \infty$, G is w-β-drawable if and only if it is s-β-drawable. Furthermore, G is w-$[1]$-drawable if and only if it is s-$[1]$-drawable.*

Sketch of proof. The if-part of both statements is trivial by Property 1. For the only-if-part of the first statement (the argument for the second is analogous), consider an allowed value of β and suppose G is w-β-drawable with w-β-drawing Γ. We show that Γ is also an s-β-drawing. Consider the graph G' that has an s-β-drawing with the same set of points for the vertices as Γ. From Property 2 we have that $G \subseteq G'$. From [3] we have that in the above interval of β values, s-β-drawable graphs are planar; hence, G' is planar. The conclusion follows from the maximality of G.

Further consideration of planar triangulated graphs reveals a connection between weak proximity drawability and angular resolution.

Lemma 10. *Let Γ be a straight-line drawing of a planar triangulated graph G such that the angle between two adjacent edges is at most $\alpha(\beta)$, where $0 \leq \beta \leq 1$. Then Γ is a w-(β)-drawing of G.*

The *angular resolution of a straight-line drawing* is the size of the minimum angle formed by adjacent edges. The *angular resolution of a straight-line drawing algorithm* $\mathcal{A}$ is a number $\mathcal{F}$ that is the infimum of all the angular resolutions of all the drawings that $\mathcal{A}$ could construct. When inputs to $\mathcal{A}$ are restricted to various classes of graphs, examples in the literature show a dependency of $\mathcal{F}$ on the maximum degree (e.g., see [13]) or on the number of vertices of graphs (e.g., see [14]). Lemma 10 is the geometric foundation of the following theorem.

Theorem 11. *Let $\mathcal{A}$ be a straight-line drawing algorithm whose inputs are planar triangulated graphs. Let $\mathcal{F}$ be its angular resolution. We have that any drawing produced by $\mathcal{A}$ is a w-β-drawing for all values of β such that $0 \leq \beta < \sin(\pi - 2\mathcal{F})$. Furthermore, such a drawing is a w-$(\sin(\pi - 2\mathcal{F}))$-drawing.*

5 Trees

We denote by $\mathcal{T}_k$ a class of trees having maximum vertex degree at most k; $\mathcal{T}_\infty$ is the class of all trees. We also denote by $\mathcal{T}(\beta)$ the class of w-(β)-drawable trees. Similarly, $\mathcal{T}[\beta]$ is the class of w-[β]-drawable trees.

First we prove that every tree is w-(β)-drawable for $\beta \leq 2$. We begin by constructing a w-(2)-drawing for an arbitrary tree. The construction can be formulated as a linear time algorithm for real RAM. In the drawing, each point p representing a tree node has the following construction devices associated with it: an open disk $D(p)$ centered at p; an open superwedge $W^+(p)$ with vertex at the parent of p (this wedge is left undefined if p is the root); a closed subwedge $W(p)$ with vertex at p.

To *generate the children* of a point p means to compute for each child q_i its superwedge, its coordinates, its disk, and its subwedge. The construction continues in breadth-first fashion from the root. Each time the children q_i of some point p are generated, the following invariants are maintained.

1. Each superwedge $W^+(q_i)$ belongs to $W(p)$, and the superwedges of distinct children of p are disjoint.
2. Each disk $D(q_i)$ lies inside the superwedge $W^+(q_i)$, tangent to its sides.
3. Subwedge $W(q_i)$ with vertex at q_i lies inside the superwedge $W^+(q_i)$ and has sides parallel to those of the superwedge $W^+(q_i)$. (Its purpose is to contain q_i and all its descendants.)

The root of the tree is placed at the origin. If the root has $k \geq 2$ children $q_1, \ldots q_k$, these are generated by dividing the plane into k equal angle superwedges $W^+(q_i)$ with vertex at the origin. Then each q_i is placed distance 1 from the origin on the bisector of its superwedge. This determines the disks and the subwedges of the q_i. If the root has only one child q_1, then the superwedge

$W^+(q_1)$ is given vertex angle $\pi/2$ and q_1 is placed at unit distance from the origin on the bisector of its superwedge; its disk $D(q_1)$ has unit radius. Once the coordinates, disks and wedges have been determined for all points at depth 0 and 1, the construction continues in a breadth-first manner. Suppose the tree has depth d, where the root has depth 0. For each depth $i = 1$ to $d - 1$, the children of each point p at depth i are generated from p and its disk and wedges as follows (see Fig. 2). Suppose p has $k \geq 1$ children q_i, $1 \leq i \leq k$. Equally subdivide $W(p)$ into k superwedges $W^+(q_1) \ldots W^+(q_k)$. Place q_i at the intersection of the bisector of its superwedge with the boundary of $D(p)$. This determines the subwedge of p, since its sides are parallel to those of the superwedge of p, and the disk of q_i, since it is tangent to the superwedge.

Theorem 12. $T(2) = T_\infty$. *Furthermore, given a tree $T \in T_\infty$, a w-(2)-drawing of T can be computed in time proportional to the size of T in the real RAM.*

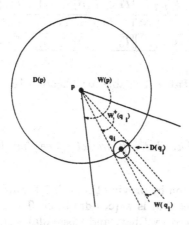

Fig. 2. Construction for trees.

We can now exploit Property 4 to extend the above result to infinitely many values of β.

Corollary 13. $T(\beta) = T_\infty$ *and* $T[\beta] = T_\infty$ *for any* $0 \leq \beta < 2$.

Surprisingly, it turns out that if $\beta = 2$ and the region of influence is a closed set, then the class of w-[2]-drawable trees does not contain trees with arbitrarily large vertex degree.

Lemma 14. $T[2] = T_5$.

The previous lemma can be generalized to values of β such that $2 < \beta \leq \infty$.

Lemma 15.

1. *Let Γ be a w-(β)-drawing of a tree for $2 < \beta \leq \infty$. Then the angle between two adjacent edges is at least $\gamma(\beta)$.*
2. *Let Γ be a w-[β]-drawing of a tree for $2 < \beta \leq \infty$. Then the angle between two adjacent edges is greater than $\gamma(\beta)$.*

From Lemma 15, an upper bound can be deduced for the maximum vertex degree of w-β-drawable trees for $2 < \beta \leq \infty$. It is worth noticing that the same lemma, and hence the same upper bounds, hold for s-β-drawable trees [3]. Thus, we can use Lemma 15 and Property 1 to import the results of [3] on the s-β-drawability of trees.

	β	$\mathcal{T}(\beta)$	$\mathcal{T}[\beta]$
1	$2 < \beta < \frac{1}{\cos(2\pi/5)}$	$T_4 \subset \mathcal{T}(\beta) \subseteq T_5$	$T_4 \subset \mathcal{T}[\beta] \subseteq T_5$
2	$\beta = \frac{1}{\cos(2\pi/5)}$	$T_4 \subset \mathcal{T}(\beta) \subset T_5$	$\mathcal{T}[\beta] = T_4$
3	$\frac{1}{\cos(2\pi/5)} < \beta < \infty$	$\mathcal{T}(\beta) = T_4$	$\mathcal{T}[\beta] = T_4$
4	$\beta = \infty$	$T_3 \subset \mathcal{T}(\beta) \subset T_4$	$\mathcal{T}[\beta] = T_3$

Table 1. Table for Theorem 16

Theorem 16. *Table 1 describes w-β-drawable trees for all values of β such that $2 < \beta \leq \infty$.*

We conclude this section by proving that it is NP-hard to determine whether a tree of maximum degree four is w-(∞)-drawable. The proof follows the NAE-3SAT paradigm introduced by Bhatt and Cosmadakis [1] and exploited for geometric graph realizability questions by Eades and Whitesides [6], [7]. First we introduce some terminology for w-(∞)-drawings. These are also called *weak open strip* drawings, as the region of influence of two points x, y is the open strip perpendicular to the line segment between them.

A weak open strip drawing is *orthogonal* if each of its edges is parallel to one of two given orthogonal direction vectors. These vectors may be assumed to be horizontal and vertical. A *normalized* orthogonal drawing is obtained from an orthogonal drawing as follows. Order the vertices in the drawing from left to right, with vertices having the same x-coordinate assigned the same place in the order. Order the vertices from bottom to top similarly. Assign each vertex a new x-coordinate equal to its order in the left-to-right order; assign each vertex a new y-coordinate equal to its order in the bottom-to-top order. The drawing resulting from this coordinatization is a normalized orthogonal drawing. Normalized drawings are weak open strip drawings if the original weak open strip drawing was orthogonal. This is *not* necessarily true in general.

A graph G is *orthogonally unique* if a) any weak open strip drawing of it must be orthogonal, and b) the set of points and edges of any normalized orthogonal drawing of G is unique up to rigid motions (i.e., combinations of reflections, rotations and translations) of the plane.

Observation 17. Suppose H is a subgraph of some graph G that has a weak open strip drawing Γ. When the part of Γ that does not represent H is discarded, what remains is a weak open strip drawing of H.

It follows from this observation that if H is orthogonally unique, then any weak open strip drawing of a graph G containing H must respect the constraints on the drawing of H.

Observation 18. Any angle formed by edges with a common endpoint in a weak open strip drawing must be $\geq \pi/2$.

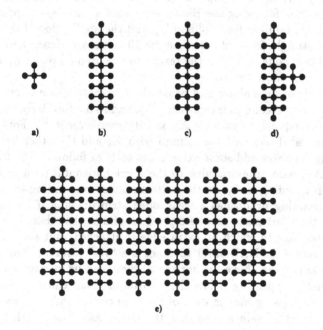

Fig. 3. Orthogonally unique graphs.

Lemma 19. *Each of the graphs a) through e) of Figure 3 is orthogonally unique and has a normalized orthogonal drawing as shown.*

Now consider an instance of NAE-3SAT consisting of m clauses $C_1 \ldots C_m$ and n variables and their complements, $X_1, \ldots X_n, X'_1, \ldots X'_n$. Each clause contains three literals, but no clause contains both a literal X and its complement X'. The instance is a "yes" instance if and only if a truth assignment can be found such that each clause contains at least one true literal and at least one false literal. We encode the instance in polynomial time in its length by designing a graph G having the form shown in part e) of Figure 3. In particular, G should have m internal vertical columns, n internal rows lying above the horizontal chain containing the crosses, and n internal rows lying below this horizontal chain.

By Lemma 19, G is orthogonally unique. Note, however, that the vertical columns can be flipped individually around the central horizontal chain without changing the drawing. Similarly, short 2-edge horizontal paths joining degree 1 vertices can be flipped around their points of attachment to long paths without changing the drawing.

Theorem 20. *To determine whether a given tree of maximum degree four has a weak open strip drawing is NP-hard.*

Sketch of proof. We modify graph G to encode a particular instance of NAE-3SAT. This is done following the Bhatt Cosmadakis paradigm. Applications of this paradigm are familiar from [6] and [7], and the application of the paradigm in this case is straight-forward once Lemma 19 is known. Hence we sketch only briefly how this is done, referring the reader to the original paper by Bhatt and Cosmadakis [1] for more details.

The first pair of rows above and below the horizontal chain of crosses represents clause C_1, the second pair represents C_2, and so on. Similarly, the left-most internal column represents variable X_1 and its complement X_1'. For the jth internal column, label one half the column with X_j and the other half with the corresponding X_j'. Now add some extra edges to G as follows. If C_i fails to contain X_j (or X_j'), add an extra edge to the short horizontal path in one of the two rows corresponding to C_i, on the half-column corresponding to X_j (or X_j'). Clearly the resulting graph has a weak open strip drawing if and only if the columns and the short paths can be flipped around so that there is at least one position in each row to which no extra edge is attached. But such flips can be found if and only if the NAE-3SAT instance is a "yes" instance. This is because half-columns appearing above the horizontal chain can be interpreted as "true", and half-columns appearing below the horizontal chain can be interpreted as "false". Hence a missing edge in each of a pair of corresponding rows above and below the horizontal chain means that the clause associated with this pair of rows contains a true literal and a false literal.

6 Open Problems

Several remaining open problems make weak proximity drawability an attractive direction of research. One class of problems concerns the use of the weak model for proximity regions other than β-regions. For example weak rectangle of influence drawings [8] could be tackled. It is easy to see that any planar graph that admits an *st*-orientation without transitive edges has a weak rectangle of influence drawing. Another problem area is to consider weak proximity models that do not allow *edges*, as opposed to vertices, to enter the proximity regions of other edges. This seems too restrictive to consider in a strong proximity model.

References

1. S. Bhatt and S. Cosmadakis. The Complexity of Minimizing Wire Lengths in VLSI Layouts. *Information Processing Letters*, **25**, 1987, pp. 263-267.
2. P. Bose, W. Lenhart, and G. Liotta. Characterizing Proximity Trees. *Algorithmica*, Special Issue on Graph Drawing (to appear).
3. P. Bose, G. Di Battista, W.Lenhart, and G. Liotta. Proximity Constraints and Representable Trees. *Graph Drawing, Proc. of the DIMACS International Workshop GD94, New Jersey, USA, October 1994*, LNCS **894**, R. Tamassia and I.G. Tollis eds., Springer-Verlag, 1995, pp. 340-351.
4. G. Di Battista, W. Lenhart, and G. Liotta. Proximity Drawability: a Survey. *Graph Drawing, Proc. of the DIMACS International Workshop GD94, New Jersey, USA, October 1994*, LNCS **894**, R. Tamassia and I.G. Tollis eds., Springer-Verlag, 1995, pp. 328-339.
5. G. Di Battista, R. Tamassia, and I. G. Tollis. Area Requirement and Symmetry Display of Planar Upward Drawings. *Discr. and Comp. Geometry*, **7**, 1992, pp. 381-401.
6. P. Eades and S. Whitesides. The Realization Problem for Euclidean Minimum Spanning Trees is NP-hard. *Proc. 10th ACM Symposium on Computational Geometry*, 1994, pp. 49-56.
7. P. Eades and S. Whitesides. Nearest Neighbor Graph Realizability is NP-hard. *Proc. Latin'95*, Valparaiso, Chile, 1995, LNCS **911**, R. Baeza-Yates, E. Goles, P. Poblete eds., Springer-Verlag, 1995, pp. 245-256.
8. M. Ichino and J. Sklansky. The Relative Neighborhood Graph for Mixed Feature Variables. *Pattern Recognition*, **18**, 1985, pp. 161-167.
9. G. Kant, G. Liotta, R. Tamassia, and I.G. Tollis. Area Requirement of Visibility Representations of Trees. *Proc.5th CCCG*, Waterloo, 1993, pp. 192-197.
10. J.M. Keil. Computing a Subgraph of the Minimum Weight Triangulation. *Comp. Geom.: Theory and Appl.*, **4**, 1994, pp. 13-26.
11. D. G. Kirkpatrick and J. D. Radke. A Framework for Computational Morphology. *Computational Geometry*, ed. G. T. Toussaint, Elsevier, Amsterdam, 1985, pp. 217-248.
12. A. Lubiw and N. Sleumer. All Maximal Outerplanar Graphs are Relative Neighborhood Graphs. *Proc. 5th CCCG*, Waterloo, 1993, pp. 198-203.
13. S. Malitz and A. Papakostas. On the Angular Resolution of Planar Graphs. *Proc. STOC*, 1993, pp. 431-437.
14. W. Schnyder. Embedding Planar Graphs on the Grid. *Proc. SODA*, 1990, pp. 41-51.
15. R. Tamassia and I. G. Tollis. A Unified Approach to Visibility Representations of Planar Graphs. *Discr. and Comp. Geometry*, **1**, 1986, pp. 321-341.
16. M. Formann, T. Hagerup, J. Haralambides, M. Kaufmann, F. T. Leighton, A. Simvonis, E. Welzl, G. Woeginger. Drawing Graphs in the Plane with High Resolution. *Proc. FOCS*, 1990, pp. 86-95.

COMAIDE: Information Visualization using Cooperative 3D Diagram Layout

David Dodson

Computer Science Department, City University,
Northampton Square, London EC1V 0HB, UK

dcd@cs.city.ac.uk

Abstract. COMAIDE is a toolkit for user-system cooperation through joint multi-focal graph browsing. It supports cooperative force-directed layout management, concurrent with dialogue handling, for heterogeneous multi-layered 3D interactive diagrams. The layout manager's intuitively 'natural' animations of multi-layered 3D graph drawings support:

1. Cooperative tidying of user-manipulable 3D layout;
2. Optional 'lucid' 3D layout optimized for a favoured viewpoint;
3. Layout annealing for good initial 3D diagram topologies;

1 Introduction

COMAIDE (Co-Operative Multilayer Application-Independent Diagram Environment) is a toolkit for cooperative Diagrammatic User Interfaces (DUIs) [15] to applications written in Prolog. It presents a simple artificial reality of 3D multi-layer node-and-link diagrams supporting multi-focal graph browsing. The nodes of the diagram occupy a set of parallel layers, each node's centre being constrained to lie in the plane of a layer.

COMAIDE diagrams are animated by an innovative force-directed layout manager, dyn, the focus of this paper. As in most force-directed graph drawing systems, dyn simulates repulsion between diagram elements to keep them spread apart, and springiness in links to promote preferred link lengths and alignments.

In dyn's normal, cooperative mode of use, the diagram is animated in a slow and intuitively predictable way which does not interfere with user manipulation of the diagram. This lets the user vary the diagram's 3D topology by dragging nodes. Each node can be dragged within a layer or to another layer by the user. dyn tidies up the result without changing the topology. This is important because the user may see ways of improving the layout that a layout manager cannot find or cannot apply without risk of causing disorientation; also purely automatic layout cannot address unpredictable or ramified user preferences. dyn also has a simulated-annealing mode for producing fresh layouts, e.g. of new diagram contents generated by an application.

Concurrently with layout animation, COMAIDE supports user-application dialogues in which an application responds to mouse actions by modifying diagram content. These dialogues can employ built-in facilities for multi-focal browsing [5, 14, 16, 9] which let both user and application expand, shrink, hide and reveal nodes. This is important in giving the user control of the allocation of screen space (external visual memory) to information whilst letting the application vary this allocation, e.g. to draw attention to important matters. User browsing manipulations incorporate a fast initial layout adjustment in the spatial neighbourhood of the affected node using the 'ductile space' metaphor [5].

Facilities for pop-up and pull-down menus are also built-in. A built-in pull-down menu allows selection of applications and control of the layout manager. Applications can also incorporate specialized layout algorithms for layout initialization as an alternative to the relatively slow but general annealing approach.

COMAIDE is a Prolog environment including sub-systems for application management, menu operations, browsing operations, layout management and diagram I/0, along with various utility sub-systems. A SICStus Prolog process incorporating these sub-systems acts as a client of City's interactive diagram graphics server, icx, which implements ICD-Edit [4, 7, 8] for X/Motif. COMAIDE is currently undergoing further development in the EU CUBIQ project. The version outlined here uses the client-server protocol ADI 1.6.1b detailed in [7].

Section 2 outlines depth perception aspects of COMAIDE diagrams. COMAIDE's layout manager, dyn, is presented and discussed in Sec. 3. Section 4 illustrates both a concept-demonstration of a cooperative Expert System DUI built using COMAIDE and the use of annealing for initial layout. Section 5 concludes.

2 Depth Perception in COMAIDE Diagrams

COMAIDE's graphics server, ICD-Edit, offers a limited 3-dimensionality, '2¾D', in which nodes appear upright and face forwards irrespective of the variable 3D orientation of a diagram. Moreover, ICD-Edit nodes are simply flat rectangles and the links between them are straight lines with optional arrowheads. These constraints allow fast diagram manipulation without 3D hardware but also limit the cues available for depth perception, which thus need special attention.

Naturally, ICD-Edit uses hidden object removal, but this only indicates the sign of the depth difference wherever a node overlaps a node or link,

Perspective is used too. Brookes [1] observes that perspective is very effective as a depth cue when parallel lines and right angles abound in the 3D model. A more detailed analysis for the case of 3D node-and-link diagrams is offered here:

1. Perspective projection is not *in itself* a depth cue.
2. Recognizable spatial arrangements of diagram elements are good depth cues; perspective usually enhances, and in particular disambiguates, this effect.
3. The more parallel lines and right angles there are in the spatial arrangement, the better its depth cueing effect.
4. Significant uniformity of sizes in the model, e.g. amongst nodes or amongst distances between nodes, is needed for perspective to aid depth cueing.

ICD-Edit's optional 'rocking' motion, inspired by SemNet[1], approximates a sinusoidal rotational oscillation with just three co-maintained images. This often seems disturbing at first, but seems generally acceptable. It very usefully depth-cues ICD-Edit diagrams which richly populate 3D space.

COMAIDE diagrams use parallel rectangular layers visualized by 3D bounding boxes and perspective viewing to exploit the potential for depth cueing in ICD-Edit diagrams. Though not shown here, intensity cueing is supported too.

3   The Layout Manager, dyn

As a layout *manager*, dyn runs as a background activity, repeatedly adjusting diagram layout. The successive adjustments approximate the motion of the mutually-repulsive springy diagram elements in a viscous medium. Links are assumed to have negligible mass. Newton's second law applies to each node:

$$\mathbf{F} = m\dot{\mathbf{V}} \tag{1}$$

where m is the mass of the node and $\dot{\mathbf{V}}$ (i.e. $d\mathbf{V}/dt$) is its acceleration vector. The force vector $\mathbf{F}$ is the sum of the motive force on the node and the drag due to viscosity. Using a simple model of viscosity, neglecting node size, this gives:

$$m\dot{\mathbf{V}} = \mathbf{F}_{\text{MOTIVE}} - k\mathbf{V} \tag{2}$$

where k is a viscosity coefficient. The option of treating nodes as having negligible mass is usually taken, so that the motion equation degenerates to:

$$\mathbf{V} = \mathbf{F}_{\text{MOTIVE}}/k \tag{3}$$

In this case, in each time step, movement is proportional to applied force — a common feature of force-directed graph drawing following its use in [10].

Fig. 1 summarizes dyn's layout animation algorithm. Table 1 lists control parameters which can be varied using pull-down menus while dyn is running.

3.1   Motive Forces in dyn

Nine different types of motive-force interaction are considered, as follows.

3D Node-boundary repulsion: Each node is repelled inwards by each face of a model-axis-aligned cuboid [6] by a force increasing linearly from zero at the cuboid's centre to 1 at a distance of bdry_rim_width from the boundary and then to max_repulsion for a node touching, crossing or outside the boundary.

3D Node-node repulsion: Each node pair within a layer undergoes inverse square law repulsion subject to a force limit of max_repulsion and a distance limit of repel_horizon.

3D Node-link repulsion: For each node, for each link which is not a link of that node and which either lies in the same layer as the node or crosses that

---

[1] SemNet [11] demonstrated fully 3D, mainly viewpoint-navigated knowledge browsing with scale-driven recursive decomposition of spatial cluster nodes)

```
For each node n: Set V(n) {the velocity of n} to [0,0,0];
While ts>0 {where ts is the assumed size of time step in seconds}:
  <<< FORCE Computation: >>>
  For each node n: Set F(n) {the motive force on n} to [0,0,0];
  For each node n: Add its boundary repulsion forces to F(n), also
  For each force couple between two diagram elements:
      For each member E of the pair of elements:
          If E is a node n: Add the relevant force to F(n)
          else (E is a link from node n1 to node n2):
              Add the relevant forces to F(n1) and F(n2);
  <<< MOTION Computation: >>>
If inertia>0 then:
   tss := ts/5;
   For each node n, repeat 5 times:
          Posn(n) := Posn(n) + V(n)*tss;
          V(n) := V(n) + (F(n)-V(n)*viscosity)*tss/(mass(n)*inertia);
else For each node n:
          V(n) := F(n)/viscosity;
          Posn(n) := Posn(n) + V(n)*ts;
viscosity := min( end_viscosity, viscosity*(100+anneal_rate)/100 );
```

Fig. 1. Outline of dyn's layout management algorithm

Parameter	Description
bdry_rim_width	Distance at which node-boundary repulsion is 1
n_n_3_repel_factor	Coefficient of 3D node-node repulsion
n_l_3_repel_factor	Coefficient of 3D node-link repulsion
l_l_3_repel_factor	Coefficient of 3D link-link repulsion
repel_horizon	Distance limit of repulsion between diagram elements
max_repulsion	Limit on each 3D repulsion force
perpendicity	Stiffness coefficient of between-layer links
preferred_length	Ideal forward size of in-layer links
fwd_rigidity	Forward stiffness coefficient of in-layer links
planar_field	Stiffness coefficient of 'magnetic' link alignment
n_n_2_repel_factor	Coefficient of 2D node-node repulsion
n_l_2_repel_factor	Coefficient of 2D node-link repulsion
max_2d_repulsion	Limit on each 2D repulsion force
viscosity	Ratio of drag force to velocity (negated)
anneal_rate	% increase in viscosity per layout step
end_viscosity	Limit to increase of viscosity
inertia	Ratio of inertia to mass

Table 1. dyn's Control Parameters

layer, inverse square law repulsion occurs between them subject to a force limit of max_repulsion and a distance limit of repel_horizon.

3D Link-link repulsion: Each pair of links such that one is in a layer which the other crosses or both cross the same gap between layers experience inverse square law repulsion acting along the line of shortest distance between the links, subject to a force limit of max_repulsion and a distance limit of repel_horizon, provided they would not be closer if extended to infinity.

Between-layer link alignment stiffness: Each link from a node in one layer to a node in another exerts a pair of forces, one on each node, of magnitude perpendicity*tan(angle from the perpendicular), seeking to make the link perpendicular to the layers. The effect is analogous to barycentering [17], but generalized to 3D.

In-layer link alignment stiffness: Each link in a layer with a forward (i.e. upward) direction is considered to be 'magnetic' if planar_field exceeds zero.

A non-magnetic link exerts a force pair seeking, with a linear elastic stiffness of fwd_rigidity, to make it preferred_length long in its current orientation.

A 'magnetic' link exerts a force pair seeking alignment in the layer's forward direction, D. The force components in D are zero if the length component in D is preferred_length, and otherwise reflect a linear elastic stiffness of fwd_rigidity. The force components in the perpendicular direction S within the layer are zero if the length component in S is zero, and otherwise reflect a stiffness proportional to planar_field. This crude but fast approximation of magnetism seems to avoid instability and to have intuitive appeal and a barycentering-like effect [17].

2D Node-node and node-link repulsion: These weak 2D repulsions are somewhat similar to their 3D counterparts, but arise in the display plane, ignoring depth differences between nodes. Their effects are, respectively, to minimise node overlap and to resist node-link crossings in the 2D display.

3.2 Force, Mass and Heat in Nature and in dyn

COMAIDE animates the diagram approximately as if suspended in a thick fluid which continually promotes equilibrium by extracting energy from it. Low-energy equilibrium states in force-directed layout have no kinetic energy because they are static, and have low potential energy in force interactions, implying good satisfaction of the separation and alignment preferences modelled by the force interactions, averaged over the diagram. The motion and heat that would be induced in a real fluid is ignored, but compared to most force-directed graph drawing systems, e.g. [13, 2, 12], dyn's dynamics are relatively natural.

dyn's universal inertial constant, inertia, controls the magnitude of the effect of mass, i.e. delay in response of motion to force and tendency to overshoot. When inertia is zero, diagram layout typically seems to converge faster with similar or perhaps better results, though no formal comparison has been made. Moreover, the animation seems more open to intuitive prediction when mass effects do not need to be considered. Indeed, naturally speaking, the more viscosity dominates mass (i.e. the smaller velocities are than they would be in the absence of viscosity), the smaller the effect of mass on trajectory.

In practice, in the course of a time step, diagram elements tend to approach each other more closely than their initial energy would allow in reality. This happens because the crude simulation algorithm does not foresee the escalation of repulsion between them in the course of their approach. Compared with a hypothetical perfectly natural animation the effect is to provide a source of random mechanical energy. A positive viscosity is needed just to counterbalance this effect. Adopting the analogy between diagram structure and molecular structure, an increase in energy equates to heating. Thus a low viscosity in dyn tends to produce heating, whilst a high viscosity tends to produce cooling.

3.3   Normal and Annealing Modes of Layout Management

dyn's 'normal' mode of cooperative operation occurs when diagram energy is low enough for diagram elements to succeed in repelling each other. In other words, between one force computation and the next, no diagram element (node or link) passes through another due to motion so fast that their high mutual repulsion when close together is not sampled — or if sampled, is insufficient to prevent through-passing. This equates to the ordinary reality of solid objects. In some sense, the 3D topology of the diagram remains constant.

At high energy levels, the diagram moves chaotically with lots of through-passing. This is somewhat analogous to the behaviour of fluids, particularly as the positions of nodes are much less constrained by the links between them.

The term annealing traditionally meant the slow cooling of a metal such that the boundaries between its crystals have time to migrate, relative to its atoms, to a particularly low energy state associated with large crystal size and with high ductility. Through metaphor, it can be now used for analogous slow reductions of energy which tend to yield particularly low-energy equilibrium states. Dropping the qualification 'simulated' from annealing seems particularly valid when, as in graph drawing, reality is being created more than simulated.

Unlike many simulated annealing systems, such as [2], dyn is deterministic, being controlled by increasing viscosity rather than by explicitly reducing the probability of random energy increase. Annealing is used to solidify the diagram into a low-energy 3D topology. The diagram can initially be melted with low viscosity and/or by 'crushing' it, which positions each node at the centre of its layer, creating high initial potential energy. If the diagram 'explodes' excessively, 'sandbags' can be switched on to limit node positions to a cuboid model space.

3.4   3D and 2D Layout

Diagram annealing in dyn using 3D forces leads to fairly good 3D layouts, although some cooperative user input is often helpful once the diagram is solid. These layouts are not however optimized for a particular direction of viewing. To maintain familiar visual context and aid fast recognition of familiar content, the diagram needs to be arranged appropriately for some principal viewing direction. dyn achieves this using 2D node-node and node-link repulsions which arise in the 2D display projection of the diagram. By making these 2D repulsions small compared to the 3D repulsions, they minimize node-node and node-link overlap in the projected view, without significantly impairing the quality of the

3D layout. The resulting layout is good both in 3D and in the 2D-displayed view. The 2D forces also tend to maintain the 2D topology of a solid diagram and are not useful to consider until the diagram is solid, especially as they currently take longer to compute than the 3D ones.

Gas, liquid and solid are familiar thermodynamic phases of everyday substances. VIM diagrams have three thermodynamic phases too: the *fluid* phase, the ordinary *solid* phase, and a third phase obtained with added 2D repulsions which can add perspicuity to solid diagrams. This third phase is the *lucid* phase.

Clearly there is much more to be gained from the correspondence between materials science and 3D force-directed graph drawing than is uncovered here.

4 Sample Results

Note that when COMAIDE presents the diagrams shown below, ICD-Edit's rocking motion makes them look far more 3-dimensional. This makes them look less tangled and resolves structural ambiguities.

4.1 Cooperative Browsing Supported by Cooperative Layout

This example shows VIM [9], a visual expert system prototype, running in CO-MAIDE. The underlying expert system shell is IM1 [3].

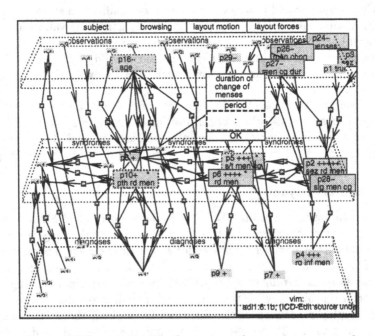

Fig. 2. VIM on first selecting a question to be answered

Figure 2 shows a view of a 3D overview diagram comprising 89 nodes and c.100 links. It shows a small IM1 knowledge base of 42 particulars (the oblong nodes) and 47 rules (the square nodes and their links). The user has entered, as the initial input, the proposition that the patient complains of amenorrhoea (particular p1, top right front). As a result, IM1 has updated various weights of evidence (visually encoded mainly with '+' and '−' characters) by forward rule application (i.e. with the data flow shown by arrowheads). It has then also updated 'investigative importances' (visually encoded by oblong-node border density) by reverse rule application. VIM has used these updates to estimate the pertinence (i.e. desired salience) of the affected nodes and has updated the diagram accordingly. dyn has then executed two cycles of layout adjustment, chiefly spreading out enlarged nodes to reduce overlap. Finally the user has clicked on the 'men cg dur' node causing a data entry pop-up to appear.

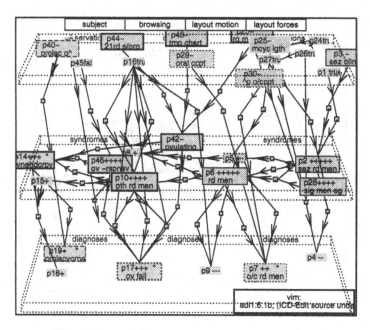

Fig. 3. Later in the same VIM consultation

Figure 3 shows a later stage in the consultation. The layout manager has executed a further 12 layout adjustment cycles. Concurrently, the user has:

1. Provided 3 answers to questions;
2. Rearranged three nodes to release a snag which inhibited layout improvement, without altering the 3D topology of the diagram;
3. De-selected most of the 'tiny' particulars from view (Ideally VIM would keep impertinent particulars hidden, but at the time of writing its automatic browsing actions vary node salience but not node selection.)

Node de-selection can be reversed by clicking on the 'link stubs' which point in the direction of omitted detail. Each such click reveals a linked rule node, together with any other links from that rule to particulars shown in the diagram.

dyn currently takes almost two minutes per animation step for the full lucid VIM diagram shown in Fig. 2, using lists for vectors in SICStus Prolog on a Sun SPARCclassic. Software tuning and quadrupled hardware power are expected to yield an acceptable speed of 1 to 5 seconds per cycle for such a diagram.

4.2 Cooperative Annealing

Fig. 4 shows a diagram which has been crushed and then given 29 annealing steps with an initial viscosity of 0.2, rising 2% per step. All of the other parameters have their default values, with 2D forces off. The diagram solidified within 11 annealing steps, and moved very little after 20 steps. Each layout step took about 2 seconds, including Prolog, ICD-Edit and X server execution time.

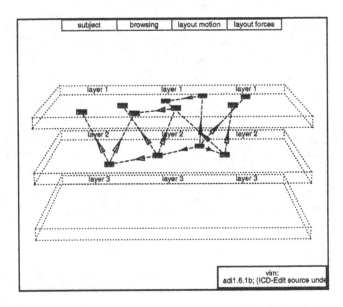

Fig. 4. A Diagram after 29 Annealing Steps

After repositioning one node by simple dragging, 10 more annealing steps produced the layout shown in Fig. 5, in which the dragged node is emphasized. dyn was stopped to aid this, as node dragging was not suppressing repositioning by dyn at that time, making dragging prone to failure in small diagrams.

The right hand half of the diagram was then manually repositioned to make the diagram *roughly* symmetrical, again with dyn turned off. After about ten

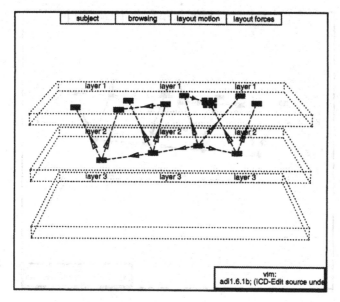

Fig. 5. After 39 Steps

more layout cycles, the node velocities were below 1 pixel per cycle. Fig. 6 shows the layout a minute later. A further 250 steps produced no noticeable result except for a very slight clockwise rotation as viewed from above.

5 Conclusion

COMAIDE demonstrates a lot of support for user-application cooperation in joint multi-focal graph browsing, illustrating key benefits of 3D in this context. The key ingredient is a cooperative layout manager, dyn, which animates multi-layered 3D graph drawings in a fairly graceful, subjectively natural and intuitively comprehensible way. Animation by dyn supports:

1. Cooperative management of 3D layout, in tidying layouts without changing their topology while allowing user manipulation;
2. Optional combined 3D and 2D 'lucid' layout for a favoured viewpoint;
3. Layout annealing, finding good 3D diagram topologies.

dyn's novelty rests largely on its rich model of diagram structure and forces:

1. Both sorts of diagram element (links as well as nodes) repel each other.
2. Links within planes are treated distincity from links between planes.
3. An optional pseudo-magnetic force promoting link alignment within a plane is modelled as an orthogonal pair of simple spring forces.
4. 2D forces which arise in the display plane are used to gently coerce good 3D layouts into being good 2D ones from the chosen viewpoint.

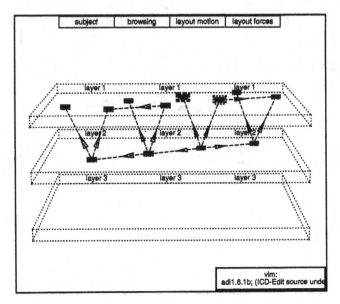

Fig. 6. After another 5 Drags and 40 Steps

5. The force computation ignores hidden diagram elements and the motion computation ignores any component of force on a node normal to its layer.

Further details including references to related work can be found using the World Wide Web URL: http://web.cs.city.ac.uk/research/dig/digpapers.html

6 Acknowledgements

Funding by EPSRC (GR/G56478, Cooperative KBS Browing), EU COPERNICUS (project 10979, CUBIQ) and City University is gratefully acknowledged.

References

1. FP Brooks. Grasping Reality Through Illusion - Interactive Graphics Serving Science. In SIGCHI Bulletin (special issue), ACM/Addison Wesley. (*CHI'88 Proceedings*, Washington, May 1988).
2. R Davidson and D Harel. Drawing Graphs Nicely Using Simulated Annealing. Tech. Report, The Weizmann Institute of Science, Rehovot, Israel, 1989; revised 1992,1993. To appear in Communications of the ACM.
3. DC Dodson and AL Rector. Importance-driven control of diagnostic reasoning. In MA Bramer, editor, *Research and Development in Expert Systems*. Cambridge University Press, 1984. (Proceedings of Expert Systems 84, Warwick, December 1984).

4. DC Dodson, LH Reeves and RB Scott. ICD-EDIT: - A Server for $2\frac{3}{4}$D Interactive Connection Diagram Graphics with Prolog Clients. Technical report TCU/CS/1995/1, Dept. of Computer Science, City University. 9pp. (Poster, Graph Drawing 94, 10-12 October 1994, Princeton, New Jersey). World-Wide Web (colour): http://web.cs.city.ac.uk/research/dig/95/TCU_1995_1/p1.html

5. DC Dodson. TRIVIAL: Refocusing in Cooperative Diagrams with Ductile Space. Tech. report TCU/CS/1995/2, Computer Science Dept., City University. 10pp. (Poster, Graph Drawing 94, 10-12 October 1994, Princeton, New Jersey). WWW (colour): http://web.cs.city.ac.uk/research/dig/95/TCU_1995_2/p2.html

6. DC Dodson. N-Dimensional RSPs, Right Multilayered Diagrams and Prolog. Technical report TCU/CS/1995/3, Dept. of Computer Science, City University. 9pp. (Poster, Graph Drawing 94, 10-12 October 1994, Princeton, New Jersey). WWW (colour): http://web.cs.city.ac.uk/research/dig/95/TCU_1995_3/p3.html

7. DC Dodson, LH Reeves and RB Scott (1995) ICD-Edit's Client-Server protocol: ADI Version 1.6.1. Technical report TCU/CS/1995/x, Dept. of Computer Science, City University, 1995. World-Wide Web: http://web.cs.city.ac.uk /informatics/cs/research/dig/95/adi1_6_1/sp.ps

8. WWW ICD-Edit page: http://web.cs.city.ac.uk/research/dig/icd-edit.html

9. DC Dodson, JA Secker, RB Scott and LH Reeves. VIM: 3D Co-operative Diagrams as KBS Surfaces. To appear in MA Bramer, editor, *Research and Development in Expert Systems XII*. (Proc. Expert Systems 95, Cambridge, Dec. 1985). WWW (colour): http://web.cs.city.ac.uk/research/dig/95/es95/es95.ps (Oct 95).

10. P Eades. A Heuristic for Graph Drawing. *Congressus Numerantium*, 42:149-160, May 1984.

11. KM Fairchild, SE Poltrock and GW Furnas. SemNet: Three Dimensional Graphic Representations of Large Knowledge Bases. In R Guidon, editor, *Cognitive Science and its Application for Human Computer Interaction*. Lawrence Erlbaum, 1988.

12. A Frick, A Ludwig and H Mehldau. A Fast Adaptive Layout Algorithm for Undirected Graphs. In R Tamassia and IG Tollis, editors, *Graph Drawing*. Lecture Notes in Computer Science 894, Springer, 1995. (Proc. DIMACS International Workshop, GD94, Princeton, New Jersey, USA, October 1994).

13. T Fruchterman and E Reingold. Graph Drawing by Force-Directed Placement. *Software-Practice and Experience*, 21(11):1129-1164, 1991.

14. K Kaugars, J Reinfelds and A Brazma. A Simple Algorithm for Drawing Large Graphs on Small Screens. In R Tamassia and IG Tollis, editors, *Graph Drawing*. Lecture Notes in Computer Science 894, Springer, 1995. (Proc. DIMACS International Workshop, GD94, Princeton, New Jersey, USA, October 1994).

15. T Lin and P Eades. Integration of Declarative and Algorithmic Approaches for Layout Construction. In R Tamassia and IG Tollis, editors, *Graph Drawing*. Lecture Notes in Computer Science 894, Springer, 1995. (Proc. DIMACS International Workshop, GD94, Princeton, New Jersey, USA, October 1994).

16. EG Noik. Encoding Presentation Emphasis Algorithms for Graphs. In R Tamassia and IG Tollis, editors, *Graph Drawing*. Lecture Notes in Computer Science 894, Springer, 1995. (Proc. DIMACS International Workshop, GD94, Princeton, New Jersey, USA, Oct. 1994).

17. K Sugiyama, S Tagawa and M Toda. Methods for Visual Understanding of Hierarchical System Structures. *IEEE Trans. Systems, Man and Cybernetics*, 11(2):109-125, February 1981.

Vertex Splitting and Tension-Free Layout

P. Eades[1]* and C. F. X. de Mendonça N.[2]**

[1] Dept. of Computer Science
Univ. of Newcastle NSW Australia
[2] Depto. de Ciência da Comp.
UNICAMP SP Brasil

Abstract. In this paper we discuss the "vertex splitting" operation. We introduce a kind of "spring algorithm" which splits vertices to obtain better drawings. We relate some experience with the technique.

1 Introduction

Suppose that $G = (V, E)$ is a graph and $w : E \rightarrow R^+$ assigns a weight to each edge of G. Graph drawing functions may be required to draw G so that the Euclidean distance between two vertices is the same as the weight of the edge between them. For example a weighted graph model is proposed in [1, 9] to represent email relationships: if two users u_1 and u_2 exchange email frequently then the weight on the edge (u_1, u_2) is small; otherwise the weight is large. To visualize these relationships, we draw the graph so that the Euclidean distance between two vertices represents the closeness of the relationship between two users. Further applications of weighted graph drawings are given [3, 4, 6].

The problem of drawing a weighted graph so that the Euclidean distance between vertices conforms to prespecified edge weights is NP-hard (see [7]). The heuristic approach of [1] uses a kind of "spring algorithm" [2] as a heuristic.

This paper takes a different approach to the problem. Suppose that vertex a is required to be close to vertex b, and b is required to be close to c, but a is required to be far from c. The triangle inequality makes this situation impossible to represent. The operation presented in this paper resolves this conflict by creating two copies of the vertex b, so that a may appear close to one copy of b and c may appear close to the other copy of b.

Intuitively, a vertex v may be "split" by making two copies v_1 and v_2 and attaching the edges incident with v to either v_1 or v_2. The operation is illustrated in Figure 1.

This simple operation is commonly used in manual layout: it has been observed in diagrams used in theorem proving systems, prerequisite diagrams for course structures, diagrammatic representations of metro systems and computer networks, and in circuit schematics. In each of these cases, a small number of vertices are split to change the graph a little to make it amenable to layout.

* Supported partially by a grant from LAC-FAPESP and by The Dept. of Comp. Sci. Univ. of Newcastle
** Supported partially by a grant from LAC-FAPESP and Proj. Integrado CNPq

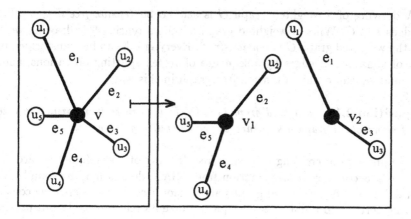

Fig. 1. The splitting operation

2 Preliminaries

A *graph* consists of a finite set V of vertices and a finite set E of edges, where each edge is an unordered pair of vertices of G. A vertex u is said to be *adjacent* to a vertex v if uv is an edge of G; in this case, the edge uv is said to be *incident* with u and v. The *neighbor* (or incident) set N_u of a vertex u is the set of all edges incident with u.

A *splitting* operation replaces v by two vertices v_1 and v_2, and partitions N_v into two nonempty disjoint sets N_{v_1} and N_{v_2}.

A *straight line drawing* of a graph $G = (V, E)$ is a function $D : V \to R^2$ that associates a position $D(v)$ to each vertex v of V. Since all drawings in this paper are straight line drawings we omit the term "straight line".

A *weight* $w(e)$ is a non-negative real value associated with an edge e of a graph. A *weighted graph* $G = (V, E, w)$ consists of a graph $G = (V, E)$ and a weight $w(e)$ for each $e \in E$.

Suppose that $G = (V, E, w)$ is a weighted graph. The *tension* in an edge $e = vu$ from a vertex v incident with e in a drawing D of G is a vector $\mathbf{t}_{ve}$ whose magnitude is the difference between the edge weight and the Euclidean distance between the two endpoint vertices, and whose direction is from $D(v)$ to $D(u)$. That is,

$$\mathbf{t}_{ve} = \frac{D(v) - D(u)}{d(v, u)}(d(v, u) - w(vu))$$

where $d(v, u)$ denotes the Euclidean distance between $D(u)$ and $D(v)$. Note that $\mathbf{t}_{ue} = -\mathbf{t}_{ve}$. A vertex v of G is *at equilibrium* if the sum of the tensions in the edges from v is zero, that is, if

$$\sum_{e \in N(v)} \mathbf{t}_{ve} = 0.$$

The drawing D is *at equilibrium* if every vertex is at equilibrium.

A drawing of a weighted graph G is said to be *tension-free* if $\mathbf{t}_{ve} = 0$ for all edges e of G. When a weighted graph admits a tension-free drawing we say that the weighted graph G is *tension-free*[3]. Every graph can be transformed into a set of nonadjacent edges by a sequence of vertex splitting operations, and in particular we can obtain a tension-free graph in this way.

Proposition 2.1 *A weighted graph $G = (V, E, w)$ can be transformed in a tension free weighted graph a sequence of vertex splitting operations.*

The problem of creating a tension free drawing of a graph is NP-hard. The proof uses a complex transformation from 3SAT; details may be found in [7]. In practice very few weighted graphs are tension-free. In this paper we consider splitting vertices to preprocess a graph to make its tension-free layout possible. Our approach is based on the spring system of Kamada and Kawai [4, 5].

3   The TensionSplit algorithm

First we review Kamada's spring algorithm [5]. The graph structure is simulated by a set of particles (for the vertices) in a plane where these particles are connected by springs (for edges and/or paths). An optimization method is applied to find a state of locally minimum energy of this system; this minimum energy state corresponds to the final drawing. The locally minimum energy state is found by repeating two steps: a *local minimization* step, and a *rearrange* step. The first step moves a vertex to a position where it is at equilibrium. In the second step some overlappings and crossings may be avoided by swapping vertices in pairs. In both steps the main goal is to reduce potential energy. The two steps are applied one after the other until the rearrange step does not change the layout.

Kamada's approach is be modified in two ways to form procedure Tension-Split in Figure 2. Our algorithm performs the local minimization step for *all* the vertices; and a modified rearrange step is used to perform splitting operations. Intuitively, the algorithm works in a very similar way to the Kamada algorithm, but when a vertex splits when it becomes "critical", that is, it has too much tension on it.

We now give details of the steps in procedure TensionSplit.

The first step partitions the edge set E into parts $E_1, E_2, \ldots, E_r$, where each E_i is the edge set of a biconnected component.

Step 2(a) is a implemented the the same way as the Kamada algorithm.

For step 2(b) we need to define some terms. For each vector $\mathbf{t}_{v,e}$ we define a *split line* which consists of a straight line through v perpendicular to the direction of the vector. Figure 3 displays a split line for the vector $\mathbf{t}_{v,e}$ (in this example, all edge weights are zero). This split line divides the plane into two semi-planes. The point where the vertex v is located divides the split line into two semi-lines.

---

[3] Our choice of terminology here is slightly nonstandard but it is motivated by the algorithm which follows.

procedure TensionSplit;

1. *divide the graph into biconnected components;*
2. *repeat until there is no critical vertex:*
   (a) *run the local minimization on all vertices until the drawing is at equilibrium;*
   (b) *if a critical vertex exists then*
       i. *choose a critical vertex v for which the the tension T_v on v is greatest;*
       ii. *split v into v_1 and v_2;*
       iii. *partition N_v into $N_{v_1} = N_v \cap R_{v,e}$ and $N_{v_2} = N_v \cap L_{v,e}$;*
       iv. *run the local minimization step on vertices v_1 and v_2;*
   (c) *run the rearrange step for all vertices of G;*

Fig. 2. The Tension Split algorithm

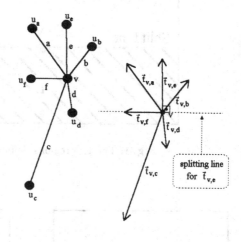

Let R denote the region of the plane defined by one semi-plane and one semi-line, and let L denote the complement of R. Figure 4 displays the two regions. We define a partition of N_v into edge sets $R_{v,e}$ and $L_{v,e}$ as follows. An edge $e = vw$ is in $R_{v,e}$ if w lies in R; and e is in $L_{v,e}$ if w lies in L. Figure 5 displays the partitions $R_{v,e}$ and $L_{v,e}$ for the graph in Figure 3.

A split line is *valid* if there is at least one biconnected component E_i for which both $E_i \cap R_{v,e}$ and $E_i \cap L_{v,e}$ are nonempty. Figure 6 displays:

Fig. 3. A sample of a split line

(a) a valid and an invalid split line for $\mathbf{t}_{v,f}$, and (b) an invalid split line for $\mathbf{t}_{v,e}$.

For each valid split line $\mathbf{t}_{v,e}$ we define the *tension for this split line* as

$$\mathbf{T}_{v,e} = \sum_{f \in R_{v,e}} \mathbf{t}_{v,f}.$$

Since the graph drawing is at equilibrium after step 2(a), we could replace $R_{v,e}$ by $L_{v,e}$ in the definition of $\mathbf{T}_{v,e}$ above. The *tension* T_v in *vertex* v is the maximum value of $|\mathbf{T}_{v,e}|$ over all valid split lines. Our algorithm uses a constant τ corresponding to the value of the minimum permissible tension. This constant can be explicitly changed by the user. A vertex is *critical* if $T_v > \tau$.

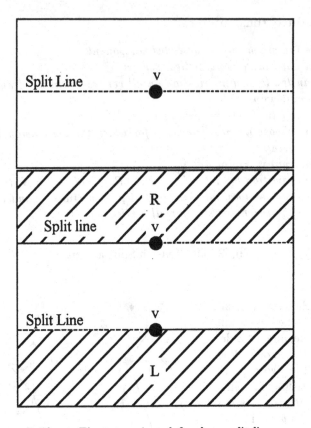

Fig. 4. The two regions define by a split line

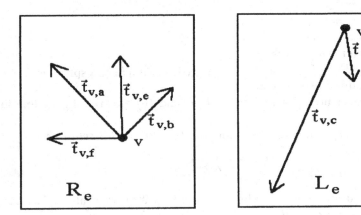

Fig. 5. The partition of the vectors into two sets

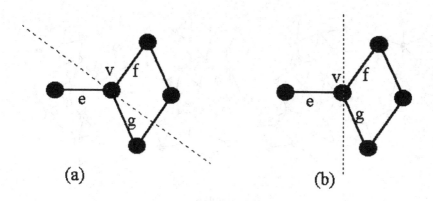

Fig. 6. A valid and an invalid split-lines

4 Samples

Some typical samples of splitting operations performed by the TensionSplit algorithm are described below.

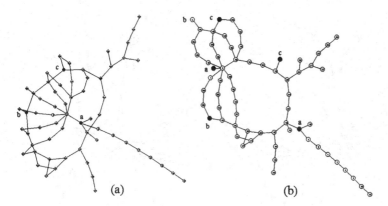

Fig. 7. example 1

Figures 7 and 8, are graphs for which spring algorithms perform poorly. Figure 7(a) displays the layout without splittings. Figure 7(b) displays the layout with three splittings (indicated by the letters a, b and c). Figure 8(a) displays the layout without splittings. Figure 8(b) and (c) display an intermediate stage layouts of the Tension Split algorithm with one splitting (indicated by the letter a) and two critical vertices (indicated by b and c). Figure 8(d) displays the final drawing with very little tension.

The diagram in Figure 9(a) is a schema from a commercial database design. It is drawn by a standard spring technique. Some node overlaps and crossings

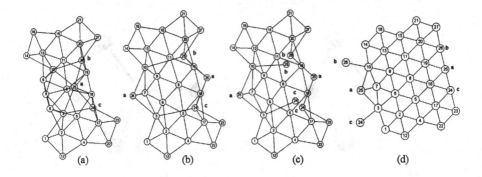

Fig. 8. example 2

appear. Algorithm TensionSplit gives better diagram in Figure 9(b) with only one vertex split (indicated by a surrounding ellipse). This sample shows a very good application of TensionSplit algorithm.

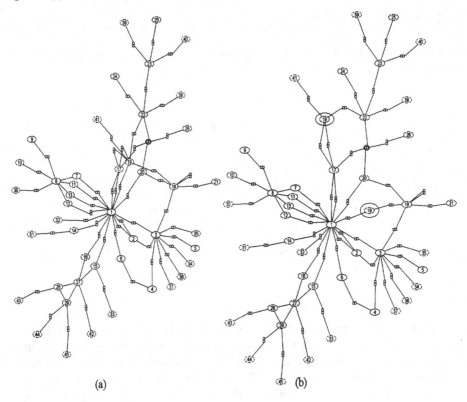

Fig. 9. example 3

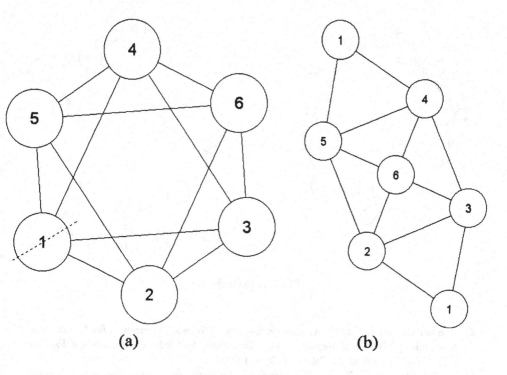

Fig. 10. example 4

Figures 10, 11 and 12 are general graphs. For each figure, a relatively small value of critical tension was used. The splitting operations are displayed. For all layouts the algorithm gives considerable symmetry. the resulting drawings are attractive: the symmetry is not bad and the edge length tends to 1.

5   Conclusion

The TensionSplit algorithm produces a layout which is near tension- free and symmetric. However, it has some problems which are inherited from Spring System [2]: it is relatively slow, and does not resolve edge crossings. For a more comprehensive approach to vertex splitting which takes edge crossings into account, see [8].

References

1. P. Eades, W. Lai, and X. Mendonça. A Visualizer for E-mail Trafic. In *4th Int. Conf. Proc. Pacific Graphics'94 / CADDM'94*, pages 64–67, 1994.
2. P. D. Eades. A Heuristic for Graph Drawing. *Congr. Numer.*, 42:149–160, 1984.
3. T. Kamada. *Visualizing Abstract Objects and Relations*. World Scientific, 1989.

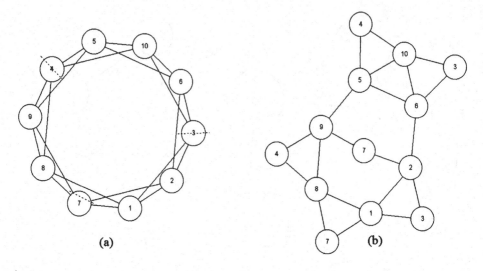

Fig. 11. example 5

4. T. Kamada and S. Kawai. Automatic Display of Network Structures for Human Understanding. Technical Report 88-007, Department of Information Science Faculty of Science, University of Tokyo, Tokyo, 1988.
5. T. Kamada and S. Kawai. An algorithm for drawing general undirected graphs. *Information Processing Letters*, 31:7–15, 1989.
6. X. Lin. *Analysis of Algorithms for Drawing Graphs*. PhD thesis, University of Queensland, Department of Computer Science, University of Queensland, 1992.
7. C. F. X. Mendonça. A Layout System for Information System Diagrams. Technical Report 94-01, Department of Computer Science, University of Newcastle, Australia, April 1994.
8. C. F. X. Mendonça and T. A. Halpin. Automatic Display of NIAM Conceptual Schemas Diagrams. Technical Report 209, Department of Computer Science, University of Queensland, Australia, July 1991.
9. Wei Lai. *Building Iteractive Diagram Applications*. PhD thesis, University of Newcastle, 1993.

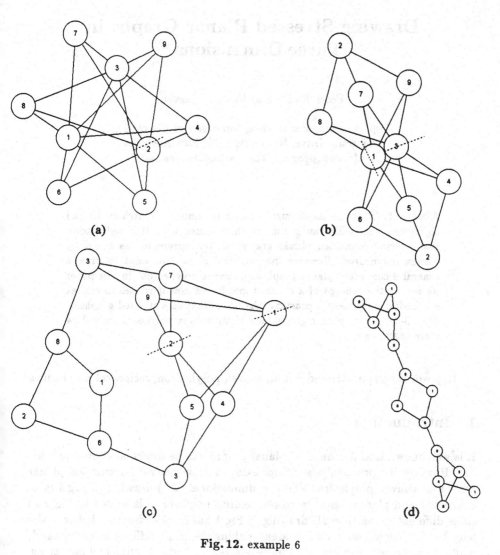

Fig. 12. example 6

Drawing Stressed Planar Graphs in Three Dimensions *

Peter Eades and Patrick Garvan

Dept. of Computer Science, University of Newcastle,
University Drive, Newcastle NSW 2308 Australia
{eades,pgarvan}@cs.newcastle.edu.au

Abstract. There is much current interest among researchers to find algorithms that will draw graphs in three dimensions. It is well known that every 3-connected planar graph can be represented as a strictly convex polyhedron. However, no practical algorithms exist to draw a general 3-connected planar graph as a convex polyhedron. In this paper we review the concept of a stressed graph and how it relates to convex polyhedra; we present a practical algorithm that uses stressed graphs to draw 3-connected planar graphs as strictly convex polyhedra; and show some examples.

Key words: graph, stressed graph, convex polyhedron, reciprocal polyhedron

1 Introduction

It is well known that 3-connected planar graphs can be drawn as convex polyhedra. However, no practical algorithms exist to draw general 3-connected planar graphs as convex polyhedra. The two-dimensional (2D) drawing in Fig.1 is 3-connected and planar, and the corresponding polyhedron is drawn in Fig.2 as three different views. The 2D drawing in Fig.1 has large differences between the lengths of its edges and in its face areas making the graph difficult to understand, while the polyhedron in Fig.2 has more nearly equal edge lengths and face areas. The ratio of the shortest distance between vertex positions to the diameter of the set of of vertex positions is 0.01 for the 2D drawing in Fig.1, and 0.09 for the three-dimensional (3D) drawing in Fig.2. We say that a planar graph is *cluttered* if this ratio is low or *easy-to-look-at* if it is high.

A method for drawing any graph in 3D using straight line edges such that no pair of edges cross is presented in [2]. This method does not restrict the drawing to the integer grid. In [3] a method is provided for drawing any graph in 3D using straight line edges such that no pair of edges cross. All vertices are located on the integer grid and, for a graph of n vertices, the required grid size is $n \times 2n \times 2n$ which is shown to be optimal to within a constant. Both [2] and [3]

---

* This paper appears as Technical Report 95-02, Department of Computer Science, University of Newcastle, Newcastle NSW 2308, Australia.

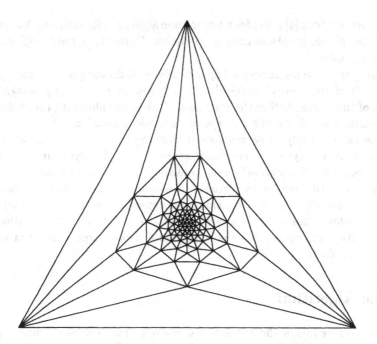

Fig. 1. Example of a cluttered 2D drawing of a planar graph.

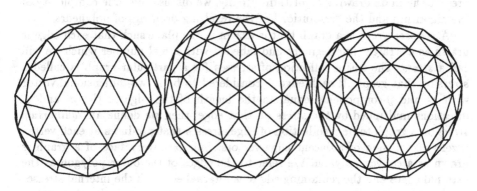

Fig. 2. Three views of the 3D polyhedron corresponding to Fig.1 as produced by Algorithm 1. The polyhedron is isomorphic to a three frequency geodesic sphere.

use the aesthetic criterion that no two edges should cross, and this is important when producing easy-to-look-at graphs. There are many other criteria that may be considered, including resolution and symmetry.

The following is a well-known theorem by Steinitz [7]:

Theorem 1 (Steinitz) *A graph G is the skeleton of a polyhedron P if and only if it is planar and 3-connected.*

This theorem guarantees that any 3-connected planar graph can be drawn as the vertices and edges of a convex polyhedron.

Some recent work [4] provides a linear-time algorithm for realizing 3-connected triangulated planar graphs as convex polyhedra. However, [4] only deals with triangulated graphs.

In this paper, we are concerned specifically with drawing 3-connected planar graphs as "strictly convex" polyhedra. A polyhedron P is *strictly convex* if the skeleton of the convex hull of the vertices of P is isomorphic with the skeleton of P. In section 2 we review the concept of "stressed graphs" and how they relate to polyhedra. We apply these concepts to drawing arbitrary 3-connected planar graphs as convex polyhedra. The time complexity of the algorithm is $O(n^{3/2})$, and the worst case "resolution" of the resulting drawings is at least exponential in n, where n is the number of vertices. In section 3 we show some examples of drawings as produced by the algorithm. Our experience with drawing polyhedra using Algorithm 1 has revealed a useful addition to our original algorithm. We conclude with some open problems in section 4. Note: some of the concepts in this paper are from [9].

2 The Algorithm

In this section we review the concept of stressed graphs (see for example [9]) and their relation to polyhedra. We give a definition of "reciprocal" polyhedra and present the main drawing algorithm. Finally, we discuss the time complexity of the algorithm and the "resolution" of the resulting drawings of polyhedra.

A *stressed graph* is a graph that is drawn in the plane such that each edge is drawn as a straight line segment and labelled with a real number which we will call a *stress*. The entire collection of stresses for a particular graph is called a *stress* on that graph. If each edge is considered to be a two-dimensional vector with x- and y-components equal to the length of the edge in the x- and y-directions multiplied by the stress on that edge, then we define an *equilibrium stress* on a graph to be a collection of edge stresses that produces at each vertex a zero vector sum of adjacent edge-vectors. Suppose every face of a graph is drawn as a convex polygon. We refer to the edges of the outer polygon as the external edges and the remaining edges as internal edges. If the internal stresses are positive while the outer stresses are negative, then an equilibrium stress on the graph is called a *convex equilibrium stress*. For convenience, a *restricted equilibrium stress* on a graph is defined as a collection of stresses where the inner stresses are positive and in equilibrium and no restrictions are placed on the outer stresses. [Note: Not every drawing of a 3-connected planar graph has a convex equilibrium stress.]

An example of a restricted equilibrium stressed graph is a Tutte [13] drawing of a 3-connected planar graph. This well known algorithm chooses a face of the graph and draws it as a convex polygon on the plane. The position of any internal vertex is then defined as the barycentre (or average) of the positions of its adjacent vertices. Fig.1 is an example of a Tutte drawing. It is clear that a collection of stresses for a such a drawing where each internal stress is $+1$ will be a restricted convex stress on the graph. [Note: Although Tutte's algorithm positions the internal vertices with the exact coordinates appropriate for a restricted equilibrium stress where each internal stress is +1, a simple variation of the algorithm will allow any collection of internal stresses to be used, as long as they are positive. Given such a collection, this variation on Tutte's algorithm will position each internal vertex so that the internal stresses are in equilibrium at each internal vertex, regardless of the shape or size of the outer polygon.]

We now describe the well known correspondence between convex stressed graphs and convex polyhedra first noted by Maxwell [11]. Consider an upturned bowl-shaped convex polyhedral surface Σ sitting on the plane H ($z = 1$) (see Fig.3(a)).

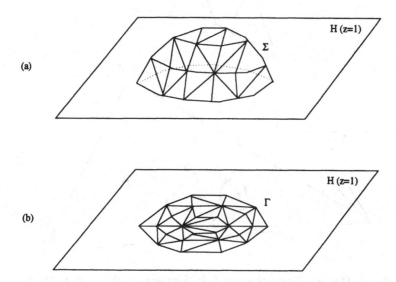

Fig. 3. (a) Upturned bowl shaped convex polyhedral surface Σ sitting on the plane H. (b) The 3-connected planar graph Γ resulting from an orthogonal projection of Σ down onto H. This projection is equivalent to flattening the bowl from above.

Let Γ be the drawing of the 3-connected planar graph resulting from an orthogonal projection of Σ down onto H (see Fig.3(b)), and ω be an edge label (or stress) for an edge of Γ defined using the following *stress equation*:

$$\omega = \frac{\varepsilon(r,s)(a_s(p_\bullet) - a_r(p_\bullet))}{[p_i, p_j, p_\bullet]}$$

where:

- ω is the stress for the edge e of Γ joining vertices v_i and v_j (see Fig.4);
- $\varepsilon(r, s)$ is ± 1 according to the orientation of faces f_r and f_s whose intersection defines e — that is, $\varepsilon(r, s) = +1$ if in an anticlockwise ordering of the vertices around f_r, v_i precedes v_j, and $\varepsilon(r, s) = -1$ otherwise;

- p_i and p_j are the coordinates of v_i and v_j in Γ;
- a_r and a_s are piecewise-affine functions mapping, respectively, faces f_r and f_s of Γ up to the corresponding facets of Σ;
- p_* (a reference point) is any point on H that is not collinear with any edge of Γ;
- $[p_i, p_j, p_*]$ is the scalar triple product of p_i, p_j, and p_*, and may be calculated by the determinant $|p_i, p_j, p_*|$, where p_i, p_j, and p_* are written as column triples.

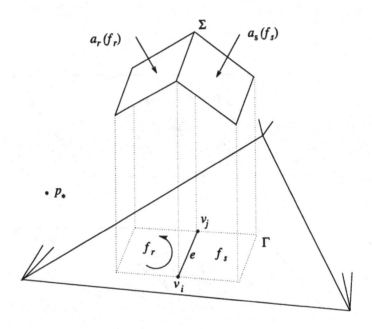

Fig. 4. Diagram illustrating the definitions used in the stress equation.

Informally, Γ records the skeletal structure of Σ, as projected onto H, and ω records the shape of Σ in the third dimension. A 3-connected plane graph Γ with an equilibrium stress ω may be written (ω, Γ).

Given the above definitions, the following theorem formally describes the bijectivity:

Theorem 2 ([9]) *Suppose (ω, Γ) is a stressed graph in H and Σ is the corresponding polyhedral surface sitting over Γ. Then Σ is the boundary of a convex polyhedron if and only if Γ and ω are convex.*

Given a polyhedron P, a *dual polyhedron* P^* is a polyhedron with a skeleton that is the graph-theoretic dual to the skeleton of P. A specific type of dual polyhedron is called a "reciprocal". Given a reference sphere S with equation $x^2 + y^2 + z^2 = r^2$, every plane $ax + by + cz = r^2$ has a reciprocal point (a, b, c)

defined with respect to S. Similarly, every point (a, b, c) (except the origin) has a reciprocal plane $ax + by + cz = r^2$ defined with respect to S. If P is a polyhedron, then a *reciprocal* polyhedron P^* with respect to S is defined as that polyhedron with vertices that are reciprocal to the facet-planes of P, and with facet-planes that are reciprocal to the vertices of P. Further, if P is convex and contains the origin, then the reciprocal polyhedron P^* will also be convex. (For more information see [1] and [8].)

Algorithm 1   *Input:* A planar embedding of a 3-connected planar graph G
Output: A strictly convex polyhedron P

1. If G does not contain a C_3 (a cycle of three edges), then replace G by its graph-theoretic dual G^*.
2. Choose any C_3 in G and draw it as a triangle on the plane $z = 1$.
3. Using Tutte's algorithm [13] draw the remaining vertices and edges of G producing a planar drawing Γ that has a restricted convex equilibrium stress. Calculate the remaining three external stresses on G [9] producing a convex equilibrium stressed graph.
4. Perform a breadth-first-search (BFS) of the face-lists of G, starting at the outerface. When visiting each face, use the stress equation to calculate the plane-equation for the corresponding facet of the polyhedron P.
5. If necessary, replace P with a convex reciprocal P^* of P.

□

Theorem 3 *Algorithm 1 draws 3-connected planar graphs as strictly convex polyhedra.*

Proof: We shall prove the correctness of the algorithm step by step.

Step 1: If a 3-connected planar graph does not contain a C_3 (a triangle), then its graph-theoretic dual will contain at least eight triangles. Therefore, if G does not contain a triangle, then its graph-theoretic dual G^* will contain a triangle.

Step 2: Step 1 guarantees that G will contain a triangle. Step 3, the Tutte drawing, requires that the outerface is a convex face. A triangular face will always be convex.

Step 3: If the outer face is convex, a Tutte drawing of a 3-connected planar graph will always be a convex plane drawing of the graph. That is, each face of the graph will be convex. Further, by definition of a Tutte drawing, the internal stresses will all be $+1$. In [9] it is shown that external equilibrium stresses for such a drawing (ie one with an outer polygon that is a triangle) can always be calculated and are unique. Thus, we have a convex equilibrium stressed graph.

Step 4: Given a convex equilibrium stressed graph, there exists a strictly convex polyhedron defined by the stress equation with a skeleton that is isomorphic with G. Using this equation, we can produce a strictly convex polyhedron.

Step 5: Every polyhedron has at least one reciprocal polyhedron. If a polyhedron is convex, then it is possible to find a reciprocal polyhedron that is also convex. Furthermore, the skeleton of a polyhedron is dual to the skeleton of a reciprocal of that polyhedron. □

Theorem 4 *Algorithm 1 requires $O(n^{3/2})$ time.*

Proof: Step 3 requires $O(n^{3/2})$ time to perform a Tutte drawing [12]. Every other step requires at most $O(n)$ time. Therefore the overall time complexity is $O(n^{3/2})$. $\square$

Unfortunately, Algorithm 1 does not always produce drawings with good "resolution", where *resolution* refers to the ratio of the smallest distance between vertex positions to the diameter of the set of vertex positions.

Lemma 5 *The worst case resolution of a Tutte drawing is $\Omega(k^n), k > 1$.*

Proof: Consider a Tutte drawing of the following 3-connected planar graph:

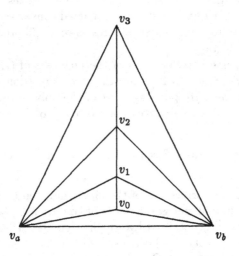

If v_a and v_b are each defined to have a y-coordinate of zero ($y_a = y_b = 0$); and v_0 is defined to have a y-coordinate of 1 ($y_0 = 1$), then, by definition of a Tutte drawing, v_1 will have a y-coordinate of $3y_0$ ($y_1 = 3y_0$). Now, $y_2 = 4y_1 - y_0$, and similarly for y_3. In this way, the maximum y-coordinate for a similar graph of n vertices can be found by solving the recurrence $y_i = 4y_{i-1} - y_{i-2}$; where $y_0 = 1$ and $y_1 = 3$. Using standard techniques [10], the solution to this recurrence equation shows that $y_n = \Theta(k^n)$, where k is a constant and $k > 1$. Therefore, the worst case resolution of a Tutte drawing is $\Omega(k^n), k > 1$. $\square$

Theorem 6 *Algorithm 1 produces a drawing of a polyhedron with at least exponential worst case resolution.*

Proof : By Lemma 4, the resolution of a Tutte drawing is at least exponential in the worst case, therefore the resolution of the drawing produced by Algorithm 1 will be at least exponential in the worst case. $\square$

3 Examples

In this section we will show some examples of drawings as produced by Algorithm 1, and suggest a possible extension of the existing algorithm so as to produce more sphere-like polyhedra. Unlike Fig.2, the drawings of polyhedra in this section are 2D projections of vertices and edges and do not explicitly show the facets. However, we have attempted to choose projections that clearly illustrate the shapes of the polyhedra. In some instances, we have drawn the polyhedron as a stereoscopic-pair, where the left drawing is the view as seen by the left eye and the right drawing is the view as seen by the right eye. A 3D effect may be produced using these pairs by looking at a point behind the page, thus causing the two drawings to overlap. An inverted image may also be produced by crossing one's eyes in order to overlap the drawings.

Figures 5 and 6 show drawings of the skeleta of the Platonic solids and the corresponding polyhedra as produced by Algorithm 1. The resulting polyhedra are isomorphic with the Platonic solids but are not regular. (For more information on the Platonic solids see [1].)

In Fig.6, the polyhedra are the direct result of a reciprocation operation (i.e. the calculation of a reciprocal polyhedron), since the skeleta of the cube and the dodecahedron have no C_3 faces. The polyhedron in drawing 6.II(a) has facets of nearly equal area and shape, as does the polyhedron in drawing 6.II(b). This near-regularity of facet area and shape is in most part due to the reciprocation step at the end of Algorithm 1. Applying *two* suitable reciprocation operations to each of the polyhedra in column II of Fig.5 will have the same effect. That is, it is possible to apply a double reciprocation to these polyhedra in such a way that the large facets will be shrunk and the small facets will be expanded. The polyhedra in 5.III are the result of such a double reciprocation.

The drawings in 5.III are nicer than those drawings in column 5.II. The most notable attributes are the approximate spherical shape of the polyhedra, and the relatively uniform facet areas and edge lengths.

This suggests changing the last step of Algorithm 1 to perform a suitable double reciprocation in those cases when the dual was not constructed. That is, when the input graph G contains at least one C_3.

Figure 7.II(a) shows a drawing of a polyhedron isomorphic to a twin cube. This polyhedron is strictly convex, where a more intuitive drawing may only be weakly convex.

Figure 7.II(b) shows the drawing of a polyhedron isomorphic to a pentagonal prism. This polyhedron is strictly convex, however no two adjacent facets are at right angles to each other.

Figure 8.III shows the drawing of a polyhedron as produced by Algorithm 1 with a skeleton that is isomorphic to the octagonal mesh shown in Fig.8.I. Figure 8.III is a bad drawing since there are large facets and many small facets clustered together. Figure 8.II is the polyhedron created by Algorithm 1 prior to the reciprocation step. This polyhedron has many long thin facets clustered down one side, and so reciprocation about a single internal point results in a cluster of small facets in the final polyhedron.

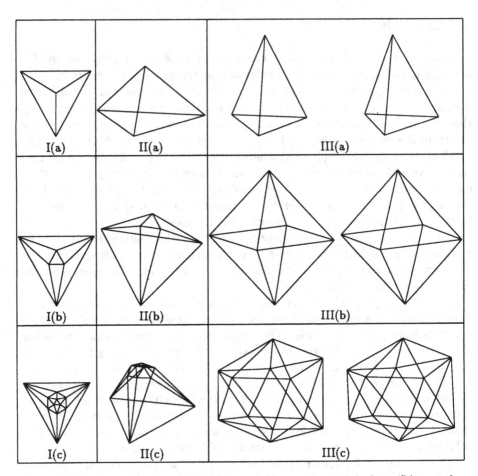

Fig. 5. (I) 2D planar drawings of the skeleta of (a) a regular tetrahedron, (b) a regular octahedron, and (c) a regular icosahedron; (II) the corresponding polyhedra as produced by Algorithm 1; and (III) stereoscopic-pair drawings of the polyhedra resulting from a double reciprocation. Note that these polyhedra consist entirely of triangular facets.

In [5] there are drawings of some 3-connected planar graphs in 2D, and in some cases the result is similar to a projection of a 3D polyhedron into 2D. In [6] there are drawings of some 3-connected planar graphs in both 2D and 3D, however the resulting drawings are not polyhedral in the sense that for any particular face of a graph the vertices of that face are not necessarily drawn so that they lie on the same plane in 3D.

As mentioned in the introduction, Fig.1 is the skeleton of a 3 frequency geodesic sphere, and Fig.2 shows three views of the corresponding polyhedron as produced by Algorithm 1. The line-hiding utilised in Fig.2 emphasises the fact that Algorithm 1 draws 3-connected planar graphs as strictly convex polyhedra, where each face of the graph is drawn in 3D as a single facet.

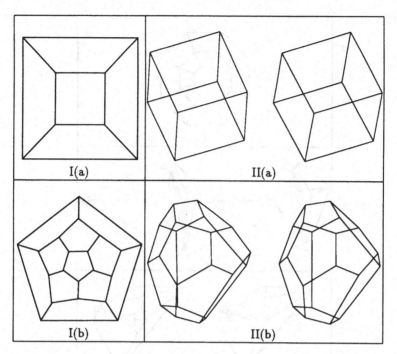

Fig. 6. (I) 2D planar drawings of the skeleta of (a) a regular hexahedron (or cube) and (b) a regular dodecahadron; and (II) stereoscopic-pair drawings of the corresponding polyhedra as produced by Algorithm 1. Note that these polyhedra consist entirely of non-triangular facets.

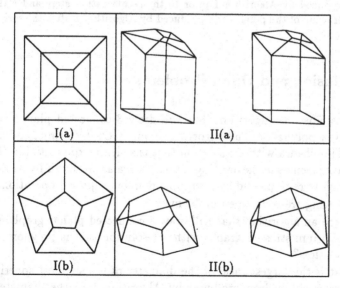

Fig. 7. (I) 2D planar drawings of the skeleta of (a) a twin cube and (b) a pentagonal prism; and (II) stereoscopic pair drawings of the corresponding polyhedra as produced by Algorithm 1.

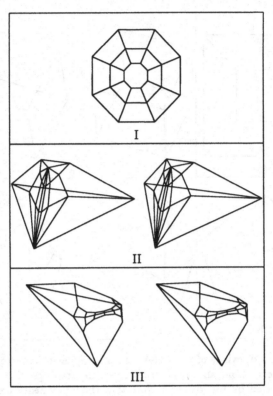

Fig. 8. (I) 2D planar drawing of an octagonal mesh; (II) the corresponding *dual* polyhedron as produced by Algorithm 1 prior to the reciprocation step; and (III) stereoscopic-pair drawing of the polyhedron produced by Algorithm 1 after the reciprocation step.

4   Conclusion and Open Problems

We have described an algorithm that will draw 3-connected planar graphs as strictly convex polyhedra. This algorithm requires $O(n^{3/2})$ time, and produces drawings of polyhedra with a resolution that is at least exponential in the worst case. We have given example drawings of polyhedra as produced by Algorithm 1, some of which were improved by a suitable double reciprocal operation.

The following are some open problems:

(i) Is there an algorithm that will draw 3-connected planar graphs in 2D as convex equilibrium stressed graphs with a resolution that is polynomial in the number of vertices?

(ii) What is the largest ratio of the diameter of the smallest spherical shell that encloses a polyhedron produced by Algorithm 1 to the diameter of the largest spherical shell enclosed by the same polyhedron, where the two spherical shells are concentric? Such sphericity of a polyhedron is a good aesthetic for identifying nice drawings of polyhedra.

(iii) Where should vertex labels be placed after drawing a 3-connected planar graph as a polyhedron? One possibility is to draw the dual of such a graph as a convex polyhedron and place labels on the facets. The facet adjacencies of such a drawing correspond exactly to the vertex adjacencies of the original graph.

(iv) What is the best way to display a three-dimensional graph? In particular, what is the best way to display a three-dimensional graph on a static two-dimensional medium such as paper? Possibilities include red/green anaglyphs and stereoscopic pairs.

Acknowledgements

We would like to thank Bob Cohen, Mike Goodrich, and Mike Houle for their useful comments on earlier versions of this paper.

References

1. H. S. M. Coxeter. *Regular Polytopes.* Dover, NY, 1973.
2. I. Cahit. Drawing the Complete Graph in 3-D with Straight Lines and Without Crossings. *Bulletin of the ICA*, Vol 12, Sep 94.
3. R. F. Cohen, P. Eades, T. Lin, and F. Ruskey. Three-Dimensional Graph Drawing. Proc. Graph Drawing 94. *Lecture Notes in Computer Science.* Volume 894, 1-11. Springer-Verlag, Berlin, 1995.
4. G. Das, M. T. Goodrich. *On the Complexity of Approximating and Illuminating Three-Dimensional Convex Polyhedra.* To appear, WADS 1995.
5. R. Davidson and D. Harel. *Drawing Graphs Nicely Using Simulated Annealing,* Technical Report CS89-13, Department of Applied Mathematics and Computer Science, The Weizmann Institute of Science, Rehovot, Israel, July 1989.
6. T. M. J. Fruchterman and E. M. Reingold. Graph Drawing by Force-directed Placement. *Software – Practice and Experience*, vol. 21(11), 1129-1164 (Nov 1991).
7. B. Grünbaum. *Convex Polytopes.* Wiley, NY, 1967.
8. B. Grünbaum and G. C. Shephard. Duality of Polyhedra. In *Shaping Space: A Polyhedral Approach.* Birkhauser, Boston, 1988.
9. J. E. Hopcroft and P. J. Kahn. A Paradigm for Robust Geometric Algorithms. *Algorithmica* (1992) 7:339-380.
10. C. L. Liu. *Introduction to Combinatorial Mathematics.* McGraw-Hill, NY, 1968.
11. J. C. Maxwell. On Reciprocal Figures and Diagrams of Forces. *Phil. Mag.* S. 4. vol. xxvii. (1864) pp.250-261
12. T. Nishizeki and N. Chiba. Planar Graphs: Theory and Algorithms. *Annals of Discrete Mathematics*, Volume 32. North-Holland, Netherlands, 1988.
13. W. T. Tutte. How to Draw a Graph. *Proc. London. Math. Soc.* (3), 13 (1963), 743-768.

Graph-Drawing Contest Report

Peter Eades[1] and Joe Marks[2]

[1] Department of Computer Science
University of Newcastle
University Drive – Callaghan
NSW 2308, Australia
E-mail: eades@cs.newcastle.edu.au
[2] Mitsubishi Electric Research Laboratories, Inc.
201 Broadway
Cambridge, MA 02139, U.S.A.
E-mail: marks@merl.com

Abstract. This report describes the the Second Annual Graph Drawing Contest, held in conjunction with the 1995 Graph Drawing Symposium in Passau, Germany. The purpose of the contest is to both monitor and challenge the current state of the art in graph-drawing technology.

1 Introduction

Text descriptions of three attributed graphs were made available on the World-Wide Web (at http://www.uni-passau.de/agenda/gd95/competition.html and http://www.cs.brown.edu/calendar/gd95/) for this year's contest. The graph attributes included both a *name* and *type* attribute for each vertex. An effective graph drawing had to communicate not only the edge connections between vertices, but also the vertex-attribute values. Thus the main judging criterion was one of information visualization. A secondary criterion was the degree to which the drawing was generated automatically, that is, without manual intervention.

Approximately 40 graphs were submitted by the contest deadline. The emphasis on information visualization and the nature of the graphs resulted in a very eclectic mix of drawings. The winners were selected by a panel of judges, and are shown below.

2 Winning submissions and honorable mentions

2.1 Graph A

Graph A models the architecture of a computer chip and was based on the hand-drawn original shown in Figure 1 [1]. The winning drawing was submitted by Georg Sander of Universität des Saarlandes (sander@cs.uni-sb.de), and is shown in Figure 2.[3] The manual steps needed to produce this drawing included a

---

[3] Several of the figures, including Figure 1, use color for enhanced effectiveness. To obtain a reprint of this article in color, please contact one of the authors.

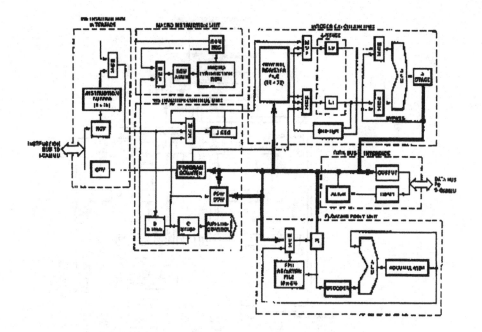

Fig. 1. Original hand-drawn copy of Graph A.

partitioning of the graph into subgraphs, adjustment of the level assignments of the nodes, and selection of various rendering parameters. Layout was computed automatically using the author's VCG tool in four seconds on a Sparc ELC workstation.

Two other drawings of Graph A received honorable mention. Figure 3 contains the drawing of Paulis Ķikusts (paulis@cclu.lv) and Pēteris Ručevskis (rpeteris@cclu.lv) from the University of Latvia. The drawing in Figure 4 was submitted by Thomas Kamps, Jörg Kleinz, and Thomas Reichenberger ([kamps, kleinz, reichen]@darmstadt.gmd.de) from IPSI, GMD Darmstadt.

2.2 Graph B

Graph B represents a collection of global symbols (e.g., functions, types, files, etc.) for a small part of a large X program. The name and type attribute values for this graph's vertices are particularly unwieldy.

The winning drawing for Graph B, shown in Figure 5, was submitted by Falk Schreiber and Carsten Friedrich ([schreibe, friedric]@fmi.uni-passau.de) of Universität Passau. Their strategies for coping with the awkward types and names included the use of color to convey vertex-type information, rotation of the drawing prior to text-label placement to avoid excessive label overlaps, rendering the edges in grayscale to enable visible overprinting of text labels, and some manual adjustment of label placements. Layout was performed using a Sugiyama-style algorithm.

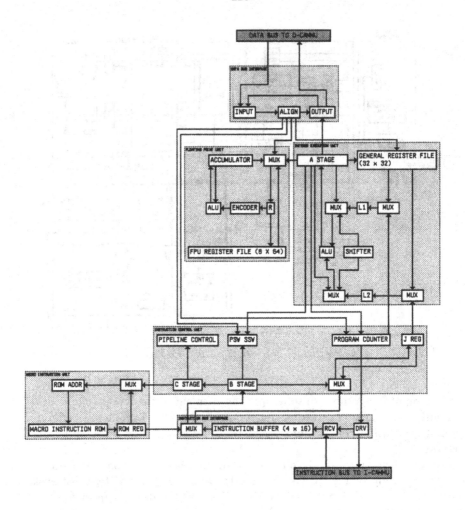

Fig. 2. Winner, Graph A.

Honorable mentions for Graph B went to Georg Sander of Universität des Saarlandes (**sander@cs.uni-sb.de**) for the drawing in Figure 6 and to Vladimir Batagelj and Andrej Mrvar ([**vladimir.batagelj, andrej.mrvar]@uni-lj.si**) from the University of Ljubljana for the drawing in Figure 7. In the latter drawing, the text-label problem was finessed by using a legend (not shown) to decode a coloring and numbering of nodes.

2.3 Graph C

Unlike Graphs A and B, Graph C was contrived without reference to a real-world application. However, what it lacks in verisimilitude it makes up for in difficulty, because it is a planar graph that is especially hard to draw well.

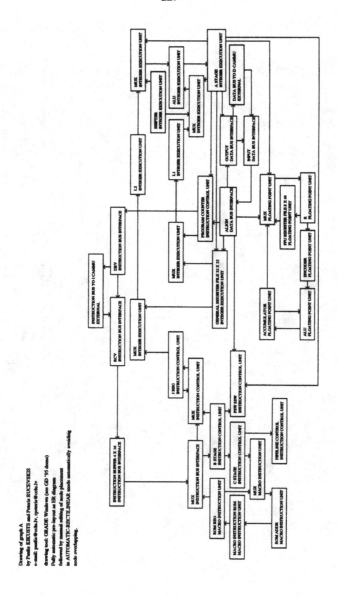

Fig. 3. Honorable mention, Graph A.

Two joint winners were declared for Graph C. They are shown in Figures 8 and 9. The former drawing was submitted by Vladimir Batagelj and Andrej Mrvar ([vladimir.batagelj, andrej.mrvar]@uni-lj.si) from the University of Ljubljana. They used an energy-minimization approach to compute the layout, and manual editing to reposition some nodes. The other winner was submitted by Paulis Ķikusts (paulis@cclu.lv) and Pēteris Ručevskis (rpeteris@cclu.lv) from the University of Latvia. Their approach to layout uses a blend of local

228

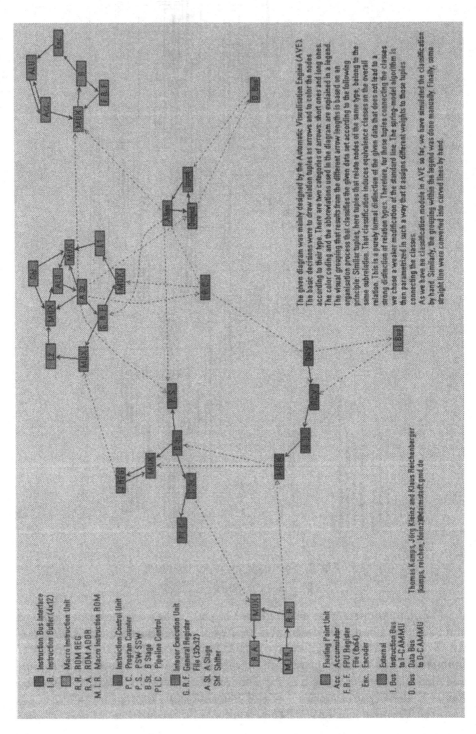

Fig. 4. Honorable mention, Graph A.

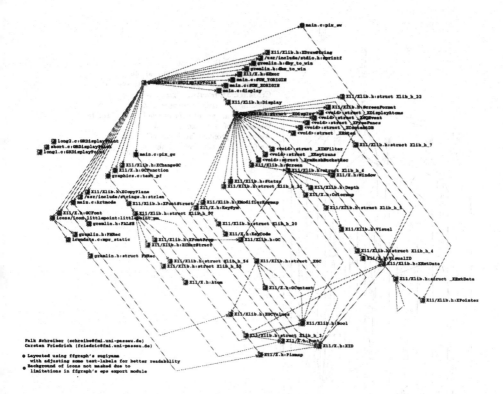

Fig. 5. Winner, Graph B.

placement heuristics, supplemented by some manual editing.

An honorable mention for Graph C was awarded to the entry submitted by Harald Lauer (lauer@informatik.uni-tuebingen) of Universität Tübingen. It is shown in Figure 10.

3 Acknowledgments

Sponsorship for this contest was provided by AT&T Bell Labs, Mitsubishi Electric Research Labs, Tom Sawyer Software, Universität Passau, and Volksbank Passau. Stephen North, Yanni Tollis, and Sue Whitesides assisted with the judging. Stephen North also provided the data for Graph B.

References

1. W. Hollingsworth, H. Sachs, and A. Smith. The CLIPPER processor: Instruction set architecture and implementation. *CACM*, 32(2):212, February 1989.

230

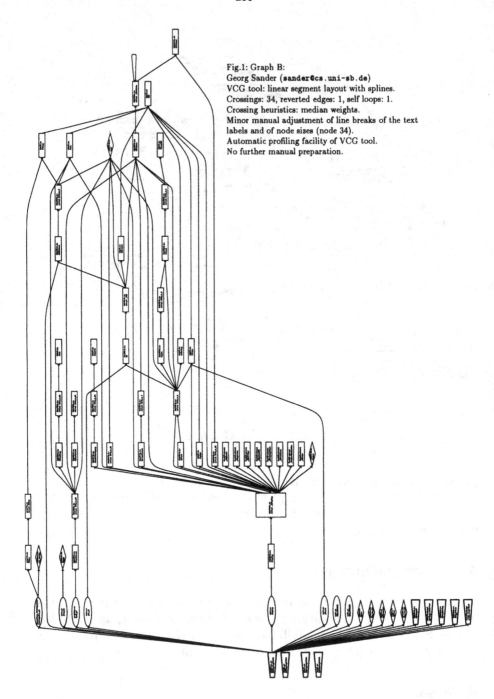

Fig.1: Graph B:
Georg Sander (**sander@cs.uni-sb.de**)
VCG tool: linear segment layout with splines.
Crossings: 34, reverted edges: 1, self loops: 1.
Crossing heuristics: median weights.
Minor manual adjustment of line breaks of the text
labels and of node sizes (node 34).
Automatic profiling facility of VCG tool.
No further manual preparation.

Fig. 6. Honorable mention, Graph B.

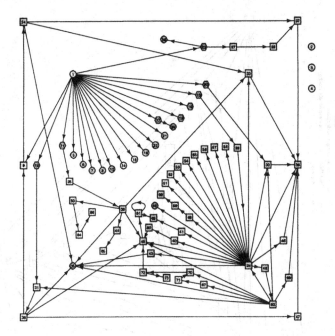

Fig. 7. Honorable mention, Graph B.

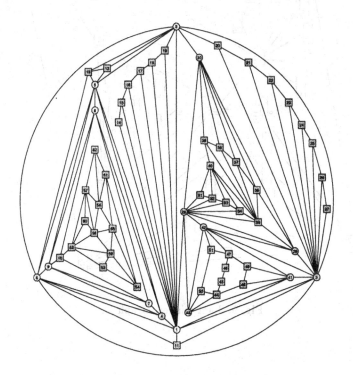

Fig. 8. Joint winner, Graph C.

232

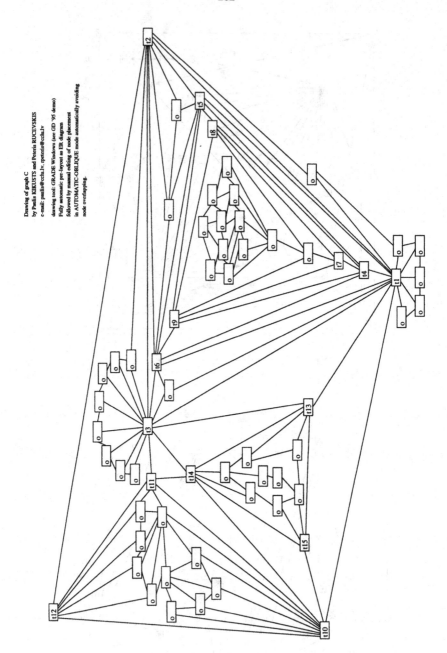

Fig. 9. Joint winner, Graph C.

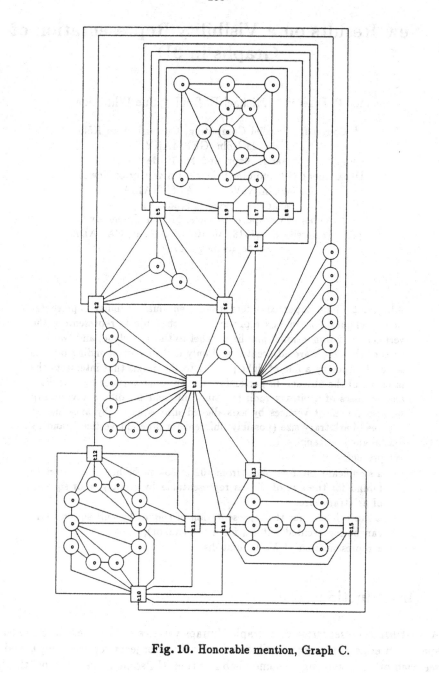

Fig. 10. Honorable mention, Graph C.

New Results on a Visibility Representation of Graphs in 3D

Sándor P. Fekete[1] *, Michael E. Houle[2], Sue Whitesides[3] **

[1] Center for Parallel Computing, Universität zu Köln
D–50923 Köln, GERMANY
sandor@zpr.uni-koeln.de
[2] Department of Computer Science, University of Newcastle
Callaghan NSW 2308, AUSTRALIA
mike@cs.newcastle.edu.au
[3] School of Computer Science, McGill University
3480 University St. #318, Montréal, Québec, CANADA
sue@opus.mcgill.ca

Abstract. This paper considers a 3-dimensional visibility representation of cliques K_n. In this representation, the objects representing the vertices are 2-dimensional and lie parallel to the x, y-plane, and two vertices of the graph are adjacent if and only if their corresponding objects see each other by a line of sight parallel to the z-axis that intersects the interiors of the objects. In particular, we represent vertices by unit discs and by discs of arbitrary radii (possibly different for different vertices); we also represent vertices by axis-aligned unit squares, by axis-aligned squares of arbitrary size (possibly different for different vertices), and by axis-aligned rectangles.

We present:

- a significant improvement (from 102 to 55) of the best known upper bound for the size of cliques representable by rectangles or squares of arbitrary size;
- a sharp bound for the representation of cliques by unit squares (K_7 can be represented but K_n for $n > 7$ cannot);
- a representation of K_n by unit discs.

1 Introduction

A *visibility representation* of a graph G maps vertices of G to sets in Euclidean space. An edge u, v occurs in G if and only if the objects representing u and v see each other according to some visibility rule. (In some investigations, the "if and only if" condition is relaxed to "only if".)

---

* Parts of this work were done while staying at SUNY Stony Brook, supported by NSF Grants ECSE-8857642 and CCR-9204585.

** Written in part while visiting the Center for Parallel Computing and INRIA-Sophia Antipolis. Research supported by NSERC and FCAR.

Application areas such as VLSI routing and circuit board layout have stimulated considerable research on visibility representations in R^2. See for example [6], [9], [12], and [13]. Recently, interest has developed in finding good 3-dimensional visualizations of graphs. See for example [4], [8], [10]. It is also of interest to develop geometric graph theory in higher dimensions.

This paper continues the study of a particular visibility representation, studied for example in [2], [3], [11] and [1], in which the objects representing vertices are 2-dimensional sets parallel to the x, y-plane. An edge u, v occurs in G if and only if the objects representing u and v see each other along a line of sight parallel to the z axis. This line of sight must intersect the interiors of the objects; hence legitimate lines of sight are extensible to tubes of small radius whose ends lie inside the objects. Furthermore, since G has an edge u, v if and only if u and v are mutually visible, the graph G is recoverable from the geometry of the representation. Throughout the paper, we use the term *representation* to refer to this specific model. Also, when we refer to representation by *arbitrary* squares or discs, we mean that the size of these objects is allowed to vary with the vertex represented. As in [2] and [3], rectangles are understood to be axis-aligned and disjoint.

In [2] the objects representing vertices are arbitrary rectangles. In this paper, we consider unit squares and discs, and arbitrary squares, rectangles and discs. We focus exclusively on the representation of cliques by these objects.

Constructions for squares can be viewed as constructions for arbitrary rectangles. Any upper bound on the size of the largest clique K_n that can be represented by arbitrary rectangles is also an upper bound on the size of the largest K_n that can be represented by arbitary squares or by unit squares. The current best bound on the size of the largest K_n that can be represented by arbitrary rectangles is 102 (see [2]), which also holds for representation by arbitrary squares and unit squares. Here we lower this bound from 102 to 55 for arbitrary rectangles, and hence for arbitrary squares and unit squares.

The results derived in this paper are as follows:

- K_{56} has no representation by arbitrary rectangles.
- K_7 (and hence K_n for all $n \leq 7$) has a representation by unit squares, but no K_n for $n \geq 8$ has a representation by unit squares.
- Any K_n has a representation by unit discs.

The rest of the paper is organized as follows. Section 2 discusses representations by arbitrary squares and rectangles, Sect. 3 considers representations by unit squares, and Sect. 4 considers representations by arbitrary discs and unit discs.

2 Squares and Rectangles

The question of determining the largest complete graph K_n that can be represented by arbitrary rectangles is addressed in [2] and [3], which give an upper bound of $n \leq 102$ and a lower bound of $n \geq 20$. The construction for K_{20} that appears in [3] uses arbitrary squares. In the following, we describe how to improve the upper bound from $n \leq 102$ to $n \leq 55$.

We consider sequences of n rectangles lying parallel to the x, y-plane in R^3, and ordered by increasing z-coordinate. We call a sequence *valid* if its associated visibility graph is K_n. Consider the projections of all the rectangles in a valid sequence onto the x, y-plane. Because each two must intersect and the objects are axis-aligned rectangles, application of a Helly-type theorem shows that the intersection of all the projections must be non-empty. Thus we can choose a common point O (henceforth regarded as the origin) belonging to the interior of each of the projections. To simplify the notation, we do not distinguish between a rectangle and its projection onto the x, y-plane; the meaning will be clear from the context.

Each rectangle R in a valid sequence can be described in terms of the perpendicular distances from O to each of its sides. Instead of giving the x, y-coordinates of the vertices of R, we describe R as a 4-tuple (E_r, N_r, W_r, S_r) whose coordinates give, respectively, the distances from $O \in R$ to the east, north, west and south sides of R.

We can assume without loss of generality that no two rectangles of a valid sequence share the same value on any of the four coordinates E, N, W, S. Hence we can assume that each coordinate value of each of the n rectangles is an integer in the range $[1, n]$ without changing the visibility relationships among the rectangles.

Consider two rectangles $A = (E_a, N_a, W_a, S_a)$ and $B = (E_b, N_b, W_b, S_b)$ in a valid sequence, and denote by $A \cap B$ the intersection of their projections onto the x, y-plane. Then $A \cap B$ contains O, and the coordinates of $A \cap B$ are $E_{A \cap B} = \min\{E_a, E_b\}$, $N_{A \cap B} = \min\{N_a, N_b\}$, $W_{A \cap B} = \min\{W_a, W_b\}$ and $S_{A \cap B} = \min\{S_a, S_b\}$. We say that a corner of $A \cap B$ is *free* if it is not covered by any of the projections of rectangles occuring between A and B in the sequence.

Suppose A and B are rectangles in a valid sequence. Then since O belongs to all the rectangles, at least one of the corners of $A \cap B$ must be free. This is because any rectangle that covers a corner also covers O and hence an entire quadrant of $A \cap B$. Thus if $A \cap B$ had no free corner, it would be covered by the union of the intervening rectangles that cover at least one corner of $A \cap B$.

The northeast corner of $A \cap B$ is not covered by a particular rectangle $R = (E_r, N_r, W_r, S_r)$ between A and B if and only if the Boolean expression $E_r < \min\{E_a, E_b\}$ OR $N_r < \min\{N_a, N_b\}$ is true. Similar conditions hold for the other three corners.

We summarize: the rectangles A and B can see each other if and only if one of the following conditions F holds *simultaneously for all* the rectangles R between A and B:

FC$_{ne}(A, B)$: northeast is free, i.e. $(E_r < \min\{E_a, E_b\}$ OR $N_r < \min\{N_a, N_b\})$;

FC$_{nw}(A, B)$: northwest is free, i.e. $(N_r < \min\{N_a, N_b\}$ OR $W_r < \min\{W_a, W_b\})$;

FC$_{sw}(A, B)$: southwest is free, i.e. $(W_r < \min\{W_a, W_b\}$ OR $S_r < \min\{S_a, S_b\})$;

FC$_{se}(A, B)$: southeast is free, i.e. $(S_r < \min\{S_a, S_b\}$ OR $E_r < \min\{E_a, E_b\})$.

Now we give a definition that is needed in the following discussions. Given a valid sequence of n rectangles $(E_1, N_1, W_1, S_1), \ldots, (E_n, N_n, W_n, S_n)$, we denote by V_E, V_N, V_W, V_S the sequences of integers obtained by projecting the rectangles onto their E, N, W and S coordinates, respectively.

The next definition and lemma provide the key tool in our analysis.

Definition 1. A sequence of distinct integers will be called *unimaximal* if it has exactly one local maximum.

Lemma 2. *For all $m > 1$, in every sequence of $\binom{m}{2} + 1$ distinct integers, there exists at least one unimaximal subsequence of length m. On the other hand, there exists a sequence of $\binom{m}{2}$ distinct integers that has no unimaximal subsequence of length m.*

This result arises from the Erdős-Szekeres Theorem (1935), whose pigeon-hole proof was given by [7]. Lemma 2 is attributed by F. R. K. Chung [5] to V. Chvátal and J. M. Steele, among others.

Lemma 3. *In a representation of K_5 by five rectangles, with no other rectangles present, it is impossible that both sequences V_N and V_S are unimaximal.*

Proof. Suppose both V_N and V_S were unimaximal. Then $N_r > \min\{N_a, N_b\}$ and $S_r > \min\{S_a, S_b\}$ must hold for all rectangles A, B, and R between A and B. Now consider the conditions **FC**$_{ne}$, **FC**$_{nw}$, **FC**$_{sw}$, **FC**$_{se}$. For **FC**$_{ne}(A, B)$ to be true, it must be the case that $E_R < \min\{E_a, E_b\}$ for all R between A and B. The same is true for **FC**$_{se}(A, B)$. Similarly, for **FC**$_{nw}(A, B)$ or **FC**$_{sw}(A, B)$ to be true, W_r must be less than $\min\{W_a, W_b\}$. Hence the free corner conditions reduce to the following. One of the two possibilities $(W_r < \min\{W_a, W_b\})$ or $(E_r < \min\{E_a, E_b\})$ holds simultaneously for all rectangles R between A and B. This means that all rectangles A and B can see each other along a line of sight with y-coordinate 0. By intersecting the arrangement of five rectangles with the x, z-plane, we get an arrangement of 5 line segments in this plane that all see each other. This contradicts the fact that only planar graphs can be represented by vertical visibility of horizontal line segments in a plane. (See [12] and [13] for results on such representations in the plane.) □

Theorem 4. *No complete graph K_n has a representation by arbitrary rectangles for $n \geq 56$.*

Proof. Suppose we had a representation of K_n with $n \geq 56$. Lemma 2 implies that V_N has a unimaximal subsequence V_N' of length 11. Consider the associated

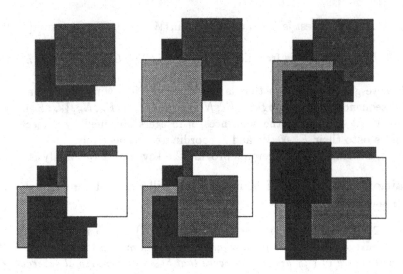

Fig. 1. Representing K_7 by unit squares.

subsequence V_S' of length 11. It follows again from Lemma 2 that there is a subsequence V_S'' of length 5 that is unimaximal. Remove the rectangles not associated with the subsequence. This destroys no visibility lines, so the five remaining rectangles represent K_5. However, both V_S'' and its corresponding subsequence V_N'' are unimaximal. This contradicts Lemma 3.

$\square$

3 Unit Squares

While it is still a challenge to narrow the gap between the known upper and lower bounds for representation of K_n by rectangles or general squares, we will see in the following that the largest possible n for which K_n has a representation by squares of equal size is $n = 7$.

Theorem 5. K_7 *has a representation by unit squares.*

Proof. See Figure 1. The six phases indicate how the seven squares are placed on top of each other. $\square$

Theorem 6. K_8 *does not have a representation by unit squares.*

Proof. We did the proof by a computer search that enumerates all maximal valid sequences of squares. The search begins with a single square. It proceeds in a depth-first-search manner, examining all the ways to add a new square above the squares in the valid sequence currently considered. It tests validity by checking the free corner conditions of the previous section. Whenever a valid extension is discovered, an immediate attempt is made to extend it further.

To be more specific, consider a valid sequence, not necessarily maximal, of m squares of the same "unit" size. As before, without loss of generality, we may describe these squares as 4-tuples (E, N, W, S). The coordinate positions give rise to four sequences V_E, V_N, V_W, V_S of distinct integers in the range $[1, m]$. Because all the squares have the same size, the E and W coordinates of any given square must sum to $m + 1$, and similarly for the N and S coordinates. Hence V_E determines V_W, and V_N determines V_S. Since the sequence is valid, the free corner conditions given in the previous section must hold for each pair of squares in the sequence.

Now suppose we want to enumerate all the possible ways to position a new unit square above the existing m squares. Each new sequence will be described by 4-tuples whose coordinates are integers in the range $[1, m + 1]$. There are $m + 1$ choices for the E coordinate of the new square and $m + 1$ choices for its N coordinate. Since these choices determine W and S, there are in total $(m + 1)^2$ possibilities. These may be considered by lexicographic order of the (E, N) pair of coordinates of the new square, from $(E, N) = (1, 1)$ to $(E, N) = (m + 1, m + 1)$.

Note that although the squares in the old sequence do not change position in any way, their coordinates must change to make room for the coordinates of the new square. For example, if the new square has $E = 3$, then all old squares with an E coordinate of 3 or greater must increase the old E coordinate by 1.

To check whether a new sequence is valid, we need only check the free corner conditions of the previous section for pairs of squares involving the new square.

Our program enumerates all valid maximal sequences of unit squares as follows. It begins with a valid sequence of length $m = 1$, described in the E, N, W, S coordinate system by the single 4-tuple $(1, 1, 1, 1)$. Then it carries out a depth-first-search as follows. When processing a valid sequence of length m, it considers in lexicographic order the possibilities for adding a new square above the existing ones. If a sequence of length $m + 1$ proves to be valid, it is recursively processed immediately.

Since every prefix of a valid sequence is valid, and since the number of valid sequences is finite (recall that $m \leq 55$), this search eventually discovers all valid sequences.

Using this depth-first-search strategy, we were able to generate all maximal valid sequences in three minutes running time on a SPARC 1. There were 2064 of these maximal valid sequences, all of which had length 7.

$\square$

4 Discs

After the results on rectangles and squares, we examine the situation for discs. We show that any K_n can be represented.

Theorem 7. *Any K_n can be represented by unit discs.*

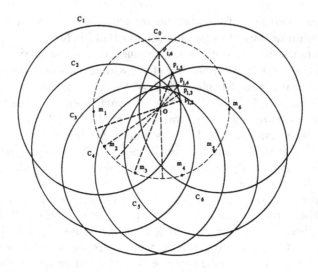

Fig. 2. Representing K_n by n unit discs.

Proof. See Figure 2. Let $\frac{1}{2} < r < 1$ and consider the circle c_0 of radius r centered at the origin O. Now pick n points $m_1 = (x_1, y_1, 0), \ldots, m_n = (x_n, y_n, 0)$ equally spaced and in counter-clockwise order on an arc of c_0 of length less than half the circumference of c_0. Without loss of generality, assume that all $y_i < 0$. Let $M_i = m_i + (0, 0, i)$. Let C_i be the unit disc parallel to the x, y-plane with center M_i, and let c_i be its projection onto the x, y-plane. We claim that the C_i form a representation for K_n.

It suffices to show that for each i, $1 \le i < n$, all the discs above C_i can see C_i, i.e., C_j can see C_i whenever $j > i$.

Consider the points p_{ij} obtained by selecting from the two intersection points of the boundaries of c_j and c_i the one that has positive y-coordinate. (Figure 2 illustrates this for $i = 1$.) Consider some fixed i and the set of indices $j > i$. Observe that p_{ij} lies on the perpendicular bisector of the line segment $[m_i, m_j]$. This bisector runs through 0. As j increases, so does the angle between the x-axis and $[0, p_{ij}]$. This means that for $j > i$, the p_{ij} appear in counter-clockwise order around the boundary of c_i. Therefore none of the points $P_j = p_j + (0, 0, j)$ on discs C_j, $i < j \le n$, are blocked from the view of the boundary of disc C_i. Obviously if $j = i + 1$, then C_i and C_j are mutually visible along lines of sight with x, y-projection in $c_i \cap c_{i+1}$. Now suppose $j > i + 1$. Because $c_{i+1} \cup c_{j-1}$ contains c_k for $i + 1 \le k \le j - 1$, C_i and C_j are visible along lines of sight with x, y-projection in $c_i \cap c_j \cap c'_{i+1} \cap c'_{j-1}$, where the prime denotes complement. For $j = i + 2$, this figure is bounded by three circular arcs, and for $j > i + 2$, by four circular arcs. $\quad\square$

Acknowledgement: We would like to thank Günter Rote for a significant simplification of the proof of Lemma 3.

References

1. H. Alt, M. Godau and S. Whitesides. Universal 3-dimensional visibility representations for graphs. See elsewhere in these proceedings.
2. P. Bose, H. Everett, S. Fekete, A.Lubiw, H. Meijer, K. Romanik, T. Shermer and S. Whitesides. On a visibility representation for graphs in three dimensions. *Proc. Graph Drawing '93*, Paris (Sèvres), 1993, pp. 38-39.
3. P. Bose, H. Everett, S. Fekete, A. Lubiw, H. Meijer, K. Romanik, T. Shermer and S. Whitesides. On a visibility representation for graphs in three dimensions. Snapshots of Computational and Discrete Geometry, v. 3, *eds.* D. Avis and P. Bose, McGill University School of Computer Science Technical Report SOCS-94.50, July 1994, pp. 2-25.
4. R. Cohen, P. Eades, T. Lin and F. Ruskey. Three-dimensional graph drawing. *Proc. Graph Drawing '94*, Princeton NJ, 1994, Lecture Notes in Computer Science LNCS #894, Springer-Verlag, 1995, pp. 1-11.
5. F. R. K. Chung. On unimodal subsequences. *J. Combinatorial Theory*, Series A, v. 29, 1980, pp. 267-279.
6. A. Dean and J. Hutchison. Rectangle visibility representations of bipartite graphs. *Proc. Graph Drawing '94*, Princeton NJ, 1994. Lecture Notes in Computer Science LNCS #894, Springer-Verlag, 1995, pp. 159-166.
7. J. M. Hammersley. A few seedlings of research. *Proc. 6th Berkeley Symp. Math. Stat. Prob.*, U. of California Press, 1972, pp. 345-394.
8. H. Koike. An application of three-dimensional visualization to object-oriented programming. *Proc. of Advanced Visual Interfaces AVI '92*, Rome, May 1992, v. 36 of World Scientific Series in Computer Science, 1992, pp. 180-192.
9. E. Kranakis, D. Krizanc and J. Urrutia. On the number of directions in visibility representations of graphs. *Proc. Graph Drawing '94*, Princeton NJ, 1994, Lecture Notes in Computer Science LNCS #894, Springer-Verlag, 1995, pp. 167-176.
10. J. Mackinley, G. Robertson and S. Card. Cone trees: animated 3d visualizations of hierarchical information. *Proc. of the SIGCHI Conf. on Human Factors in Computing*, 1991, pp. 189-194.
11. K. Romanik. Directed VR-representable graphs have unbounded dimension. *Proc. Graph Drawing '94*, Princeton, NJ, 1994, Lecture Notes in Computer Science LNCS #894, Springer-Verlag, 1995, pp. 177-181.
12. R. Tamassia and I. Tollis. A unified approach to visibility representations of planar graphs. *Discrete Comput. Geom.* v. 1, 1986, pp. 321-341.
13. S. Wismath. Characterizing bar line-of-sight graphs. *Proc. ACM Symp. on Computational Geometry*, 1985, pp. 147-152.

Generalized Fisheye Views of Graphs*

Arno Formella and Jörg Keller

Universität des Saarlandes, FB 14 Informatik
Postfach 151150, 66041 Saarbrücken, Germany

Abstract. Fisheye views of graphs are pictures of layouted graphs as seen through a fisheye lens. They allow to display, in one picture, a small part of the graph enlarged while the graph is shown completely. Thus they combine the features of a zoom—presenting details— and of an overview picture—showing global structure. In previous work the part of the graph to be enlarged—the focus region—was defined by a focus point. We generalize fisheye views such that the focus region can be defined by a simple polygon and show efficient algorithms to compute generalized fisheye views. We present experimental results on two applications where generalized fisheye views are advantageous: travel planning and ray tracing.

1 Introduction

Graphs are a common data structure in computer programs. Graph layout, i.e. the science to display a graph, has become important to visualize the underlying data sets and their relations. Focusing on one or several regions in a two dimensional layout is also important in other applications, e.g. CAD–systems. Interesting features in layouted graphs or generally in two dimensional images are both the global structure and the local structure in special regions. If the data set to be displayed is large, a picture showing the whole graph only allows to assess the global structure. Details can be obtained by zooming into some part of the graph, but global information is lost.

Fisheye views of graphs—the name is taken from the similarity to viewing a picture through a magnifying lens—try to combine both features. The node positions of the layouted graph are transformed such that a part of the layouted graph is displayed enlarged, but that the graph is still completely visible.

Fisheye views can be computed by defining a focus point and transforming node positions with respect to their euclidean distance from this focus point. The distance function can also be some other relation such as the length of the shortest path between the nodes. The different types are called graphical and logical fisheye views, respectively. A recent paper by Sarkar and Brown [3] gives a survey.

---

* This research was partly supported by DFG (German Science Foundation) under SFB 124 – TP D4. The first author is partly supported by Universidade de Vigo/Xunta de Galicia, the second author is partly supported by DFG through a habilitation fellowship.

Often however, one is not only interested in details in one part of the graph, but in several parts or even a whole region of the graph. The first demand can be accomplished by multiple focus points. The second demand can be accomplished by having one or several focus regions instead of a focus point. Misue and Sugiyama [2] used focus regions but restricted themselves to rectangles. Sarkar et. al. [4] allow only convex polygons. We will derive fisheye views based on a focus region from transformations based on a focus point. In contrast to Sarkar et. al. we allow arbitrary simple polygons, and our implementation is more efficient because we need not iterate. Moreover, the magnifying process is easy to reverse, so that editing in the distorted view is possible.

The remainder of this paper is organized as follows. In Sect. 2, we review graphical fisheye views as described in [3] and we derive general properties of the transformation function. In Sect. 3, we extend these transformations to focus regions formed by simple polygons, we describe our implementation and give performance results. In Sect. 4, we discuss the use of fisheye views on our applications, travel planning and ray tracing. Section 5 concludes the paper.

2 Computation of Fisheye Views

In a layouted graph, each node v is given a position $p(v)$ in the plane respective to an orthogonal coordinate system. Edges (v, w) are represented by straight lines from $p(v)$ to $p(w)$. All nodes and edges of the graph are positioned within the *frame*, an axis-parallel rectangle.

If we define a *focus point* f within the frame, then the position of each node v can be represented as

$$p(v) = f + \mathbf{a}_{v,f} \ .$$

A fisheye view of a graph moves each node v from $p(v)$ to $p'(v)$ by applying a transformation function g to the vector $\mathbf{a}_{v,f}$ such that v is moved "away" from f but does not leave the frame:

$$p'(v) = f + g(\mathbf{a}_{v,f}) \ .$$

In [3], two possibilities for g are presented: cartesian and polar transformation.

2.1 Cartesian Transformation

We consider a vertical line through f that partitions the plane into two half-planes. We define a horizontal ray starting in f such that $p(v)$ and the ray are in the same halfplane. Analogously, we consider a horizontal line through f and define a vertical ray starting in f. Let i_x and i_y be the intersections of these rays with the frame. Then $\mathbf{a}_{v,f}$ can be uniquely represented as

$$\mathbf{a}_{v,f} = \alpha_x \cdot (i_x - f) + \alpha_y \cdot (i_y - f) \ ,$$

where α_x and α_y are in $[0 : 1]$. An example is given in Fig. 1(a).

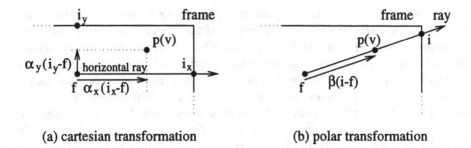

(a) cartesian transformation   (b) polar transformation

Fig. 1. Representation with respect to focus f

We define the *cartesian transformation function* g_c with the help of a bijective *distortion function* $h : [0 : 1] \rightarrow [0 : 1]$ as

$$g_c(\mathbf{a}_{v,f}) = h(\alpha_x) \cdot (i_x - f) + h(\alpha_y) \cdot (i_y - f) \ .$$

We adopt from [3] the function $h(x) = (d+1)x/(dx+1)$, where $d \geq 0$ controls the amount of movement. More general, a function h should satisfy the following requirements, also partly given in [3]:

$h(0) = 0$: The focus point itself is not moved.

$h(1) = 1$: Points on the frame are not moved.

$h(x) > x$ for all $x \in (0 : 1)$: All other points are moved away from f.

$h(x)$ **be strictly monotonous increasing**: Points cannot overtake during the transformation, i.e. points being closer to f before the transformation must be closer afterwards as well.

$(h(x) - x)/x$ **be strictly monotonous decreasing**: The closer a point is to f, the farther away should it be moved. The moving distance is taken relative to the point's distance to f before the transformation:

$h(x) - x$ is the absolute distance a point is moved during the transformation. By division through x, this is set in relation to the distance between the point and f before the transformation.

The above function satisfies these requirements. Another possibility would be $h_2(x) = \sin(\pi/2 \cdot x)$. It is obtained by considering fisheye views as projections of hemispheres onto planes.

2.2 Polar Transformation

Consider a ray through $p(v)$ that starts in f as shown in Fig. 1(b). Let i be the intersection of the ray and the frame. Then $\mathbf{a}_{v,f}$ can be uniquely represented as

$$\mathbf{a}_{v,f} = \beta \cdot (i - f) \ , \tag{1}$$

(a) cartesian transf.   (b) polar transf.

Fig. 2. Example graph after transformation

where $\beta \in [0 : 1]$. We define the *polar transformation function* g_p as

$$g_p(\mathbf{a}_{v,f}) = h(\beta) \cdot (i - f) .$$

For both transformations, one can prove that no point will disappear. For cartesian transformations, one can prove that horizontal and vertical edges keep their orientation. Figure 2 shows a graph forming a regular 20×20-grid. It is transformed with $d = 3$, the focus point is marked with a filled square. All further pictures showing transformations will use $d = 3$ to have comparable results. The influence of d has already been analyzed by others [3].

3 Extension to Focus Polygons

For many applications, a simple focus point is not satisfying; e.g. for travel planning with graphs that represent road maps, one might want to focus on the whole area around the chosen route. Thus instead of a focus point, one needs a *focus region* surrounded by a simple closed curve. We will restrict to using polygons that completely lie within the frame.

To define transformations based on polygons, we will first deal with nodes that are positioned within the polygon. To derive transformations for nodes outside the focus polygon, we will define focus points on the polygon and apply the techniques from the previous section. For both transformations to be developed, the case of a focus point is obtained if polygons consist of only one point.

3.1 Scaling Polygons

While in the previous section the area around the focus point was enlarged, here the focus region itself should be enlarged, but it should not be "deformed". Hence, we will scale the focus region. To do this, we have to define a center point c and a scaling factor γ. Then, any node v that is positioned within the polygon can be represented as

$$p(v) = c + \mathbf{a}_{v,c}$$

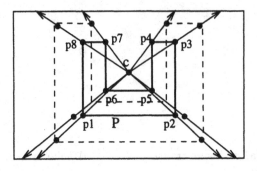

Fig. 3. Computation of the scaled polygon

and will be moved to

$$sc(p(v)) = c + \gamma \cdot \mathbf{a}_{v,c} \ .$$

For a polygon P with n corners $p_1, \ldots, p_n$, we define the center point c as the simple barycentric combination

$$c = \frac{1}{n} \cdot \sum_{i=1}^{n} p_i \ ,$$

For a convex polygon, c will lie inside the polygon.

The scaling factor must guarantee that the scaled polygon still lies within the frame, and that the amount of scaling is somehow related to the transformation function g. As in (1), we represent each corner p_i of the polygon by a parameter β_i with respect to the center point c. The scaling factor γ then is computed as

$$\gamma = \min \{ h(\beta_i)/\beta_i \ : \ 1 \le i \le n \text{ and } \beta_i \ne 0 \} \ .$$

The term $h(\beta_i)/\beta_i$ represents the factor by which corner p_i would be scaled from center c when moved by a polar transformation with focus point c. The minimum guarantees that the scaled corners and thus the scaled polygon completely lies within the frame. Figure 3 shows an example.

Scaling of the focus polygon P to $P' = sc(P)$ induces the following: we use P to determine parameters α_x and α_y or β for nodes v with position $p(v)$, and we use the scaled polygon P' to compute the new position $p'(v)$.

3.2 Cartesian Transformation

For each node v that is positioned outside P, we define two focus points $f_{v,x}$ and $f_{v,y}$, such that a line through $p(v)$ and $f_{v,x}$ ($f_{v,y}$) is horizontal (vertical).

We first assume that the horizontal line through $p(v) = (x_v, y_v)$ intersects P in $i_1 = (x_1, y_v), \ldots, i_t = (x_t, y_v)$. Then $f_{v,x}$ is chosen to be the intersection point (x_j, y_v) closest to $p(v)$. If P is convex, then $t = 2$ and $p(v)$ will be either

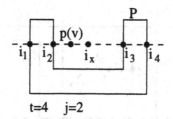

Fig. 4. Transformation with a non-convex polygon

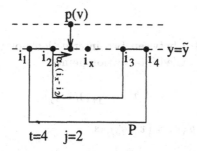

Fig. 5. Computation of α_x by a second horizontal line

left of i_1 or right of i_2. Then we can define i_x exactly as in Sect. 2.1 to be the intersection of a horizontal ray from $f_{v,x} = (x_j, y_v)$ and the frame, and

$$x_v = x_j + \alpha_x \cdot |i_x - f_{v,x}| \ .$$

However, if P is not convex, then t might be a multiple of two and v might be positioned between two intersection points i_j and i_{j+1}. An example is given in Fig. 4. We choose as focus point the intersection point that is closer to $p(v)$. This property must also hold after the transformation. Thus, we choose i_x to be the point on the horizontal line with equal distance to both i_j and i_{j+1}:

$$i_x = ((x_j + x_{j+1})/2, y_j) \ .$$

If the horizontal line through $p(v)$ does not intersect P, we choose the closest horizontal line $y = \tilde{y}$ that intersects P. We replace $p(v) = (x_v, y_v)$ by $(x_v, \tilde{y})$ and compute $f_{v,x}$, i_x and α_x for this point. The procedure is illustrated in Fig. 5. In exactly the same way, we compute $f_{v,y}$, i_y and α_y.

To compute $p'(v) = (x'_v, y'_v)$, we use the scaled polygon P'. We first scale $f_{v,x} = (x_j, y_j)$ to $f'_{v,x} = sc(f_{v,x}) = (x'_j, y'_j)$. If i_x was a point on the frame, i'_x is a point on the same edge of the frame, only the position on that edge is adjusted such that a line between i'_x and $f'_{v,x}$ is horizontal.

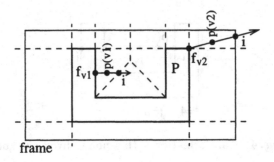

Fig. 6. Voronoi diagram of a simple polygon

If i_x lay in the center between two intersection points, then i'_x lies in the center between the two scaled intersection points:

$$i'_x = ((x'_j + x'_{j+1})/2, y'_j) \ .$$

Now, we can define $p'(v) = (x'_v, y'_v)$ as

$$x'_v = x'_j + h(\alpha_x) \cdot |i'_x - f'_{v,x}| \ ,$$
$$y'_v = y'_j + h(\alpha_y) \cdot |i'_y - f'_{v,y}| \ .$$

3.3 Polar Transformation

For each node v outside P, we define its focus point f_v to be the point on P with minimum distance to $p(v)$. This might either be a corner of P or the intersection point of one edge of P and the perpendicular line through $p(v)$. As the minimum distance property must also hold after the transformation, we have a problem in defining the intersection i. While we could use the center between two intersections with P in cartesian transformation, we need more notation here.

The *voronoi diagram* $VD(P)$ of a simple polygon P partitions the plane outside the polygon in regions such that all points in one region are closer to one edge or corner of the polygon than to all other edges and corners. We make all regions finite by intersecting $VD(P)$ with the frame. We denote by $VR(x, P)$ the curve that encloses the voronoi region (with respect to polygon P) that contains point x. Now we can represent $p(v)$ as

$$p(v) = f_v + \alpha \cdot (i - f_v) \ ,$$

where i is the intersection of the ray r from f_v through $p(v)$ and $VR(p(v), P)$. Figure 6 gives an example.

To compute $p'(v)$ we use once again the scaled polygon P'. We scale f_v to $f_v' = sc(f_v)$, and we compute i' to be the intersection of a ray from f_v', parallel to r, and $\text{VR}(f_v', P')$. Then we obtain

$$p'(v) = f_v' + h(\alpha) \cdot (i' - f_v') \; .$$

3.4 Implementation

We use scanlines to decide whether a point is located within the polygon or within a voronoi region, and to compute intersections with the polygon in cartesian transformations. To avoid recomputation of intersections between scanlines and polygons, we use a software cache to store scanlines and intersection points. We tested the usefulness of the software cache on a regular 30×30–grid and a graph with 900 randomly positioned nodes, using a rectangular polygon. For the grid, hit ratios are 98 % (cartesian) and 95 % (polar). For the random graph, hit ratios drop to 45 % and 0 %, respectively. This indicates that caching is mostly useful for regular graph layouts and for cartesian transformation.

We benchmarked our prototype implementation on a SPARCstation2. The user times to transform a 512×512–grid with respect to a rectangle are 30 s for cartesian and 57 s for polar transformation. The times with respect to a ring with 64 corners are 44 s and 206 s, respectively. Cartesian transformation is much faster due to caching effects and because computations are simpler (no intersections between arbitrarily oriented lines and polygons). Polar transformation time increases with the number of voronoi regions, i.e. polygon corners. This could be improved by a better search strategy to find the voronoi region of a point.

The overall performance is between 1.14 and 7.86 Milliseconds per transformed node. Hence, for graphs with a few hundred nodes, response times are still acceptable for interactive systems.

4 Applications

We illustrate the concept of focus regions on two applications, travel planning and ray tracing. Further applications where the concept of a dynamical viewport is of great value are CAD–Systems. Here, for instance, a designer needs to place wires in a certain region without loosing the global view of the long wires.

4.1 Travel Planning

Figure 8(a) shows the motorways in the southern part of Germany. Suppose we plan a travel from Nürnberg to Munich but are not sure whether to take the direct route or whether to visit Regensburg. With a single focus point, one can only concentrate on Nürnberg or on Munich (see Fig. 8(b) and (c)). However, the route between the towns is deformed. With the concept of a focus region, we can simply put a pentagon around the possible routes (see Fig. 8(d)). Cartesian and polar transformation with respect to this rectangle are shown in Fig. 8(e) and (f).

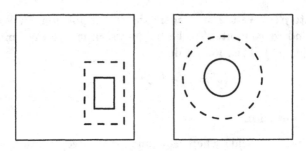

Fig. 7. Polygons to transform the example protein

The cartesian transformation in Fig. 8(e) seems better for us, because the surrounding motorways reflect more the original picture. This, however, may well be a matter of taste. Figures 8(b) and (c) show cartesian transformations for the same reason. In a complete travel planning environment, one would assume that each node and each edge has a certain priority. A node or edge is only visible if its priority supersedes a threshold that depends on the distance from the focus region. Then, in Fig. 8(e) and (f), smaller towns and roads along the main route would appear. The details of such features are described in [3].

4.2 Ray Tracing

Fisheye lens effects are also of interest in computer generated pictures. One of the main techniques to generate photorealistic pictures on a computer is ray tracing [1]. The basic idea is to have, in a 3–dimensional space, a camera position z, a view plane with pixel points, and a scene consisting of objects. From z, a ray is sent through each pixel. An intensity value for that pixel is obtained by computing intersections with objects and applying illumination models.

To have the impression of a fisheye lens effect on the scene, we simply apply a transformation with the inverse function h^{-1} to all pixel points on the view plane before calculating the primary rays through the pixel points. As h is bijective, this is always possible.

We incorporated our transformation code into a ray tracing program. Figure 9(a) shows the protein pdb2sni from the PDB[2] data base to illustrate the effect of the distortion on curved surfaces. We use the two polygons shown in Fig. 7. The circle is a ring with 64 corners. Figures 9(b) and (c) show the cartesian and polar transformations of the protein with respect to the rectangle. Figures 9(d) and (e) show the transformations with the ring.

An advantage of the polar transformation are fewer deformations of spheres close to the focus region. A disadvantage, however, are the deformations of spheres near the corner of the picture. Here, spheres get a corner. The reason for this is the rectangular form of the frame.

---

[2] Brookhaven Protein Data Bank

The concept of fisheye lenses introduces a new kind of special effect which can be used for zooming–in, zooming–out, image distortion, image morphism, etc., especially when animations are produced by an artist.

5   Conclusion

We have shown how to extend fisheye views based on a single focus point to views based on simple polygons. We have presented efficient algorithms and data structures to implement these views, and we have given performance results of a prototype implementation. We have tested our concepts on two applications: travel planning with computer generated road maps, and ray tracing. By allowing concave polygons, we are also able to handle multiple focus regions.

The prototype can still be extended in several ways. For maps, it might be useful to implement other features of fisheye views, e.g. presenting additional information like town names and population depending on the distance from the focus region. This extension is fairly simple given the representation with parameters (α_x, α_y) or β [3]. Another possible extension is hierarchical application, i.e. a lens within a lens.

In ray tracing, it might be helpful to use an animated fisheye lens in animated films. This allows to inspect three–dimensional close ups without loosing the overview. Further, a dynamic viewport might be useful in many two–dimensional display tools.

It is possible to extend cartesian transformation to arbitrary simple closed curves. Only the computation of intersection points between the curve and scan-lines has to be changed. Scaling of the curve can be realized by approximation through a large number of points on the curve and computation of center and scaling factor with respect to these points.

Extension of polar transformation to arbitrary curves is more difficult because when given a point p, one has to find the point of the curve closest to p.

Acknowledgements

We would like to thank H.-P. Lenhof for providing the protein description.

References

1. Glassner, A. S. (Ed.): An Introduction to Ray Tracing (Academic Press, 1989)
2. Misue, K., Sugiyama, K.: Multi-viewpoint perspective display methods: formulation and application to compound graphs. Human Aspects in Computing, Proc. 4th Int.l Conf. on Human-Computer Interaction (Elsevier, 1991) 834–838
3. Sarkar, M., Brown, M. H.: Graphical fisheye views. Comm. ACM **37**(12) (1994) 73–84
4. Sarkar, M., Snibble, S. S., Tversky, O. J., Reiss, S. P.: Stretching the rubber sheet: a metaphor for viewing large layouts on small screens. Proc. Symp. User Interface Software and Technology (ACM, 1993) 81–91

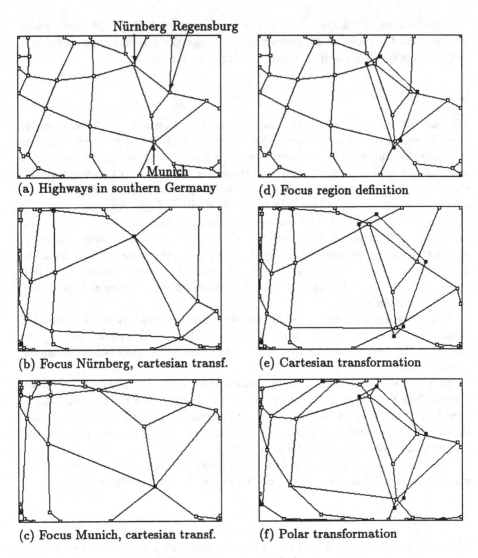

(a) Highways in southern Germany

(b) Focus Nürnberg, cartesian transf.

(c) Focus Munich, cartesian transf.

(d) Focus region definition

(e) Cartesian transformation

(f) Polar transformation

Fig. 8. Travel planning with focus region

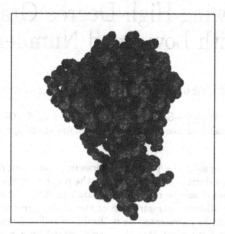

(a) Protein

(b) Cartesian transf. with rectangle

(d) Cartesian transf. with ring

(c) Polar transf. with rectangle

(e) Polar transf. with ring

Fig. 9. Protein visualization

Drawing High Degree Graphs
with Low Bend Numbers

Ulrich Fößmeier   Michael Kaufmann

Universität Tübingen, Wilhelm-Schickard-Institut, Sand 13, 72076 Tübingen, Germany,
email: foessmei / mk@informatik.uni-tuebingen.de

Abstract. We consider the problem of drawing plane graphs with an arbitrarily high vertex degree orthogonally into the plane such that the number of bends on the edges should be minimized. It has been known how to achieve the bend minimum without any restriction of the size of the vertices. Naturally, the vertices should be represented by uniformly small squares. In addition we might require that each face should be represented by a non-empty region. This would allow a labeling of the faces. We present an efficient algorithm which provably achieves the bend minimum following these constraints. Omitting the latter requirement we conjecture that the problem becomes NP-hard. For that case we give advices for good approximations. We demonstrate the effectiveness of our approaches giving some interesting examples.

1 Introduction

Embedding planar graphs into a grid while optimizing quality parameters like area, edge lengths or bend number is not only a challenging combinatorial problem, but viewed as a graph drawing problem, it has many direct practical applications.

A drawing of a graph in the plane roughly consists of a representation of the vertices by geometric objects (circles, squares, rectangles, lines, ...) and an assignment of the vertices to geometric positions in the plane. The edges are represented by curves between the objects representing two vertices; in orthogonal representations they form a continous sequence of horizontal and vertical line segments embedded in the plane. Intersections of a horizontal and a vertical line segment belonging to the same curve are called *bends*. The goal in graph drawing is the production of *nice* drawings: The representation of the graph should be simple, the incidence structure should be easily readable, planar graphs should be drawn planar, the vertices should be distributed nicely in the drawing. There are many other requirements, which might be even contradicting.

Orthogonal drawings of planar graphs have been extensively studied in the last decade. There are plenty of applications like VLSI-design [10], entity relationship and data flow diagrams in Software Engineering (e.g. [5, 12, 2, 14, 1]) or visualization of interactions inside of molecular structures in Astrophysics [3]. In some of them multiple edges are necessary, so we have to handle multigraphs. Many results and algorithms have been worked out [4], one of the most important is Tamassia's algorithm from [16]: How to draw a 4-planar graph orthogonally with the minimum number of bends preserving a given planar representation. (A 4-planar graph has a vertex degree of at most 4. A planar representation is given by a fixed cyclic order of the incident edges of each vertex.) We call a planar graph together with a planar representation a *plane* graph.

It makes sense to preserve the planar representation, since very often a (non orthogonal) drawing is given which should have a certain familiarity with the orthogonal layout. Moreover an efficient algorithm for the problem without this restriction cannot be expected since finding a bend minimum solution over all planar representations is NP-hard [8].

In most of the applications it is necessary to visualize planar graphs with a vertex degree of more than four; this aspect cannot naturally be captured in a usual orthogonal drawing, so we have to look for extensions of Tamassia's approach.

Approaches like to proceed to k-gonal grids if the maximum degree is k or to split each vertex of large degree into cluster of many subvertices of degree 3 or 4 turn out to be useless or to be at most good heuristics which may terribly fail in some cases. Our target is a provably good quality of the drawing under the commonly used standards.

Our criterion of optimum is the number of bends. Note that there is always a representation without any bends if we demand no other qualities of the drawing: A *visibility representation* e.g. [17, 13] satisfies all requirements established so far. But in such drawings the size of the vertices may grow independently of the vertex degree: Fig. 1a) shows an example where the size of a vertex grows arbitrarily whereas the degree of the vertex v is a small constant (in the example: 5).

Drawings like Fig. 1a) do not fulfill our wishes in many applications. The size of the vertices should be determined by the degree and not by the structure of the graph. More concrete our model is as follows: We use a grid with uniform distance λ between the grid lines. The size of the vertices should be smaller than λ, and their centers should be placed on the intersections of the grid lines; this ensures that no vertex is intersected by any grid line except of those defining its position and consequently two vertices can never intersect. Let max be the maximal number of edges being incident to a single side of any vertex. We require every vertex to be a square with side length $(2 \cdot max - 1)$. The motivation for the factor 2 is given in section 4. We choose λ to be twice the side length of the vertices. This leaves enough space to route the edges.

With these requirements we ensure a clear and well-arranged look of the drawing. Note that for each vertex v, no two different adjacent vertices can be connected by straight lines emanating from the same side of v.

Segments of edges being incident to the same side of some vertex may run very close together (i.e. the distance between two of them is much smaller than λ) and so it is possible to draw a face in such a way that every point inside of the face has a tiny distance to some bounding edge of the face (see face C in Fig. 1b). We call such faces *empty faces*.

We consider two models: In the first model we allow empty faces and call the corresponding drawings 'Planar Orthogonal Drawings with Equal Vertex Size' (*podevs*). An example can be seen in Fig. 1b. Trying to meet the general podevs-requirements, we noticed that the possible representation of the faces by empty regions (face C in Fig. 1b) does not fit in our efficient approach. We conjecture that in general the bend-minimization problem for podevs is NP-hard. Therefore in our second model we additionally require that at least the non-trivial faces (faces with at least three adjacent vertices) should have a non-empty region. An optimal drawing with this constraint is given in Fig. 1c). We call such drawings 'Planar Orthogonal Drawings with Equal Vertex Size and Non-Empty Faces' (*podevsnef*). Note that podevsnef-drawings always allow a (reasonable) labeling of the faces whereas this is impossible in the case of faces which are represented by an empty region.

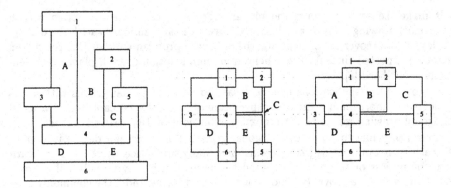

Fig. 1.

a) A visibility representation

b) An optimal solution (podevs)

c) An optimal solution allowing labeling of the faces (podevsnef)

The rest of this paper is organized as follows: In section 2 we perform the first modifications on Tamassia's algorithm and receive some kind of two-dimensional visibility representation. In section 3 the main algorithm (for drawings like in Fig. 1c)) is given. We incorporate the requirements of small uniform square sizes for each vertex and of non-zero area for the faces. In section 4 we shortly describe the postprocessing transformation from the topological to the geometric layout, which is called compaction. Some remarks on further improvements (heuristics for empty faces, compact visibility representations using the approach in section 2) conclude the paper. Many proofs and details are omitted and can be found in [7].

2   Nearly Orthogonal Representations

The basis of our data structure is the *orthogonal representation* defined in [16]: Given a graph $G = (V, E)$ and a planar representation for it; an orthogonal representation for G and its planar representation is a set of lists $H(f)$, one for every face f of G, whose elements are triples $((u, v), s, a)$, where

— (u, v) is an edge of G,

— s is a binary string, and

— a is an integer in the set $\{90, 180, 270, 360\}$.

If f is an internal face the edges in $H(f)$ appear in clockwise order and counterclockwise otherwise. The string $s[(u, v)]$ describes the bends of the edge (u, v): The kth bit of $s[(u, v)]$ represents the kth bend that appears at the right side of (u, v), as it is encountered when going along (u, v). The binary symbols 0 and 1 represent angles of 90 and 270 degrees, respectively. $a[(u, v)]$ specifies the angle formed in face f at the vertex v by the edge (u, v) and the following edge in $H(f)$.

For our purposes we have to modify this concept and use *nearly orthogonal representations*. We assign $0°$-values to angles between two parallel edges being incident to the same vertex at its same, because in this case the representation remains consistent such that the sum of the interior angles of a polygon with k edges is equal to $(k - 2) \cdot 180°$. So in a nearly orthogonal representation $a[(u, v)]$ is a number in the set $\{0, 90, 180, 270, 360\}$.

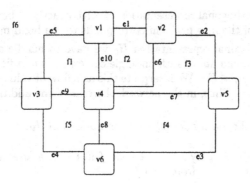

Fig. 2. A Nearly Orthogonal Representation

Fig. 2 shows a drawing for the graph of Fig. 1 with edge identifiers. The nearly orthogonal representation of the example in Fig. 2 is:

$f_1 = ((e_9, \varepsilon, 90), (e_5, 0, 90), (e_{10}, \varepsilon, 90));$
$f_2 = ((e_{10}, \varepsilon, 90), (e_1, \varepsilon, 90), (e_6, 0, 90));$
$f_3 = ((e_6, 1, 90), (e_2, 0, 90), (e_7, \varepsilon, 0));$
$f_4 = ((e_7, \varepsilon, 90), (e_3, 0, 90), (e_8, \varepsilon, 90));$
$f_5 = ((e_4, 0, 90), (e_9, \varepsilon, 90), (e_8, \varepsilon, 90));$
$f_6 = ((e_1, \varepsilon, 180), (e_5, 1, 180), (e_4, 1, 180), (e_3, 1, 180), (e_2, 1, 180)).$
Note the angle of $0°$ in the face f_3.

In [16] Tamassia describes a 1:1-correspondence between an orthogonal representation H of a graph G and a flow in some network N_H defined as follows: $N_H = (U, A, s, t, b, c)$ where
$b : A \to \mathbb{R}^+$ is a nonnegative capacity function,
$c : A \to \mathbb{R}$ is a cost function,
U (the *nodes* of the network) $= \{s\} \cup \{t\} \cup U_V \cup U_F$, where s and t are the *source* and the *sink* of the network, U_V contains a node for every vertex of G and U_F contains a node for every face of G,
A (the *arcs* of the network) contains

a) arcs from s to nodes v in U_V with cost 0 and capacity $4 - deg(v)$;
b) arcs from s to nodes f in U_F, where f represents an internal face of G with $deg(f) \le 3$; these arcs have cost 0 and capacity $4 - deg(f)$; $deg(f)$ for a face f always denotes the number of edges in the list $H(f)$;
c) arcs from nodes f in U_F representing the external face or representing internal faces f with $deg(f) \ge 5$ to t; these arcs have cost 0 and capacity $deg(f) - 4$ if f is an internal face and capacity $deg(f) + 4$ for the external face;
d) arcs of cost 0 and capacity ∞ from nodes v in U_V to nodes f in U_F, if v is incident to an edge of $H(f)$;
e) arcs of cost 1 and capacity ∞ from a node f in U_F to a node g in U_F, whenever the faces f and g of G have at least one common edge.

Every flow unit on an arc between two faces stands for a bend on an edge between these faces. The flow on the arcs in d) defines the angles of H: If $x_{v,f}$ is the flow from the node $v \in U_V$ to the node $f \in U_F$ then the angle at vertex v in face f is $(x_{v,f} + 1) \cdot 90°$. Every feasible flow of value $\Sigma_u b(s, u) = \Sigma_w b(w, t)$ with cost B

corresponds to an orthogonal representation H with exactly B bends. Thus the cost minimum solution of the flow problem corresponds to the bend minimum drawing.

In a nearly orthogonal representation H_0 we have to handle angles of degree 0. According to the formula above such an angle corresponds to a flow of value -1 from some $v \in U_V$ to some $f \in U_F$. We interpret this as a flow of value +1 in the opposite direction, from f to v. Thus, in the network there are some additional arcs:

f) arcs of cost 0 and capacity $deg(v) - 4$ from nodes v in U_V to t, if $deg(v) \geq 5$; and

g) arcs of cost 0 and capacity 1 from a node f in U_F to a node v in U_V, whenever there is an arc of type d) from v to f.

Fig. 3 shows the network and the flow for the drawing of Fig. 2; only arcs with nonzero flow are drawn. Note the arcs from face f_3 to vertex v_4 and from vertex v_4 to t.

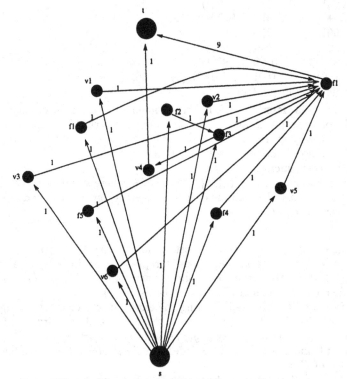

Fig. 3. Network and Flow for the drawing of Fig. 2

The flow in Fig. 3 is not optimal: Since arcs of type d) as well as arcs of type g) have cost zero, it is always possible to establish a zero cost flow in the network. This flow leads to a drawing resembling a visibility representation drawing without any bends, but with very large vertices (cf. Fig. 1a).

We discuss the quality of such drawings and possible extensions in section 5. In the next section we modify the network in order to get drawings with vertices that are not much larger than necessary.

3 A Bend-Minimizing Algorithm

We shortly recall the requirements for the model given in Section 1. We want to generate drawings like the one in Fig. 1b or 1c respectively, that means: All vertices have square shape and the same size and the centers of the squares lie on grid points of a grid where the unit distance is twice the length of a square side.

We already mentioned that at most one edge being incident to a certain side of some vertex can be a straight line. The only exception are a group of consecutive multiple edges between two vertices v and w, i.e. multiple edges arising in consecutive order in the adjacency lists of v and w, thus defining faces of degree two. We call this (these) edge(s) without bend the *middle edge(s)*.

Moreover, all edges on the same side of v at the left of the middle edge (counterclockwise) are required to bend to the left and the edges at the right of the middle edge (clockwise) bend to the right. We call these bends *vertex bends*, because they have nothing to do with the topological structure of the graph, but they are necessary only because of the vertex degree. For illustration see Fig. 4a).

For the next considerations we assume that there is only one middle edge in each direction. We consider two models:
a) The general *podevs*   and
b) the *podevsnef*, where we demand every face to have a non-zero area.

At first we discuss the difference between the two models; with *k-face* we denote a face f with $\deg(f) = k$.

Lemma 1 *[7] Every k-face with $k \geq 4$ can be represented by a non-empty region.*

So the significant difference between our models is to allow or to forbid triangles with zero area. Our algorithm for a podevsnef is based on the following

Lemma 2 *[7] Every $0°$-angle of a podevsnef has a unique corresponding $270°$-bend.*

Using these observations we describe an algorithm to compute a podevsnef: A flow on an arc of type g) from $f \in U_F$ to $v \in U_V$ is allowed if and only if there is a flow on an arc of type e) from some $g \in U_F$ to f where g is one of the two faces on the other sides of the edges e_1 and e_2 which define the face f at the vertex v (this expresses the correspondence between the $0°$-angle and the $270°$-bend). We model this situation by replacing the arcs of type g) by arcs of type h) going directly from face g to node v: For every edge e being incident with vertex v and neighboring faces f and g. there are arcs of type h) from f to v and from g to v. Arc (g, v) stands for the combination of arc (g, f) of type e) and arc (f, v) of type g). Thus type h) arcs have capacity 1 and cost 1. Note that there are two arcs of type h) from a face f to a node v (v belonging to f): They correspond to the two type e) arcs belonging to the two edges of f incident to v.

After having solved the Min-Cost-Flow-problem we re-insert the arcs (f, v) and (g, f) instead of (g, v) (with the corresponding flow values) and compute the nearly orthogonal representation as described in section 2.

Not every feasible flow in the network described so far has a corresponding drawing: Let e_i and e_j be two consecutive edges in the adjacency list of some vertex v and f_i, f_j and f_k the resulting faces such that f_j lies between e_i and e_j (cf. Fig. 4b)).

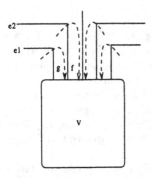

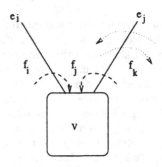

Fig. 4. a) Vertex Bends

b) Forbidden Combinations

We have to avoid two cases:

(a) Since there are no negative angles, the flow from a face into a node must be restricted to 1: it is not allowed that the flows on the arcs of type h) from f_i to v over e_i and from f_k to v over edge e_j are simultaneously equal to 1 (dashed lines in Fig. 4b)).

(b) Further, since all edges at the left of the middle edge(s) must have their vertex bend to the left and all edges at the right of the middle edge(s) must have their vertex bend to the right, it is forbidden to have a flow of value 1 from f_i to v over edge e_i and simultaneously a flow of value 1 from f_j to v over the same edge e_i (see dotted lines in Fig. 4b)).

Although it is easy to avoid any of the forbidden cases by usual means of flow problems, we cannot guarantee both conditions simultaneously. So we realize one of the conditions (we choose condition (a)) using the capacity restrictions of the flow problem and the other one by punishing it with extraordenarily high cost; thus we establish the arcs of type h) not directly, but use the following construction: Let v be a node in U_V and $f_{i_1}, \dots, f_{i_k}$ an ordered list of the faces around the vertex v in the graph (e.g. in clockwise order); let $e_{i_1}, \dots, e_{i_k}$ be the edges that separate these faces such that e_{i_j} separates face $f_{i_{j-1} \bmod k}$ and face f_{i_j}. See Fig. 5 for illustration (for $k = 3$).

Then we add for every edge e_{i_j} being incident to v two nodes $H^l_{e_{i_j}}$ and $H^r_{e_{i_j}}$, where $H^l_{e_{i_j}}$ ($H^r_{e_{i_j}}$) corresponds to the arc of type h) crossing edge e_{i_j} in clockwise (counterclockwise) order around v; further $H_{f_{i_j}}$ are new nodes in the network for every face f_{i_j}.

New arcs are:

– Arcs with capacity 1 and cost $2c + 1$ (c having a suitable value) from f_{i_j} to $H^r_{e_{i_j}}$ and to $H^l_{e_{i_{j+1} \bmod k}}$, i.e. to the edges of the graph corresponding to the arcs of type h) starting in face f_{i_j} and crossing a bounding edge of this face.

- Arcs with capacity 1 and cost 0 from the nodes $H_{f_{i_j}}$ to node v; the arcs of these two types replace the arcs of type h).

- Arcs with capacity 1 and cost 0 from node $H_{e_{i_j}}^l$ to node $H_{f_{i_j}}$ and from node $H_{e_{i_{j+1} \bmod k}}^r$ to node $H_{f_{i_j}}$, i.e. from two auxiliary nodes for two neighboured edges to the auxiliary node for the face between them. These arcs guarantee condition (a).

- Arcs with capacity 1 and cost $-c$ from node $H_{e_{i_j}}^l$ to node $H_{e_{i_j}}^r$ and vice versa. Every pair of such arcs defines a cycle of cost $-2c$ and thus a cost minimum path from a node f_{i_j} to a node v has cost $2c+1-2c=1$ corresponding to one necessary bend as in the case at the arcs of type h). Every time a second flow unit passes one of the nodes $H_{e_{i_j}}^l$ or $H_{e_{i_j}}^r$, the arcs with negative cost are already satisfied and thus the path from f_{i_j} to v has cost $2c+1$.

Fig. 5 shows the construction of this part of the network. All capacities are 1 and all costs not indicated are 0.

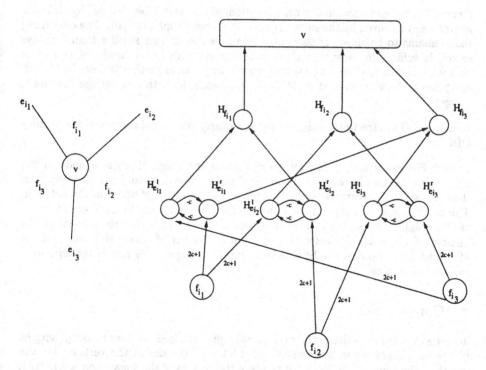

Fig. 5. The construction for the arcs of type h)

Now we can formulate the algorithm that computes the nearly orthogonal representation:

Algorithm

(1) Establish the network as described above;
(2) Solve the Min-Cost-Flow problem;
(3) Re-insert the arcs of type h) instead of the auxiliary construction and replace them by arcs of type e) and g);
(4) Compute the bends and angles of the drawing using the value of the flow on arcs of type d), e) and g).

Note that we have to solve a Min-Cost-Flow problem in a network with negative cycles. But that does not cause any troubles here, since all these cycles are known and have length 2; so we can use a standard augmentation algorithm where the cost-minimum paths are determined by (a slightly modified version of) Dijkstra's algorithm.

Unfortunately we do not have a 1:1-correspondence between a feasible max-flow in the network and a feasible drawing; but if we choose c large enough, we can state the following

Lemma 3 *For every feasible max-flow in the network with cost $b < c$ there is a corresponding drawing with exactly b bends.*

Proof: Taking into account the considerations about vertex bends (see Fig. 4a)) the lemma can be proved by the same arguments as those applied in [16]. The only thing that remains to be shown is that a feasible max-flow of cost smaller than c always exists. It suffices to show that there is a drawing with $b < c$ bends, since in this case it is easy to construct a flow with cost b. In [7] an algorithm is described which computes a drawing with at most $2m \leq 6n$ bends, so with $c = 6n$ the Lemma is correct. ◊

Lemma 4 *The algorithm above computes a nearly orthogonal representation in time $O(n^2 \log n)$.*

Proof: For every vertex and for every face of the graph there is a node in the network; further we have auxiliary nodes: Two for every edge and one for every face; thus planarity of the graph implies linearity of the number of nodes in the network. The number of arcs of type a), b), c) and f) is proportional to the number of vertices of the graph, the number of arcs of type d), e), g) and h) is proportional to the number of edges. So the network has a linear number of nodes and arcs and the Min-Cost-Flow problem can be solved in time $O(n^2 \log n)$. The rest of the algorithm runs in linear time. ◊

4 Compaction

To get a drawing from the nearly orthogonal representation, we have to assign lengths to the edge segments in a consistent way; let s be the size of the vertices, i.e. the length of the square side. We want to place the centers of the squares on a grid with unit distance $\lambda = 2s$; then the distance between two vertices is at least as large as the vertex itself. So we have to guarantee that the edge lengths are concurring with this constraint, in particular that the length of every straight edge is a multiple of s. Tamassia [16] describes a linear time algorithm that solves the corresponding problem for a vertex degree of at most 4, and we want to use a similar technique. We replace every vertex v by a square of $8s$ small vertices (see Fig. 6), where the small vertex in the middle of each side (marked with an m in Fig. 6) will be incident to the (or one of the) middle edge(s) in this direction; there are enough small vertices for the rest of the edges being incident to v.

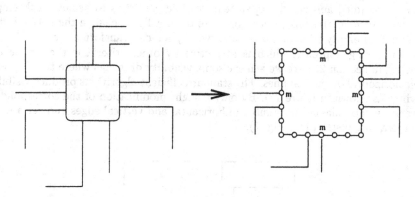

Fig. 6. Replacing a Large Vertex by Small Vertices.

Note that a square of $4s$ small vertices would not be enough: Let e.g. s be equal to 3 and $e = (u, v)$ be a vertical middle edge such that e is the rightmost edge at the bottom side of u and the leftmost edge at the top side of v. Then it would be impossible to draw e without any bend, if the centers of the two squares representing u and v should have the same x-coordinate.

So the size of the vertices is $2s \times 2s$ instead of a possible size of $s \times s$; but in the latter case we would have drawings with more bends, because we could not guarantee that we can draw middle edges without bends.

For the resulting graph (after the construction described here) we run a variant of the compaction algorithm of [16] to compute coordinates for the vertices and thus the final drawing.

Summarizing the results of the sections 3 and 4 we formulate

Theorem 1 *Our algorithm together with the compaction algorithm computes a pode-vsnef with the minimum number of bends in $O(n^2 \log n)$ time.*

5   Concluding Remarks and Discussion

This work has been motivated by the two figures at the end of the paper taken from the doctoral dissertation of Petra Mutzel [11]. She has convinced us to work on this specific model: On the left side the original picture from the paper on Astrophysics, on the right side the hand-made layout after the maximum planarization step. The third picture shows a bend-minimum podevsnef produced by an implementation of our Algorithm using GraphEd [9] (the non-planar edge between the vertices HCO^+ and CH was deleted by hand since the algorithm can only handle planar graphs). Another interesting example drawn as podevsnef and shown at the end is the planar graph from the competition in last year's GD-conference.

Our algorithm works well for podevs restricted to have non-empty faces. Trying to omit this restriction we saw that the case of empty triangles is the only problematic case. We are currently working on an NP-hardness proof for the general problem. Running several examples we get the impression that a preference of those (computably hard) configurations lead to clear and understandable drawings, since the edges are bundled and clearly separated. Possible heuristics to achieve such empty faces are to eventually flip corners, or more drastically to change the network flow problem if an empty triangle can be achieved by a local transformation.

A promising approach which is attractive by other criteria is the simple one described in section 2. Here we achieve some visibility drawings where the edges are either horizontal or vertical lines. The standard efficient algorithms produce visibility drawings as shown in figure 1a). By only a slight modification of the corresponding network we can balance the number of horizontal and vertical edges and get a much better drawing instead (see Fig. 7).

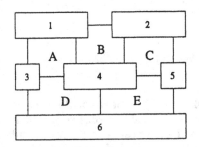

Fig. 7. A more attractive 'visibility representation'

Aspects for further research are to improve the used area by local transformations maintaining the minimum bend number and to give bounds for the used area. Another is to allow different vertex sizes not depending on the graph structure, but depending on the size of some text labels to be written inside.

Acknowledgement: We wish to thank Harald Lauer, Petra Mutzel and Roberto Tamassia for helpful discussions and comments on earlier versions of this paper.

References

1. Batini, C., E. Nardelli, and R. Tamassia, *A Layout Algorithm for Data-Flow Diagrams*, IEEE Trans. on Software Engineering, Vol. SE-12 (4), pp. 538-546, 1986.
2. Batini, C., M. Talamo, and R. Tamassia, *Computer Aided Layout Of Entity-Relationship Diagrams*, The Journal of Systems and Software, Vol. 4, pp. 163-173, 1984.

265

3. H.K.B. Beck, H.-P. Galil, R. Henkel, and E. Sedlmayr: *Chemistry in circumstellar shells, I. Chromospheric radiation fields and dust formation in optimcally thin shells of M-giants*, Astron. Astrophys. 265 (1992) 626–642.

4. Di Battista G., P. Eades, R. Tamassia and I.G. Tollis, *Algorithms for Automatic Graph Drawing: An Annotated Bibliography*, Tech.Rep., Dept.of Comp.Sc., Brown Univ., 1993.

5. Di Battista, E. Pietrosanti, R. Tamassia and I.G. Tollis, *Automatic Layout of PERT Diagrams with XPERT*, Proc. IEEE Workshop on Visual Lang. (VL'89), 171-176, 1989.

6. Di Battista, G., L. Vismara, *Angles of Planar Triangular Graphs*, Proc. of the 25th ACM Symposium on the Theory of Computing, San Diego, California, 1993.

7. Fößmeier, U., and M. Kaufmann, *Drawing High Degree Graphs with Low Bend Numbers*, Technical Report WSI-95-21, Univ. Tübingen 1995.

8. Garg, A. and R. Tamassia, *On the Computational Complexity of Upward and Rectilinear Planarity Testing*, Proc. of GD '94, Princeton, 1994.

9. Himsolt, M., *Konzeption und Implementierung von Grapheneditoren*, Doctoral Dissertation, Passau 1993.

10. Lengauer, Th., *Combinatorial Algorithms for Integrated Circuit Layout*, Teubner/Wiley & Sons, Stuttgart/Chichester, 1990.

11. Mutzel, P., *The Maximum Planar Subgraph Problem*, Doctoral Dissertation, Köln 1994.

12. Reiner, D., et al., *A Database Designer's Workbench* in Entity-Relationship Approach, ed. S. Spaccapietra, pp. 347-360, North-Holland, 1987.

13. Rosenstiehl, P., and R.E. Tarjan, Rectilinear planar layouts and bipolar orientations of planar graphs, *Discrete and Comp. Geometry* 1 (1986), pp. 343–353.

14. Protsko, L.B., P.G. Sorenson, J.P. Tremblay, and D.A. Schaefer, *Towards the Automatic Generation of Software Diagrams*, IEEE Trans. on Software Engineering, Vol. SE-17 (1), pp. 10-21, 1991.

15. Storer, J.A., *The node cost measure for embedding graphs in the planar grid*, Proc. 12th ACM Symposium on the Theory of Computing, 1980, pp. 201-210.

16. Tamassia, R., *On Embedding a Graph in the Grid with the Minimum Number of Bends*, SIAM Journal of Computing, vol. 16, no. 3, 421 - 444, 1987.

17. Tamassia, R., and I.G. Tollis, *A unified approach to visibility representations of planar graphs*, Discr. and Comp. Geometry 1 (1986), pp. 321–341.

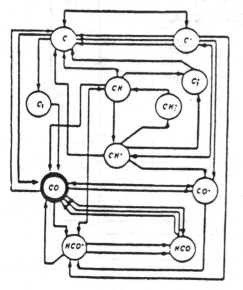
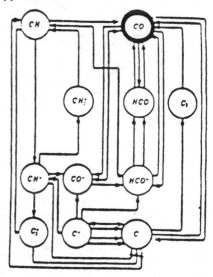

The layouts for the astrophysics-example from [11]

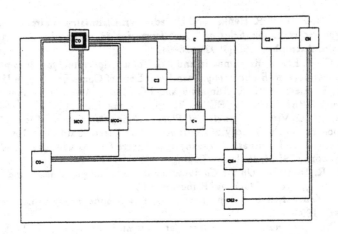

The podevsnef for the astrophysics-example

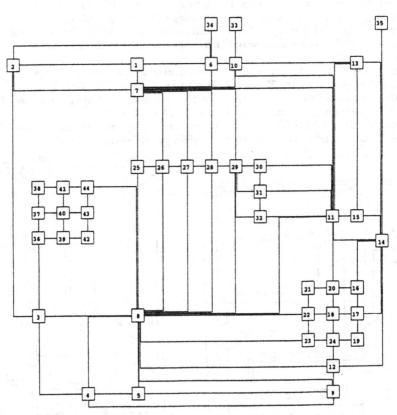

The podevsnef for the planar competition graph from GD'94

The Drawing of Configurations

Harald Gropp

Mühlingstr.19, D-69121 Heidelberg, Germany

Abstract. The drawing of configurations and other linear hypergraphs is discussed. From their historical and geometrical context it is quite natural to denote hyperedges of vertices as lines, i.e. to position the points (vertices) in the plane such that those points which form hyperedges are collinear in the plane (or as close to collinear as possible). This is a new concept in the area of hypergraph drawing. However, in mathematics it has been used for more than 100 years. The exact drawing of configurations is mainly based on the realization of matroids and techniques in computer algebra.

1 Introduction and Notation

1.1 Graphs and Geometry

The topic of graph drawing is interesting not only from the point of view of finding nice drawings of graphs for all kinds of applications. It is, of course, the fascinating meeting of two mathematical theories, geometry and graph theory.

Geometry is one of the oldest mathematical theories which reached a high level already more than 2000 years ago. It was and perhaps is still dominated by the axiomatic way of Euclid. The appearance of noneuclidean geometries in the 19th century extended the meaning of geometry within mathematics. However, in graph drawing the main task is still to display graphs in 2-dimensional (or 3-dimensional) geometry.

The development of graph theory started from many different applications in a rather different way. Many important roots can be found in sciences like physics and chemistry in the 19th century whereas graph theory was established as a theory of its own around 1960.

This paper being in the section *Graphs defined by geometry* discusses the drawing of hypergraphs in general and configurations in particular. Whether these combinatorial structures are regarded to be defined by geometry or whether they define new kinds of geometry (finite and noneuclidean) depends on the point of view.

These "geometric" structures contain only a finite number of points (and hence lines). It may even occur that there is no line connecting two points which clearly contradicts to the axioms of Euclid. This makes it possible that many new geometric structures arise which are "counterexamples" to classical or more modern theorems of euclidean geometry (Pappus or Sylvester-Gallai).

The problem of drawing linear hypergraphs is to obtain (of course) nice drawings of those structures which exist in euclidean geometry and to produce reasonable drawings of the "new" structures although they have "noneuclidean" properties.

1.2 Some Definitions

Definition 1.1 *A hypergraph $H = (V, E)$ consists of a set of vertices V and a set of hyperedges H where each hyperedge is a non-empty subset of V. A hypergraph is called k-uniform if the size of each hyperedge is k, r-regular if each vertex belongs to exactly r hyperedges, and linear if the intersection of two different hyperedges is at most one vertex.*

For further information on hypergraphs, see e.g. [1].

Definition 1.2 *A configuration (v_r, b_k) is a finite incidence structure with v points and b lines such that*
(1) there are k points on each line and r lines through each point, and
(2) two different points are connected by a line at most once.

A symmetric configuration (v_k, v_k) is shortly denoted by v_k.
From the point of view of hypergraph theory it is easy to describe configurations as follows.

Remark 1 *A configuration (v_r, b_k) is a linear r-regular k-uniform hypergraph with v vertices and b hyperedges.*

1.3 Hypergraph Drawing

In the business of graph drawing the drawing of hypergraphs plays a specific role. For example, a 3-uniform hypergraph with v vertices has subsets of size 3 as edges. Perhaps the smallest very interesting hypergraph is shown in Fig.1. It has 7 vertices and 7 hyperedges (the 6 lines and the "curved" line).

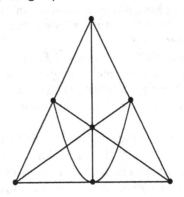

Fig.1.The unique configuration 7_3

The usual drawing of edges (of size 2) in a graph is the (linear) segment between its two vertices. The main question is how to position the vertices (as points in the plane or on a grid etc.) in order to obtain a more or less nice drawing. It depends on the special purpose of the drawing whether the focus is on aspects of symmetry or readability for example.

In the case of hypergraphs the problem becomes much more difficult since through 3 or more given points there is no line (in general). There is a certain class of hypergraphs with a history of more than 100 years, the *configurations* (v_r, b_k). Even their drawing problem was already attacked more than 100 years ago. It should be mentioned that the role of configurations as hypergraphs is just one interesting relation of configurations and graphs (see [5]).

1.4 Configurations and Their History

Hence the question arises: If configurations can be described as particular hypergraphs so easily why is there a special name for them ? The answer is quite easy: Configurations are much older than hypergraphs. While the term of a hypergraph was introduced around 1960 the first definition of configurations was already given in 1876 in the book *Geometrie der Lage* of T. Reye at a time when even graph theory had not yet found an established place within mathematics.

It is not only true that configurations are older than graphs and hypergraphs. Thus certain graph-theoretic results were already obtained earlier in the language of configurations. It is also this *geometric* language of configurations which raises the question of drawability rather naturally. The objects of configurations are called points and lines. However, the language of graph theory uses more abstract terms like vertex and edge. Moreover, the roots of graph theory are very different ones, from logic and topology to physics and chemistry.

Configurations were "born" in geometry. The paragraph where Reye defines configurations is illustrated by a drawing of the famous configuration of Desargues (see Fig. 2, not identical with the drawing in Reye's book). By the way, a picture of the Desargues configuration is shown on the cover of the *Proceedings of Graph Drawing '93* [3] edited by the *Centre d'analyse et de mathématique sociales*.

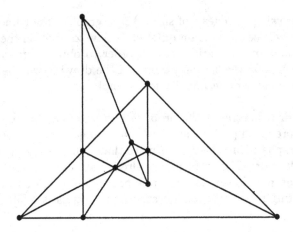

Fig.2.The Desargues configuration 10_3

In the beginning, only those configurations which were drawn in the plane were accepted as "geometric" configurations, the others were called "schematic". Quite soon, however, configurations were regarded as combinatorial structures defined by the axioms given above. It turned out that the configuration 7_3 (the hypergraph of Fig. 1) cannot be drawn with straight lines only. Since all its lines contain 3 points such a drawing would be a contradiction to the theorem of Gallai (which was, however, proved later). In particular, many configurations v_3 with $v \leq 12$ were constructed (combinatorially) and also drawn as points and lines in the plane. Not all the drawings were correct (compare Fig. 3 below).

A very remarkable result is due E. Steinitz. In his dissertation of 1894 [13] he proved that there is always a drawing for a configuration v_3 with straight lines and at most one "quadratic" line (compare Fig. 1).

It is not possible here to describe the history of configurations and their drawings in detail. The interested reader is referred to [4] and [8] for further information. However, I hope to have clarified why the drawing of configurations is not only an interesting topic itself. I think it plays a key role in the drawing of hypergraphs and similar combinatorial structures, because of its long tradition and its geometrical context.

2   The Different Goals of Drawing

As already mentioned, it is quite clear that in the drawing of graphs the edges should be realized by straight lines or curves like segments of lines, circles etc. Of course, there remains a lot of discussion how to position the vertices in order to obtain a good, nice, readable etc. drawing of the graph.

In the case of hypergraphs the main strategy until now has been to denote the hyperedge by enclosing the corresponding vertices by a circle or ellipse like curve. The reason is probably that the main root of hypergraph theory is the theory of subsets of sets. For these mathematical structures a Venn diagram is the usual tool of drawing pictures.

There is not much literature on drawing of hypergraphs available. The only software for drawing hypergraphs which I have seen until now is the Hypergraph Drawing and Optimization System (HDOS) of Vitaly Voloshin of the State University of Moldova [15]. This HDOS draws hypergraphs as indicated above (Venn like diagrams).

However, since graphs are also hypergraphs, it provides nice drawings of graphs, e.g. of the Petersen graph. Since the Petersen graph has at least 3 nice drawings it is a challenging task for every graph drawing software system. Moreover, the HDOS also computes a lot of graph-theoretical parameters, e.g. the new upper chromatic number, see [14].

In this paper the strategy of drawing will be to describe hyperedges by (straight) lines, at least as far as possible. If the hypergraph is linear, i.e. two hyperedges never intersect in more than one vertex, this aim can be reached perhaps. The result of Steinitz shows that at least for configurations n_3 it can nearly be reached. Of course, for non-linear hypergraphs the strategy can be bad since in euclidean geometry there is only one line through 2 given points.

2.1 Realizations over the Real or Rational Numbers

While the mathematicians of the last century were mainly interested in obtaining nice pictures of their configurations, quite recently in the theory of matroids the question of realizability has been discussed from a more theoretical point of view. By generalizing the concept of independence in linear algebra it was asked whether a given system of subsets can be realized by a matrix such that dependent sets correspond to matrices of determinant 0. For the following purpose it is not necessary to define matroids here explicitly. For further details on matroids, see e.g. [11].

The problem can be explained in the language of matrices as follows. Given a matrix with 3 rows and n columns with entries in a certain field (e.g. $\mathbb{R}$). Identify the set $\{1, ..., n\}$ with the set of n columns of the matrix. Determine all 3-by-3 subdeterminants of the matrix. If this determinant is 0 call the set of 3 numbers corresponding to the chosen columns to be a line or hyperedge, otherwise not. This is an algebraic definition of a combinatorial (or geometric) structure.

Vice versa, this problem is closely related to the drawing of hypergraphs. For example, for configurations v_3 the problem of realization can be defined as follows.

Definition 2.1 *A coordinatization or realization of a configuration v_3 with point set P and line set L over a field K is a mapping from P to K^3 such that for all distinct $i, j, k \in P$ $det(x_i, x_j, x_k) = 0$ if and only if i, j, k are collinear in the configuration.*

For our purposes the field will be the field of real or rational numbers. For details and references concerning this realization problem the reader is referred to [8]. As an example the following configuration 11_3 and their realization matrix are given.

Its configuration lines are:

1,2,4; 2,3,5; 3,4,6; 4,5,7; 5,6,8; 6,7,9; 7,8,10; 8,9,11; 1,9,10; 2,10,11; 1,3,11.

The 3-by-11 matrix

$$\begin{vmatrix} 1 & 1 & 0 & 0 & 1 & 0 & 17 & 1 & 1 & 37 & 4 \\ 0 & 0 & 1 & 0 & 1 & -17 & 17 & 51 & 2 & 18 & -9 \\ 0 & 1 & 0 & 1 & 1 & 100 & 185 & 285 & 5 & 45 & 0 \end{vmatrix}$$

has the property that a 3-by-3 subdeterminant with columns i, j, k is 0 if and only if i, j, k is one of the 11 lines given above.

In order to obtain a drawing in the 2-dimensional plane a suitable projection of $\mathbb{R}^3$ to $\mathbb{R}^2$ has to be chosen. What suitable means depends on the further properties of the drawing which we require.

2.2 Linear Drawings

At first, it has to be checked whether a configuration can be realized over the real numbers. These realizations have been constructed for all configurations v_3 with $v \leq 12$ for which this is possible. There are exactly 3 of them which are not realizable over $\mathbb{R}$, the unique configuration 7_3, the unique configuration 8_3, and one configuration 10_3 (see Fig. 3).

E.g. realizations of 8 of the other 9 configurations 10_3 are published in [2]. The Desargues configuration is omitted since its drawing does not cause any problems. The Theorem of Desargues tells us how to draw it, at least on an "infinite" part of paper.

For the convenience of the reader these 8 matrices are printed below. Of course, drawings of these configurations would be nicer, but these matrices are the basis of the drawings and shorten the length of this paper.

$$\begin{vmatrix} 0 & 12 & 4 & 1 & -1 & -4 & 17 & 0 & 0 & 1 \\ -3 & 6 & -1 & -1 & -2 & 1 & 0 & 1 & 0 & -1 \\ 4 & -9 & 1 & 0 & 3 & 3 & 0 & 0 & 1 & 1 \end{vmatrix}$$

$$\begin{vmatrix} -552 & 48 & -3600 & 2 & 1 & 8 & 0 & 0 & 0 & 1 \\ 529 & -42 & 3450 & 4 & 0 & -7 & 4 & 1 & 0 & -1 \\ 0 & 117 & -3450 & 71 & 0 & .8 & 71 & 0 & 1 & 1 \end{vmatrix}$$

$$\begin{vmatrix} -49 & 483 & 49 & 3 & 23 & 1 & 1 & 0 & 0 & 1 \\ 0 & 252 & 21 & 32 & 12 & -1 & 0 & 1 & 0 & -1 \\ -9 & 108 & 9 & 3 & 3 & 0 & 0 & 0 & 1 & 1 \end{vmatrix}$$

$$\begin{vmatrix} -10 & 12 & 0 & 10 & 1 & 18 & 0 & 2 & 0 & 1 \\ -6 & 36 & 36 & 6 & 0 & 18 & 1 & -2 & 0 & -1 \\ 6 & 0 & 9 & 9 & 0 & 270 & 3 & 1 & 1 \end{vmatrix}$$

$$\begin{vmatrix} 89 & -4 & 0 & 4 & 11 & 0 & 10 & 10 & 1 \\ 7832 & 11 & 88 & 85 & 88 & 0 & 1 & -1 & 0 & -1 \\ 979 & -11 & 8 & 11 & 8 & 0 & 0 & 0 & 1 & 1 \end{vmatrix}$$

$$\begin{vmatrix} -3 & -270 & 0 & 3 & 1 & 0 & -18 & 30 & 1 \\ 4 & 45 & 2 & 2 & 0 & 1 & 3 & -3 & 0 & -1 \\ -4 & -99 & 0 & 4 & 4 & 0 & 0 & -18 & 18 & 1 & 1 \end{vmatrix}$$

$$\begin{vmatrix} -2 & 0 & 672 & 16 & 2 & 1 & 0 & 20 & 1 \\ -3 & 2 & -322 & 3 & 3 & 0 & 1 & -2 & 0 & -1 \\ 3 & -21 & 5376 & 16 & 16 & 0 & 0 & 21 & 1 & 1 \end{vmatrix}$$

$$\begin{vmatrix} -60 & 500 & 0 & 4 & 1 & 60 & 0 & 16 & 0 & 1 \\ 160 & -400 & 16 & 16 & 0 & -160 & 1 & -16 & 0 & -1 \\ 0 & 525 & 9 & 9 & 0 & 60 & 0 & 21 & 1 & 1 \end{vmatrix}$$

Afterwards a projection yields the required picture. The list of further properties which the drawing should have is quite similar to those for graphs like a certain minimum distance of points, minimum angle between lines, good distribution of points in the picture etc.

2.3 Approximative Drawings

From a more practical point of view it is not that bad if a configuration cannot be realized over the reals. Mathematically, this is a remarkable result but it should not discourage us. If the main task of the drawing is to show clearly what exactly the hyperedges or lines are an approximative drawing is suitable as well.

Figure 3 was produced in a paper of S. Kantor of 1881 [10] where he published drawings of all 10 configurations 10_3. However, it was proved a few years later by E. Schroeter and after that several times by other mathematicians that the configuration 10_3 in Fig. 3 is the only one which is not realizable over the reals nor over any other field. In fact, the given drawing is not correct. It is not clear whether Kantor himself knew about it.

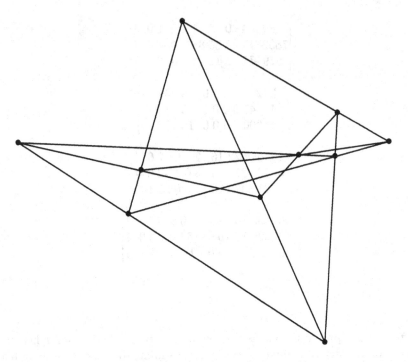

Fig.3.The non-realizable cfz. 10_3

But who cares ? It is not so easy to find out which of these 10 "lines" is no line at all. More important everybody will find this drawing very helpful to have an easy description of the hyperedges (or lines or subsets) of the hypergraph.

2.4 Schematic Drawings

Even if there is one "curved" line contained in the picture it does not really disturb the drawing as we can see in Fig. 1. The above mentioned result of Steinitz makes it reasonable to look for such drawings of configurations v_3 and to try to prove similar results for other classes of combinatorial structures.

If the number of "curved" lines is relatively small, such a picture is still much clearer than a picture using Venn diagrams.

2.5 Nice Drawings

Apart from the above topics the problem remains to find a nice linear drawing if there is one at all. Perhaps a nice drawing containing a few "curved" lines is more acceptable than an ugly linear drawing.

3 Other Combinatorial Structures

The question of drawing hypergraphs is not restricted to configurations. They are just the best examples to introduce the topic because of their historical and geometric context. For many purposes in mathematics as well as in applications it is desirable to describe a certain structure by a drawing rather than by giving a list of points and lines.

3.1 Linear Hypergraphs

If the hypergraph is linear but not necessarily uniform or regular this does not change the problem discussed above very much. We can read through the above paragraphs and follow the same strategy again. In particular, it should be investigated which linear hypergraphs can be drawn with straight lines only or with at most "quadratic" line in the sense of Steinitz.

3.2 (r,1)-Designs and Linear Spaces

Perhaps the problem with linear hypergraphs is that although they are defined in books on hypergraphs so far they have not been investigated very much in detail.

A certain subclass, however, linear spaces and $(r, 1)$-designs have been enumerated for small numbers of points and are an interesting data base. In the language of hypergraphs a linear space is a hypergraph where through two given points there is always a common hyperedge. An $(r, 1)$-design is an r-uniform linear space.

Recently all small linear spaces and $(r, 1)$-designs have been constructed. For further information on all $(r, 1)$-designs with at most 13 points see [6], [7], and [9]. Moreover, all linear spaces with at most 11 points have been determined (see [12]).

4 Conclusion

In my opinion the strategy to draw hypergraphs or at least linear hypergraphs by displaying the hyperedges as lines (as far as possible) will lead to drawings which in most cases will be better readable than traditional drawings. Moreover, this strategy is closely related to an interesting theoretical background in geometry, combinatorics, and algebra having its roots back in the 19th century. For further details the reader has to be referred to the references below.

Of course, the drawing of hypergraphs is much more difficult than graph drawing. Perhaps this is the reason why so far it has not been attacked seriously. In particular, for lines of size greater than 3 the problem becomes very hard. For these hypergraphs only very few results have been obtained until now. I hope that this paper will start a more intensive research on the drawing of hypergraphs, from a theoretical and a practical point of view.

References

1. C. Berge, Hypergraphs, Amsterdam-New York- Oxford-Tokyo (1989)
2. J. Bokowski, B. Sturmfels, Computational Synthetic Geometry, Springer LNM 1355, Berlin-Heidelberg-New York (1989)
3. G. DiBattista, P. Eades, H. de Fraysseix, P. Rosenstiehl, R. Tamassia (eds.), Graph Drawing '93 Proceedings, September 1993, Paris
4. H. Gropp, On the history of configurations, Internat. Symp. on Structures in Math. Theories, ed. A.Díez, J.Echeverría, A.Ibarra, Universidad dei País Vasco, Bilbao (1990) 263-268
5. H. Gropp, Configurations and graphs, Discrete Math. **111** (1993) 269-276
6. H. Gropp, Configurations and $(r, 1)$-designs, Discrete Math. **129** (1994) 113-137
7. H. Gropp, Graph-like combinatorial structures in $(r, 1)$-designs, Discrete Math. **134** (1994) 65-73
8. H. Gropp, Configurations and their realization (to appear)
9. H. Gropp, The $(r, 1)$-designs with 13 points (submitted to Discrete Math.)
10. S. Kantor, Die Configurationen $(3, 3)_{10}$, Sitzungsber. Akad. Wiss. Wien, math.-naturwiss. Kl. **84** (1881) 1291-1314
11. J.G. Oxley, Matroid theory, Oxford-New York-Tokyo (1992)
12. C. Pietsch, On the classification of linear spaces of order 11, J. Combin. Des. 3 (1995), 185-193
13. E. Steinitz, Über die Construction der Configurationen n_3, Dissertation Breslau (1894)
14. V.I. Voloshin, On the upper chromatic number of a hypergraph, Australasian J. Comb. **11** (1995) 25-45
15. V.I. Voloshin, Hypergraph Drawing and Optimization System, preprint

Upward Drawings on Planes and Spheres

Extended Abstract for *Graph Drawing '95*

20 – 22 September 1995

S. Mehdi Hashemi[1], Andrzej Kisielewicz[2] and Ivan Rival[3]

[1] Department of Mathematics
University of Ottawa
Ottawa K1N 6N5 Canada
shashemi@csi.uottawa.ca

[2] Mathematical Institute
University of Wroclaw
Wroclaw, Poland
kisiel@math.uni.wroc.pl

[3] Department of Computer Science
University of Ottawa
Ottawa K1N 6N5 Canada
rival@csi.uottawa.ca

Abstract. Although there is a linear time algorithm to decide whether an ordered set has an upward drawing on a surface topologically equivalent to a sphere, we shall prove that the decision problem whether an ordered set has an upward drawing on a sphere itself is NP-complete. To this end we explore the surface topology of ordered sets highlighting especially the role of their saddle points.

Introduction

The search for an efficient upward planarity testing algorithm for ordered sets[4] is a longstanding problem, much sought after by theoreticians of graphical data structures. It has always been a mystery how upward planarity testing for orders could be so difficult if its undirected companion, planarity-testing for graphs is so easy. In spite of well-known linear time algorithms for graph planarity testing (e.g. [**Hopcroft** and **Tarjan** 1974]), it is not even self-evident that there is a finite algorithm for upward planarity testing.

For many years progress has been slow.

- *Planar lattices are dismantlable* [**Baker, Fishburn** and **Roberts** 1971]
- *Algorithmics of planar lattices* [**Kelly** and **Rival** 1975].
- *Planarity-testing for lattices using graph planarity-testing* [**Platt** 1976].
- *Straight lines for planar upward drawings* [**Kelly** 1987] .
- *Bipartite planar upward drawings* [**di Battista, Liu** and **Rival** 1990].

---

[4] In fact, by adjoining subdivision points as needed we may just as well consider *directed acyclic graphs.*

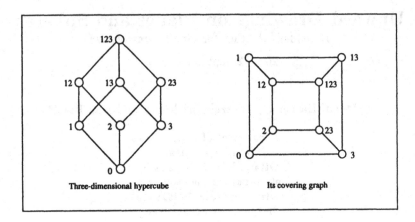

Fig. 1. The three-dimensional hypercube is a nonplanar ordered set (with planar covering graph).

- *Planar ordered sets of width two* [**Czyzowicz**, **Pelc** and **Rival** 1990].
- *Planar ordered sets with bottom* [**Hutton** and **Lubiw** 1991].
- *Planar triangle-free graphs have planar orientations* [**Kisielewicz** and **Rival** 1993].

Recently, [**A. Garg** and **R. Tamassia** (1995)] made a major breakthrough. By transforming the *NOT-ALL-EQUAL-3-SAT* decision problem into an auxiliary flow decision problem with integer coordinates, and then providing yet another transformation to upward planarity, they proved that upward planarity-testing is NP-complete.

The results presented here are based on our work of recent years whose twofold purpose is to understand upward drawings of orders through the two-dimensional surfaces in $\mathbf{R}^3$ on which they may be embedded (without crossing edges), and conversely, to better understand two-dimensional surfaces in $\mathbf{R}^3$ by means of the upward drawings that fit on them (cf. [**Hashemi** and **Rival** (1994)].

Our main result bears on this problem.

Upward Sphericity Testing

INSTANCE Given an ordered set P.
QUESTION Does P have an upward drawing on the sphere $\{(x, y, z) : x^2 + y^2 + z^2 = 1\}$ without the crossing of edges?

Although much is already written about planarity for graphs and ordered sets, there is much less about sphericity. Indeed, apart from bits and pieces in the topological graph theory and differential geometry literature there is virtually a blank about the algorithmics of sphericity.

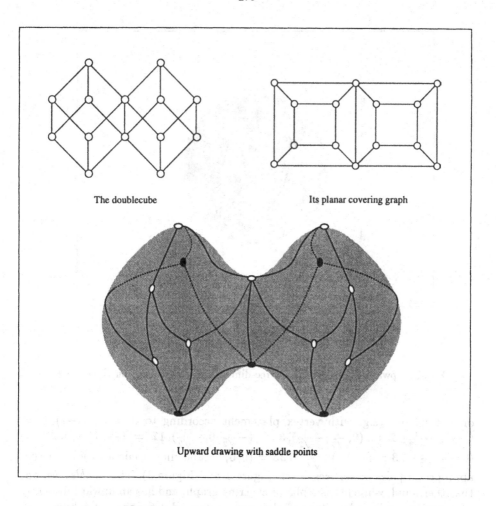

The doublecube

Its planar covering graph

Upward drawing with saddle points

Fig. 2. Upward drawing of the doublecube on a surface of genus zero.

Like *upward planarity testing* the complexity of *upward sphericity testing* seems far from obvious. Indeed, according to [**Ewacha, Li** and **Rival (1991)**] — and unlike *upward planarity testing*, the decision problem whether an ordered set has an *upward drawing* on a surface of genus zero (that is, a *topological* sphere) is itself polynomial. The reason is that its planar covering graph can be "lifted" to an upward drawing on a surface of genus zero. On the other hand, it is our main result that *upward sphericity* is still *NP-complete*.

The two-dimensional cube $\{0, 1, 2, 12\}$ has an upward drawing on the plane, (e.g. $0 = (1, 0, 0), 1 = (0, 1, 0), 2 = (2, 1, 0), 12 = (1, 2, 0)$) : the two-dimensional cube is a *planar* ordered set. The three-dimensional cube (cf. Figure 1) is a nonplanar ordered set.

The three-dimensional cube, although nonplanar, has an upward drawing

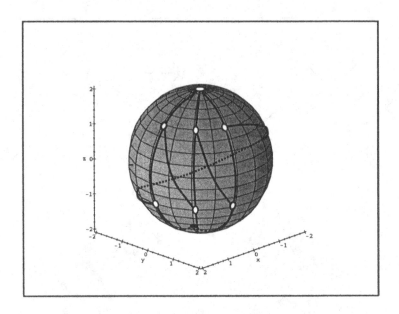

Fig. 3. Upward drawing of the three-dimensional hypercube on the sphere.

on the sphere, (e.g. with vertex placement according to $0 = (0,0,-1), 1 = (\frac{1}{\sqrt{2}}, 0, -\frac{1}{\sqrt{2}}), 2 = (0, \frac{1}{\sqrt{2}}, -\frac{1}{\sqrt{2}}), 3 = (-\frac{1}{\sqrt{2}}, 0, -\frac{1}{\sqrt{2}}), 12 = (\frac{1}{\sqrt{2}}, 0, \frac{1}{\sqrt{2}}), 23 = (0, \frac{1}{\sqrt{2}}, \frac{1}{\sqrt{2}}), 13 = (-\frac{1}{\sqrt{2}}, 0, \frac{1}{\sqrt{2}}), 123 = (0,0,1)$, and monotonic arcs joining the appropriate pairs of vertices — cf. Figure 1 and Figure 3). The *doublecube*, on the other hand, while it has a planar covering graph, and has an upward drawing on a surface with no handles or holes, has no upward drawing on a sphere: any smooth surface in $\mathbf{R}^3$ on which the doublecube has an upward drawing has a *saddle point* (cf. Figure 2).

Theorem 1. *Upward sphericity testing is NP-complete.*

The proof, in part, runs parallel to this important recent result of [**A. Garg** and **R. Tamassia** (1995)]. Despite initial appearances, though, we see no way to derive our theorem, directly, or by way of corollary, from it.

Theorem 2. *Upward planarity testing is NP-complete.*

In the process, we shall derive a proof (which seems to us simpler and more transparent) based on

Exact Cover By 3-Sets

INSTANCE Given a set X with $|X| = 3q$ and a collection C of 3-element subsets of X.

QUESTION Does C contain an exact cover for X, i.e., a subcollection $C' \subseteq C$ such that every element of X occurs in exactly one member of C'?

Both the present proof, and the earlier one of [**A. Garg** and **R. Tamassia** **(1995)**], rely on "gadgets" that they call *tendrils* and *wiggles*, in our jargon *spirals*. An example, $S(1,0)$ a spiral with a "frame" illustrated in Figure 4, has a planar covering graph and yet no upward drawing on a plane. Moreover, although it does have an upward drawing on a smooth surface of genus one, such a surface must always have a saddle point. The argument is based on this technical result which seems to be of independent interest.

Theorem 3. *Let P be an ordered set, let S be a smooth surface in $\mathbf{R}^3$ on which P has an upward drawing. Let F be a face of this upward drawing and let m stand for the number of paths of vertices of this face F, all members of which are interior, or all members of which are extremal. If*

$$interior\,(F) > \; extremal\,(F) + 2m$$

then S must have a saddle point.

Although upward planarity testing is difficult, it may be that, if it is known that an ordered set already has an upward drawing on a genus zero surface then there is an efficient procedure to decide whether it has an upward drawing on a plane, too.

Upward Planarity Testing of Spherical Ordered Sets

INSTANCE Given an ordered set P with upward drawing on a sphere.
QUESTION Does P have an upward drawing on a plane?

Problem 4. Is **Upward Planarity Testing of Spherical Ordered Sets** polynomial?

Our analysis here has made no distinction between one or more saddle points.

m-Saddle Point Surfaces

INSTANCE Given an ordered set P with planar covering graph and a nonnegative integer s.
QUESTION Does P have an upward drawing on a smooth surface of genus zero with at most s saddle points?

Problem 5. Is the s-saddle point problem NP-complete?

In spite of the intractability of *upward spherical testing* it is still of interest to know just which ordered sets have an upward drawing on a sphere.

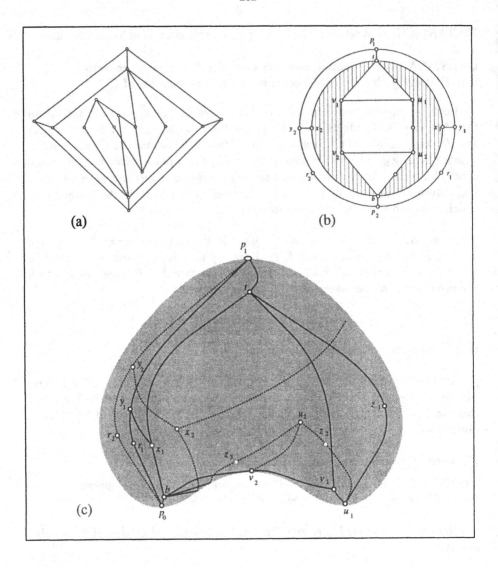

Fig. 4. (a) The Spiral $S(1, 0)$. (b) The planar covering graph of $S(1, 0)$ (c) An upward drawing of $S(1, 0)$ on a surface with a saddle point.

Problem 6. Characterize those ordered sets which have an upward drawing on a sphere.

The bridge between the *NP-completeness* of *upward planarity testing* and *upward sphericity testing* is built over *spirals*. In effect, a *spiral* is spherical if and only if it is planar.

Problem 7. Characterize those ordered sets for which sphericity implies planarity.

Strategy of the Proof

Spherical Ordered Sets at a Glance

What is the difference between the sphere $\{(x, y, z) : x^2 + y^2 + z^2 = 1\}$ and an arbitrary homeomorph of it, that is, any smooth (compact and closed) surface in $\mathbf{R}^3$ of genus zero?

In a word, *saddles*.

An ordered set is *spherical* if it has an upward drawing on the unit sphere such that no two edges cross and all edges are monotonic paths with respect to the positive z-axis, the northerly direction (although the precise direction is arbitrary). [**S. Foldes, I. Rival** and **J. Urrutia (1992)**] showed that an ordered set is spherical if it has a *top*, a *bottom*, and its covering graph is planar. Actually, *an ordered set is spherical if and only if, by adjoining new edges as needed, it can be extended to a directed acyclic graph, with top and bottom, and with planar covering graph.* (This echoes a characterization of planar ordered sets in terms of *planar lattices* [**D. Kelly (1987)**] or, equivalently, *st-graphs* [**G. di Battista** and **R. Tamassia (1988)**].)

Fix an upward drawing (without crossing of its edges) of an ordered set P on a surface S. To any covering edge $a \succ b$ in P we associate two values, $+$ to that end of the covering edge outgoing from a and $-$ to that end of it incoming to b. In this way, every covering edge acquires two values and, for every element $a \in P$, every incident covering edge associates, in this way, a sign ($+$ or $-$), to a. A minimal element of P will acquire all $-$ values and a maximal element all $+$ values. In general, to every element there is associated a circular sequence of $+$'s and $-$'s corresponding to a clockwise orientation in a neighbourhood about it. What is of importance is the number of alternations of $+$'s and $-$'s. Thus, we call an element *ordinary* just if this sequence consists of an interval of $+$'s and an interval of $-$'s — one alternation. An *extremal* element, that is, a maximal or a minimal element has no alternations at all. If an element's circular sequence has two or more alternations then it must be a *saddle point* — the surface on which the ordered set is drawn cannot be spherical.

On the boundary $v_1, v_2, \ldots, v_m$ of a face F we consider all vertices which are extremal with respect to F, that is, all vertices v_i such that either $v_{i-1} \succ v_i$ and $v_{i+1} \succ v_i$, or else, $v_{i-1} \prec v_i$ and $v_{i+1} \prec v_i$. Such a vertex is *extremal* if all of its neighbours are either larger or all are smaller. We call it *interior* with respect to this face F if it has a neighbour v, not on the boundary of this face, with opposite sign, that is, $v \prec v_i$ if $v_{i-1} \succ v_i$ and $v_{i+1} \succ v_i$, while $v \succ v_i$ if $v_{i-1} \prec v_i$ and $v_{i+1} \prec v_i$.

An algorithmic analysis of a graph on a surface ultimately entails a triangulation of it. *Any face which itself (as an ordered subset) contains a* top *and* bottom *can be triangulated without increasing the alternation about any of its vertices.* The essential conclusion to which all of these observations lead is this.

An ordered set has an upward drawing on a sphere if and only if its covering graph has a triangulation with no saddle point at all.

Reduction Gadgets

The *NP-completeness* reduction is based on standard techniques of "component design" (cf. [**M. R. Garey** and **D. S. Johnson (1979)**]).

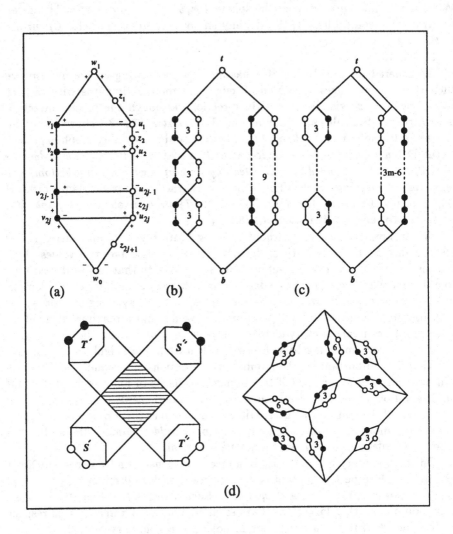

Fig. 5. (a) The spiral S_j (b) Q (c) $P(m)$ (d) R

Here is the system of gadgets that are used in the reduction.

Loosely speaking, our aim is to construct a directed acyclic graph P such that "flipping" its components corresponds to an instance of **Exact Cover By 3-Sets**.

— *Spirals* S_j, j a positive integer, constitute the "flipping" components.

- We collect four spirals at a time to build an ordered set (we call Q) to represent the elements of the set X. More precisely, Q consists of three copies of S_3 and one S_9. (The point is that Q has a spherical upward drawing if and only if either all of the three spiral components are "flipped" or none at all are "flipped".) For every member $c \in C$ we associate a copy of Q and "flipping" the components shall correspond to choosing the member c to be an element of the desired exact cover.
- Construct ordered sets $P(m)$, with $m + 1$ spirals, m of which are copies of S_3 and one S_{3m-6}, intended to represent members of C, such that $P(m)$ has a spherical upward drawing if and only if exactly one of the spiral sets is "flipped". (For every element $x \in X$ there is a suitable copy of $P(m)$ and "flipping" a component corresponds to choosing the member of C to cover the element x.
- We construct "communication" edges in P to ensure that choices in the Q-components and the $P(m)$-components agree. Because these communication edges may cross, we replace each by a "crossover" R.

In summary, the directed acyclic graph is constructed in polynomial time and has a spherical upward drawing if and only if the corresponding instance of **Exact Cover By 3-Sets** has a positive solution.

References

K. A. Baker, P. C. Fishburn, and F. S. Roberts (1971) Partial orders of dimension 2, *Networks* (**2**), 11–28.

G. di Battista, W.-P. Liu and I. Rival (1990) Bipartite graphs, upward drawings, and planarity, *Inform. Proc. Letters* **36**, 317–322.

G. di Battista and R. Tamassia (1988) Algorithms for plane representations of acyclic digraphs. *Theoretical Computer Science* **61**, 175–198.

J. Czyzowicz. A. Pelc and I. Rival (1990) Planar ordered sets of width two, *Math. Slovaca* **40** (4), 375–388.

K. Ewacha, W. Li, and I. Rival (1991) Order, genus and diagram invariance, *ORDER* **8**, 107–113.

S. Foldes, I. Rival and J. Urrutia, Light sources, obstructions, and spherical orders, *Discrete Math.* **102**, 13–23.

M. R. Garey and D. S. Johnson (1979) *Computers and Intractability : A Guide to the Theory of NP-completeness.* Freeman,

A. Garg and R.Tamassia (1995) On the computational complexity of upward and rectilinear planarity testing, *Lecture Notes in Computer Science* (**894**) (eds. R. Tamassia and I. G. Tollis), pp. 286 – 297.

A. Garg and R.Tamassia (1995) Upward planarity testing, *ORDER* (**12(2)**).

S. Mehdi Hashemi and I. Rival (1994) Upward drawings to fit surfaces, in *Orders, Algorithms, and Applications* (*ORDAL* '94) (eds. V. Bouchitté and M. Morvan), *Lecture Notes in Computer Science* **831**, Springer pp. 53–58.

J. Hopcroft and R. E. Tarjan (1974) Efficient planarity testing, *J. Ass. Comp. Mach.* **21** (4), 549–568.

M. D. Hutton and A. Lubiw (1991) Upward planar drawing of single source acyclic digraphs, *Proc. 2nd A.C.M./S.I.A.M. Symposium Discrete Appl. Math.*, pp. 203–211.

D. Kelly (1987) Fundamentals of planar ordered sets, *Discrete Math.* **63**, 197–216.

D. Kelly and I. Rival (1975) Planar lattices, *Canad. J. Math.* **27**, 636–665.

A. Kisielewicz and I. Rival (1993) Every triangle-free planar graph has a planar upward drawing, *ORDER* **10**, 1–16.

A. Lempel, S. Even and I. Cederbaum (1967) An algorithm for planarity testing of graphs, in *Theory of Graphs, International Symposium*, Rome (1966) (P. Rosenstiehl ed.), Gordon and Breach, pp. 215–232.

C. R. Platt (1976) Planar lattices and planar graphs, *J. Comb. Th. Ser. B* **21**, 30–39.

K. Reuter and I. Rival (1991) Genus of orders and lattices, in *Graph-Theoretic Concepts in Computer Science* (R. Möhring ed.), *Lect. Notes Comp. Sci.* **484**, pp. 260–275.

I. Rival (1993) Reading, drawing, and order, in *Algebras and Orders* (I. G. Rosenberg ed.), Kluwer.

C. Thomassen (1989) Planar acyclic oriented graphs, *ORDER* **5**, 349–361.

Grid Embedding of 4-Connected Plane Graphs

Xin He

Department of Computer Science, State University of New York at Buffalo, Buffalo, NY 14260. The work was partially supported by NSF grant CCR-9205982.

Abstract. A straight line grid embedding of a plane graph G is a drawing of G such that the vertices are drawn at grid points and the edges are drawn as non-intersecting straight line segments. In this paper, we show that, if a 4-connected plane graph G has at least 4 vertices on its exterior face, then G can be embedded on a grid of size $W \times H$ such that $W + H \leq n$, $W \leq (n+3)/2$ and $H \leq 2(n-1)/3$, where n is the number of vertices of G. Such an embedding can be computed in linear time.

1 Introduction

A *straight line grid embedding* of a n-vertex planar graph is a drawing where the vertices are located at distinct grid points, and the edges are represented by straight line segments. Such embeddings on reasonably small grids are very useful in visualizing planar graphs on graphic screens [5]. Wagner [19], Fáry [6], and Stein [17] independently showed that every planar graph has a straight line embedding. Since then, many embedding algorithms have been reported (e.g. [18, 1]). The earlier algorithms all suffer two serious drawbacks, as noted in [7]. First, they require high-precision real arithmetic, and therefore cannot be used even for a graph of moderate size. Second, in the drawings produced by these algorithms, the ratio of the largest distance to the smallest distance between vertices are so large (exponential in n) that it is very difficult to view those drawings on graphic screens. In view of these drawbacks, Rosenstiehl and Tarjan [12] posed the problem of computing a straight line embedding on a grid of polynomial size. Schnyder [14] proved that every planar graph has a straight line embedding on a $(2n-4) \times (2n-4)$ grid. Independently, de Fraysseix, Pach, and Pollack showed that every planar graph has an embedding on a $(2n-4) \times (n-2)$ grid, which can be computed in $O(n \log n)$ time [7]. The running time of their algorithm was improved to $O(n)$ [4]. Schnyder proved the existence of an embedding on an $(n-2) \times (n-2)$ grid [13] and gave an $O(n)$ time algorithm to compute such an embedding [15]. Schnyder's algorithm can be implemented in parallel in $O(\log n \log \log n)$ time with $n/\log n$ processors on a PRAM [8]. It was shown in [9] that every 3-connected planar graph G can be embedded on a $(2n-4) \times (n-2)$ grid such that all internal faces of G are convex. The grid size of such embedding is reduced to $(n-2) \times (n-2)$ in [2, 16].

There exists a plane graph G such that, for any straight line grid embedding of G, each dimension of the grid needs to be at least $\lfloor 2(n-1)/3 \rfloor$ even if the other dimension is allowed to be unbounded [7, 3]. It has been conjectured that every planar graph can be embedded on a $2n/3 \times 2n/3$ grid. This conjecture remains open. Chrobak and Nakano showed that every planar graph has a straight line embedding on a $(\lfloor 2(n-1)/3 \rfloor) \times (4\lfloor 2(n-1)/3 \rfloor - 1)$ grid [3].

For the grid embedding problem, we can assume that all internal faces of G are triangles. (If not, we can triangulate G and remove the added edges after an embedding is obtained.) If all internal faces of G are triangles, it is an *internally triangulated plane* graph. If the external face of G is also a triangle, then G is a *triangulated plane* graph. A *non-empty triangle* of G is a triangle containing some vertices in its interior. In this paper we show that if G is 4-connected and has at least 4 vertices on its external face, then the above mentioned $(2n/3 - 1) \times (2n/3 - 1)$ lower bound on grid size does not hold. We will prove:

Theorem 1. *Every n vertex 4-connected plane graph G with at least 4 vertices on the external face has a straight line embedding on a $W \times H$ grid such that $W + H \leq n$, $W \leq (n+3)/2$ and $H \leq 2(n-1)/3$. Such an embedding can be computed in linear time.*

In Theorem 1, we assume G is given by its adjacency list representation, where the neighbors of each vertex are given in clockwise order of embedding. The 4-connectivity of G can be checked in linear time [10]. Every such a graph G can be internally triangulated so that it has no non-empty triangles. From now on, we only consider such internally triangulated plane graphs. The present paper is organized as follows. In Section 2, we review some definitions and describe a *generic shift algorithm* in [3], which is the basis of our algorithm. In Section 3, we present our algorithm. In section 4, we prove Theorem 1.

2 Preliminaries

The embedding algorithms in [7, 4, 3, 9] are based on the following concept [7].

Definition 2. Let $G = (V, E)$ be a triangulated plane graph and $\pi = v_1, v_2 \ldots v_n$ an ordering of V such that the edge (v_1, v_2) is on the external face of G. Let G_k be the subgraph of G induced by $v_1, \ldots, v_k$ and C_k be the external face of G_k. π is called a *canonical ordering* of G if the following hold for $k = 3, \ldots, n$:

(co1) G_k is 2-connected and all internal faces of G_k are triangles.

(co2) C_k contains the edge (v_1, v_2).

(co3) If $k < n$, then v_{k+1} is in the external face of G_k and all neighbors of v_{k+1} in G_k belongs to C_k.

(co4) If $k < n$, then v_k has at least one neighbor v_j with $j > k$.

Every triangulated plane graph has a canonical ordering [7]. We will use $\prec$ to denote the linear order of the canonical ordering. Fig 1 shows a canonical ordering of G. By the *contour* of G_k we mean its external face $C_k = (w_1 = v_1, w_2, \ldots, w_m = v_2)$. For a given k ($3 \leq k \leq n - 1$), let $w_p, \ldots, w_q$ be the neighbors of $v = v_{k+1}$ in C_k. When we add v to G_k, the edges (w_p, v) and (v, w_q) become contour edges. We call (w_p, v) a *forward edge* and (v, w_q) a *backward edge*. All vertices and edges that disappear from the contour when we add v are said to be *covered* by v. We denote $ind_v(w_i) = i - p + 1$ and call it the *index* of w_i

with respect to v. The *in-degree* $deg^-(v)$ of v is the number of children of v in C_k, that is, $deg^-(v) = q - p + 1$.

A contour vertex w_i $(1 < i < m)$ is called a *valley vertex*, if $w_{i-1} \succ w_i \prec w_{i+1}$; a *peak vertex*, if $w_{i-1} \prec w_i \succ w_{i+1}$. A vertex $v_{k+1} \neq v_1, v_2, v_3$ is called: *forward-oriented*, if $w_p \prec w_{p+1} \prec \ldots \prec w_{q-1} \prec w_q$; *backward-oriented*, if $w_p \succ w_{p+1} \succ \ldots \succ w_{q-1} \succ w_q$; *crossing-valley*, if it covers a valley vertex w_r $(p < r < q)$; *crossing-peak*, if it covers a peak vertex w_r $(p < r < q)$.

We next describe a *generic shift algorithm* in [3] which is the basis of the algorithms in [2, 3, 4, 7, 9] and our algorithm. Given G with canonical ordering π, the algorithm works as follows: We add vertices one at a time according to π. At each step, the contour C_k satisfies certain *contour invariants* that involve restrictions on the slopes of contour edges. When adding a vertex v_{k+1}, we determine its location in the grid and, if necessary, shift some vertices of G_k to the right in order to preserve the contour invariants. We maintain a set $U(v)$ of vertices for each $v \in V$. $U(v)$ contains the vertices located "under" v that need to be shifted whenever v is shifted. The *shift operation* on a contour vertex w_j, denoted by $shift(w_j)$, is achieved by increasing the x-coordinate of each $u \in \cup_{i=j}^m U(w_i)$ by 1.

Generic Shift Algorithm:

Initially, place v_1, v_2, v_3 at the points $(0,0), (2,0), (1,1)$, respectively. Let $U(v_i) = \{v_i\}$ $(1 \leq i \leq 3)$.

Suppose G_k $(3 \leq k \leq n-1)$ has been embedded, and we are about to add $v = v_{k+1}$. Let $w_p, \ldots, w_q$ be the children of v in the contour C_k. Define: $U(v_{k+1}) = \{v_{k+1}\} \cup_{i=p+1}^{q-1} U(w_i)$. Apply $shift(w_i)$ to some of $w_1, \ldots, w_m$ (possibly none), so that afterwards there exists at least one point (x', y') satisfying the following conditions, and that placing v at (x', y') preserves the contour invariants.

Definition 3. Generic shift conditions:
(gs1) $x(w_p) \leq x' \leq x(w_q)$;
(gs2) (x', y') is above C_k, i.e. the half line $\{(x', z) | z \geq y'\}$ does not intersect C_k;
(gs3) all vertices $w_p, \ldots, w_q$ are visible from (x', y').

Place v at an arbitrary point (x', y') satisfying these conditions.

Theorem 4. [3] *For all choices of shift operations and vertex coordinates, as long as (gs1), (gs2), (gs3) are satisfied, the Generic Shift Algorithm produces a correct straight line grid embedding.*

3   Drawing Algorithm

Our algorithm crucially depends on the following theorem (proved in [10]).

Theorem 5. *Let G be a triangulated plane graph whose external face $\{v_1, v_2, v_n\}$ is the only non-empty triangle. Then G has a canonical ordering π satisfying the conditions (co1), (co2), (co3), and the following:*

(co4') *Each v_k ($k \leq n-2$) has at least **two** neighbors v_j with $v_j \succ v_k$; v_{n-1} has one neighbor v_n with $v_n \succ v_{n-1}$.*

Moreover, π can be computed in linear time.

Lemma 6. v_n *is the only crossing-peak vertex of G with respect to π.*

Proof. Toward a contradiction, suppose v_{k+1} ($k+1 < n$) is a crossing-peak vertex with respect to π. Let $w_p, \ldots, w_q$ be the children of v_{k+1} in C_k. Then v_{k+1} covers a peak vertex w_r ($p < r < q$). So v_{k+1} is the only neighbor of w_r with $w_r \prec v_{k+1}$. This contradicts the condition (co4'). $\square$

Let G be an $(n-1)$ vertex internally triangulated plane graph with no non-empty triangles (obtained from triangulating a 4-connected plane graph). In order to apply Theorem 5 to G, we add a new vertex v_n in the external face F and connect v_n to all vertices on F such that $\{v_1, v_2, v_n\}$ is the external face. The resulting graph G^+ is called the *extended graph* of G. Since G is 4-connected, the only non-empty triangle of G^+ is $\{v_1, v_2, v_n\}$. In the following, we will discuss the embedding of G^+ by using the canonical ordering π of G^+ satisfying Theorem 5, with the understanding that v_n needs not be embedded.

Let n_f and n_b denote the number of forward- and backward-oriented vertices in G^+, among $v_4, \ldots, v_n$. Since v_n is a peak vertex, we have $n_f + n_b \leq n - 4$. Without loss of generality, we assume $n_f \leq n_b$. (If not, we vertically "flip" G^+ and swap v_1 and v_2. A forward-oriented vertex in G^+ is a backward-oriented vertex in the flipped graph and vice versa).

Direct each edge $e = (u, v)$ of G^+ from u to v, if $u \prec v$. Denote the resulting directed graph by $\overrightarrow{G^+}$. Our algorithm needs a special canonical ordering π_{left} of G^+ obtained as follows. The first three vertices of π_{left} are v_1, v_2, v_3. Suppose the vertex v_k of π_{left} has been defined. Consider a vertex v not in G_k^+. If all incoming neighbors of v are in the contour C_k, v is called a *candidate vertex*. The incoming neighbors of a candidate vertex form a contiguous interval in C_k. Define v_{k+1} of π_{left} to be the candidate vertex whose interval is the leftmost in C_k. This completes the description of π_{left}. Clearly, π_{left} is a canonical ordering of G^+, which is called the *leftmost canonical ordering*. We use $\prec_{left}$ to denote the order defined by π_{left}. (Fig 1 shows the leftmost canonical ordering.)

We need the following concepts introduced in [3]. Each vertex v ($v \neq v_1, v_2$) of G^+ is classified as either *stable* or *unstable*. With each v, we associate a sequence $DC(v)$ of vertices called its *domino chain*, and a vertex $dom(v)$ called its *dominator*. They are defined as follows. For v_n, $DC(v_n) = (v_n)$, $dom(v_n)$ is undefined, and v_n is stable. Consider $v = v_{k+1}$ ($2 \leq k \leq n - 2$). Let u be the leftmost child of v in C_k. Let z be the vertex that covers the edge (u, v). Then:

Definition 7. Domino chain and dominator:
(dc1) If $ind_z(v) = 2$, then $DC(v) = (v)$, $dom(v) = z$ and v is unstable.
(dc2) If $ind_z(v) \geq 4$, then $DC(v) = (v)$, $dom(v) = z$ and v is stable.
(dc3) If $ind_z(v) = 3$ and $DC(z) = (z_1, \ldots, z_i, z)$, then $DC(v) = (z_1, \ldots, z_i, z, v)$ and $dom(v) = dom(z)$. Also, v is stable if and only if z is stable.

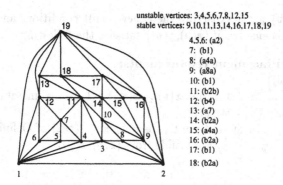

Fig. 1. The leftmost canonical ordering and the grid embedding of G.

As in [3], an unstable vertex of in-degree 2 is called a *room-shift vertex*. The intuition is that a stable vertex can be placed above its leftmost child, while an unstable vertex must be placed at least one x-coordinate to the right of its leftmost child. If v is a room-shift vertex, this can result in putting v directly above its rightmost child w and violating the contour invariants. In this case, we have to shift w to the right in order to "make room" for v.

Example: In Fig 1, $DC(14) = (18, 14)$, $dom(14) = 19$. $DC(9) = (10, 9)$; $dom(9) = 11$. $DC(3) = (3)$; $dom(3) = 4$.

The slope of an contour edge $e = (w_i, w_{i+1})$ is denoted by *slope(e)*. If $slope(e) = 0$, then e is *horizontal*; If $0 < slope(e) < +\infty$, then e is *upward*; If $slope(e) = +\infty$ or $-\infty$, then e is *vertical*; If $-\infty < slope(e) < 0$, then e is *downward*. We are now ready to describe our algorithm. It is a version of the generic shift algorithm. Our contour invariants are as follows:

Definition 8. Contour invariants:
(ci1) $x(w_1) \leq x(w_2) \leq \ldots \leq x(w_{m-1}) \leq x(w_m)$.
(ci2) Each forward edge (w_{i-1}, w_i) is either horizontal, or upward, or vertical. If w_i is unstable, then (w_{i-1}, w_i) must be horizontal or upward.
(ci3) Each backward edge (w_{i-1}, w_i) is either downward, or horizontal.

Lemma 9. *Let $w_p, \ldots, w_q$ be the children of $v = v_{k+1}$ in the contour C_k. Assume the contour invariants hold. Then $x(w_p) < x(w_{p+1})$. If $deg^-(v) > 2$ and v is unstable, then $x(w_{p+1}) < x(w_{p+2})$.*

Proof. If $e_1 = (w_p, w_{p+1})$ is a backward edge, then e_1 cannot be vertical. If e_1 is a forward edge, then w_{p+1} is unstable and e_1 cannot be vertical. In either case, we have $x(w_p) < x(w_{p+1})$. Suppose $deg^-(v) > 2$ and v is unstable. If $e_2 = (w_{p+1}, w_{p+2})$ is a backward edge, then e_2 cannot be vertical. If e_2 is a forward edge, then since v is unstable and w_{p+2} is the third child of v, w_{p+2} is unstable. Hence e_2 cannot be vertical. In either case, $x(w_{p+1}) < x(w_{p+2})$. $\square$

Suppose we are about to add $v = v_{k+1}$ $(3 \leq k \leq n - 2)$. Our algorithm must perform shift operation on some contour vertex (if necessary), and place

v at a point $(x(v), y(v))$ satisfying the generic shift conditions and the contour invariants. This is ensured if $(x(v), y(v))$ satisfies the following:

Definition 10. Placement requirements:
(pr1) $x(v) < x(w_q)$.
(pr2) $x(v) \geq x(w_p)$, if v is stable. $x(v) \geq x(w_p) + 1$, if v is unstable.
(pr3) $y(v) \geq \max\{y(w_p), y(w_q)\}$.
(pr4) $(x(v), y(v))$ is located above C_k, as defined in (gs2) of Definition 3.
(pr5) all vertices $w_p, \ldots, w_q$ are visible from $(x(v), y(v))$.

The placement rules for v depend on: (a) the in-degree of v; (b) v is stable or unstable; (c) v is forward-oriented, backward-oriented, or crossing-valley; (d) the slope of the edges covered by v. Let W and H denote the width and the height of the current grid. First consider the case $deg^-(v) = 2$.

Case (a1): $deg^-(v) = 2$, v is unstable and forward-oriented, (w_p, w_q) is horizontal. v is placed as in Fig 2 (a1): $x(v) = x(w_p) + 1$ and $y(v) = y(w_p) + 1$. If $x(v) = x(w_q)$ perform $shift(w_q)$. W is increased by ≤ 1. H is increased by ≤ 1.

Case (a2): $deg^-(v) = 2$, v is unstable and forward-oriented, (w_p, w_q) is upward. v is placed as in Fig 2 (a2): $x(v) = x(w_p) + 1$ and $y(v) = y(w_q)$. If $x(v) = x(w_q)$ then perform $shift(w_q)$. W is increased by ≤ 1. H is unchanged.

Case (a3): $deg^-(v) = 2$, v is unstable and backward-oriented, the edge (w_p, w_q) is horizontal. v is placed as in Fig 2 (a3): $x(v) = x(w_p) + 1$ and $y(v) = y(w_p) + 1$. If $x(v) = x(w_q)$ then perform $shift(w_q)$. W is increased by at most 1. H is increased by at most 1.

Case (a4a): $deg^-(v) = 2$, v is unstable and backward-oriented, (w_p, w_q) is downward. v is placed as in Fig 2 (a4a): $x(v) = x(w_p) + 1$ and $y(v) = y(w_p)$. If $x(v) = x(w_q)$ perform $shift(w_q)$. W is increased by ≤ 1. H is unchanged.

Case (a4b): Same as (a4a). Fig 2 (a4b) shows an alternative rule: $x(v) = x(w_p) + 1$ and $y(v) = y(w_p) + 1$. If $x(v) = x(w_q)$ then perform $shift(w_q)$. W is increased by at most 1. H is increased by at most 1.

Case (a5): $deg^-(v) = 2$, v is stable and forward-oriented, (w_p, w_q) is horizontal. v is placed as in Fig 2 (a5): $x(v) = x(w_p)$ and $y(v) = y(w_p) + 1$. W is unchanged. H is increased by at most 1.

Case (a6): $deg^-(v) = 2$, v is stable and forward-oriented, the edge (w_p, w_q) is upward. v is placed as in Fig 2 (a6): $x(v) = x(w_p)$ and $y(v) = y(w_q)$. W is unchanged. H is unchanged.

Case (a7): $deg^-(v) = 2$, v is stable and backward-oriented, (w_p, w_q) is horizontal. v is placed as in Fig 2 (a7): $x(v) = x(w_p)$ and $y(v) = y(w_p) + 1$. W is unchanged. H is increased by at most 1.

Case (a8a): $deg^-(v) = 2$, v is stable and backward-oriented, (w_p, w_q) is downward. v is placed as in Fig 2 (a8a): $x(v) = x(w_p) + 1$ and $y(v) = y(w_p)$. If $x(v) = x(w_q)$ then perform $shift(w_q)$. W is increased by ≤ 1. H is unchanged.

Case (a8b): Same as (a8a). Fig 2 (a8b) is an alternative placement of v: $x(v) = x(w_p)$ and $y(v) = y(w_p) + 1$. W is unchanged. H is increased by ≤ 1.

Next consider the case $deg^-(v) \geq 3$. By Lemma 6 and the contour invariants, v's children are embedded as a *V-shaped* polygonal line P. The vertex among w_p and w_q with larger y-coordinate is called the *high-end* vertex of v. Let (w_i, w_{i+1})

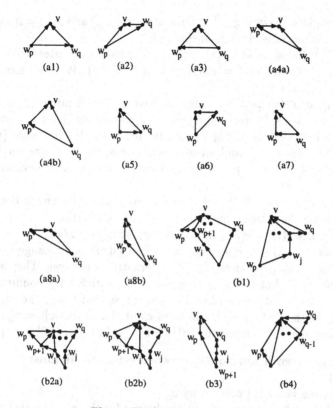

Fig. 2. The placement rules.

be the last downward edge in P. (If no such edge exists, let $w_i = w_p$). Let (w_{j-1}, w_j) be the first upward or vertical edge in P. (If no such edge exists, let $w_j = w_q$). Consider any point $(x(v), y(v))$ such that:

(i) $x(w_i) \leq x(v) < x(w_j)$, if either (w_{j-1}, w_j) or (w_j, w_{j+1}) is vertical; $x(w_i) \leq x(v) \leq x(w_j)$, if neither (w_{j-1}, w_j) nor (w_j, w_{j+1}) is vertical.

(ii) $y(v) \geq \max\{y(w_p), y(w_q)\} + 1$, if the edge adjacent to the high-end vertex of v is horizontal; $y(v) \geq \max\{y(w_p), y(w_q)\}$, if the edge adjacent to the high-end vertex of v is not horizontal.

It is easy to see that such a point satisfies (pr3), (pr4), and (pr5). In the following placement rules, the coordinate $x(v)$, $y(v)$ satisfies the conditions (i), (ii), (pr1) and (pr2). Note that **no** shift operation is needed for any contour vertex.

Case (b1): $deg^-(v) \geq 3$, the edge covered by v and adjacent to the high-end vertex of v is horizontal. v is placed as in Fig 2 (b1) (two examples are shown): Let $y(v) = \max\{y(w_p), y(w_q)\} + 1$ and determine $x(v)$ as follows. (a) v is stable: Let $x(v) = x(w_i)$. (b) v is unstable: If $w_i = w_p$, let $x(v) = x(w_p) + 1$. If $w_i \neq w_p$, let $x(v) = x(w_i)$. W is unchanged. H is increased by at most 1.

Case (b2a): $deg^-(v) \geq 3$; the edge covered by v and adjacent to the high-end vertex of v is not horizontal; w_p is the high-end vertex of v; and $x(w_j) \geq$

$x(w_p)+2$. v is placed as in Fig 2 (b2a): Let $y(v) = y(w_p)$ and $x(v) = \max\{x(w_p)+1, x(w_i)\}$. W is unchanged. H is unchanged.

Case (b2b): Same as in (b2a). Fig 2 (b2b) shows an alternative rule: Let $y(v) = y(w_p) + 1$, and $x(v) = \max\{x(w_p) + 1, x(w_i)\}$. W is unchanged. H is increased by at most 1.

Case (b3): $deg^-(v) \geq 3$; the edge covered by v and adjacent to the high-end vertex of v is not horizontal; w_p is the high-end vertex of v; and $x(w_j) = x(w_p) + 1$. v is placed as in Fig 2 (b3). In this case, the first edge (w_p, w_{p+1}) covered by v is downward and all other edges covered by v are vertical. So v must be stable. Let $y(v) = y(w_p) + 1$ and $x(v) = x(w_p)$. W is unchanged. H is increased by at most 1.

Case (b4): $deg^-(v) \geq 3$; the edge covered by v and adjacent to the high-end vertex of v is not horizontal; w_q is the only high-end vertex of v. v is placed as in Fig 2 (b4): Let $y(v) = y(w_q)$. If v is stable, let $x(v) = x(w_i)$. If v is unstable, let $x(v) = \max\{x(w_p) + 1, x(w_i)\}$. W is unchanged. H is unchanged.

In the cases (a4), (a8), and (b2), two alternatives are given. They are chosen as follows: Suppose that $v = v_{k+1}$ ($k + 1 < n - 1$) satisfies the conditions of the rule (a4) (or (a8) or (b2), resp.) Let u be the vertex that covers the edge (w_p, v). Let z_1, z_2, z_3 be the first, second, and third child of u. If the following conditions hold, we must place v by using the rule (a4b), (or (a8b) or (b2b), resp.)

Definition 11. Avoid-horizontal-forward-edge conditions:
(1) $deg^-(u) \geq 3$;
(2) (w_p, v) is the last edge covered by u;
(3) either of the following two conditions hold: (3a) the edge (z_1, z_2) is upward; or (3b) $z_3 \neq v$, the edge (z_2, z_3) is upward, and $y(z_1) \leq y(w_p)$.

Remark 1: The rules (a4a), (a8a) and (b2a) are the only rules that create horizontal forward edges. By using the rules (a4b), (a8b) and (b2b), this can be avoided. More precisely, if v is placed by using the rules (a4a), (a8a) or (b2a), the edge (w_p, v) is horizontal and v is not a peak point (see Fig 3(1)). If v is placed by using the rules (a4b), (a8b) or (b2b), the edge (w_p, v) is either upward or vertical, and v is a peak point (Fig 3 (2)).

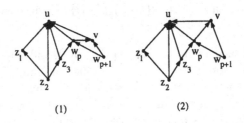

(1)   (2)

Fig. 3. The avoid-horizoantal-forward-edge conditions.

Remark 2: By the definition of π_{left}, z_1, z_2, z_3 are embedded before v. So the avoid-horizontal-forward-edge conditions can be checked when v is embedded.

For a vertex v satisfying the conditions of the rule (a4) (or (b2), respectively), if the avoid-horizontal-forward-edge conditions do not hold, then we place v by using the rule (a4a) (or (b2a), respectively). For a vertex v satisfying the conditions of the rule (a8), if the avoid-horizontal-forward-edge conditions do not hold, v is called a *free-lance* vertex. A free-lance vertex can be placed by using either the rule (a8a) or (a8b). We use this freedom to adjust the height and the width of the embedding as follows. Let n_r be the number of room-shift vertices. Let $d = (n/2-1) - n_r$. We place the first (at most) d free-lance vertices by using the rule (a8a). All other (if any) free-lance vertices are placed by using the rule (a8b). This completes the description of our algorithm.

Example: The graph G in Fig 1 is embedded by using our algorithm. (Recall $v_n = 19$ is not embedded). The rule used for each vertex is as indicated. The free-lance vertex 9 is placed by using the rule (a8a). The vertex 11 is placed by using the rule (b2b) since it satisfies the avoid-horizontal-forward-edge conditions.

4 Bounding the Grid Size

Let W and H be the width and the height of the final grid. If v is placed by using the rule (ai) ($i = 1, 2, 3, 4a, 4b, 5, 6, 7, 8a, 8b$) or the rule (b$i$) ($i = 1, 2a, 2b, 3, 4$), we call v an (ai) or a (bi) vertex. Let a_i be the number of (ai) vertices and b_i be the number of (bi) vertices. A vertex that is the first one reaching a new y-coordinate is called a *height-increasing* vertex. A height-increasing vertex must be either an (ai) vertex (for $i = 1, 3, 4b, 5, 7, 8b$), or a (bi) vertex (for $i = 1, 2b, 3$). A height-increasing vertex of type (ai) or (bi) is called an (a'i) or a (b'i) vertex. Let a_i' and b_i' be the number of (a'i) and (b'i) vertices, respectively. We have:

$$H = 1 + a_1' + a_3' + a_{4b}' + a_5' + a_7' + a_{8b}' + b_1' + b_{2b}' + b_3' \tag{1}$$

If the placement of v increases W, v is a *width-increasing* vertex. A width-increasing vertex must be an (ai) vertex for $i = 1, 2, 3, 4a, 4b, 8a$. Let a_i'' ($i = 1, 2, 3, 4a, 4b, 8a$) be the number of (ai) vertices that are **not** width-increasing:

$$W = 2 + (a_1 + a_2 + a_3 + a_{4a} + a_{4b} + a_{8a}) - (a_1'' + a_2'' + a_3'' + a_{4a}'' + a_{4b}'' + a_{8a}'') \tag{2}$$

Since v_1, v_2, v_3 are neither (ai) nor (bi) vertices and v_n is not embedded, we have:

$$a_1 + a_2 + a_3 + a_{4a} + a_{4b} + a_5 + a_6 + a_7 + a_{8a} + a_{8b} + b_1 + b_{2a} + b_{2b} + b_3 + b_4 = n - 4 \tag{3}$$

Bound on $H + W$: From equations (1), (2), and (3), we have:
$$\begin{aligned}
W + H &= 3 + (a_1 + a_2 + a_3 + a_{4a} + a_{4b} + a_5' + a_6 + a_7' + a_{8a} + a_{8b}' + b_1' + b_{2a} \\
&\quad + b_{2b}' + b_3' + b_4) + (a_1' + a_3' + a_{4b}') - (a_6 + b_{2a} + b_4) - \\
&\quad (a_1'' + a_2'' + a_3'' + a_{4a}'' + a_{4b}'' + a_{8a}'') \\
&\leq (n-1) + (a_1' + a_3' + a_{4b}') - (a_6 + b_{2a} + b_4) - \\
&\quad (a_1'' + a_2'' + a_3'' + a_{4a}'' + a_{4b}'' + a_{8a}'') \tag{4}
\end{aligned}$$

Let D denote the set of (a1), (a3), (a4b) vertices; D' the set of (a'1), (a'3), (a'4b) vertices; and J the set of (a6), (b2a), (b4) vertices. Thus $|D'| = a_1' + a_3' + a_{4b}'$

and $|J| = a_6 + b_{2a} + b_4$. A vertex in D' may increase both W and H by 1. A vertex in J increases neither W nor H. We will show $|D'| \leq |J|$. Define:

$K = \{v | v$ is unstable and (w_p, v) is upward, where w_p is the leftmost child of $v\}$

Note that $D' \subseteq D \subseteq K$. For each $v \in K$, we define a sequence of vertices $S(v) = (x_0 = v, x_1, \ldots, x_k, u)$ (possibly $k = 0$) such that:

(1) For each i ($1 \leq i \leq k$), x_i is an (a2) vertex and covers the edge (w_p, x_{i-1}).

(2) u is either an (a6), (b4), (b2a), or (b2b) vertex and covers the edge (w_p, x_k).

$S(v)$ is defined as follows. Start with $S = (x_0 = v)$. Suppose that x_i has been defined. Let u be the vertex that covers the edge (w_p, x_i), where w_p is the leftmost child of x_i. There are three cases:

Case 1: $deg^-(u) = 2$ and u is unstable. Define $x_{i+1} = u$, and continue. Note that (w_p, x_i) is upward. By our rules, x_{i+1} must be an (a2) vertex.

Case 2: $deg^-(u) = 2$ and u is stable. Let $k = i$ and u be the last vertex of $S(v)$ and we are done. (Since (w_p, x_k) is upward, u must be an (a6) vertex).

Case 3: $deg^-(u) \geq 3$. Let $k = i$ and u be the last vertex of $S(v)$ and we are done. Since u covers the edge (w_p, x_k) and x_k is unstable, there are two subcases:

Case 3A: $ind_u(x_k) = 2$. Since the edge (w_p, x_k) is upward, u is forward-oriented (Fig 4(1)). By the avoid-horizontal-forward-edge conditions, the last edge covered by u is either upward or vertical. So u is a (b4) vertex.

Case 3B: $ind_u(x_k) = 3$ and u is unstable. Let $e_1 = (t_1, w_p)$ and $e_2 = (t_2, t_3)$ be the first and the last edge covered by u (see Fig 4(2)).

(i) Suppose $deg^-(u) > 3$ and $y(t_1) \leq y(t_2)$. By the avoid-horizontal-forward-edge conditions, e_2 is upward or vertical and t_3 is the only high-end vertex of u. So u is placed as a (b4) vertex.

(ii) Suppose $deg^-(u) > 3$ and $y(t_1) > y(t_2)$. If e_2 is horizontal, then t_1 is the only high-end vertex of u and e_1 is not horizontal. If e_2 is upward, then neither e_1 nor e_2 is horizontal. So u is either a (b2a), or (b2b), or (b4) vertex.

(iii) Suppose $deg^-(u) = 3$. Then $(w_p, x_k) = (t_2, t_3)$. Depending on which of t_1 and t_3 is the high-end vertex of u, u is either a (b2a), (b2b) or a (b4) vertex.

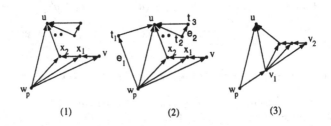

(1)　　　　　　　(2)　　　　　　　(3)

Fig. 4. The last vertex u in the sequence $S(v)$.

Note that in $S(v) = \{x_0 = v, x_1, \ldots, x_k, u\}$, the vertices $x_1, \ldots, x_k$ (if any) are uniquely determined by v. On the other hand, a vertex u can be the last vertex of $S(v_1)$ and $S(v_2)$ for two distinct vertices $v_1, v_2 \in K$, where u satisfies Case 3A for $S(v_1)$ and Case 3B for $S(v_2)$. (See Fig 4(3)).

We construct a directed forest F such that the following hold: (a) The node set of F is a subset of the vertices of G^+; (b) The set of leaf nodes in F is the set D; (c) Each non-root internal node of F is a (b4) vertex, and has exactly two children; (d) Each root node of F is either a (b2b) vertex or in J, and has exactly one child.

The forest F is constructed as follows. We keep a set $Q \subseteq K$. For each $v \in Q$, we find the sequence $S(v)$ and define $parent(v)$ to be the last vertex u of $S(v)$. When a vertex v is put in Q, $parent(v)$ is identified. At any moment, for each vertex $v \in Q$, $parent(v)$ is the root of a tree. Initially, let $Q = D$.

By the remark above, at most two vertices v_1, v_2 in Q can have the same parent u. In this case, u must be unstable and the edge (w_p, u) (where w_p is the leftmost child of u) is upward. (See Fig 4 (3)). So u is a (b4) vertex and is also in K. In this case, we remove v_1, v_2 from Q and put u into Q.

Repeat this until all vertices in Q have distinct parents. Now the parents of the vertices in Q are the roots of the trees of the forest F to be constructed. It is easy to check that F satisfies all above conditions.

Consider a tree T in F. Let $Leaf(T)$ and $Int(T)$ be the number of leaf nodes and internal nodes in T, respectively. Clearly, $Leaf(T) = Int(T)$. Let $D(T)$, $D'(T)$, and $J(T)$ be the number of nodes in T that are in D, D', and J, respectively. We will show $D'(T) \leq J(T)$.

If the root of T is in J, then $D'(T) \leq D(T) = Leaf(T) = Int(T) = J(T)$. Suppose the root of T is a (b2b) vertex. Let v be the leaf node in T that has the smallest y-coordinate among all leaf nodes in T. Then v is not a height-increasing vertex. Thus v is in D but not in D'. Hence, $D'(T) \leq D(T) - 1 = Leaf(T) - 1 = Int(T) - 1 = J(T)$.

Since each vertex in D' corresponds to a distinct leaf node in F and $D'(T) \leq J(T)$ holds for every tree T in F, we have: $|D|' \leq |J|$. By (4), this gives:

$$W + H \leq (n-1) - (a_1'' + a_2'' + a_3'' + a_{4a}'' + a_{4b}'' + a_{8a}'') \leq (n-1) \quad (5)$$

Bound on the width W: We first bound n_r (the number of room-shift vertices). Each room-shift vertex v is associated with a vertex $dom(v)$. It was shown in [3] that, for two distinct room-shift vertices v_1 and v_2, $dom(v_1) \neq dom(v_2)$. By the definition of the dominator, only the forward-oriented vertices and v_n can be dominators of room-shift vertices. By our assumption on canonical ordering, we have $n_f \leq n_b$. Hence:

$$n_r = a_1 + a_2 + a_3 + a_{4a} + a_{4b} \leq n_f + 1 \leq (n-4)/2 + 1 = n/2 - 1$$

Thus $d = (n/2 - 1) - n_r \geq 0$. Our algorithm places at most d free-lance vertices by using the rule (a8a). So $a_{8a} \leq d$. From equation (2), we have:

$$W \leq 2 + (a_1 + a_2 + a_3 + a_{4a} + a_{4b}) + a_{8a} \leq 2 + n_r + (n/2 - 1) - n_r = n/2 + 1 \quad (6)$$

Bound on the heigh H: First suppose the number of free-lance vertices is at least $d = (n/2 - 1) - n_r$. Then our algorithm places $a_{8a} = d$ of them by using the rule (a8a). So $a_1 + a_2 + a_3 + a_{4a} + a_{4b} + a_{8a} = n_r + d = n/2 - 1$. By (2) and (5), we have:

$$H \leq (n-1) - (a_1'' + a_2'' + a_3'' + a_{4a}'' + a_{4b}'' + a_{8a}'') - W = (n-1) - (2 + a_1 + a_2 + a_3 + a_{4a} + a_{4b} + a8a) = (n-1) - (n/2 + 1) < 2(n-2)/3 \tag{7}$$

Next suppose the number of free-lance vertices is less than d. Then no free-lance vertices are placed by using the rule (a8b). So all (a8b) vertices satisfy the avoid-horizontal-forward-edge conditions.

Let A denote the set of (a'1), (a'3), (a'5), (a'7), and (b'1) vertices; B the set of (a'4b), (a'8b), (b'2b) vertices; and C the set of (b'3) vertices. From equation (1), we have: $H = 1 + |A| + |B| + |C|$. For each $v \in A \cup B \cup C$, we define a vertex $mate(v)$ as follows. Consider a vertex $v \in A$. v covers a horizontal edge (x, y) with y-coordinate $y(v) - 1$. (See Fig 5(1)). Define $mate(v)$ to be the vertex among x and y that is not a height-increasing vertex.

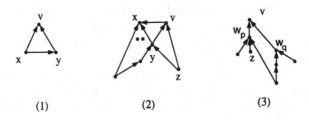

(1)   (2)   (3)

Fig. 5. The definition of the mate vertices.

Consider a vertex $v \in B$. Let y be the first child of v. Let x be the vertex that covers the edge (y, v) (Fig 5(2)). v is a peak point and (x, v) is horizontal. So x is not height-increasing. Define $mate(v) = x$.

Consider a vertex $v \in C$. Let w_p and w_q be the leftmost and the rightmost child of v, respectively (Fig 5(3)). By the rule (b3), we have $y(w_p) \geq y(w_q)$. By the definition of π_{left}, we have $w_p \prec_{left} w_q$. Thus there is a vertex z such that $z \prec_{left} w_p \prec_{left} w_q$ and $y(z) = y(w_q)$. So w_q is not height-increasing. Define $mate(v) = w_q$.

Define: $Mate = \{mate(v) \mid v \in A \cup B \cup C\}$. From the definition, it is easy to check that: (i) If $x \in Mate$, then x is not in $A \cup B \cup C$. (ii) If $x = mate(v)$ for a vertex $v \in C$, then x cannot be $mate(v')$ for any $v' \neq v$. (iii) A vertex x can be the mate vertex for at most two vertices: one $v \in A$ and another $v' \in B$. Hence each vertex in $Mate$ corresponds to at most two vertices in $A \cup B \cup C$. Thus:

$$|A| + |B| + |C| \leq 2|Mate|$$

The vertices v_1, v_2, v_3, v_n are neither in $A \cup B \cup C$ nor in $Mate$. Thus: $|A| + |B| + |C| + |Mate| \leq (n-4)$. So $|A| + |B| + |C| \leq 2(n-4)/3$. From (1), this gives:

$$H = 1 + (|A| + |B| + |C|) \leq 1 + 2(n-4)/3 < 2(n-2)/3 \tag{8}$$

Recall that G^+ has n vertices, and the original graph G has $n - 1$ vertices. So the bounds in (5), (6), (7), (8) imply the bounds on size in Theorem 1.

Implementation of the algorithm: The graph G^+ can be constructed from G in linear time. The canonical ordering π can be computed in $O(n)$ time [10]. The leftmost canonical ordering π_{left} can be computed from π in linear time by using the method in [11]. After π_{left} is known, the algorithm can be implemented by using the method in [3]. So the algorithm takes linear time. This completes the proof of Theorem 1.

References

1. N. Chiba, T. Yamanouchi, and T. Nishizeki, Linear algorithms for convex drawings of planar graphs, in Progress in Graph Theory, J. A. Bondy and U. S. R. Murty (eds.), 1982, pp.153–173.

2. M. Chrobak and G. Kant, Convex grid drawings of 3-connected planar graphs, Tech. Rep. RUU-CS-93-45, Dept. of Comp. Sci. Utrecht University, 1993.

3. M. Chrobak and S. Nakano, Minimum-width grid drawings of planar graphs, in Proc. Workshop on Graph Drawing'94, Princeton, NJ, Oct. 1994.

4. M. Chrobak and T. Payne, A linear time algorithm for drawing planar graphs on a grid, TR UCR-CS-89-1, Dept. of Math. and Comp. Sci., UC at Riverside, 1989.

5. G. Di Battista, P. Eades, R. Tamassia, and I. G. Tollis, Algorithms for drawing graphs: an annotated bibliography, Comput. Goem. Theory Appl. Vol 4, 1994, pp. 235-282.

6. I. Fáry, On straight line representation of planar graphs, Acta. Sci. Math. Szeged, 11, 1948, pp. 229–233.

7. H. de Fraysseix, J. Pach and R. Pollack, How to draw a planar graph on a grid, *Combinatorica* 10, 1990, pp. 41–51.

8. M. Fürer, X. He, M. Y. Kao, and B. Raghavachari, $O(n \log \log n)$-work parallel algorithms for straight line grid embeddings of planar graphs, SIAM J. Disc. Math 7(4), 1994, pp. 632-647.

9. G. Kant, Drawing planar graphs using the *lmc*-ordering, in *Proc. 33th Ann. IEEE Symp. on Found. of Comp. Science*, Pittsburgh, 1992, pp. 101-110.

10. G. Kant and X. He, Two algorithms for finding rectangular duals of planar graphs, in Proc. 19th on Graph-Theoretic Concepts in CS, 1993, LNCS 790, pp. 396-410.

11. F. Preparata and R. Tamassia, Fully dynamic techniques for point location and transitive closure in planar structures, in Proc. 29th FOCS, 1988, pp. 558-567.

12. P. Rosenstiehl and R. Tarjan, Rectilinear planar layouts and bipolar orientations of planar graphs, Discrete & Computational Geometry 1, 1986, pp. 343-353.

13. W. Schnyder, Embedding planar graphs on the grid, Abs. AMS 9, 1988, p. 268.

14. W. Schnyder, Planar graphs and poset dimension, Orders 5, 1989, pp. 323-343.

15. W. Schnyder, Embedding planar graphs on the grid, in Proc. of the 1st Annual ACM-SIAM Symp. on Discrete Algorithms, 1990, pp. 138-147.

16. W. Schneider and W. Trotter, Convex drawings of planar graphs, Abstracts of AMS 13 (5), 1992.

17. S. K. Stein, Convex maps, in Proc Amer Math Soc, Vol. 2, 1951, pp. 464–466.

18. W. T. Tutte, How to draw a graph, Proc. London Math. Soc. 13, 1963, pp. 743–768.

19. K. Wagner, Bemerkungen zum Vierfarben problem, Jahresbericht Deutsch Math 46, 1936, pp. 26–32.

Recognizing Leveled-Planar Dags in Linear Time*

Lenwood S. Heath** and Sriram V. Pemmaraju***

1 Introduction

Let $G = (V, E)$ be a directed acyclic graph (dag). A *leveling* of G is a function $lev : V \to \mathbf{Z}$ mapping the nodes of G to integers such that $lev(v) = lev(u) + 1$ for all $(u, v) \in E$. G is a *leveled dag* if it has a leveling. If $lev(v) = j$, then v is a *level-j node*. Let E_j denote the set of arcs in E from level-j nodes to level-$(j + 1)$ nodes. Without loss of generality, we may assume that the image of *lev* is $\{1, 2, \ldots, m\}$ for some m. Let $V_j = lev^{-1}(j)$ denote the set of level-j nodes. Each V_j is a *level* of G. The leveling partitions V into the levels $V_1, V_2, \ldots, V_m$, and according we denote G as $G = (V_1, V_2, \ldots, V_m; E)$.

Let ℓ_j denote the vertical line in the Cartesian plane $\ell_j = \{(j, y) \mid y \in \mathbf{R}\}$, where $\mathbf{R}$ is the set of reals. Suppose G has a planar embedding in which all nodes in V_j are placed on ℓ_j and each arc in E_j, where $1 \leq j < m$, is drawn as a straight line segment between lines ℓ_j and ℓ_{j+1}. Then this planar embedding is called a *directed leveled-planar embedding* of G. Figure 1 shows a directed leveled-planar embedding of a dag. A dag is called a *leveled-planar dag* if it has a directed leveled-planar embedding.

In this paper we present a linear time algorithm for the problem of determining if a given dag has a directed leveled-planar embedding. Our algorithm uses a variation of the PQ-tree data structure introduced by Booth and Lueker [2]. One motivation for our algorithm is that it can be extended to recognize 1-queue dags, thus answering an open question in [6]. Combinatorial and algorithmic results related to queue layouts of dags and posets can be found in [4, 7, 5]. Our algorithms also contrasts leveled-planar undirected graphs and leveled-planar dags, since the problem of recognizing leveled-planar graphs has been shown to be NP-complete by Heath and Rosenberg [8]. Another motivation comes from the importance of the above problem in the area of graph drawing. Our result extends the work of Di Battista and Nardelli [1], Chandramouli and Diwan [3], and Hutton and Lubiw [9]. These authors assume solve the problem assuming certain restrictions on the given dag and leave the general problem open.

The organization of the rest of the paper is as follows. Section 2 discusses the nature of the problem and outlines our approach. Section 3 defines the data

* This research was partially supported by National Science Foundation Grant CCR-9009953.
** Department of Computer Science, Virginia Tech, Blacksburg, VA 24061-0106, heath@cs.vt.edu.
*** Department of Computer Science, University of Iowa, Iowa City, IA 52242-1316, sriram@cs.uiowa.edu.

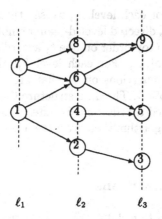

ℓ_1 ℓ_2 ℓ_3

Fig. 1. A leveled-planar dag.

structures (PQ-trees and collections) that we need to represent sets of permutations of nodes in a particular level. Section 4 defines the operations we use to restrict or combine sets of permutations. Section 5 presents our linear time algorithm for recognizing leveled-planar dags.

2   The Problem

It is easy to check whether a dag is leveled in linear time. Therefore, without loss of generality, we may assume that $G = (V_1, V_2, \ldots, V_m; E)$ is a connected, leveled dag, and we wish to determine whether G has a directed leveled-planar embedding.

Suppose G has a directed leveled-planar embedding $\mathcal{E}$. For each j, where $1 \leq j \leq m$, $\mathcal{E}$ determines a total order $\leq_j$ on V_j given by the bottom to top order of the nodes on ℓ_j. Conversely, if a total order $\leq_j$ on V_j is given for each j, then it is easy to check whether those total orders witness a directed leveled-planar embedding of G. It suffices to check that there are no two arcs (u, v) and (x, y) such that $lev(u) = lev(x) = j$, $u <_j x$, and $y <_{j+1} v$. In Figure 1, the total orders are given by $1 <_1 7$, $2 <_2 4 <_2 6 <_2 8$, and $3 <_3 5 <_3 9$.

The problem of recognizing whether a connected leveled dag G is a leveled-planar dag is then equivalent to determining whether there are total orders on all the levels that are witness to a leveled-planar embedding of G. Let G_j denote the subgraph of G induced by $V_1 \cup V_2 \cup \cdots \cup V_j$. (Note that, unlike G, G_j is not necessarily connected.) Each total order on V_j can be thought of as a permutation on V_j. Moreover, for each j, there is a set of permutations Π_j that contains exactly the total orders on V_j that occur in witnesses to directed leveled-planar embeddings of G_j. So to recognize whether G is a leveled-planar dag, we need only compute Π_m and check that it is nonempty. Our basic approach to doing this efficiently is to perform a left-to-right sweep processing the levels in the order $V_1, V_2, \ldots, V_m$. For each level V_j, we say that a permutation π of the nodes in V_j is a *witness* to a directed leveled-planar embedding of G_j if the nodes in V_j appear in a bottom to top order on line ℓ_j according to π in some directed leveled-planar embedding of G_j. For each level V_j, the algorithm constructs a representation of all permutations on V_j that are witness to some directed leveled-planar embedding of G_j. The data structure that we use for maintaining sets of permutations is called a *collection*. So after processing $V_1, V_2, \ldots, V_j$, we have a collection C_j. The algorithm then processes V_{j+1} and uses C_j to construct the next collection C_{j+1}.

3 PQ-trees and Collections

In order to define the collection data structure precisely, we need the PQ-tree data structure of Booth and Lueker [2] to represent sets of permutations. A PQ-tree T for a set S is a rooted tree that contains three types of nodes: leaves, P-nodes, and Q-nodes. The leaves in T are in one-one correspondence with the elements of S. The set S is called the *yield* of T, denoted YIELD(T). The PQ-tree T represents permutations of YIELD(T) according to the following rules: (a) The children of a P-node may be permuted arbitrarily, (b) The children of a Q-node must occur in the given order or in the reverse order. As a special case, the empty PQ-tree ϵ represents the empty set of permutations. The set of permutations represented by T is denoted by PERM(T). The *yield* YIELD(r) of a node r in T is the yield of the subtree rooted at r. Without loss of generality, we may assume that every P-node has 3 or more children and that every Q-node has 2 or more children. A *collection* is a finite set of PQ-trees with pairwise disjoint yields.

For any PQ-tree T and connected leveled-planar dag F with k levels, for some $k \geq 0$, we say that T *represents* F if and only if PERM(T) is the set of all permutations of the level-k nodes in F that witness some leveled-planar embedding of F. For each level V_j, our algorithm maintains a collection C_j satisfying the property stated in the following theorem.

Theorem 1. *For each j, $1 \leq j \leq m$, and for each connected component F in G_j, there is a corresponding PQ-tree $T[F]$ in C_j that represents F.*

Since $G = G_m$ is connected, the above theorem implies that C_m contains a single PQ-tree $T[G]$, that represents G. So C_m contains a non-empty PQ-tree if and only if G has a directed leveled-planar embedding. Thus the goal of our algorithm is to compute C_m. The proof of Theorem 1 is inductively established in the following description of the algorithm.

The algorithm initializes $C_1 = \{\text{leaf } v \mid v \in V_1\}$. Thus for $j = 1$ (the base case), the correspondence claimed in Theorem 1 is trivially true. The algorithm then proceeds to inductively construct $C_2, C_3, \ldots, C_m$ in that order. As an inductive hypothesis, we assume that Theorem 1 holds for some $j \geq 1$.

In order to construct C_{j+1} from C_j, we assume that some information is maintained in each non-leaf node of a PQ-tree in C_j and one additional piece of information is maintained at the root of a PQ-tree in C_j. Let F be any connected component of G_j. By the inductive hypothesis, $T[F]$ is the PQ-tree in C_j that represents F. For any subset S of the set of nodes in V_j that belong to F, define MEETLEVEL(S) to be the greatest $d \leq j$ such that $V_d, \ldots, V_j$ induces a dag in which all nodes of S occur in the same connected component. For example, in Figure 1, MEETLEVEL($\{3,5\}$) = 1 and MEETLEVEL($\{5,9\}$) = 2. Note that if $|S| > 1$, then MEETLEVEL(S) < j. For a Q-node q in $T[F]$ with ordered children $r_1, r_2, \ldots, r_t$, maintain in node q integers denoted ML(r_i, r_{i+1}), where $1 \leq i < t$, that satisfy ML(r_i, r_{i+1}) = MEETLEVEL(YIELD(r_i)∪YIELD(r_{i+1})). For a P-node p in $T[F]$, maintain in node p a single integer denoted ML(p) that satisfies

$$\text{ML}(p) = \text{MEETLEVEL}(yield(p)).$$

Let S be any subset of the set of nodes in V_j that belong to F. Now define LEFTLEVEL(S) to be the smallest d such that F contains a node in V_d. We always have LEFTLEVEL(S) $\leq$ MEETLEVEL(S) and inequality is possible. At the root of $T[F]$, maintain a single integer denoted LL($T[F]$) satisfying

$$\text{LL}(T[F]) = \text{LEFTLEVEL}(\text{YIELD}(T[F])).$$

When our algorithm computes the collection C_{j+1} from C_j, it also maintains the values of ML and LL in the PQ-trees in C_{j+1}. Note that since every PQ-tree in C_1 is a leaf, ML values are not defined, while LL(T) = 1 for each tree $T \in C_1$.

4 Operations

We have described our data structure and now describe two simple operations on PQ-trees that serve as building blocks of the algorithm that constructs C_{j+1} from C_j.

1. ISOLATE(T, x), where T is a PQ-tree and $x \in$ YIELD(T). This operation returns a PQ-tree T' such that (a) The root of T' is a Q-node with x as its first or last child. (b) PERM(T') is the subset of permutations in PERM(T) in which x is either the first or the last element. If YIELD(T) = $\{x\}$, then ISOLATE(T, x) returns a PQ-tree that is just the single leaf x. If there is no permutation in PERM(T) that has x as its first or last element, then ISOLATE(T, x) returns ϵ.

2. IDENTIFY(T, x, y, z), where T is a PQ-tree, $x, y \in$ YIELD(T), $x \neq y$, and $z \notin$ YIELD(T). Let P be the subset of permutations in PERM(T) in which x and y appear consecutively. Let P' be obtained from P as follows: If P contains the permutation $a, \ldots, b, x, y, c, \ldots, d$, then put in P' the permutation $a, \ldots, b, z, c, \ldots, d$, obtained by replacing x, y by z. The operation IDENTIFY(T, x, y, z) returns a PQ-tree T' such that PERM(T') = P'. Note that P' may be empty, in which case $T' = \epsilon$.

The operation ISOLATE(T, x) is a special case of Booth and Lueker's RE-DUCE operation and can be implemented as follows. Let r be the root of T. If

$x = r$, then ISOLATE(T, x) simply returns T. Otherwise, there are two cases based on whether x is a child of r or not.

Base Case: x is a child of r. If r is a Q-node and x is not its first or last child, then there are no permutations in PERM(T) with x at the end or at the beginning, so the operation returns ϵ. If r is a Q-node and x is either the first or the last child of r, then nothing needs to be done and the operation simply returns T. If r is a P-node, then T is transformed as shown in Figure 2.

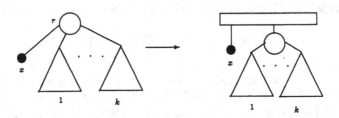

Fig. 2. The transformation of T in the base case of ISOLATE(T, x).

Inductive Case: x is not a child of r. Let T' be the subtree rooted at a child of r whose yield contains x. Let $T'' = $ ISOLATE(T', x). If $T'' = \epsilon$, then ISOLATE(T, x) also returns ϵ. Otherwise, replace T' by T''. The root of T'' is a Q-node with x as either its first or its last child. If r is a P-node, perform the transformation on T shown in Figure 3. If r is a Q-node and T'' is not the

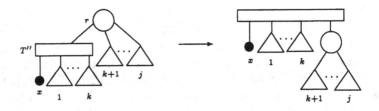

Fig. 3. The transformation of T in the inductive case of ISOLATE(T, x) when r is a P-node.

first or the last child of r, then the algorithm returns ϵ; otherwise, perform the transformation on T shown in Figure 4. The running time of ISOLATE(T, x) is proportional to the depth of x in T.

The operation IDENTIFY(T, x, y, z) can be implemented in the following four steps.

Step 1. Locate r, the node in T that is the least common ancestor of x and y.

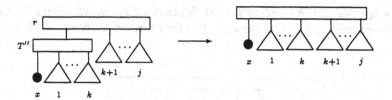

Fig. 4. The transformation of T in the inductive case of ISOLATE(T, x) when r is a Q-node.

Step 2. Let T_1 and T_2 be subtrees of T rooted at a children of r such that $x \in$ YIELD(T_1) and $y \in$ YIELD(T_2). Let $T_1' =$ ISOLATE(T_1, x) and $T_2' =$ ISOLATE(T_2, y). If either $T_1' = \epsilon$ or $T_2' = \epsilon$, then IDENTIFY(T, x, y, z) returns ϵ. Otherwise, replace T_1 by T_1' and T_2 by T_2'. The root of T_1' (respectively, T_2') is a Q-node with x (respectively, y) being the first or the last child of the root.

Step 3. This step depends on whether r is a P-node or a Q-node.

(a) r is a P-node. T is transformed as shown in Figure 5.

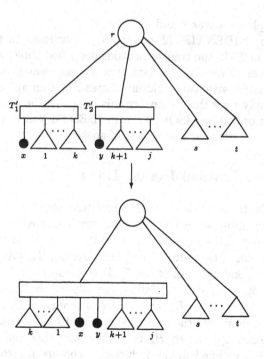

Fig. 5. The transformation of T in IDENTIFY(T, x, y, z) when r is a P-node.

(b) r is a Q-node. If the subtrees T_1 and T_2 are not adjacent children of r, then the operation returns ϵ. Otherwise, T is transformed as shown in Figure 6.

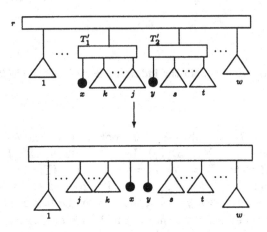

Fig. 6. The transformation of T in IDENTIFY(T, x, y, z) when r is a Q-node.

Step 4. Leaf z replaces leaves x and y.

The running time of IDENTIFY(T, x, y, z) is proportional to the sum of the depths of x and y in T. In the transformations described above we have ignored several special cases caused by the fact that a transformation might lead to the birth of a P-node with two children. Instead of dealing with these cases separately, we simply note that whenever this happens, the P-node is replaced by a Q-node. The operations ISOLATE and IDENTIFY also update ML and LL values. Details are omitted due to lack of space.

5 Recognizing Leveled-Planar Dags

We are now ready to describe how our algorithm constructs C_{j+1} from C_j. To understand the intuition behind the construction, imagine that through a sequence of simple operations (to be described later) dag G_j is transformed into dag G_{j+1}. This yields a sequence of dags $H_1, H_2, \ldots, H_k$ with $G_j = H_1$ and $G_{j+1} = H_k$. Correspondingly, collection C_j is transformed into collection C_{j+1} via a sequence of operations that mimic those applied to H_s, $1 \leq s < k$. This yields a sequence of collections $D_1, D_2, \ldots, D_k$, where $C_j = D_1$ and $C_{j+1} = D_k$. In what follows, we show that the operation applied to D_s, for each s, $1 \leq s < k$, mimics the operation applied to H_s in such a way that the correspondence between G_j and C_j, claimed in the induction hypothesis, also exists between G_{j+1} and C_{j+1}.

We use the following three operations to transform G_j into G_{j+1}:

/* Algorithm for transforming dag G_j into dag G_{j+1} */

$H:=G_j$;

/* GROWTH PHASE */
for all connected components F in H **do**
       for all level-j nodes u in F **do**
           $N_u := \{v[u] \mid (u, v) \in E_j\}$;
           Replace F in H by GROW(F, u, N_u);

/* MERGE PHASE */
for all pairs of level-$(j + 1)$ nodes $(v[X], v[Y])$ in H **do**
       if $v[X]$ and $v[Y]$ belong to the same connected component F **then**
           Replace F in H by MERGE1$(F, v[X], v[Y], v[X, Y])$
       else if $v[X]$ and $v[Y]$ belong to components F_1 and F_2 **then**
           Replace F_1 and F_2 in H by MERGE2$(F_1, F_2, v[X], v[Y], v[X, Y])$;

/* CLEANUP PHASE */
Relabel each node $v[X]$ in H as v;
Add all the level-$(j + 1)$ sources in G to H;

$G_{j+1}:=H$;

Fig. 7. Transforming G_j into G_{j+1}

1. GROW(F, u, S), where F is a connected, leveled-planar dag with k or $k + 1$ levels, for some $k \geq 0$, u is a level-j node in F, and S is a set of nodes not in F. The operation returns the dag obtained by adding arcs (u, v) for all $v \in S$, to F.
2. MERGE1(F, u, v, w), where F is a connected, leveled-planar dag with k levels, for some $k \geq 0$, u and v are distinct level-$(j + 1)$ nodes in F, and w is a node not in F. The operation returns the dag obtained by identifying nodes u and v in F and replacing the resulting node by w.
3. MERGE2(F_1, F_2, u, v, w), where F_1 and F_2 are connected, leveled-planar dags, each with k levels, for some $k \geq 0$, u is a level-$(j + 1)$ node in F_1 and v is a level-$(j + 1)$ node in F_2, and w is a node not in F_1 or F_2. The operation returns the dag obtained by identifying nodes u and v in F and replacing the resulting node by w.

Figure 7 shows the algorithm that uses the above operations to transform G_j into G_{j+1}. In this algorithm, H is initialized to G_j and then transformed into G_{j+1} by repeatedly applying the three operations described above in three *phases*. In the GROWTH PHASE, to each level-j node in H with outdegree equal to d, d out-neighbors are attached. Thus after the GROWTH PHASE, each level-$(j+1)$ node v in G with p in-neighbors is represented by p copies in H, each copy having in-degree equal to 1. Note that these p copies have labels $v[u_1], v[u_2], \ldots v[u_p]$,

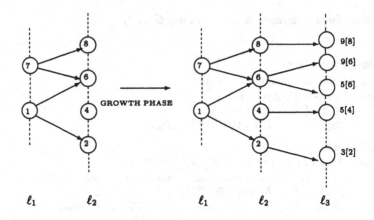

Fig. 8. Illustration of the GROWTH PHASE.

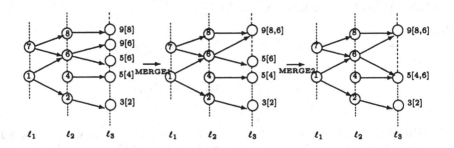

Fig. 9. Illustration of the MERGE PHASE.

where $u_1, u_2, \ldots, u_p$ are the p in-neighbors of v in G. Figure 8 illustrates the GROWTH PHASE applied to G_2, the subgraph of the dag in Figure 1, induced by the first two levels.

We extend the $v[u]$ notation to include $v[X]$, where X is a sequence of distinct level-j nodes that are all adjacent to v; we think of X as a set of nodes, while $v[X]$ is a single level-$(j+1)$ node that is adjacent to each $u \in X$. In the MERGE PHASE, each pair of level-$(j+1)$ nodes $(v[X], v[Y])$, where $X \cap Y = \emptyset$, is merged into a single node with label $v[X, Y]$. Note that $v[X]$ and $v[Y]$ may belong to the same connected component or to different connected components and accordingly the operations MERGE1 or MERGE2 are used. Figure 9 illustrates the MERGE PHASE applied to the dag resulting from the GROWTH PHASE shown in Figure 8. In the CLEANUP PHASE, each level-$(j+1)$ node in H with label $v[X]$ is relabeled v so as to match its name in G_{j+1}. For example, the dag obtained after the MERGE PHASE in Figure 9 contains nodes 9[8,6], 5[4,6], and 3[2] which are relabeled 9, 5, and 3 in the CLEANUP PHASE. Finally level-$(j+1)$ sources in G are added to H. This completes the transformation of H into G_{j+1}.

Having described the sequence of operations that transforms G_j into G_{j+1}, we now describe the parallel sequence of operations that transforms C_j into C_{j+1}. As mentioned earlier, the operations on collections closely mimic the operations on dags and so corresponding to each of the operations GROW, MERGE1, and MERGE2 that operate on connected leveled-planar dags, we define the following three operations that operate on PQ-trees:

1. GROW(T, u, S), where T is a PQ-tree, $u \in$ YIELD(T), and $S \cap$ YIELD$(T) = \emptyset$.
2. MERGE1(T, u, v, w), where T is a PQ-tree, $u, v \in$ YIELD(T), and $w \notin$ YIELD(T).
3. MERGE2(T_1, T_2, u, v, w), where T_1 and T_2 are PQ-trees, $u \in$ YIELD(T_1), $v \in$ YIELD(T_2), and $w \notin$ YIELD$(T_1) \cup$ YIELD(T_2).

Each of these operation returns a PQ-tree. To see how these operations are applied to collections, let us suppose that a collection D is initialized to C_j. Imagine that when F is replaced by GROW(F, u, N_u) in H in the GROWTH PHASE, the corresponding PQ-tree $T[F]$ is replaced by GROW$(T[F], u, N_u)$ in D. Similarly, in the MERGE PHASE, when F is replaced by MERGE1$(F, v[X], v[Y], v[X, Y])$, then the corresponding PQ-tree $T[F]$ is replaced by the PQ-tree MERGE1$(T[F], v[X], v[Y], v[X, Y])$. Finally, when F_1 and F_2 are replaced by the dag

$$\text{MERGE2}(F_1, F_2, v[X], v[Y], v[X, Y]),$$

then the corresponding PQ-trees $T[F_1]$ and $T[F_2]$ are replaced by

$$\text{MERGE2}(T[F_1], T[F_2], v[X], v[Y], v[X, Y]).$$

Corresponding to the relabeling of the level-$(j+1)$ nodes in H in the CLEANUP PHASE, all the leaves of trees in D are similarly relabeled and corresponding to the addition of the level-$(j + 1)$ sources to H, PQ-trees that just contain a leaf are added to D. This completes the construction of C_{j+1}. Thus the transformation of C_j into C_{j+1} can also be viewed as proceeding in three distinct phases: GROWTH PHASE, MERGE PHASE, and CLEANUP PHASE. Our task now is to describe the three operations on PQ-trees mentioned above and to prove their correctness.

1. **GROW(T, u, S).** This operation returns a PQ-tree T' obtained from T as follows. If $S = \emptyset$, then T' is obtained by deleting u from T. Otherwise, let $S = \{v_1, v_2, \ldots, v_k\}$, for some $k \geq 1$. If $k = 1$, then T' is obtained by replacing u by the leaf v_1. If $k = 2$, then T' is obtained by replacing u by a Q-node q whose children are the leaves v_1 and v_2. Then the information ML$(v_1, v_2) = j$ is inserted at q. If $k > 2$, then T' is obtained by replacing u by a P-node p whose children are the leaves $v_1, v_2, \ldots, v_k$. Then the information ML$(p) = j$ is inserted at the new P-node. The operation leaves the LL value of T unchanged, that is, LL$(T') =$ LL(T). The correctness of GROW(T, u, S) is embodied in the following lemma, which follows from the discussion.

Lemma 2. *After the* GROWTH PHASE, *for each connected component F in H, there is a PQ-tree $T[F]$ in D that represents F.*

2. **MERGE1**(T, u, v, w). This operation simply calls IDENTIFY(T, u, v, w) and returns the PQ-tree obtained. The correctness of MERGE1(T, u, v, w) is embodied in the following lemma.

Lemma 3. *Suppose that F is a leveled-planar dag with j levels. Further suppose that T is a PQ-tree that represents F. If $F' =$ MERGE1(F, u, v, w) and $T' =$ MERGE1(T, u, v, w), then T' represents F.*

3. **MERGE2**(T_1, T_2, u, v, w). Without loss of generality, suppose that $LL(T_1) \leq LL(T_2)$. The trees T_1 and T_2 are merged in two steps. In the first step, the PQ-tree T_2 is attached to T_1 at an appropriate location. The resulting tree, T_3, contains the leaves u and v. In the second step, these two leaves in T_3 are identified into one leaf w using the operation IDENTIFY. The two steps are discussed in detail below.

Step 1. Attaching T_2 to T_1. Start with the leaf u in T_1 and proceed upward in T_1 until a node r' and its parent r are encountered such that:

1. r is a P-node with $ML(r) < LL(T_2)$. T_3 is obtained by attaching T_2 as a child of r in T_1.
2. r is a Q-node with ordered children $r_1, r_2, \ldots, r_t$, $r' = r_1$, and $ML(r_1, r_2) < LL(T_2)$. T_3 is obtained by replacing r_1 in T_1 with a Q-node q having two children, r_1 and the root of T_2. The case where $r' = r_t$ and $ML(r_{t-1}, r_t) < LL(T_2)$ is symmetric.
3. r is a Q-node with ordered children $r_1, r_2, \ldots, r_t$, $r' = r_i$, for some i satisfying $1 < i < t$, and both $ML(r_{i-1}, r_i) < LL(T_2)$ and $ML(r_i, r_{i+1}) < LL(T_2)$. T_3 is obtained by replacing r_i in T_1 with a Q-node q having two children, r_i and and root of T_2.
4. r is a Q-node with children $r_1, r_2, \ldots, r_t$, $r' = r_i$, $1 < i < t$, and

$$ML(r_{i-1}, r_i) < LL(T_2) \leq ML(r_i, r_{i+1}).$$

   T_3 is obtained by attaching T_2 as a child of r between r_{i-1} and r_i. The case where

$$ML(r_i, r_{i+1}) < LL(T_2) \leq ML(r_{i-1}, r_i)$$

   is symmetric.
5. r' is the root of T_1. In this case, construct T_3 by making its root a Q-node q with two children, r' and the root of T_2.

Step 2. Identifying u and v in T_3. Return IDENTIFY(T_3, u, v, w).
Steps 1 and 2 described above also update ML and LL values. Details are not presented due to lack of space. The following lemma establishes the correctness of MERGE2(T_1, T_2, u, v, w).

Lemma 4. *Suppose that F_1 and F_2 are connected leveled-planar dags and T_1 and T_2 are PQ-trees that represent F_1 and F_2 respectively. If $F = \text{MERGE2}(F_1, F_2, u, v, w)$ and $T = \text{MERGE2}(T_1, T_2, u, v, w)$, then T represents F.*

This completes the discussion of our algorithm and establishes this theorem.

Theorem 5. *A leveled-planar dag can be recognized in linear time.*

The time complexity follows from an amortized analysis that we sketch here. It suffices to show that the time complexity of all the MERGE1 and MERGE2 operations together is linear. Each arc in the dag G is allocated three credits; since G is planar, the total number of credits is linear. In MERGE1, there are two paths in the dag involved in forming a new face; since each arc is in two faces, two credits from each arc in the face pays for the MERGE1. In MERGE2, the work is proportional to the height of the shorter dag; there is always a path in the dag that has a credit on each arc. We conclude that the allocated credits are sufficient and that the time complexity is linear.

References

1. Giuseppe Di Battista and Enrico Nardelli. Hierarchies and planarity theory. *IEEE Transactions on Systems, Man, and Cybernetics*, 18:1035–1046, 1988.
2. Kellogg S. Booth and George S. Lueker. Testing for the consecutive ones property, interval graphs, and graph planarity using PQ-tree algorithms. *Journal of Computer and System Sciences*, 13:335–379, 1976.
3. Mahadevan Chandramouli and A. A. Diwan. Upward numbering testing for triconnected graphs (an extended abstract). Accepted at Graph Drawing 95.
4. Lenwood S. Heath and Sriram V. Pemmaraju. Stack and queue layouts of posets. Technical Report 93-06, University of Iowa, 1993. Submitted.
5. Lenwood S. Heath and Sriram V. Pemmaraju. Stack and queue layouts of directed acyclic graphs: Part II. Technical Report 95-06, University of Iowa, 1995. Submitted.
6. Lenwood S. Heath, Sriram V. Pemmaraju, and Ann Trenk. Stack and queue layouts of directed acyclic graphs. In William T. Trotter, editor, *Planar Graphs*, pages 5–11, Providence, RI, 1993. American Mathematical Society.
7. Lenwood S. Heath, Sriram V. Pemmaraju, and Ann Trenk. Stack and queue layouts of directed acyclic graphs: Part I. Technical Report 95-03, University of Iowa, 1995. Submitted.
8. Lenwood S. Heath and Arnold L. Rosenberg. Laying out graphs using queues. *SIAM Journal on Computing*, 21(5):927–958, 1992.
9. Michael D. Hutton and Anna Lubiw. Upward planar drawing of single source acyclic digraphs. In *Proceedings of the 2nd Symposium on Discrete Algorithms*, pages 203–211, 1991.

Contact Graphs of Curves
(extended abstract)

Petr Hliněný

Dept. of Applied Mathematics, Charles University,
Malostr. nám. 25, 118 00 Praha 1, Czech republic
(E-mail: hlineny@kam.ms.mff.cuni.cz)

Abstract. Contact graphs are a special kind of intersection graphs of geometrical objects in which we do not allow the objects to cross but only to touch each other. Contact graphs of simple curves (and line segments as a special case) in the plane are considered. Several classes of contact graphs are introduced and their properties and inclusions between them are studied. Also the relation between planar and contact graphs is mentioned. Finally, it is proved that the recognition of contact graphs of curves (line segments) is NP–complete (NP–hard) even for planar graphs.

1 Introduction

The intersection graphs of geometrical objects have been extensively studied for their many practical applications. Formally the *intersection graph* of a set family $\mathcal{M}$ is defined as a graph G with the vertex set $V(G) = \mathcal{M}$ and the edge set $E(G) = \{\{A, B\} \subseteq \mathcal{M} \,|\, A \neq B, A \cap B \neq \emptyset\}$. Probably the first type studied were interval graphs (intersection graphs of intervals on a line), owing to their applications in biology, see [14],[1]. We may also mention other kinds of intersection graphs such as the intersection graphs of chords of a circle (circle graphs [2]), of boxes in the space [15], of curves or line segments in the plane [3],[16],[11],[12].

A special type of geometrical intersection graphs—the *contact graphs*, for which we do not allow the geometrical objects to cross but only to touch each other, are considered here. Unlike the general intersection graphs, in this field only a few results are known. There is a nice old result of Koebe [10] about representations of planar graphs as contact graphs of circles in the plane. In [5] a similar result about contact graphs of triangles is proved. The contact graphs of line segments are considered in [4],[6] and [17]. It is proved that every bipartite planar graph is a contact graph of vertical and horizontal line segments [4], and for contact graphs of line segments of any direction, with contact of 2 segments in one contact point, a characterization is given in [17].

We follow the ideas of intersection graphs of curves and of contact graphs of segments, and generalize the definition to contact graphs of simple curves in the plane. We also allow a contact of more than 2 curves in a point and for such contact points we distinguish one-sided and two-sided contacts. We define classes of contact graphs and study the inclusions among them and their properties such as the maximal clique and the chromatic number. Then we show some relations of contact graphs and planar graphs, and using them we prove the main result—that the recognition of contact graphs of curves (line segments) is NP–complete (NP–hard) even for planar graphs, while the similar question for planar triangulations is solvable in polynomial time.

For complete proofs of the presented results refer to [9].

2 Curve Contact Representations

2.1 Definitions

Simple curves of finite length (Jordan curves) in the plane are considered. Each curve has two *endpoints* and all of its other points are called interior points; they form the *interior* of the curve. We say that a curve φ *ends (passes through)* in a point X if X is an endpoint (interior point) of φ.

Definition. A finite set $\mathcal{R}$ of curves in the plane is called a *curve contact representation* of a graph G if interiors of any two curves of $\mathcal{R}$ are disjoint and G is the intersection graph of $\mathcal{R}$. The graph G is called the *contact graph* of $\mathcal{R}$ and denoted by $G(\mathcal{R})$. A curve contact representation $\mathcal{R}$ is said to be a *line segment contact representation* if each curve of $\mathcal{R}$ is a line segment.
A graph H is called a *contact graph of curves* (*contact graph of line segments*) if there exists a curve contact representation (line segment contact representation) $\mathcal{S}$ such that $H \cong G(\mathcal{S})$.

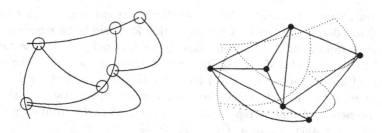

Fig. 1. An example of a curve contact representation of a graph

A curve contact representation is called simply a *representation*, a contact graph of curves simply *contact graph*. Any subset $\mathcal{S} \subseteq \mathcal{R}$ is called a *subrepresentation* of $\mathcal{R}$. A point C of the plane is said to be a *contact point* of a representation $\mathcal{R}$ if it is contained in at least two curves of $\mathcal{R}$. The *degree* of a contact point C in $\mathcal{R}$ is the number of curves of $\mathcal{R}$ containing C, a contact point of degree k is called a *k–contact point*. Note that for any k–contact point

C of a representation $\mathcal{R}$ either all k curves of $\mathcal{R}$ containing C end in C or one curve is passing through C and the other $k - 1$ curves end in C. We say an endpoint of a curve is *free* if it is not a contact point.

In Figure 1 an example of a curve contact representation and its contact graph is given. For a better view of the representation we introduce the following convention for drawing the contact representations: Each contact point is emphasized by a circle around it and the curves of this contact point are drawn into the circle but not necessarily touching each other.

Fig. 2. The difference between one-sided and two-sided contact points

For contact representations there is an important difference in contact points with a curve passing through (see Figure 2)—whether the other curves of this contact point are only on one side of the passing curve or on both sides of it. We may formally define a *one-sided contact point* as a contact point C in which either all of its curves end or there exists a curve ϱ passing through C such that for all other curves $\sigma_1, \ldots, \sigma_{k-1}$ ending in C the cyclic order of the curves outgoing from C is $\varrho, \varrho, \sigma_1, \ldots, \sigma_{k-1}$. We say a contact point is *two-sided* if it is not one-sided. It is obvious that any 2–contact point is one-sided, but from the contact degree 3 there exist graphs that have a contact representation and cannot be represented without two-sided contact points.

A representation $\mathcal{R}$ is said to be a *k–contact representation* if each contact point of $\mathcal{R}$ has degree at most k. A representation $\mathcal{R}$ is said to be *simple* if each pair of curves from $\mathcal{R}$ has at most one common contact point. A representation $\mathcal{R}$ is said to be *one-sided* if each of its contact points is one-sided. The same definitions of a k–contact or one-sided representations are applied for line segment representations. It is clear that every line segment representation is simple. All these properties of contact representations are transferred to contact graphs, and we refer to contact graphs as k-contact, simple or one-sided in the obvious sense. Unless explicitly stated otherwise, we will consider only one-sided contact representations. Therefore by representation we will mean one-sided representation, and we will say a two-sided representation otherwise. Similarly, we will consider one-sided contact graphs by default.

2.2 Simple Results

For a description of a curve contact representation we define the following tool: The *incidence graph* of a representation $\mathcal{R}$ (denoted by $I(\mathcal{R})$) is a directed graph, whose vertices correspond to curves and contact points of $\mathcal{R}$ and each vertex of a curve is connected with all contact points that lie on it. The edge is oriented from curve to contact point iff the point is an endpoint of the curve. We understand here a directed graph as an orientation of undirected graph, i.e. strictly without multiple edges. An example of the incidence graph of a representation is presented in Figure 3.

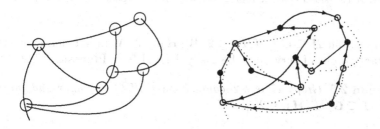

Fig. 3. An example of the incidence graph of a representation

From the definitions it follows that the contact graph of a representation is fully determined by its incidence graph, while the opposite is, of course, not true. We say that two representations are *similar* if their incidence graphs are isomorphic. It is not difficult to show the following lemma that enables us to handle a curve contact representation easier and to describe it using finite (polynomial) space.

Lemma 2.1. *For each two-sided representation $\mathcal{R}$ there exists a two-sided representation S similar to $\mathcal{R}$, so that each curve from S is a piecewise linear curve, consisting of linear number of segments. Additionally, if $\mathcal{R}$ is one-sided, then S can be also chosen one-sided.*

For one-sided contact representations another description, which is using the incidence graph of a representation, is proposed next.

Proposition 2.2 *For a graph G there exists a contact representation $\mathcal{R}$ such that $G \cong I(\mathcal{R})$ iff G is a planar directed graph and its vertices can be divided into two independent set $V(G) = A \cup B$ so that the outdegrees in A are at most 2, the outdegrees in B are at most 1 and the total degrees in B are at least 2.*

We omit proofs of this technical results, they may be found in [9]. To show the difference between one-sided and two-sided contact representations, we present the representation in Figure 4. It is a two-sided contact representation with non-planar incidence graph, and its contact graph has no one-sided contact representation (for example by Proposition 2.4).

The following characterization of the 2–contact graphs of line segments is given in [17]: Graph G is a 2-contact graph of line segments iff G is planar and

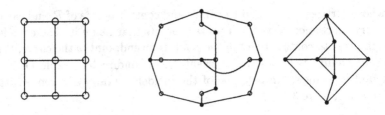

Fig. 4. A two-sided contact representation with non-planar incidence graph

for each subgraph $H \subseteq G$, $|E(H)| \leq 2 \cdot |V(H)| - 3$. A similar characterization of 2–contact graphs of curves can be easily derived from Proposition 2.2.

Proposition 2.3 *Graph G is a 2–contact graph iff G is planar and for each subgraph $H \subseteq G$, $|E(H)| \leq 2 \cdot |V(H)|$.*

It also follows from Proposition 2.2 that one-sided 3–contact graphs of curves are planar too. However, there is probably no such nice characterization as the previous ones, due to the results presented in Section 5.

Proposition 2.4 *If G is a contact graph of a one-sided 3–contact representation $\mathcal{R}$, then G is planar. Moreover, there exists a planar drawing of G such that for each 3–contact point X of curves $u, v, w \in \mathcal{R}$ the triangle u, v, w forms a face.*

3 Classes of Contact Graphs

Various classes of contact graphs of curves or line segments, with bounds on contact degrees and simplicity, are defined. Remember that only one-sided contact representations are considered here. The inclusions among the classes are described in Theorem 1, see the diagram in Figure 5.

Definition. For an integer $k \geq 2$, we denote by $CONCUR$ $(k\text{--}CONCUR)$ the class of all contact (k–contact) graphs of curves, by $SCONCUR$ $(k\text{--}SCONCUR)$ the class of all simple contact (simple k–contact) graphs of curves, and by $CONSEG$ $(k\text{--}CONSEG)$ the class of all contact (k–contact) graphs of line segments.

Theorem 1. *All the inclusions among contact graph classes described in Figure 5 are strict and no other inclusion holds.*

Sketch of proof. All inclusions shown in Figure 5 are obvious from definitions. The equalities $2\text{--}SCONCUR = 2\text{--}CONCUR$ and $3\text{--}SCONCUR = 3\text{--}CONCUR$ follow from the fact (not proved here) that any 3–contact representation may be rearranged to be simple.

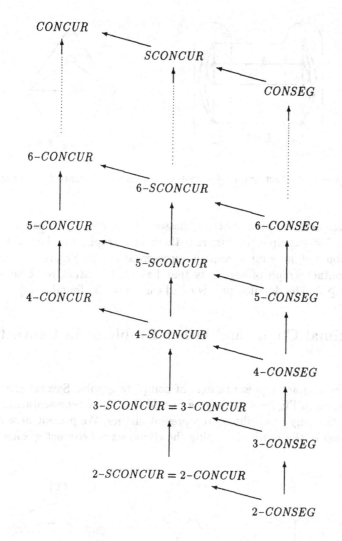

Fig. 5. The inclusions between classes of contact graphs

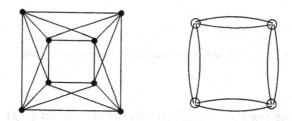

Fig. 6. A 4-contact graph that has no simple contact representation

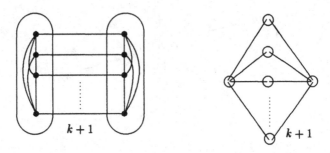

Fig. 7. A $(k+1)$–contact graph of segments that has no k–contact representation

The differences between distinct classes are proved by constructing special graphs. For example, in Figure 6 there is a graph that has a 4–contact representation but no simple contact representation. In Figure 7 a scheme of a $(k+1)$–contact graph of segments that has no k–contact representation, is shown for $k \geq 3$. The detailed proofs of all cases may be found in [9].

4   Maximal Clique and Other Problems in Contact Graphs

We study the contact representations of complete graphs. Several examples of them are shown in Figure 8. It may be proved that these representations are, in some sense, the only possibilities to represent cliques. We present here only the weaker version of the result, bounding the clique size of contact graphs.

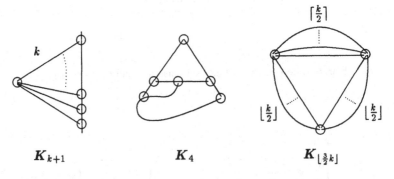

Fig. 8. Contact representations of complete graphs

Theorem 2. *The complete graphs contained in the contact graph classes are*

1. for every $k \geq 2$, $K_m \in k\text{-}CONSEG$ iff $m \leq k+1$,

2. *for every* $k \geq 3$, $K_m \in k\text{-}SCONCUR$ *iff* $m \leq k+1$,
3. *for every* $k \geq 3$, $K_m \in k\text{-}CONCUR$ *iff* $m \leq \lfloor \frac{3}{2}k \rfloor$,
4. *specially* $K_m \in 2\text{-}CONCUR = 2\text{-}SCONCUR$ *iff* $m \leq 4$.

Corollary 4.1. *For* $k \geq 3$, *the maximal clique size of a* k*-contact graph, simple* k*-contact graph, is bounded by* $\lfloor \frac{3}{2}k \rfloor$, $k+1$.

Sketch of proof. We include here only the proof of the easy case of simple contact representations. The whole proof is in [9].

Let us suppose that there exists a simple k–contact representation $\mathcal{R}$ of the graph K_{k+2}, $k \geq 3$. If there is no contact point of degree at least 3 in $\mathcal{R}$, we get a 2–contact representation of K_5, a contradiction to Proposition 2.4. Otherwise we take a contact point X of degree at least 3 and curves $\varrho_1, \varrho_2, \varrho_3$ containing X. Because the contact degree of X is at most k, there exist two curves σ_1, σ_2 not containing X. Then $\varrho_1, \varrho_2, \varrho_3, \sigma_1, \sigma_2$ form a 3–contact subrepresentation of K_5 (in contact points distinct from X there may be only one of the curves $\varrho_1, \varrho_2, \varrho_3$), again a contradiction to Proposition 2.4.

Many graph problems that are hard in the general case, can be solved quickly for special intersection graphs. For example, it is easy to find the chromatic number, maximal clique or independent set of an interval graph, using the simplicial decomposition of it. We show, based on the previous result, that the maximal clique of a contact graph can be found in polynomial time if the contact representation is given.

Proposition 4.2 *There exists a polynomial algorithm that for given contact representation of a graph* G *finds the maximal clique of* G, *while the INDEPENDENT SET and the 3-COLOURABILITY problems remain NP–complete for contact graphs (2–contact graphs) even when the contact representation is given.*

Further we show that the contact graphs are "almost perfect", i.e. their chromatic number is bounded by a linear function of the maximal clique size. However, an infinite sequence of contact graphs, for which the chromatic number grows faster than the maximal clique, may be constructed.

Proposition 4.3 *For any contact graph* G, $\chi(G) \leq 2 \cdot \omega(G)$. *There exists a contact graph* H_m *with* $\omega(H_m) = m$ *and* $\chi(H_m) \geq m + \lceil \frac{m-1}{4} \rceil$, *for any integer* m.

5 Recognition of Contact Graphs

The problem to decide, whether a given graph can be represented as an intersection graph of specified objects, is important in studying the intersection graphs. The decision version of the problem is called the *recognition* of intersection graphs (of a special kind). For the interval graphs a simple characterization is given in [14], and a more efficient algorithm for their recognition is in [1]. Circle graphs (intersection graphs of chords of a circle) may be mentioned [2] as other kind of intersection graphs that can be quickly recognized, but the algorithm is not easy. On the other hand, the recognition of intersection graphs of curves in the plane is proved to be NP–hard [11], moreover graphs with at least exponential complex representations [13] are known in that case.

The complexity of recognizing the contact graphs of curves or line segments is considered here, especially for planar graphs.

5.1 Contact Representations of Planar Triangulations

Firstly we present an important lemma that is used to disprove existence of certain contact representations of planar graphs.

Lemma 5.1. *Let $\mathcal{R}$ be a two-sided 3–contact representation of a graph G containing f free endpoints of curves. Then the representation $\mathcal{R}$ must contain at least $(|E(G)| - 2 \cdot |V(G)| + f)$ 3–contact points forming non-neighbouring triangles in G (two triangles are said to be neighbouring if they have a common edge).*

We already know by Proposition 2.3 that the recognition of 2–contact graphs is polynomial—the edge number condition may be checked using the polynomial algorithm for network flows, and the planarity is also known to be polynomial. There are also other results on representing planar graphs as contact graphs that follow.

From [4] it is known that every bipartite planar graph is a 2–contact graph of segments. A planar triangulation is a planar graph that has all faces, including the outer face, triangles. In [6], representations of planar triangulations by contacts of segments are considered: A 4–connected planar triangulation is a 3–contact graph of segments iff it is 3–colourable (and this condition can be checked in linear time).

We study the curve contact representations of planar triangulations. From Lemma 5.1 the following statement is derived:

Theorem 3. *There exists a polynomial algorithm that for a given planar triangulation decides whether it is a 3–contact graph, and finds the representation if exists.*

In further constructions we need a special graph that has a simple 3–contact representation, but no contact representation in which some curve has a free endpoint (an endpoint of a curve is called free if it is not a contact point). This graph is presented in Figure 9, we denote it by $\mathcal{E}$. The property of having no free endpoint in any contact representation follows from Lemma 5.1— $\mathcal{E}$ has only 16 non-neighbouring triangles. It is used to "eat" an endpoint of a curve in a contact representation—if any vertex v of it is adjacent to some other vertex w (of another graph), the only way to represent the edge $\{v, w\}$ is to use one endpoint of the curve w in the contact point of v, w.

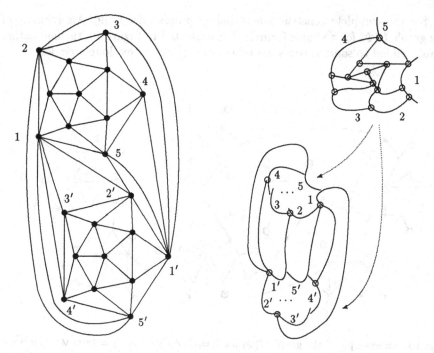

Fig. 9. The end-eating graph $\mathcal{E}$ and a scheme of its contact representation

5.2 NP–completeness of Recognizing 3–contact Graphs

Unlike the planar triangulations, the problem of deciding whether a given general planar graph is a 3–contact graph, is NP–complete. The recognition of (two-sided) contact graphs clearly belongs to NP from Lemma 2.1. For the NP–completeness reduction of our problem we use the PLANAR 3–SAT problem, that is defined as a special case of the SAT problem (a formula Φ with a set variables V and a set of clauses C) for which the bipartite graph F, $V(F) = C \cup V$, $E(F) = \{xc : x \in c \text{ or } \neg x \in c\}$, is planar with degrees of all vertices bounded by 3.

Given a formula Φ (of the PLANAR 3–SAT problem), we construct a graph $R(\Phi)$ that has a contact representation iff the formula Φ is satisfiable. In the construction each variable and each clause of Φ are replaced by special graphs, then clauses are connected with their variables by connectors. The positive and negated occurrences of a variable are distinguished by connecting to clauses using different terminals of the variable graph. For clauses with less than 3 variables special false terminators are used. The variable and clause graphs are constructed using the end-eating graph, and from Lemma 5.1 it is derived that the only possible representations of the variable graph reflects the values 0 and 1 of the variable, and the clause graph is representable iff at least one of its literals is true.

For the complete construction including proofs refer to [9]. An example of the graph $R(\Phi)$ for a simple formula Φ is presented in Figure 10, the end-eating graphs added to some vertices are schematically drawn by double circles.

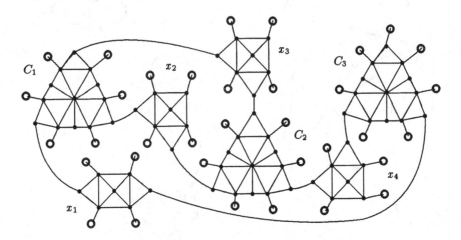

Fig. 10. An example of the graph $R(\Phi)$ for $\Phi = C_1 \wedge C_2 \wedge C_3$, $C_1 = (\neg x_1 \vee \neg x_2 \vee \neg x_3)$, $C_2 = (x_2 \vee x_3 \vee \neg x_4)$, $C_3 = (\neg x_1 \vee x_4)$

Lemma 5.2. *For a formula Φ of a given instance of the PLANAR 3–SAT problem the graph $R(\Phi)$ is planar, and is a one-sided 3–contact graph if Φ is satisfiable but has no two-sided contact representation if Φ is not satisfiable.*

Theorem 4. *The recognition of contact graphs (simple contact graphs, k–contact graphs for $k \geq 3$) is NP–complete.*
The recognition of two-sided contact graphs (two-sided k–contact graphs for $k \geq 3$) is NP–complete.
The recognition of contact graphs is NP–complete even within the class of planar graphs.

Using more involved methods, a similar reduction can be constructed for contact graphs of line segments. However, we do not know whether the recognition of contact graphs of segments belongs to NP. Thus it is proved:

Theorem 5. *The recognition of contact graphs (k–contact graphs for $k \geq 3$) of segments is NP–hard.*
The recognition of two-sided contact graphs (two-sided k–contact graphs for $k \geq 3$) of segments is NP–hard.
The recognition of contact graphs of segments is NP–hard even within the class of planar graphs.

References

1. K. S. Booth, G. S. Lucker, *Testing for the consecutive ones property, interval graphs, and graph planarity using PQ-tree algorithms*, J. Comp. Systems Sci. 13 (1976), 255–265.

2. A. Bouchet, *Reducing prime graphs and recognizing circle graphs*, Combinatorica 7 (1987), 243–254.

3. G. Ehrlich, S. Even, R.E. Tarjan, *Intersection graphs of curves in the plane*, J. of Comb. Theory Ser. B 21 (1976), 8–20.

4. H. de Fraysseix, P.O. de Mendez, J. Pach, *Representation of planar graphs by segments*, 63. Intuitive Geometry (1991), 110–117.

5. H. de Fraysseix, P.O. de Mendez, P. Rosenstiehl, *On triangle contact graphs*, Combinatorics, Probability and Computing 3 (1994), 233–246.

6. H. de Fraysseix, P.O. de Mendez, to appear.

7. M.R. Garey, D.S. Johnson, Computers and Intractability, W.H. Freeman and Company, New York 1979.

8. F. Gavril, *Algorithms for a maximum clique and maximum independent set of a circle graph*, Networks 4 (1973), 261–273.

9. P. Hliněný, *Contact graphs of curves*, KAM Preprint Series 95–285, Dept. of Applied Math., Charles University, Czech rep., 1995.

10. P. Koebe, *Kontaktprobleme der konformen Abbildung*, Berichte über die Verhandlungen der Sächsischen, Akad. d. Wiss., Math.–Physische Klasse 88 (1936), 141–164.

11. J. Kratochvíl, *String graphs II: Recognizing string graphs is NP-hard*, J. of Comb. Theory Ser. B 1 (1991), 67–78.

12. J. Kratochvíl, J. Matoušek, *Intersection graphs of segments*, J. of Comb. Theory Ser. B 2 (1994), 289–315.

13. J. Kratochvíl, J. Matoušek, *String graphs requiring exponential representations*, J. of Comb. Theory Ser. B 2 (1991), 1–4.

14. C.B. Lekkerkerker, J.C. Boland, *Representation of finite graphs by a set of intervals on the real line*, Fund. Math. 51 (1962), 45–64.

15. S. Roberts, *On the boxicity and cubicity of a graph*, in "Recent Progresses in Combinatorics" 301–310, Academic Press, New York, 1969.

16. F.W. Sinden, *Topology of thin film RC-circuits*, Bell System Techn. J. (1966), 1639–1662.

17. C. Thomassen, presentation at Graph Drawing '93, Paris, 1993.

On Representations of Some Thickness-Two Graphs

Extended Abstract

Joan P. Hutchinson
Department of Mathematics
Macalester College
St. Paul, MN 55105, U.S.A.
hutchinson@macalstr.edu

Thomas Shermer
Department of Computer Science
Simon Fraser University
Burnaby, BC, Canada
shermer@cs.sfu.ca

Andrew Vince
Department of Mathematics
University of Florida
Gainesville, FL 32611, U.S.A.
vince@math.ufl.edu

ABSTRACT. This paper considers representations of graphs as *rectangle-visibility graphs* and as *doubly linear graphs*. These are, respectively, graphs whose vertices are isothetic rectangles in the plane with adjacency determined by horizontal and vertical visibility, and graphs that can be drawn as the union of two straight-edged planar graphs. We prove that these graphs have, with n vertices, at most $6n - 20$ (resp., $6n - 18$) edges, and we provide examples of these graphs with $6n - 20$ edges for each $n \geq 8$.

1. Introduction

A *thickness-two* graph G is one whose edge set can be partitioned into two planar graphs, each on one copy of the vertex set of G. These graphs are of theoretical interest and arise in a multitude of applications. For example, it is an NP-complete problem to determine whether a graph has thickness two [9], and the upper bound on their chromatic number is known only to lie between 9 and 12 [12, 4, 6]. These graphs arise in models for printed circuit boards [5, 6] and in VLSI design and layout [18] in which all connections are either horizontal or vertical and so divide naturally into two planar layers. Visibility representation of these graphs is of increasing interest; see, for example, [11, chap.7].

We study thickness-two graphs and their representations (and non-representations) as *rectangle-visibility graphs* and as *doubly linear graphs*. We show that not all thickness-two graphs have these representations. Specifically we show that the most (edge) dense thickness-two graphs are neither rectangle-visibility nor doubly linear graphs, though these graph representations are ubiquitous among thickness-two graphs of lower density.

A *bar-visibility graph* [8, 19] is one whose vertices can each be represented by a closed horizontal line segment in the plane, having pairwise disjoint relative interiors, with two vertices adjacent in the graph if and only if the corresponding segments are *vertically visible*. Two segments are considered *vertically visible* when there is a nondegenerate rectangle R such that R intersects only these two segments and the horizontal sides of R are subsets of these two segments. (For variations on this definition, see [11, 14].) Clearly a bar-visibility graph is planar. Not all planar graphs are bar-visibility graphs since the latter are characterized as those planar graphs for which there is a planar embedding with all cut vertices on a common face [14, 19].

A natural two-dimensional analogue is that of a *rectangle-visibility graph*, a graph whose vertices can each be represented by a closed rectangle in the plane with sides parallel to the axes, having pairwise disjoint interiors, with two vertices adjacent in the graph if and only if the corresponding rectangles are vertically or horizontally visible (with horizontal visibility defined analogously to vertical). Note that the bands of visibility may overlap and cross. Every planar graph is a rectangle-visibility graph [8], and it is clear that every rectangle-visibility graph has thickness at most two. Even more, a rectangle-visibility graph is the union of two bar-visibility graphs. Our main result on these graphs is that a rectangle-visibility graph with n vertices has at most $6n - 20$ edges, as distinguished from thickness-two graphs which have at most $6n - 12$ edges. (The latter fact follows from Euler's formula for planar graphs, which implies that a planar graph with n vertices has at most $3n - 6$ edges.) Thickness-two graphs can have as many as $6n - 12$ edges; we show also that for every $n > 7$ there is a rectangle-visibility graph with $6n - 20$ edges. Using similar methods, in [2, 3] it is proved that a bipartite rectangle-visibility graph has at most $4n - 12$ edges.

It is a consequence of a classical theorem of Steinitz on polyhedra (see [13]) that every planar graph G has a *linear* or *straight-line* embedding in the plane. This means that

(1) every edge is a straight line segment,

(2) no vertex lies in the interior of an edge, and

(3) edges do not cross.

(Results on rectilinear drawings, which satisfy only (1) and (2), are obtained in [15].) If, instead of property (3), we require that

(3') G can be partitioned into two subgraphs, each without edge crossings,

then G is called *doubly linear*. Again it is clear that doubly linear graphs have thickness at most two. We prove that a doubly linear graph with n vertices has at most $6n - 18$ edges, and for each $n > 7$ we have an example of a doubly linear graph with $6n - 20$ edges, two edges less than our upper bound. We give examples of doubly linear graphs that are not rectangle-visibility graphs and conjecture that every rectangle-visibility graph is doubly linear.

Since it is known that the problem of recognizing thickness-two graphs is NP-complete, it may be difficult to obtain a complete characterization of rectangle-visibility or of doubly linear graphs. These concepts come from [5, 8], and from the Workshop on Visibility Representations, McGill University Bellairs Research Institute, held in February, 1993. Complete details of this work are included in [7].

2. Examples of rectangular and doubly linear representations

Some examples of these representations are shown in the figures. Figures 1 and 2 show a rectangle-visibility and a doubly linear representation, respectively, of the complete graph K_8.

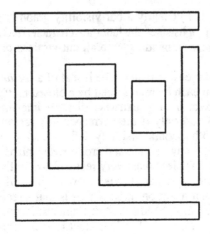

Fig. 1. A rectangle-visibility representation of K_8.

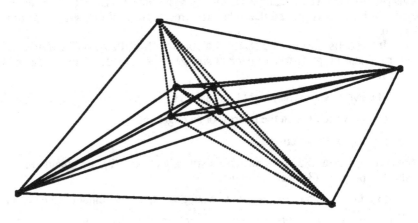

Fig. 2. A doubly linear representation of K_8.

This is the largest complete graph so representable since K_9 has thickness three [1, 16]. It is not hard to add another vertex, adjacent to six others, to each of these representations to obtain K_9 minus two edges; it can be arranged for these missing edges to be either mutually incident or nonincident. These graphs and K_8 have $6n - 20$ edges, $n = 9, 8$, respectively. K_9 minus one edge $(K_9 - e)$ has thickness two with $6n - 19$ edges [17]. Theorem 3 will show that $K_9 - e$ is therefore not a rectangle-visibility graph, though it is the union of two bar-visibility graphs. We conjecture that $K_9 - e$ is not doubly linear.

Figure 3 shows a rectangular representation of $K_{5,5}$ plus four edges, and Figure 4 shows a closely related doubly linear representation of $K_{5,5}$. Figure 3 can be extended to a rectangular representation of $K_{5,6}$ plus edges by adding a long rectangle along the left side, and Figure 4 can be similarly extended to a doubly linear representation of $K_{5,6}$. In [2, 3] it is shown that $K_{p,q}$ with p and q at least 5 is not a rectangle-visibility graph (and that $K_{5,5}$ minus an edge and $K_{5,5}$ plus an edge are rectangle-visibility graphs). Thus $K_{5,5}$ and $K_{5,6}$ are doubly linear graphs, but not rectangle-visibility graphs. These examples point up an essential difference between the two classes of graphs: namely, that although a subgraph of a doubly linear graph is also doubly linear, the same is not true for rectangle-visibility graphs.

OPEN QUESTION. Is there a rectangle-visibility graph that is not doubly linear?

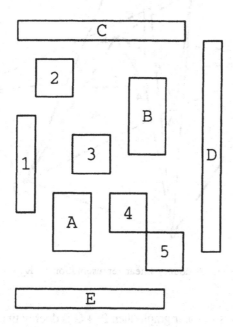

Figure 3. A rectangle-visibility representation of $K_{5,5}$ plus four edges.

It is not difficult to obtain, for every n, a rectangle-visibility and a doubly linear representation of the join of K_4 and P_n and the join of K_4 and C_n, where P_n and C_n are, respectively, the path and the cycle on n vertices. The *join* of two disjoint graphs G and H is the disjoint union of these two graphs together with an edge joining vertices g and h, for each vertex g of G and vertex h of H, and is denoted $G+H$. In these examples K_4 cannot be replaced by K_5 for n > 12 since these graphs would contain $K_{5,13}$ which, by Euler's formula, has thickness at least 3.

In [7] we prove the following two propositions on representations of joins. Proposition 1 gives another family of graphs that have a rectangular representation, and Proposition 2 proves these are also doubly linear. Note that, as long as G contains a cycle, P_2+G is not planar since it contains a homeomorph of K_5.

PROPOSITION 1. If G is a 2-connected planar graph or, more generally, a bar-visibility graph, then P_2+G is a rectangle-visibility graph.

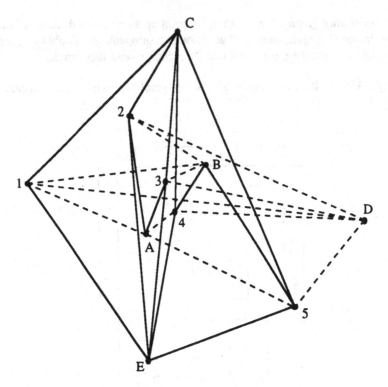

Fig. 4. A doubly linear representation of $K_{5,5}$.

PROPOSITION 2. If G is a planar graph, then $P_2 + G$ is doubly linear.

Additional rectangular examples are found in Figures 5 and 6, and related doubly linear graphs are described in a sketch of the proof of Theorem 8.

3. Edge bounds and densities for rectangle-visibility graphs

In [7] we determine a tight bound, $6n - 20$, on the maximum number of edges in a rectangle-visibility graph with n vertices. In addition we examine the possible edge-densities for rectangle-visibility graphs and for graphs that do not have a rectangle-visibility representation. More precisely, if m, the number of edges, is greater than $6n - 20$, then no rectangle-visibility representation is possible; for essentially all $m \leq 6n - 20$ there exist both examples of graphs that have a rectangle-visibility representation and examples of graphs that do not. In particular, we give an example of a rectangle-visibility graph with n vertices and m edges for each m with $0 \leq m \leq 6n - 20$ and an example of a connected rectangle-visibility graph for each m with $n - 1 \leq m \leq 6n - 20$ (except for two pairs (m, n)).

THEOREM 3. A rectangle-visibility graph on $n \geq 5$ vertices has at most $6n - 20$ edges.

SKETCH OF PROOF. Let G be a rectangle-visibility graph with representation R*. Let R be a rectangle in R* and define N(R) to be the set of rectangles in R* that intersect with positive area the one-way infinite band of all points "north" of R. Select R_1 to be a rectangle R with N(R) empty and with the greatest y-coordinate for its bottom. Note that if R' is visible to R_1 horizontally, then N(R') is empty; otherwise there is another rectangle with N(R) empty and y-coordinate larger than R_1's for its bottom. Move R_1 northward, above the rest of the configuration, and expand it horizontally until it is wider than the whole representation. The new R_1 has retained all its previous visibilities and may have gained some.

We repeat this process of rectangle movement and expansion to the south, east, and west. The resulting configuration is bordered by four rectangles as in Figure 1. The resulting rectangle-visibility graph G' contains G as a subgraph. Using Euler's formula for planar graphs, upper bounds on the number of horizontal and vertical edges of G' are $3n - 12$ and $3n - 8$, respectively, because the topmost and bottommost rectangles of G' have horizontal degree zero and the rightmost and leftmost rectangles have vertical degree two. Thus G has at most $6n - 20$ edges. QED

COROLLARY 4. Let G' with $n' \geq 5$ vertices be a subgraph of a rectangle-visibility graph G. Then G' has at most $6n' - 20$ edges.

See [3] for similar proofs that a bipartite rectangle-visibility graph and a bipartite subgraph of a rectangle-visibility graph with n vertices have at most $4n - 12$ edges. Bipartite rectangle-visibility examples with at most $4n - 16$ edges are given also for each $n \geq 7$, and for each $n \geq 16$ bipartite graphs with n vertices and $4n - 12$ edges are known that are subgraphs of rectangle-visibility graphs [10].

We also show that the bound of Theorem 3 is best possible for all $n \geq 8$. (For $n \leq 8$, as noted before, the complete graphs give the best possible bound.) Figures 5 and 6 show rectangular representations with $6n - 20$ edges and $n = 16$ and 17 vertices, respectively, and Theorem 5 states that similar graphs can be constructed for all $n \geq 8$.

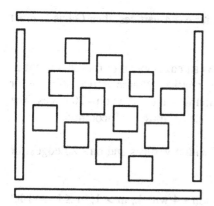

Fig. 5. A rectangular representation with $n = 3 \cdot 4 + 4$.

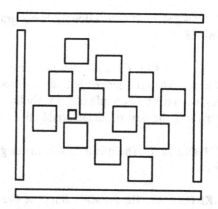

Fig. 6. A rectangular representation with $n = 13 + 4$.

THEOREM 5. There is a rectangle-visibility graph with n vertices and $6n - 20$ edges for each $n \geq 8$.

Rectangle-visibility graphs with fewer edges are also possible, as seen in the next result.

THEOREM 6. With the exception of the cases $(n, m) = (6, 16)$ and $(n, m) = (7, 22)$, the following holds for all $n \geq 4$:

(a) For each m with $0 \leq m \leq 6n - 20$, there is a rectangle-visibility graph with n vertices and m edges.

(b) For each m with $n - 1 \leq m \leq 6n - 20$, there is a connected rectangle-visibility graph with n vertices and m edges.

The exceptions arise since a simple graph with n vertices has at most $n(n-1)/2$ edges. Otherwise these graphs are constructed from those of Theorem 5 unioned either with a graph with no edges or with a path (for the connected case.)

Families of graphs with $n' \geq 9$ vertices and $m' \geq 35$ edges that are not representable by rectangles can also be found. For any $0 \leq m \leq 6n - 20$ for which there is a rectangle-visibility graph G with n vertices and m edges, form the disjoint union of G with $K_9 - e$ to obtain a graph with $m + 35$ edges and $n + 9$ vertices. By Corollary 4, the new graph is not a rectangle-visibility graph since $K_9 - e$ has more than $6n - 20$, $n = 9$, edges. Connected graphs G together with $K_9 - e$ plus an adjoining edge similarly give connected examples.

4. Edge bounds and densities for doubly linear graphs

We also present parallel results for doubly linear graphs; however, our examples, related to those of Theorem 5 and having $6n - 20$ edges, come only within two of the edge-bound $6n - 18$ of Theorem 7.

THEOREM 7. If G is a doubly linear graph with $n \geq 4$ vertices, then G has at most $6n - 18$ edges.

In fact, for $n > 4$, the proof establishes an edge-bound of $6n - 20$ except in the case when the convex hull of the embedding consists of three vertices. Except for K_4, we have no example of a doubly linear graph with more than $6n - 20$ edges. The graphs of the next result are doubly linear analogues of those of Theorem 5.

THEOREM 8. There is a doubly linear graph with n vertices and $6n - 20$ edges for each $n \geq 8$.

SKETCH OF THE PROOF. Suppose that $n = km + 4$ with $k, m > 1$, $m \leq k$, and let $q \geq 2$ be an integer such that $k \leq qm$. Let S be the grid points in the rectangle

$$T = \{(x, y) : 0 \leq x \leq k - 1, 0 \leq y \leq m - 1\},$$

and let

$$a = (-4k, -m), b = (4k, 3m), c = (8k, 3m - 10qm), \text{ and } d = (-k, 3m + 13qm).$$

The vertices a, b, c, and d are joined by straight edges to form a K_4 with no edge intersecting the rectangle T.

We form two sets R and B of edges, each a linear triangulation. Let R have the edges

> {s, s + (1, 0)}, {s, s + (1, 1)}, {s, s + (2, 1)} for all s in S (and when the second vertex is an element of S);
>
> {a, (x, y)} for x = 0 or x = 1 or y = 0;
>
> {b, (x, y)} for x = k − 1 or x = k − 2 or y = m − 1;
>
> the edges of the K_4 formed by a, b, c, and d; and
>
> {c, (k − 1, 0)}.

Thus a is connected to the left two columns of vertices and to the bottom vertices of T, and b is connected to those on the top and in the two rightmost columns.

Let B have the edges

> {s, x + (0, 1)}, {s, s + (−1, 1)}, {s, s + (−1, 2)} for all s in S (and when the second vertex is in S);
>
> {c, (x, y)} for x = k − 1 or y = 0 or y = 1;
>
> {d, (x, y)} for x = 0 or y = m − 1 or y = m − 2;
>
> the edges of the K_4 except for the edge ab;
>
> {a, (0, 0)}, and {b, (k − 1, m − 1)}.

Thus c is connected to the bottom and to the right of the rectangle by straight edges (since $q \geq 2$), and d is connected to the left and to the top vertices.

It is then a routine set of slope calculations and edge counts to see that this gives a doubly linear graph with $6n - 20$ edges. QED

Since a subgraph of a doubly linear graph is doubly linear, we can achieve in a graph with n vertices any desired number of edges less than $6n - 20$. To construct families of non-doubly linear graphs one can begin as before with a specific graph that is not doubly linear and form the union with a doubly linear graph of any desired size. For example, one can begin with K_9, which is not doubly linear since it has thickness three. Or one can begin with $K_{12} - F$, the complete graph on 12 vertices minus a one-factor. Let G_1 be the graph of the icosahedron, and let G_2 be the graph on the same set of vertices with vertices adjacent if they are at distance two in G_1. In fact, G_1 and G_2 are isomorphic, and their union is $K_{12} - F$, showing the latter to have thickness two. However, $K_{12} - F$ has 12 vertices and $60 = 6n - 12$ edges, and so by Theorem 7 is not doubly linear.

ACKNOWLEDGMENTS. The authors would like to thank the participants in the Workshop on Visibility Representations, held in February, 1993, at the McGill University Bellairs Research Institute, Barbados, and especially the other workshop organizer, S. H. Whitesides, for stimulating questions and answers.

References

1. J. Battle, F. Harary, and Y. Kodama, Every planar graph with nine points has a nonplanar complement, *Bull. Amer. Math. Soc.* 68 (1962) 569–571.

2. A. Dean and J. P. Hutchinson, Rectangle-visibility representations of bipartite graphs, Extended Abstract, *Lecture Notes in Computer Science (Proc. DIMACS Workshop Graph Drawing, 1994)*, R. Tamassia and I. G. Tollis, eds., vol. 894, Springer-Verlag, 1995, 159–166.

3. _____, Rectangle-visibility representations of bipartite graphs (submitted).

4. M. Gardner, Mathematical Games, *Scientific American* 242 (Feb. 1980) 14–19.

5. M. R. Garey, D. S. Johnson, and H. C. So, An application of graph coloring to printed circuit testing, *IEEE Trans. Circuits and Systems* CAS-23 (1976) 591–599.

6. J. P. Hutchinson, Coloring ordinary maps, maps of empires, and maps of the Moon, *Math. Mag.* 66 (1993) 211–226.

7. J. P. Hutchinson, T. Shermer, and A. Vince, On representations of some thickness-two graphs (submitted).

8. D. G. Kirkpatrick and S. K. Wismath, Weighted visibility graphs of bars and related flow problems, *Lecture Notes in Computer Science (Proc. 1st Workshop Algorithms Data Struct.)*, vol. 382, Springer-Verlag, 1989, 325–334.

9. A. Mansfield, Determining the thickness of graphs is NP-hard, *Math. Proc. Cambridge Phil. Soc.* 93 (1983) 9–23.

10. H. Meijer, personal communication.

11. J. O'Rourke, *Art Gallery Theorems and Algorithms*, Oxford University Press, N.Y., 1987.

12. G. Ringel, *Färbungsproblems auf Flächen und Graphen*, Deutscher Verlag der Wissenschaften, Berlin, 1959.

13. E. Steinitz and H. Rademacher, *Vorlesungen über die Theorie der Polyeder*, Springer, Berlin, 1934.

14. R. Tamassia and I. G. Tollis, A unified approach to visibility representations of planar graphs, *Disc. and Comp. Geom.* 1 (1986) 321–341.

15. C. Thomassen, Rectilinear drawings of graphs, *J. Graph Theory* 12 (1988) 335–341.

16. W. T. Tutte, On the non-biplanar character of the complete 9-graph, *Canad. Math. Bull.* 6 (1963) 319–330.

17. _____, The thickness of a graph, *Indag. Math.* 25 (1963) 567–577.

18. J. D. Ullman, *Computational Aspects of VLSI Design*, Computer Science Press, Rockville, Md., 1984.

19. S. K. Wismath, Characterizing bar line-of-sight graphs, *Proc. 1st Symp. Comp. Geom.*, ACM (1985) 147–152.

Drawing Force-Directed Graphs Using Optigraph *

Jovanna Ignatowicz

Department of Computer Science

Brown University

`jai@cs.brown.edu`

Abstract. Optigraph is an interactive, multi-threaded tool for force-directed graph drawing. The interface allows a user to construct an arbitrary graph made of edges, free vertices and fixed vertices, and then apply circular and orthogonal spring forces to the graph to achieve an optimal graph layout. The user can step through simulations, transform the graph and vary forces at any time making the application highly interactive and educational.

1 Introduction

The primary purpose of Optigraph is to provide an implementation based on classic force-directed algorithms that encourages the user to experimentally apply various spring forces on a set of vertices and edges. The algorithms are based on work done by Eades, and Fruchterman and Reingold [1, 2]. Attracting forces are polynomial and based on the actual and ideal distances between two neighboring vertices. Repelling forces are based on the inverse square law and also consider the ideal and actual distances of two vertices, but are applied to non-neighboring vertices. An additional force is introduced to produce orthogonal drawings. It is called an orthogonal spring and redefines the ideal distance between two vertices. By this definition, two vertices are the ideal distance apart if both their x and y components are that distance apart.

2 Use of Concurrency

Optigraph achieves a high level of interactivity through the use of concurrency. It is written in Concurrent ML [4, 5], a high-level concurrent language that provides constructs to dynamically create multiple threads within one process, establish channels of communication between threads and employ various synchronization primitives. EXene [3] is the multi-threaded interface toolkit that is used on top of CML. It is unique in that the input to the interface is directed to the thread assigned to the user interface component associated with that event. Therefore,

---

* Research supported in part by the National Science Foundation under grant CCR-9423847 and by the U.S. Army Research Office under grant 34990–MA–MUR.

neither the application nor the interface is forced to wait for input to be handled by the other.

To make the best use of the concurrency at its disposal, Optigraph is organized into several parts, each of which is dominated by one or more threads. The graphics server and thread sequencer, for instance, are two infinitely looping threads. The graphics server handles requests to store, retrieve and display vertex and edge data. The thread sequencer waits for the user to start a new simulation or step through an existing one. At each step of a simulation, it spawns one thread for each vertex. These threads then calculate their new position based on their location and the location of all other vertices.

3   Results

The graphs produced by the circular spring forces greatly resemble those produced by Eades and Fruchterman and Reingold, since the algorithm is based on their work. They tend to be very symmetrical with minimal edge crossings and optimal vertex placement. The graphs produced by the orthogonal springs are orthogonal drawings that would ordinarily be produced by other methods[6].

In figure 1, an initial tree layout is shown on the left. This tree is an example of a graph constructed entirely by the user. It contains no symmetries and has varied edge lengths. The graph on the right is the result of applying circular attracting forces to the initial tree while specifying some natural distance for the edges. Although the tree is not much more symmetric, it does have uniform edge lengths. The result of adding repelling circular forces to the right graph in figure 1 is shown in figure 2. Here, the symmetries of the tree are apparent while the edge lengths stay the same. Finally, figure 3 is the result of applying orthogonal attracting and repelling forces to figure 2 to achieve an orthogonal drawing of the tree. It is interesting to note that the edge lengths of the graphs in figures 2 and 3 are longer than those in the right graph of figure 1 although the same natural distance is specified. This is due to the fact that the repelling forces of the vertices are stronger than the attracting forces of the edges. Optigraph provides a slider so that the user can modify the ratio of these two forces while they are both being applied.

Acknowledgements

I wish to thank Roberto Tamassia for his guidance and enthusiasm, Emden Gansner for introducing me to SML, CML and eXene, and Jeffrey Korn for his suggestions and technical support.

Fig. 1. Left- initial graph. Right- attracting circular forces

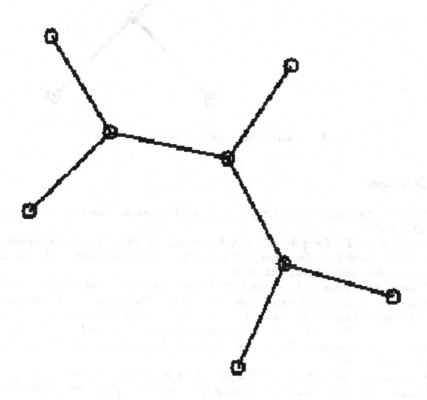

Fig. 2. attracting and repelling circular forces

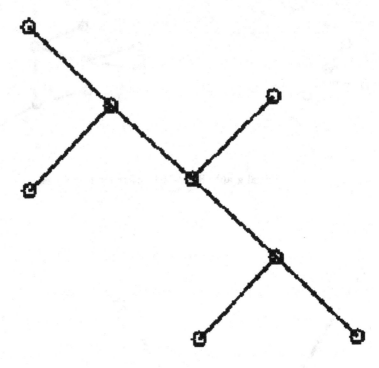

Fig. 3. attracting and repelling orthogonal forces

References

1. Eades, P., "A Heuristic For Graph Drawing", *Congressus Numerantium*, 42, pp. 149-160, 1984.
2. Fruchterman, T., Reingold, E., *Graph Drawing by Force Directed Placement*, UIUCDCS-R-90-1609, Department of Computer Science, University of Illinois at Urbana-Champaign, Urbana, IL, 1990.
3. Gansner, E. and Reppy, J., *eXene*, CMU Workshop on SML, 1991.
4. Reppy, J., *Concurrent Programming with Events - The Concurrent ML Manual*, Department of Computer Science, Cornell University, Ithaca, NY, 1990.
5. Reppy, J., "CML: A Higher-order Concurrent Language", *Proceedings of the SIGPLAN '91 Conference on Programming Language Design and Implementation*, June 1991, pp. 293-305.
6. Sugiyama, K. and Misue, K., "A Simple and Unified Method for Drawing Graphs: Magnetic-Spring Algorithm", *Proceedings of Graph Drawing '94*, 1994, pp. 365-376.

Exact and Heuristic Algorithms for 2-Layer Straightline Crossing Minimization

Michael Jünger[1]* and Petra Mutzel[2]*

[1] Institut für Informatik, Universität zu Köln, Pohligstr. 1, D-50969 Köln, Germany
[2] Max-Planck-Institut für Informatik, Im Stadtwald, D-66123 Saarbrücken, Germany

Abstract. We present algorithms for the two layer straightline crossing minimization problem that are able to compute exact optima. Our computational results lead us to the conclusion that there is no need for heuristics if one layer is fixed, even though the problem is NP-hard, and that for the general problem with two variable layers, true optima can be computed for sparse instances in which the smaller layer contains up to 15 nodes. For bigger instances, the iterated barycenter method turns out to be the method of choice among several popular heuristics whose performance we could assess by comparing the results to optimum solutions.

1 Introduction

Two layer straightline crossing minimization is receiving a lot of attention in automatic graph drawing. The problem consists of aligning the two shores V_1 and V_2 of a bipartite graph $G = (V_1, V_2, E)$ on two parallel straight lines (layers) such that the number of crossings between the edges in E is minimized when the edges are drawn as straight lines connecting the endnodes. There appears to be a general agreement that good solutions for this problem contribute to better readability of diagrams representing hierarchical organizations on two or more layers.

Let $n_1 = |V_1|$, $n_2 = |V_2|$, $m = |E|$, and let $N(v) = \{w \in V \mid e = \{v, w\} \in E\}$ denote the set of neighbors of $v \in V = V_1 \cup V_2$ in G. Any solution is obviously completely specified by a permutation π_1 of V_1 and a permutation π_2 of V_2. For $k = 1, 2$ let $\delta_{ij}^k = 1$ if $\pi_k(i) < \pi_k(j)$ and 0 otherwise. Thus π_k $(k = 1, 2)$ is uniquely characterized by the vector $\delta^k \in \{0, 1\}^{\binom{n_k}{2}}$. Given π_1 and π_2, the number of crossings is

$$C(\pi_1, \pi_2) = C(\delta^1, \delta^2) = \sum_{i=1}^{n_2-1} \sum_{j=i+1}^{n_2} \sum_{k \in N(i)} \sum_{l \in N(j)} \delta_{kl}^1 \cdot \delta_{ji}^2 + \delta_{lk}^1 \cdot \delta_{ij}^2$$

$$= \sum_{k=1}^{n_1-1} \sum_{l=k+1}^{n_1} \sum_{i \in N(k)} \sum_{j \in N(l)} \delta_{kl}^1 \cdot \delta_{ji}^2 + \delta_{lk}^1 \cdot \delta_{ij}^2.$$

* Partially supported by DFG-Grant Ju204/7-1, Forschungsschwerpunkt "Effiziente Algorithmen für diskrete Probleme und ihre Anwendungen"

It has been proven in [GJ83] that the two layer straightline crossing minimization problem is NP-hard, even if the permutation on one layer is fixed [EW94]. Therefore, a lot of effort went into the design of efficient heuristics, for the version in which one permutation is fixed as well as for the general case. Eades and Kelly [EK86] observe that the computation of true optima would be desirable in order to assess the performance of various heuristics, however, [EK86] believe that the NP-hardness of the problem renders such an experimental evaluation impractical.

In this paper, we would like to demonstrate that, if one permutation is fixed, it is indeed possible to compute the exact minima in surprisingly short computation times. In section 2, we outline our algorithm which transforms the problem to a linear ordering problem that is subsequently solved via the branch and cut method. In section 3, we give computational results that allow us to assess the performance of several popular heuristics accurately.

Assume the permutation π_1 of V_1 is fixed. For each pair of nodes $i, j \in V_2$, $i \neq j$, we define c_{ij} to be the number of crossings between edges incident with i and edges incident with j if π_2 is such that $\pi_2(i) < \pi_2(j)$. Then

$$L = \sum_{i=1}^{n_2-1} \sum_{j=i+1}^{n_2} \min\{c_{ij}, c_{ji}\}$$

is a trivial lower bound on the number of crossings. One observation in our experiments was that this trivial lower bound is surprisingly good. In section 4, we utilized this fact and the branch and cut algorithm of section 2 for the design and implementation of a program that solves the general two layer straightline crossing minimization problem to optimality.

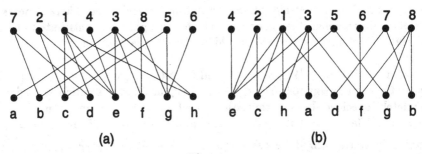

Fig. 1.

Figure 1 demonstrates that the number of crossings can indeed be considerably less if both layers can be freely permuted. The left drawing was given in [STT81] with fixed lower layer, [STT81] obtained the shown drawing with 48 crossings that we could show to be optimum. The right drawing is the optimum when both layers can be freely permuted. It has only 19 crossings.

As was to be expected, two sided crossing minimization can be done only for small instances. For large instances, we adopt the common method that consists of fixing the first layer, "optimizing" the second, fixing the found permutation of the second, "optimizing" the first, etc., back and forth, until the crossing number is not reduced anymore. We follow this iterative approach both using the heuristics of section 3 as well as the exact algorithm. The results are somewhat surprising, e.g., using the barycenter heuristic rather than exact one-sided crossing minimization yields slightly better results.

2 Branch and Cut for One Sided Crossing Minimization

The one sided straightline crossing minimization problem consists of fixing a permutation π_1 of V_1 and finding a permutation π_2 of V_2 such that the number of straightline crossings

$$C(\pi_2) = C(\delta^2) = \sum_{i=1}^{n_2-1} \sum_{j=i+1}^{n_2} \sum_{k \in N(i)} \sum_{l \in N(j)} \delta_{kl}^1 \cdot \delta_{ji}^2 + \delta_{lk}^1 \cdot \delta_{ij}^2$$

is minimized. Let

$$c_{ij} = \sum_{k \in N(i)} \sum_{l \in N(j)} \delta_{lk}^1$$

denote the number of crossings among the edges adjacent to i and j if $\pi_2(i) < \pi_2(j)$. Then

$$C(\pi_2) = C(\delta^2) = \sum_{i=1}^{n_2-1} \sum_{j=i+1}^{n_2} c_{ij}\delta_{ij}^2 + c_{ji}(1 - \delta_{ij}^2)$$

$$= \sum_{i=1}^{n_2-1} \sum_{j=i+1}^{n_2} (c_{ij} - c_{ji})\delta_{ij}^2 + \sum_{i=1}^{n_2-1} \sum_{j=i+1}^{n_2} c_{ji}.$$

For $n = n_2$, $x_{ij} = \delta_{ij}^2$ and $a_{ij} = c_{ij} - c_{ji}$ we solve the linear ordering problem

$$\text{(LO)} \quad \text{minimize} \quad \sum_{i=1}^{n_2-1} \sum_{j=i+1}^{n_2} a_{ij}x_{ij}$$

$$0 \le x_{ij} + x_{jk} - x_{ik} \le 1 \quad \text{for } 1 \le i < j < k \le n$$

$$0 \le x_{ij} \le 1 \quad \text{for } 1 \le i < j \le n$$

$$x_{ij} \in \{0, 1\} \quad \text{for } 1 \le i < j \le n.$$

If z is the optimum value of (LO), $z + \sum_{i=1}^{n_2-1} \sum_{j=i+1}^{n_2} c_{ji}$ is the minimum number of crossings.

The constraints of (LO) guarantee that the solutions correspond indeed precisely to all permutations π_2 of V_2. Furthermore, it can be shown that the "3-cycle constraints" are necessary in any minimal description of the feasible solutions by linear inequalities, if the integrality conditions are dropped. The NP-hardness of the problem makes it unlikely that such a complete linear description can be found and exploited algorithmically. Further classes of inequalities with a number of members exponential in n that must be present in a complete linear description of the feasible set, are known, and some of them can be exploited algorithmically. For the details see [GJR85].

When the integrality conditions in (LO) are dropped, only $2\binom{n}{2}$ hypercube inequalities and $2\binom{n}{3}$ 3-cycle inequalities are left that define a relaxation of (LO) which has been proven very useful in practical applications. In [GJR84a] a branch and cut algorithm for (LO) is proposed that solves this relaxation with a cutting plane approach, since writing down all 3-cycle inequalities, even though taking only polynomial space, and solving the corresponding linear program, is not practical for space reasons. Rather the algorithm starts with the hypercube constraints that are handled implicitly by the LP-solver, and iteratively adds violated 3-cycle constraints and deletes nonbinding 3-cycle constraints after an LP has been solved, until the relaxation is solved. If the optimum solution is integral, the algorithm stops, otherwise it is applied recursively to two subproblems in one of which a fractional x_{ij} is set to 1 and in the other set to 0. In [GJR84b] such a branch and cut approach could be used to find optimum linear orderings with n up to 60 in an application involving input-output matrices that are used in economic analysis. For the many details and the inclusion of further useful inequalities in the cutting plane part, see [GJR84a].

A new implementation of the algorithm is used in our computational experiments. It is written in C and uses the [CPLEX] software for solving the linear programming relaxations coming up in the course of the computation.

3 One Sided Crossing Minimization

The fact that we are able to compute optimum solutions allows us to assess the quality of various popular heuristics for one-sided two layer straightline crossing minimization experimentally. Our computational comparison includes the following heuristics: the barycenter heuristic by [STT81], the median heuristic by [EW94], the stochastic heuristic by [D94], the greedy-insert heuristic by [EK86], the greedy-switch heuristic by [EK86], and the split heuristic by [EK86].

In order to gain confidence in the correctness of our implementations, we repeated the computational tests in [EK86]. We could reproduce their results accurately. There are no published computational results for the stochastic heuristic, but a personal communication with the author [D95] confirms the correctness of our implementation.

All subsequent tables have the following columns:

- n_i: Number of nodes on layer i for $i = 1, 2$
- m: Number of edges
- Low: The trivial lower bound for the number of crossings
- Min: The minimum number of crossings (computed by the branch and cut algorithm)
- Bary: The number of crossings found by the barycenter heuristic
- Median: The number of crossings found by the median heuristic
- Stoch: The number of crossings found by the stochastic heuristic

- Gre-ins: The number of crossings found by the greedy-insert heuristic
- Gre-swi: The number of crossings found by the greedy-switch heuristic
- Split: The number of crossings found by the split heuristic

For each type of graph, three numbers are given: the average number of crossings taken over all sampled instances of this type, the relative size of this number in percentage of the minimum number of crossings, and the average running time in seconds on a SUN Sparcstation 10. All samples are generated by the program random_bigraph of the Stanford GraphBase by Knuth [K93]. The generators are hardware independent and are available from the authors so that exactly the same experiments can be run by anyone who is interested.

In Table 1, we give the results for "20+20-graphs", i.e., bipartite graphs with 20 nodes on each layer and various fixed numbers of edges chosen uniformly and independently from the set of all possible edges. Each average is taken over 100 samples. The most surprising fact is perhaps that the exact computation by the branch and cut algorithm is faster than many of the heuristics. Only the barycenter and the median heuristic are between two to four times faster than the exact algorithm. Furthermore, the table indicates, less surprisingly, that dense instances are not very interesting. The data is visualized in Figure 2.

Table 1. Results for 100 instances on 20 + 20 nodes with increasing density

n_i	m	Low	Min	Bary	Median	Stoch	Gre-ins	Gre-swi	Split
20	40	180.35	180.75	185.34	206.27	185.44	248.37	275.99	183.39
		99.78	100.00	102.54	114.12	102.60	137.41	152.69	101.46
			0.02	0.01	0.01	0.05	0.02	0.04	0.08
20	80	957.62	959.23	968.80	1051.14	970.01	1175.11	1044.14	964.35
		99.83	100.00	101.00	109.58	101.12	122.51	108.85	100.53
			0.03	0.01	0.01	0.06	0.05	0.10	0.11
20	120	2420.14	2422.32	2433.53	2564.82	2437.39	2763.72	2460.94	2428.23
		99.91	100.00	100.46	105.88	100.62	114.09	101.59	100.24
			0.03	0.01	0.01	0.07	0.10	0.16	0.16
20	160	4625.79	4627.72	4638.24	4825.06	4644.35	5098.27	4644.10	4632.17
		99.96	100.00	100.23	104.26	100.36	110.17	100.35	100.10
			0.04	0.01	0.02	0.08	0.17	0.23	0.23
20	200	7560.42	7561.88	7571.08	7817.99	7582.47	8157.86	7572.24	7566.79
		99.98	100.00	100.12	103.39	100.27	107.88	100.14	100.07
			0.05	0.02	0.02	0.09	0.24	0.31	0.31
20	240	11314.37	11315.55	11323.26	11625.54	11338.06	12033.34	11321.10	11318.68
		99.99	100.00	100.07	102.74	100.20	106.34	100.05	100.03
			0.07	0.02	0.03	0.09	0.34	0.42	0.41
20	280	15859.70	15860.35	15865.69	16225.57	15883.69	16667.12	15863.66	15861.76
		99.99	100.00	100.03	102.30	100.15	105.09	100.02	100.01
			0.09	0.03	0.03	0.10	0.45	0.52	0.53
20	320	21290.56	21290.76	21294.12	21727.43	21313.78	22116.56	21292.93	21291.56
		99.99	100.00	100.02	102.05	100.12	103.88	100.01	100.00
			0.11	0.03	0.04	0.11	0.59	0.65	0.66
20	360	27751.63	27751.69	27752.99	28257.47	27768.41	28459.57	27752.01	27751.84
		100.00	100.00	100.01	101.82	100.06	102.55	100.00	100.00
			0.14	0.04	0.04	0.12	0.74	0.81	0.80

In Table 2, we concentrate on sparse instances in which, on the average, every node has two adjacent edges. We believe that such instances are among the most interesting in practical applications. It turns out that the barycenter, the

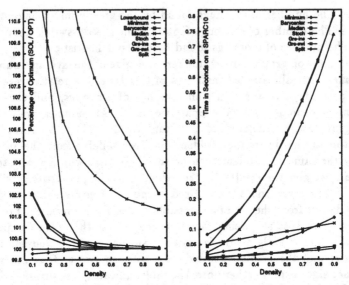

Fig. 2. Results for 100 instances on 20 + 20 nodes with increasing density

Table 2. Results for 10 instances of sparse graphs with increasing size

n_j	m	Low	Min	Bary	Median	Stoch	Gre-ins	Gre-swi	Split
10	20	37.90	38.00	38.90	45.40	38.70	46.40	50.90	38.50
		99.74	100.00	102.37	119.47	101.84	122.11	133.94	101.32
		0.00	0.00	0.00	0.00	0.01	0.00	0.01	0.02
20	40	171.70	171.90	175.70	193.70	174.90	240.80	293.60	174.70
		99.88	100.00	102.21	112.68	101.74	140.08	170.80	101.63
		0.01	0.01	0.01	0.05	0.02	0.05	0.09	
30	60	436.60	438.30	451.90	491.10	451.30	602.30	692.40	445.60
		99.61	100.00	103.10	112.05	102.97	137.42	157.97	101.67
		0.11	0.01	0.01	0.13	0.05	0.11	0.25	
40	80	761.50	765.70	785.60	856.60	782.70	1105.00	1367.50	783.20
		99.45	100.00	102.60	111.87	102.22	144.31	178.60	102.29
		0.30	0.01	0.02	0.28	0.08	0.22	0.57	
50	100	1247.30	1252.20	1279.90	1389.50	1273.20	1770.60	2200.50	1277.80
		99.61	100.00	102.21	110.97	101.68	141.40	175.73	102.04
		0.68	0.02	0.03	0.50	0.13	0.32	1.00	
60	120	1683.10	1687.60	1738.30	1890.90	1720.20	2453.10	2994.50	1736.10
		99.73	100.00	103.00	112.05	101.93	145.36	177.44	102.87
		1.09	0.03	0.04	0.83	0.18	0.61	1.67	
70	140	2465.00	2479.00	2541.30	2730.00	2522.50	3592.20	4498.80	2549.20
		99.44	100.00	102.51	110.13	101.76	144.91	181.48	102.83
		4.46	0.04	0.04	1.28	0.26	0.73	2.82	
80	160	3153.90	3172.10	3254.60	3521.60	3232.90	4583.10	5885.70	3240.60
		99.43	100.00	102.60	111.02	101.92	144.48	185.55	102.16
		6.42	0.05	0.06	1.85	0.33	0.99	4.11	
90	180	4104.00	4132.80	4233.70	4566.80	4206.80	5843.70	7331.30	4293.90
		99.30	100.00	102.44	110.50	101.79	141.40	177.39	103.90
		25.13	0.05	0.06	2.66	0.41	1.32	5.84	
100	200	5127.40	5162.70	5287.50	5728.80	5247.60	7469.90	9407.50	5333.50
		99.32	100.00	102.42	110.97	101.64	144.69	182.22	103.31
		435.51	0.06	0.08	3.35	0.49	1.45	7.56	

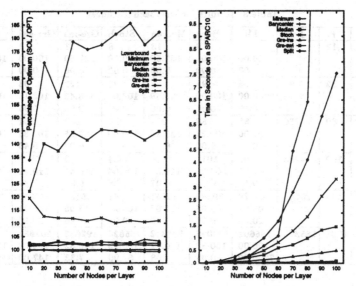

Fig. 3. Results for 10 instances of sparse graphs with increasing size

stochastic and the split heuristic perform very well in terms of quality, however, the split heuristic takes roughly the same time as the branch and cut computation up to size 80+80, whereas the barycenter heuristic obtains results of similar quality as split, but much faster (see Figure 3).

In Table 3, we repeat an experiment by Dresbach [D94] for instances defined by Warfield [W77] as follows: For $k = 3, 4, 5, 6, 7, 8$ we let $n_1 = k$, $n_2 = 2^k - 1$, and the adjacency matrix of the bipartite graph is a $n_1 \times n_2$ matrix whose rows are labelled $1, 2, \ldots, k$, whose columns are labelled $1, 2, \ldots, 2^k - 1$, and column j contains j in k-digit binary notation. Layer 1 is fixed and layer 2 is "optimised". Again, it turns out that barycenter is the fastest method with excellent quality solutions. The results of the stochastic heuristic, the barycenter and the split heuristic are very close to the optimum solution. Up to size 7+127, the branch and cut algorithm needs only moderate computation time, for the instance 8+255 it is not competitive in terms of time, but we found it surprising that such a big linear ordering instance with $n = 255$ could be solved at all. The branch and cut algorithm was the only method that found the true optima for $k \geq 6$, whereas for $3 \leq k \leq 5$, the fact that the optimum value equals the value of the trivial lower bound seems to indicate that these instances are not hard.

4 Two Sided Crossing Minimization

The trivial lower bound on the number of crossings that turned out to be excellent in our previous experiments, can obviously be adapted to partial orderings rather than complete orderings (permutations) on one of the layers. This encouraged us to devise a simple branch and bound algorithm for the general two layer straightline crossing minimization problem in which both π_1 and π_2 must

Table 3. Results for Dresbach instances

n_1	n_2	m	Low	Min	Bary	Median	Stoch	Gre-ins	Gre-swi	Split
3	7	12	8	8	8	13	8	11	8	8
				100.00	100.00	162.50	100.00	137.50	100.00	100.00
				0.00	0.00	0.00	0.02	0.00	0.00	0.02
4	15	32	95	95	95	127	95	122	98	95
				100.00	100.00	133.68	100.00	128.42	103.16	100.00
				0.00	0.00	0.00	0.03	0.02	0.05	0.07
5	31	80	756	756	758	922	756	934	804	760
				100.00	100.27	121.96	100.00	123.55	106.35	100.53
				0.03	0.00	0.03	0.18	0.08	0.40	0.43
6	63	192	4998	5002	5015	5818	5004	6023	5523	5043
				100.00	100.26	116.31	100.04	120.41	110.42	100.90
				0.73	0.05	0.07	1.38	0.38	2.87	2.65
7	127	448	29745	29778	29883	33641	29841	35152	34366	30086
				100.00	100.35	112.97	100.21	118.05	115.41	101.03
				20.50	0.17	0.20	9.02	1.98	20.20	24.30
8	255	1024	165375	165602	166098	183342	165824	192633	202957	167546
				100.00	100.30	110.71	100.13	116.32	122.56	101.17
				7200.00	0.95	1.08	67.90	7.33	147.00	189.00

be determined. Namely, we enumerate all permutations π_1 (let without loss of generality $|V_1| \leq |V_2|$, $V_1 = \{1, 2, \ldots, n\}$) as follows: Initially all $v \in V_1$ are unfixed. At depth l in a depth-first-search, $l - 1$ nodes of V_1 are fixed in positions $1, 2, \ldots, l-1$. Then the first unfixed node in the canonical ordering of V_1 is fixed at position l, and the trivial lower bound L is computed for the resulting partial ordering. If L is greater than the value of the best known solution, the next unfixed node in the canonical ordering of V_1 is fixed at position l, else we move to position $l + 1$, if $l < n$, and otherwise ($l = n$) we call the branch and cut algorithm to determine an optimum ordering of V_2 and update the best known solution, if necessary. Backtracking, i.e. moving from position l to position $l - 1$ occurs whenever the list of unfixed nodes at depth l in the enumeration tree is exhausted. Before the enumeration is entered, a heuristic solution is determined in order to initialize the best known solution. A good initial solution makes the enumeration tree smaller.

In Table 4, we use this algorithm to determine optimum solutions for 10+10 graphs with increasing edge densities, 100 samples for each type of graph. It turns out that with increasing density, the computation times increase rapidly for the minimum computation, whereas the heuristics are not very sensitive to density. All heuristics are iterated between the two layers until a local optimum is obtained, as outlined in the introduction, starting from the canonical ordering on V_1. An additional column labelled "LR-Opt" gives according results for the iterated minimum crossing computation by branch and cut, which is, remarkably, sometimes outperformed by the best iterated heuristics. For sparse instances, the minimum is much better than any of the heuristically found solutions. In Figure 4, we show an example of a 10+10 graph with 20 edges. The first drawing was found by the LR-Opt heuristic and has 30 crossings, the second by the barycenter heuristic and contains 11 crossings and the third one is the optimum solution with only 4 crossings.

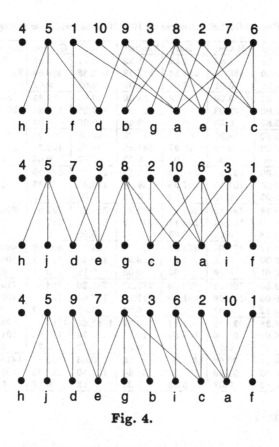

Fig. 4.

Within one hour of computation time, we can find optimum solutions for 11+11 instances with up to 80% density, 12+12 with up to 50% density, 13+13 with up to 30% density, 14+14, 15+15, 16+16 with up to 10% density.

In Table 5, we repeat the same experiment with 10 starts from random orderings of the nodes in V_1. The results show that this way a considerable performance gain for all heuristics can be achieved. LR-Opt, Barycenter and Split obtain results of similar good quality.

Tables 6 and 7 deal with the more interesting sparse instances of bigger size for which we can not compute the optimum anymore, Table 6 with canonical start, Table 7 with 10 random starts. Summarizing, the barycenter method turns out to be the clear winner, both in terms of quality as well as in terms of computation time.

Table 4. Results for 100 instances on 10 + 10 nodes with increasing density

n_i	m	Min	LR-Opt	Bary	Median	Stoch	Gre-ins	Gre-swi	Split
10	10	0.29	1.64	1.52	1.53	2.71	4.32	9.61	2.63
		100.00	565.52	524.14	527.59	934.48	1489.66	3313.79	906.90
		1.10	0.01	0.01	0.01	0.03	0.02	0.02	0.04
10	20	11.62	19.99	18.78	24.08	26.96	38.85	34.81	23.25
		100.00	172.03	161.62	207.23	232.01	334.34	299.57	200.09
		3.89	0.02	0.01	0.01	0.06	0.04	0.03	0.07
10	30	56.60	66.98	65.30	81.78	82.98	109.96	80.29	70.11
		100.00	118.34	115.37	144.49	146.61	194.28	141.86	123.87
		14.06	0.02	0.02	0.02	0.07	0.06	0.07	0.11
10	40	146.89	157.91	157.70	189.55	182.77	225.26	165.65	160.20
		100.00	107.50	107.36	129.04	124.43	153.35	112.77	109.06
		43.02	0.03	0.02	0.02	0.08	0.10	0.11	0.15
10	50	276.78	287.32	288.15	333.25	320.21	387.87	296.38	290.79
		100.00	103.81	104.11	120.40	115.69	140.14	107.08	105.06
		91.58	0.04	0.03	0.02	0.09	0.13	0.15	0.21
10	60	463.17	475.04	475.52	539.59	509.38	598.98	482.76	478.46
		100.00	102.56	102.67	116.50	109.98	129.32	104.23	103.30
		206.61	0.06	0.03	0.03	0.10	0.17	0.22	0.28
10	70	698.35	709.91	710.88	782.33	747.20	854.61	715.73	712.73
		100.00	101.66	101.79	112.03	107.00	122.38	102.49	102.06
		379.12	0.07	0.04	0.03	0.11	0.22	0.29	0.35
10	80	1008.38	1021.46	1021.44	1110.39	1051.66	1165.97	1025.84	1024.78
		100.00	101.30	101.30	110.12	104.29	115.63	101.73	101.63
		763.53	0.08	0.04	0.03	0.12	0.27	0.34	0.40
10	90	1405.57	1420.68	1421.86	1524.18	1430.86	1516.62	1423.90	1421.72
		100.00	101.08	101.16	108.44	101.80	107.90	101.30	101.15
		1549.12	0.07	0.03	0.03	0.12	0.29	0.32	0.37

5 Conclusions

The outcome of our computational experiments lead to the following conclusions.

(1) When one layer is fixed, the exact minimum crossing number can be efficiently computed in practice, so there is no real need for heuristics.

(2) In the general case, small sparse instances as they occur in applications can be solved to optimality if the smaller sized shore has up to about 15 vertices. For larger instances, the iterated barycenter method, started with a few random orderings of one layer, is clearly the method of choice among all tested methods.

6 Acknowledgements

We would like to thank Thomas Ziegler for implementing and running all heuristics in this paper except LR-Opt and Stefan Dresbach for helpful discussions concerning the stochastic heuristic.

Table 5. Results for 100 instances on 10 + 10 nodes with increasing density, 10 trials each

n_j	m	Min	LR-Opt	Bary	Median	Stoch	Gre-ins	Gre-swi	Split
10	10	0.29	0.30	0.31	0.71	0.73	2.10	3.95	0.42
		100.00	103.45	106.90	244.83	251.72	724.14	1362.07	144.83
		1.10	0.11	0.08	0.08	0.27	0.21	0.16	0.38
10	20	11.62	12.50	12.44	16.57	17.44	30.55	21.00	13.83
		100.00	107.57	107.06	142.60	150.09	262.91	180.72	119.02
		3.89	0.18	0.12	0.13	0.52	0.38	0.34	0.64
10	30	56.60	57.27	57.46	68.66	66.33	97.22	62.59	58.30
		100.00	101.18	101.52	121.31	117.19	171.77	110.58	103.00
		14.06	0.26	0.17	0.15	0.68	0.60	0.62	1.01
10	40	146.89	147.35	147.73	166.41	159.31	205.97	150.34	148.24
		100.00	100.31	100.57	113.29	108.46	140.22	102.35	100.92
		43.02	0.36	0.21	0.18	0.79	0.90	1.02	1.45
10	50	276.78	277.11	277.78	304.62	292.34	363.43	277.85	277.61
		100.00	100.12	100.36	110.06	105.62	131.31	100.39	100.30
		91.58	0.47	0.26	0.22	0.87	1.23	1.50	2.03
10	60	463.17	463.76	464.07	499.46	478.48	565.63	464.54	464.17
		100.00	100.13	100.19	107.82	103.31	122.12	100.30	100.22
		206.61	0.59	0.32	0.25	0.96	1.65	2.15	2.67
10	70	698.35	698.75	699.23	745.00	712.78	816.80	699.37	699.04
		100.00	100.06	100.13	106.68	102.07	116.96	100.15	100.10
		379.12	0.68	0.34	0.29	1.03	2.23	2.78	3.30
10	80	1008.38	1008.62	1008.88	1070.82	1018.66	1120.31	1008.96	1008.94
		100.00	100.02	100.05	106.19	101.02	111.10	100.06	100.06
		763.53	0.81	0.37	0.31	1.11	2.70	3.39	3.89
10	90	1405.57	1406.14	1406.22	1490.03	1410.31	1461.52	1406.43	1406.44
		100.00	100.04	100.05	106.01	100.34	103.98	100.06	100.06
		1549.12	0.70	0.33	0.34	1.17	2.86	3.13	3.53

Table 6. Results for 10 instances of sparse graphs

n_j	m	LR-Opt	Bary	Median	Stoch	Gre-ins	Gre-swi	Split
10	20	19.70	15.70	25.70	27.20	35.80	34.20	20.90
		0.02	0.02	0.01	0.05	0.04	0.04	0.06
20	40	73.70	72.50	79.60	132.50	170.70	237.70	91.20
		0.10	0.03	0.04	0.36	0.17	0.17	0.41
30	60	176.00	147.90	188.50	288.20	442.30	549.80	208.30
		0.48	0.10	0.09	1.18	0.49	0.48	1.33
40	80	309.80	273.30	374.20	555.70	760.60	1207.00	368.80
		1.81	0.17	0.14	2.72	0.93	0.67	3.45
50	100	457.70	392.30	561.90	824.40	1284.40	1971.20	548.10
		5.87	0.25	0.17	5.92	1.37	1.10	7.14
60	120	645.60	567.00	811.20	1219.90	1954.80	2667.90	811.10
		13.34	0.38	0.24	8.58	2.24	1.87	10.52
70	140	861.30	764.60	1146.20	1689.30	2549.30	4122.80	1032.40
		24.95	0.55	0.34	14.09	2.89	2.19	19.48
80	160	1246.10	1080.70	1481.30	2183.30	3279.40	5495.90	1467.70
		62.65	0.68	0.52	21.09	4.58	3.22	25.01
90	180	1697.70	1272.40	1848.00	2859.50	4280.00	6853.70	1762.40
		86.37	1.10	0.57	31.84	6.41	4.30	38.36
100	200	2027.30	1555.10	2084.10	3453.10	5405.00	8796.30	2209.40
		178.93	1.46	0.82	40.23	7.41	5.25	47.78

Table 7. Results for 10 instances of sparse graphs, 10 trials each

n_i	m	LR-Opt	Bary	Median	Stoch	Gre-ins	Gre-swi	Split
10	20	13.60	12.70	18.70	17.50	30.00	22.30	14.70
		0.12	0.15	0.12	0.55	0.40	0.34	0.68
20	40	51.00	48.30	59.10	89.00	150.80	163.40	63.70
		0.98	0.42	0.39	3.61	1.82	1.58	3.93
30	60	133.40	117.00	145.80	228.60	421.30	422.10	160.10
		5.55	0.96	0.76	11.48	4.59	4.18	13.13
40	80	234.10	212.40	271.40	432.80	724.50	949.80	279.90
		18.45	1.75	1.29	26.57	8.25	7.42	31.26
50	100	384.20	325.60	407.30	715.60	1245.60	1715.90	462.70
		52.01	2.79	2.06	51.33	13.59	11.60	60.80
60	120	541.10	479.90	599.90	1106.80	1909.70	2472.10	654.00
		128.12	4.38	2.93	92.08	21.97	18.27	114.07
70	140	733.20	641.30	858.00	1489.30	2514.30	3640.00	896.90
		304.08	5.79	3.82	139.95	30.18	23.22	175.83
80	160	1022.90	903.70	1145.10	1993.30	3248.70	4843.50	1169.60
		619.36	7.57	5.28	204.64	38.96	31.18	264.82
90	180	1282.50	1044.70	1323.70	2516.50	4209.10	6228.20	1466.40
		1134.67	10.81	6.55	307.44	57.13	43.19	377.72
100	200	1599.20	1313.20	1793.20	3119.40	5323.90	8145.30	1807.60
		2313.48	13.76	8.02	402.74	67.13	50.24	504.25

References

[CPLEX] CPLEX: Using the CPLEX callable library and the CPLEX mixed integer library. CPLEX Optimization Inc. (1993)

[D94] Dresbach, S.: A New Heuristic Layout Algorithm for DAGs. Derigs, Bachem & Drexl (eds.) Operations Research Proceedings 1994, Springer Verlag, Berlin (1994) 121–126

[D95] Dresbach, S.: Personal communication. (1995)

[EK86] Eades, P., and D. Kelly: Heuristics for Reducing Crossings in 2-Layered Networks. Ars Combinatoria 21-A (1986) 89–98

[EW94] Eades, P., and N.C. Wormald: Edge crossings in Drawings of Bipartite Graphs. Algorithmica 10 (1994) 379–403

[GJ83] Garey, M.R., and D.S. Johnson: Crossing Number is NP-Complete. SIAM J. on Algebraic and Discrete Methods 4 (1983) 312–316

[GJR84a] Grötschel, M., M. Jünger, and G. Reinelt: A cutting plane algorithm for the linear ordering problem. Operations Research 32 (1984) 1195–1220

[GJR84b] Grötschel, M., M. Jünger, and G. Reinelt: Optimal triangulation of large real world input-output matrices. Statistische Hefte 25 (1984) 261–295

[GJR85] Grötschel, M., M. Jünger, and G. Reinelt: Facets of the linear ordering polytope. Mathematical Programming 33 (1985) 43–60

[K93] Knuth, D.E.: The Stanford GraphBase: A Platform for Combinatorial Computing. ACM Press, Addison-Wesley Publishing Company (1993) New York

[STT81] Sugiyama, K., S. Tagawa, and M. Toda: Methods for Visual Understanding of Hierarchical System Structures. IEEE Trans. Syst. Man, Cybern., SMC-11 (1981) 109–125

[W77] Warfield, J.N.: Crossing Theory and Hierarchy Mapping. IEEE Trans. Syst. Man, Cybern., SMC-7 (1977) 505–523

Constraint–Based Spring–Model Algorithm for Graph Layout

Thomas Kamps and Joerg Kleinz
Institute for Integrated Publication and Information Systems (GMD–IPSI)
Dolivostr. 15, D–64293 Darmstadt, Germany, {kamps,kleinz}@darmstadt.gmd.de
John Read
Department of Combinatorics and Optimization
University of Waterloo, Ontario N2L 3G1, Canada, jread@orion.uwaterloo.ca

Abstract

In this paper we will discuss the question of how to develop an algorithm for automatically designing
an optimal layout of a given constraint graph. This automatic layout algorithm must observe certain
aesthetic principles, which facilitate the user's interpretation of the graph. The constraint graph is
generated by the visualization algorithm AVE (Automatic Visualisation Engine) which decides in
this case that the object relation between the objects which are to be graphically represented must be
visually realized by lines. The constraints describe how subsets of objects are geometrically represented relative to each other. We require that each individual object be of fixed size throughout the
algorithm, but we allow for each of these sizes to differ one from another. This automatic layout algorithm is developed along the lines of a (spring) force model, a method which has its roots in such
works as [Eade]. As a measure of any particular layout's fidelity to our aesthetic principles, we have
developed a function which assigns a real value to each possible layout. We have insured that this
function has good differentiability properties, in order that we may exploit gradient descent methods
to arrive at a layout that minimizes this function. Any such layout is then by definition the optimal
layout we seek. These gradient methods are integral in assuring that the algorithm we develop can
efficiently implement the aesthetic principles so that we obtain a very fast routine.

Introduction

In cooperation with Macmillan Publishers Ltd. the department PaVE (Publication and
Visualisation Environment), as part of GMD–IPSI, has developed a prototype version
of the electronic Dictionary of Art. In this application scenario an art historian, as a user
of the interactive electronic refence work, queries a semantic network for information.
The semantic network represents knowledge about the domain of art. The relations of
that network establish facts between domain objects, e.g., "Frank Lloyd Wright's *pro-
fession* was architect" or "Applied Arts is a *broader term of* Graphic Design". In order
to support the user in the process of information access we have built the system AVE
(Automatic Visualisation Engine) that is able to automatically design diagrams as a
visualisation of the query result. We refer the reader to [Reich] for detailed information
concerning the generative theory underlying AVE's approach and to [Golo] for its implementation. In the following section we present a brief summary of AVE's features
in order to explain the overall environment in which our spring–model algorithm is applied.

AVE's Approach

We follow a functional notion of aesthetics (see [Kamp94a]) and thus our approach to
automatic data visualisation is based on the idea that the diagram should mirror the inherent characteristics of the data. As mentioned above we consider data to be represented as relations between objects. We interpret the relational properties as the characteristics, or the regularities, of the data and see the aggregations of properties,
establishing the relation types, as a sensible way to type the data. To exploit data regu-

larities is in analogy to many graph layout approaches. A bibliography of algorithms that graphically realize such properties ('tree', 'symmetric', 'hierarchy', et.) as line diagrams is given in [DiBa]. What makes the difference for us are requirements coming from our application. In the semantic network there are different sorts of relation types. Not all of them are sensibly visualized using the graphical resource 'line' as can be seen from the examples presented in figure 2. In order to integrate these different visulizations we think a rich classification of the data is inevitable. To describe these relations we use the Smalltalk Frame Kit (SFK) [Fisch], an object–oriented, frame–based representation language. SFK implements the theory of binary relations in order to describe the semantics of the relations in terms of their mathematical properties. We use the following incomplete but for our purposes sufficient classification schema for binary relations:

Relation (unqualified)
    Symmetric Relation (symmetric)
    Acyclic Relation (acyclic)
       Tree Relation (tree)
       Irreflexive Order Relation (irreflexive, transitive, asymmetric)
          Irreflexive Tree Order Relation (irreflexive–order, tree)
             Discrete Linear Order Relation (irreflexive–tree–order, linear, discrete domain)
             Continuous Linear Order Relation (irreflexive–tree–order, linear, cont. domain)
       Bipartite Relation (bipartite)
       Function (functional)

figure 1: classification schema for binary relations

Our philosophy is thereby to separate data and data analysis strictly from visualization decisions. [Kosa] proposed an approach in which they define subgraphs, e.g., 'T–Shapes' and 'Hub–Shapes' 'Axial Symmetry', etc., to be the regularities. We see these regularities not as an alternative to the ones we use but as an extension (a relation which is of order relation type may contain T–Shapes). However we would not make 'Axial Symmetry' a regularitiy because this is a purely geometric property and thus should become the result of a visualization decision. As a result of this discussion our aim is, similarly to that of [Kosa], to integrate results from the field of graph drawing but also form the field of information visualization for which we only representatively mention [Mack] and [Roth] who proposed approaches that mainly dealt with the visualization of functional relationships among data.

Using the above mentioned modeling language we can qualfiy relations with appropriate relation types on the class level. It depends on the application which types are assigned to which data relations. This is in accordance with the conformance criteria given by [Lin], which say that the visualization should be adaptable to the application.

In our approach we decompose the given semantic subnet into binary relations and analyse them to find out whether their types can be refined on the instance level. It might turn out that a relation which is defined to be of order relation type on the class level is of a more specific type, say tree order, on the given instance level. This strongly affects the visualization. We will explain this below. In order to decide how to assign graphical resources to the given relations we also rank them based on their specificity in the classification schema and based on the quantitative share that a binary relation has in the overall subnet. The first criterion is justified by the assumption that a specific relation type organizes the data more effectively than a more general type. For instance, a linear order relation organizes the data in a stronger way than an irreflexive order relation because in the case of linearity all elements are pairwise comparable whereas in the other case this is not true. The second criterion makes sure that a more specific relation type does not get the best graphical resources a priori.

To generate expressive diagrams (in analogy to [Mack]) means to map the relations and their properties onto graphical relations that match these properties. One graphical relation is the 'line' relation that connects arbitrary sorts of shapes (see figure 2, examples b) and c)). This graphical resource is very general and can thus be applied to any relation type. Another one, the 'inclusion' graphical relation constructs rectangles that include each other (see example d)). It can only be applied for tree order relations because it exactly matches the tree order properties. Relative position relations, such as 'object A sits below object B', can be applied to represent discrete linear orderings, for instance lists of elements. However, we use this graphical resource also in combination with the 'line' relation in which case it constrains the possible positions of the shapes in one dimension. The "line" relation that is constrained in such a way can visualize order relation type. Together with a legend 'attributes' can be used to represent bipartite relations (see examples a) and b)) and in the more specific case of a qualitative function its values can be visualised using the 'colour' resource. 'Length of box' can be used to represent quantitative function values such as time spans (see example). These examples show that we do the graphical binding at the relation types. Although this list of graphical relations is not complete it shows, by examples, how we assign graphical resources to the data. For a more detailed discussion of the assignment of graphical means of expression to the relation types we refer the reader to [Reichen].

The assigment of graphical relations to the types is not unique. Thus, in a given situation AVE has to decide what graphical relation, among those that are expressive for a certain data relation, is assigned to it. However, this can only be done in the context of the overall subnet to be visualized because the graphical means of expression are limited and because of that the different data relations have to compete with each other in a resource allocation process (see [Golo]). This process is constructive on the relation level. The variety of diagrams which is presented in figure 2 illustrates the different possible outcome of AVE's constructive design process. Thus, in example b) the 'continuous distance' graphical resource is combined with the 'line' graphical resource and this constructs the diagram. We do not have an explicit notion of a 'time–line' template or 'line' template that have to be combined in the process.

The aim of this paper is to discuss the final positioning algorithm by which the graphical elements are displayed given that a relation is visualised using the graphical relation 'line'. In order to do that we first sketch the interface between the overall visualisation system AVE and the spring–model algorithm. Then, we outline the constraints, that may be established during the resource allocation process. They affect the size, the position and the colour of the graphical objects. In this paper we only introduce the position constraints. After that we discuss the objective function which is succeeded by a description of our gradient descent algorithm. Then we illustrate the results by presenting eight example layouts together with their computation times. Finally, we outline possible improvements.

The Interface to AVE

Assume we are given the connected graph $D = (V, E)$, where V is the set of nodes and E is the set of unordered edges $\{v_i, v_j\} \in E$ for $v_i, v_j \in V$. We further assume a set of constraints (as defined below) for the set of nodes. We let N be the total number of nodes to be represented in our graph, and in order to preserve the legibility of our layout, we

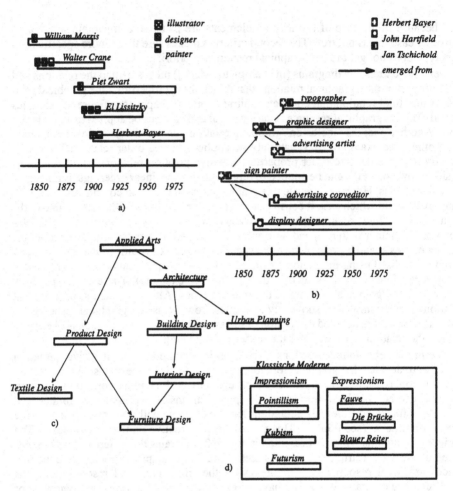

figure 2: some example diagrams

artificially assume N is less than or equal to 40 (neverthless, our algorithm may also handle graphs with a larger number of nodes). As already noted, each node (or object) is represented by a rectangle of fixed size (in contrast to taking them as points of a zero area). Thus, for each node we give three coordinates, "the left upper corner", "the right lower corner" and the "center". That is, $v_i = \left\{ \left(x_i^L, y_i^T \right), \left(x_i^R, y_i^B \right), \left(x_i^C, y_i^C \right) \right\}$ where $x_i^L \leq x_i^C \leq x_i^R$ und $y_i^B \leq y_i^C \leq y_i^T$. Each rectangle contains a core area that represents a domain object the name of which is printed above this area. We take the center of the core area as the center of the overall rectangle. Therefore, the center of the ith rectangle does not correspond in general to the midpoint of the ith rectangle. Again, we allow for the size of each rectangle to differ from one another. We define $d(v_i, v_j) \in N$ to be the topological distance between the two nodes v_i and v_j and this is simply the number of edges in the shortest path between the two nodes. Finally, we assume that the graph is to be represented in the square $\left[-\frac{1}{2}, \frac{1}{2} \right] \times \left[-\frac{1}{2}, \frac{1}{2} \right]$ with origin at $(0,0)$.

The visualization machine AVE, in which our spring model algorithm is embedded, gives a set of geometrical constraints which we now list. As we have mentioned before, the constraints we consider here are means to express data relations geometrically. Thus, they are an important means for the communication of the data and their characteristics into graphics.

Fixed Positions Certain node positions can be fixed in either the x or y components. Thus, there is a (possibly empty) subset $P \subseteq V$ such that for each $v_i = (x_i, y_i) \in P$ there exists p_i^x or $p_i^y \in \mathbb{R}$ with $x_i = p_i^x$ or $y_i = p_i^y$. The reader should note that by writing $v_i = (x_i, y_i)$ we are being intentionally vague as to what is being fixed, that is whether we are fixing x_i^L, x_i^C or x_i^C. This is certainly of little importance since we require that each individual object be of fixed size throughout the algorithm. Thus, $x_i^C - x_i^L$ and $x_i^R - x_i^C$ are both positive constants throughout.

Fixed Distances The distance between certain pairs of nodes can be fixed in either the x or the y direction. Thus, there is a (possibly empty) subset $D \subseteq V \times V$ such that for each $(v_i, v_j) \in D$ there exists d_{ij}^x or $d_{ij}^y \in \mathbb{R}$ with $x_i = x_j + d_{ij}^x$ or $y_i = y_j + d_{ij}^y$. Here $v_i = (x_i, y_i)$ and $v_j = (x_j, y_j)$.

Relative Positions A node can be restricted to lying above (below) to the right of (to the left of) another node. Thus, there is a (possibly empty) subset $R \subseteq V \times V$ such that for each $(v_i, v_j) \in R$ we have $x_i \leq x_j$ or $y_i \leq y_j$.

Orientation Two nodes can have a natural vertical or horizontal positioning relative to one another. Thus, there are (possibly empty) subsets $O_h \subseteq V \times V$ and $O_v \subseteq V \times V$ such that for each $(v_i, v_j) \in O_h$ (O_v) the orientation between v_i and v_j is horizontal (vertical). We will set $O = O_h \cup O_v$.

The Fixed Position and Fixed Distance constraints reduce the dimension of the problem from $2N$ to a smaller number depending on the magnitude of the sets P and D. The Relative Position constraint serves to restrict the domain of the objective function, which we will introduce below. The Orientation constraints are not absolute and will be incorporated in this objective function.

State of the Art in Spring Model Algorithms

[Eade] proposes a spring–model based algorithm which meets two aesthetic principles: firstly, all edges ought to have same length and secondly the layout should display as much symmetry as possible. He considers the graph that is laid out as a mechanical system in which the vertices of the graph represent steel rings and the edges represent spring forces. The input is a random initial placement of the vertices. His iteration process involves computing in each step the resultant force of the springs acting on each steel ring, which provide a force field whose action on the ring–spring system tends to move this system to a new position of lower potential energy. This is then repeated until a state of minimal energy is achieved.

The paradigm of the approach taken by Kamada [Kama] is the same as the one proposed by Eades. Here, the energy potential of the underlying mechanical system is given explicitly as an objective function

$$E = \sum_{i=1}^{n} \sum_{j=i+1}^{n} \frac{1}{2}(|p_i - p_j| - l_{ij})^2$$

that is to be minimised. In this formula $p_i = (p_i^x, p_i^y)$ and $p_j = (p_j^x, p_j^y)$ denote the positions of node i and node j and l_{ij} denotes the optimal distance between the two nodes. The aesthetic goal is that the Euclidean distance between any pair of nodes matches their topological distance. To realize this, Kamada implemented an optimization method which always finds a local minimum for E. An important difference to Eades' approach is that whenever a node is moved to another position all other nodes are frozen.

This is due to the 2n interdependent non–linear equations that they obtain from the minimisation condition (see [Kama]).

Both, Eades and Kamada, applied their algorithms to undirected graphs. The advantage of this technique is that it preserves symmetry without explicitly requiring knowledge of the automorphism group of the graph, a problem which is computationally expensive. Davidson and Harel [Davi] proposed a simulated annealing approach in which the energy potential of a system is described in terms of vertex distribution, nearness to borders, edge–lengths, and edge crossings. Fruchterman and Reingold [Fruc] presented an effective modification of Kamada's model. Sugiyama [Sugi] extended the spring model approach by introducing different sorts of magnetic forces that are applied in conjunction with the distance forces. One sensible way to apply magnetic forces is, e.g., to assign different edge types different orientations.

In contrast to the approaches we have discussed in this section our algorithm must consider a set of geometric constraints as introduced above. In our model, nodes are not just points in the plane, but rather are rectangles having varying x–and y–dimensions extents. Since our nodes are inhomogeneous, seeking a minimum distance between the centers of two nodes is not sufficient for the to minimization of overlapping. Therefore, in addition to 'distance forces', we introduce 'node–node repulsion forces' which serve to assure that the overlapping area of two nodes is minimized. Finally, we employ 'angle forces' in order to incorporate the Orientaion Constraints into the objective function.

In our first prototype [Klei] we extended Eades' approach by the additional forces introduced in the last paragraph. The algorithm we obtained this way was successful but rather slow. However, being a part of an interactive system, a major requirement any algorithm must meet is that diagrams should be laid out and displayed within interaction time. In order to satisfy this goal we now employ a gradient descent algorithm for the minimization of the objective function, which measures how well any given layout can be interpreted by the user.

The Objective Function

In the implementation of the objective function we take into account the "forces" arising due to the distance between pairs of nodes. Further, as we already mentioned we incorporate the "forces" arising from the predetermined orientation between pairs of nodes $(v_i, v_j) \in O$. The distance forces are calculated in such a way that this force is inversely proportional to the topological distance between pairs of nodes. We will adopt the following notation $\xi_{ij} = |x_i^C - x_j^C|$ and $\eta_{ij} = |y_i^C - y_j^C|$. Using this notation we can write the objective function for the 'distance forces' between the nodes v_i and v_j, similar to the formulation in [Kama], as

$$\rho_{ij}(v) = 1 - \frac{\sqrt{\xi_{ij}^2 + \eta_{ij}^2}}{d(v_i, v_j) \cdot \alpha}$$

where alpha is the optimal distance between two nodes connected by a single edge. The optimal edge length would be, in the best of all possible cases, twice as long as the "standard" node. This length is approximated by the assumed average length of the string representing the object name. This is in contradistinction to [Kama] in which this optimal length is determined from the topology of the graph.

As we have indicated, the orientation constraints impose a natural horizontal or vertical positioning on any pair of nodes in the set O. The angle between the x–axis and the line through the nodes v_i and v_j is given by the formula

$$\gamma_{ij}(v) = \begin{cases} \tan^{-1}\left(\dfrac{y_j^C - y_i^C}{x_j^C - x_i^C}\right) \in \left(-\dfrac{\pi}{2}, \dfrac{\pi}{2}\right] & \text{if } x_i^C \neq x_j^C \\ 0 & \text{if } x_i^C = x_j^C \end{cases}$$

Then the deviation angles for any pair $(v_i, v_j) \in V \times V$ is

$$\theta_{ij}^h(v) = \begin{cases} \gamma_{ij} & \text{if } (v_i v_j) \in O_h \\ 0 & \text{otherwise} \end{cases}$$

and

$$\theta_{ij}^v(v) = \begin{cases} \gamma_{ij} - \dfrac{\pi}{2} & \text{if } (v_i v_j) \in O_v \\ 0 & \text{otherwise} \end{cases}$$

These are the objective functions which, upon differentiating, generate the 'angle forces' between the nodes v_i and v_j in the vertical direction and the horizontal direction, respectively. Now in as much as our nodes carry information for the user, we must insure that they do not overlap, thereby blocking part of that information from view. In order to achieve this, we consider the following functions

$$\alpha_{ij}(v) = \begin{cases} \max\{x_i^R - x_j^L, 0\} & \text{if } x_i^L + x_i^R \leq x_j^L + x_j^R \\ \max\{x_j^R - x_i^L, 0\} & \text{if } x_j^L + x_j^R < x_i^L + x_i^R \end{cases}$$

$$\beta_{ij}(v) = \begin{cases} \max\{y_i^T - y_j^B, 0\} & \text{if } y_i^B + y_i^T \leq y_j^B + y_j^T \\ \max\{y_j^T - y_i^B, 0\} & \text{if } y_j^B + x_j^T < y_i^B + y_i^T \end{cases}$$

We take the product of α_{ij} and β_{ij} as the objective function for the 'node–node repulsion forces' between the nodes v_i and v_j and write $\psi_{ij}(v) = \alpha_{ij}(v) \cdot \beta_{ij}(v)$

Upon a closer examination it is clear that ψ_{ij} is at least as large as the square of the area of the overlap of the i-th and j–th nodes. Further, if this pair of nodes do not overlap then $\psi_{ij} = 0$. With this notation we obtain the j–th term of our objective function, given by

$$f_j(v) = \sum_{i<j}\left(\frac{1}{d(v_i, v_j)^2}\rho_{ij}^2 + \omega_1(\theta_{ij}^{h\,2} + \theta_{ij}^{v\,2}) + \omega_2\psi_{ij}\right)$$

We have multiplied the square of the deviation angles by the scaling factor ω_1, which in actual calculations was always set equal to 0.05 and we have multiplied the overlap function ψ_{ij} by the factor ω_2. In order to weaken the forces between node pairs connected over topologically long paths, we have included the factor $\dfrac{1}{d(v_i, v_j)^2}$. For the sake of keeping the formula of the partial derivatives of $f_j(v)$ simple we first define the partial derivatives for its composites (we present only the derivatives in x) as

$$\frac{\delta\rho_{ij}(v)}{\delta x_i} = \begin{cases} \dfrac{-4\rho_{ij} \cdot \xi_{ij}}{d(v_i, v_j) \cdot \sqrt{\xi_{ij}^2 + \eta_{ij}^2}} & \text{if } x_i < x_j \\ 0 & \text{otherwise} \end{cases}$$

$$
\frac{\delta\Theta_{ij}{}^{v}(v)}{\delta x_i} = \begin{cases} 2\ \dfrac{\Theta_{ij}^{v}\,(y_j^C - y_i^C)}{(x_j^c - x_i^c)^2 + (y_j^c - y_i^c)^2} & \text{if } x_i > x_j \\ 0 & \text{otherwise} \end{cases}
$$

$$
\frac{\delta\Theta_{ij}{}^{h}(v)}{\delta x_i} = \begin{cases} 2\ \dfrac{\Theta_{ij}^{h}\,(y_j^C - y_i^C)}{(x_j^c - x_i^c)^2 + (y_j^c - y_i^c)^2} & \text{if } x_i > y_j \\ 0 & \text{otherwise} \end{cases}
$$

$$
\frac{\delta\psi_{ij}(v)}{\delta x_i} = \begin{cases} a_{ij} & \text{if } x_i^L + x_i^R \le x_j^L + x_j^R \text{ and } x_i^R \ge x_j^L \\ -a_{ij} & \text{if } x_j^L + x_j^R < x_i^L + x_i^R \text{ and } x_j^R > x_i^L \\ 0 & \text{otherwise} \end{cases}
$$

Thus, we can write the partial derivative of $f_j(v)$ in the following way

$$
\frac{\delta f_j(v)}{\delta x_i} = \frac{\delta\rho_j(v)}{\delta x_i} + \frac{\delta\Theta_j^v(v)}{\delta x_i} + \frac{\delta\Theta_j^h(v)}{\delta x_i} + \frac{\delta\psi_j(v)}{\delta x_i}
$$

$$
\frac{\delta f_j(v)}{\delta y_i} = \frac{\delta\rho_j(v)}{\delta y_i} + \frac{\delta\Theta_j^v(v)}{\delta y_i} + \frac{\delta\Theta_j^h(v)}{\delta y_i} + \frac{\delta\psi_j(v)}{\delta y_i}
$$

and by summing over j we get the objective function $f(v) = \sum_{j=1}^{N} f_j(v)$ whose gradient is constructed from the partial derivatives in the obvious way.

The Algorithm

Let us now consider an algorithm which will minimize the value of the objective function as defined in the previous section. We begin by reminding the reader that the gradient descent method can be applied successfully to any smooth function f over $\mathbb{R}^n$, which grow at infinity. Given an initial starting position $\vec{p}_i$, one calculates $\nabla f(\vec{p}_i)$ and defines τ_i as the smallest positive value which minimizes the function $\varphi(\tau) := f(\vec{p}_i - \tau\nabla f(\vec{p}_i))$. We then define $\vec{p}_{i+1} = \vec{p}_i - \tau_i\nabla f(\vec{p}_i)$. Iterating over $i \in \mathbb{N}$ gives us a sequence $(\vec{p}_i)$ which converges to a (local) minimum. We have modified this theoretically satisfying model in order to speed convergence and avoid problems one might encounter due to local minima. First we let

$$
m_{j,i} = \sqrt{\left(\frac{df(\vec{p}_i)}{dx_j}\right)^2 + \left(\frac{df(\vec{p}_i)}{dy_j}\right)^2} \quad \text{for } 1 \le j \le N \text{ be the movement of node } v_j \text{ in it-}
$$

eration i. Then, similar to the approach of [Fric], we define $\mu \in [0, \frac{\pi}{4}]$ to be the opening angle for oscillation detection and we let κ_i be the angle between the vectors given by the positions of node v_j during the i–2nd and i – 1st iteration. Given these definitions, we call

$$
O_{j,i}(O_{j,i-1}, \kappa_i) = \begin{cases} O_{j,i-1} + r_0 * \cos(\kappa_i) & \text{if } |\cos(\kappa_i)| > \cos(\mu) \\ O_{j,i-1} & \text{otherwise} \end{cases}
$$

the oscillation factor of node v_j for $i > 2$ and $r_0 \in [0, 1]$. In contrast to their approach, we call the product $T_{j,i} = O_{j,i} * m_{j,i}$ of the oscillation factor with the movement of an arbitrary node v_j the temperature of this node. To give a formula for τ_i we need to define $T_{k,i} = \max\{T_{j,i} : 1 \le j \le N\}$ as the maximum temperature of the nodes in the ith iteration and from this we obtain the ratio $q_i = \dfrac{T_{k,i}}{T_{k,i-1}}$ that describes the maximum change of temperature between two consecutive iterations. Then we set

$$\tau_i = \begin{cases} \dfrac{c}{m_{k,i}} & \text{if } i \leq 2 \\ \tau_{i-1} * f & \text{otherwise} \end{cases}$$

in which $c = \frac{a}{2}$. The factor

$$f = \begin{cases} 1 + w * (1 - q_i) & \text{if } q < 1,\ w \in [0, 0.5] \\ \dfrac{1}{q_i} & \text{otherwise} \end{cases}$$

describes the correction of the τ_i with respect to the change of temperature. Such a choice of c and f, respectively, and hence τ_i have proven in practice to lead quickly to a very pleasing layout, and help control oscillations that can cause such an algorithm to become very inefficient. As a consequence, convergence is obtained if $T_{k,i} < \epsilon$ for some i which means that it depends only on the given graph and its characteristics and is not enforced by an artificial damping function nor by a limitation on the number of iterations. After the application of the above algorithm, it might be necessary to rescale the results so that the optimal layout is neither too large nor too small. We define the quantities

$$R_L = \min\{x_i^L | 1 \leq i \leq N\},\ R_R = \max\{x_i^L | 1 \leq i \leq N\},$$
$$R_T = \min\{y_i^T | 1 \leq i \leq N\},\ R_B = \max\{y_i^B | 1 \leq i \leq N\}.$$

that determine the bounding box of the graph.

Then we set $l_{max} = \max\{|R_R - R_L|, |R_T - R_B|\}$, which we use as a scaling factor. Thus, we redefine the scaling factor $s_{i+1} = s_i \cdot \dfrac{1}{l_{max}}$, for $s_0 = 1$ and translate the center of the bounding box to the origin in order to recenter the graph.

Results

In this section we will discuss example layouts that were generated by our spring–model algorithm. Along the first three diagrams we demonstrate how the semantics of the data may affect their visualization. Then, we present a couple of other example layouts some of which are taken from the literature. For each diagram we give the computation time which was measured on SPARCstation 20. The algorithm is implemented in ANSI–C. The three diagrams in figure 3 have the same topology if one abstracts from the directedness of the edges. Example a) represents a relation which is called 'work togteth- er'. It is defined on the set of 'artists' and it has the relation type Symmetric Relation. Therefore it is visualized using lines. Example b) represents the terminology relation 'broader term of' which is defined on the set of art concepts. It is qualified as an order relation on the class level but in the given instance it forms a tree–order relation. According to our model included rectangles would best visualize this graph. However, here we chose the 'arrow' resource in order to demonstrate the algorithm described in this paper. Thus, AVE decides to use the graphical relation 'below', which is defined on the y– coordinates, to constrain the positions of the nodes in the y–dimension. This way, we guarantee that the transitivity of the 'broader term of' relation is expressively visualized. In this case the Orientation constraints require the sons of a node to be verti- vally positioned relative to the father node. Example c) represents an organization chart in which additionally the nodes on the same level have a fixed zero distance in the y coordinate.

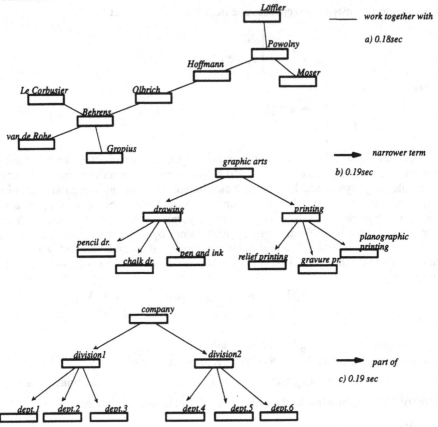

figure 3: Three graphs with the same topology but different relation types

In figure 4 we present additional examples of layouts. Example a) and b) are unconstrained which implies that only the distance forces work which is in analogy to the algorithms of Eades and Kamada. The lattice in example c) is of order relation type visualized using arrows and 'below' constraints. Example d) the x–extents are determined by time intervals and thus we obtain fixed x–positions. In example e) we have defined 'below' constraints and fixed distance constraints for hierarchic subgraph whereas the remaining subgraph is not constrained at all.

Summary and Future Work

In this paper we have presented a force directed placement routine which is an extension of the model originally proposed by [Eade]. Since we apply this routine in the context of the visualization algorithm AVE, we had to add two additional forces to the 'distance forces': the 'node–node repulsion force' that minimizes the overlapping area of two nodes and the 'angle force' that graphically realize a relative vertical (horizontal) positioning of two nodes to one another. Apart from the additional forces which we have modeled, a set of geometric constraints that are generated by AVE have to be integrated

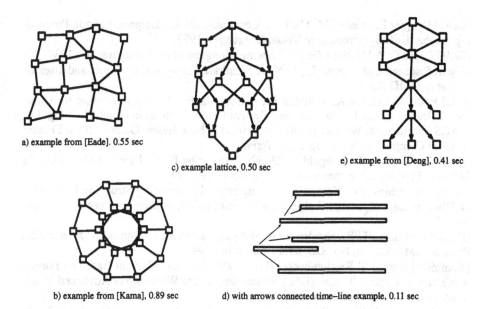

a) example from [Eade]. 0.55 sec

c) example lattice, 0.50 sec

e) example from [Deng], 0.41 sec

b) example from [Kama], 0.89 sec

d) with arrows connected time–line example, 0.11 sec

figure 4: additional examples

into the graph layout process. Since a major requirement from the application is to insure interactivity the computation time was a crucial issue for us. This led us to the construction of an objective function with good differentiability properties in order exploit gradient descent methods and thus obtain a very fast algorithm producing expressive line diagrams effectively. However, we still encountered some problems which we consider should be solved in the future.

As can be seen in figure 4 example e), patterns such as the T–Shape template and the Hub–Shape template in the approach of [Kosa] may be visually realized just by using geometric constraints, that is, without explicitly having a notion of these patterns. However, this does not mean that the algorithm always preserves such patterns, for example, in the particular case when parts of the pattern are connected to other structures which themselves work against a construction of patterns. As mentioned before we consider the integration of such structural regularities as an important extension of our approach. However, this means to extend AVE's data analysis process in order to detect such regularities in the data and to assign them visualization templates.

Another problem occurs when the node extents are very inhomogeneous. In this case the ideal length of the edges between two nodes, which now relies on the size of the ideal node's extent, is no more a good measure. For this more general case we still have find a sensible notion of distance.

References

[Davi] Davidson R. Harel D., Drawing Graphs Nicely Using Simulated Annealing. Technical Report CS89–13, Department of Applied Mathematics and Computer Sciences, Weizman Institute Of Sciences, Israel, July, 1989.

[DiBa] Di Battista G. Eades P. Tamassia R. Tollis G., Algorithms for Drawing Graphs: an Annotated Bibliograpy, Brown University, Department of Computer Science, Technical Report, 1993.

[Deng] Dengler E. Friedell M. Marks J., Constraint–Driven Diagram Layout, Proceedings of the IEEE Symposium on Visual Languages, 1993.

[Eade] Eades P., A Heuristic Graph Drawing, Congressus Numerantium, 42, 1984.

[Fisch] Fischer D. H. Rostek L., SFK: A Smalltalk Frame Kit, Concepts and Use, To appear as GMD Report.

[Fric] Frick A. Ludwig A. Mehldau H., A fast Adaptive Layout Algorithm for Undirected Graphs (Extended Abstract and System Demonstration, in Graph Drawing, DIMACS International Workshop, GD '94, Princeton New Jersey, Ovtober 1994, Lecture Notes in Computer Science, Springer Verlag.

[Fruc] Fruchterman T. Reingold E., Graph Drawing by Force–Directed Placement. In Software Practice and Experience, 21(11), November 1991.

[Golo] Golovchinsky G. Kamps T. Reichenberger K, Subverting Structure: Data–driven Diagram Layout, accepted at IEEE Visualization '95, Atlanta, October 30–November 1 1995.

[Kamp94a] Kamps T. Reichenberger K. (1994a), Automatic Layout as an Organization Process, GMD Report, No. 825, Sankt Augustin 1994.

[Kamp94b] Kamps T. Reichenberger K. (1994b), Automatic Layout Based on Formal Semantics, in Catarci, T. et.al. (Eds.), Proceedings of the Workshop of Advanced Visual Interfaces, AVI '94, June 1194, Bari, Italy.

[Kama] Kamada T. Kawai S., An Algorithm for General Undirected Graphs, in Information Processing Letters 31 (1989). Elsevier Science Publishers B.V. (North Holland).

[Kosa] Kosak C. Marks J. Shieber S., Automating the Layout of Network Diagrams with Specified Visual Organization, in IEEE Transactions On Systems. MAn and Cybernetics, Vol 24, No. 3, March 1994.

[Klei] Kleinz J., Entwicklung eines constraint–gesteuerten 2D–Positionierungsverfahrens, to appear as GMD report.

[Lin] Lin T. Eades P., Integration of Declarative and Algorithmic Approaches for Layout Creation, in Proceedings of DIMACS International Workshop, GD '94 Princeton, New Jersey, USA, October 1994, Tamassia R. Tollis G. (Eds.), Lecture Notes in Computer Science 894,1994.

[Mack] Mackinlay J. (1986) Automating the Design of Graphical Presentations of Relational Information, in ACM Transactions on Graphics, 5(2).

[Reich] Reichenberger K. Kamps T. Golovchinsky G., Towards a Generative Theory of Diagram Design, accepted at IEEE Symposium on Information Visualization (Infovis '95), Atlanta, October 31 1995

[Roth] Roth S. F. Hefley W.E. Intelligent Multimedia Presentation Systems: Research and Principles, in Mark Marbury (Ed.) Intelligent Multi–Media Interfaces, AAAI Press, 1993.

[Sugi] Sugiyama K. Misue K. A Simple and Unified Method for Drawing Graphs: Magnetic–Spring Algorithm, in Proceedings of DIMACS International Workshop, GD '94 Princeton, New Jersey, USA, October 1994, Tamassia R. Tollis G. (Eds.), Lecture Notes in Computer Science 894,1994.

Layout Algorithms of Graph-Like Diagrams for GRADE Windows Graphic Editors

Paulis ĶIKUSTS and Pēteris RUČEVSKIS

Institute of Mathematics and Computer Science
University of Latvia, 29 Rainis blvd., Riga, Latvia
paulis@cclu.lv, rpeteris@cclu.lv

Abstract. We propose a set of layout operations ensuring flexible and convenient interactive editing of communication diagrams and nested entity-relationship models having textual labels on connections. The set includes several procedures for incremental diagram layout. Tools for fully automatic layout and for direct manual painting of graphic primitives are also integrated in a single system. In this way we have filled to some extent the gap between both extremal levels of editing.

1 Introduction

GRADE Windows is a CASE tool, based on specification language GRAPES/ 4GL [1]. There are communication diagrams (CD) and entity-relationship models (ER) among system models built by GRADE Windows (Fig. 1, 2). Such diagrams are graph-like structures and consist of elements of two principal kinds: *blocks* and *connections*. The graphical notations used for ER models are those introduced by J.Martin. An additional graphic element is a notation for nested entities shown by placing descendent entities inside parent entities. Thus, we allow more general graph-like diagrams in which blocks are arbitrary nested without combinatorial restrictions for connections. Moreover, blocks and connections may be supplied with text areas called *superscriptions* which are the third principal kind of diagram elements. In such a way our ER diagrams generalize K.Sugiyama's and K.Misue's compound digraphs [2].

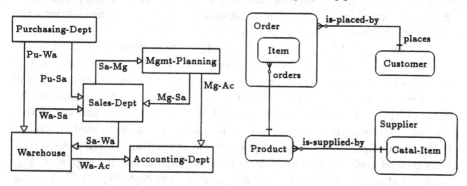

Fig. 1   Fig. 2

Creating a complex diagram using only simple painting tools is very complicated process because adding or displacing even a single block may cause a need to change the placement of all diagram elements. Fully automatic layout of graph-like structures is a well-studied problem and there are a great amount of layout algorithms for entire diagrams [3]. However, since creating a complex diagram is a gradual interactive process during which only fragments of diagram are presented, fully automatic tool may produced fragment layouts that are non-adequate for desired final layout of the whole diagram. Thus, the gap between the both extremal levels of editing deserves attention, but only few papers [4, 5] have focused on this topic.

We propose a set of layout operations ensuring flexible and convenient interactive editing of CD and ER diagrams. The set of our operations includes several procedures oriented on processing of intermediate stages of layouts. Tools for fully automatic layout and for direct manual painting of graphic primitives are also integrated in a single system. In this way we have filled to some extent the gap between the both extremal levels of editing.

2   Principles of Algorithms and Data Structures

All our editing procedures are grouped into several kinds of *editing modes*, the most important of which are AUTOMATIC and MANUAL. AUTOMATIC mode has the strongest set of geometric constraints maintained by the editors: contours of the block figures do not cross each other; block figures of non-nested blocks do not overlap; connection lines have no common segments and do not intersect CD block figures; superscriptions do neither overlap each other, nor block figures, nor connection lines. MANUAL mode has only one constraint: contours of the nested block figures do not cross each other. Fast procedures switching among the modes have also been implemented, and they are important additional elements for filling the gap between the extremal levels of editing.

In CD and ER editors geometry calculations are performed with a simplified representation of the actual graphical data: block figures and text areas are represented by their smallest enclosing rectangles, connections lines are modified accordingly. The obtained rectangles and line segments form the primary geometric data (PGD) structure (see Fig. 3 for our ER example). All newly interactively or automatically created diagram elements are inserted into the PGD-structure maintaining the constraints.

While in MANUAL mode CD and ER editors work directly with the PGD-structure, there are two secondary geometric data structures in the case of AUTOMATIC mode: H-structure and V-structure. H(V)-structure consists of rectangles of three sorts: (1) slim rectangles of width 2δ that are obtained by spreading each horizontal (vertical) segment of block rectangles and connection lines from PGD-structure by constant width δ in all four axes-parallel directions; (2) the block superscription rectangles; (3) the connection superscription rectangles pushed aside by distance δ from adjacent connection line segments. Fig. 4 shows H-structure of our ER example (dashed lines copy Fig. 3). Fast access to rectangles from H, V-structures are ensured by means of well known regular cell structure associated with them.

Our basic procedure that acts with rectangles from H-structure (V-structure) is the shifting procedure *Shift*. The procedure takes care of freeing a place of the necessary size among diagram elements: when an axes-parallel vector is given, *Shift* recursively translates all the rectangles from an adjacent rectangular area by appropriate shorter vectors, and then translates the given rectangle by the given vector.

Procedure *Shift*, together with four converting procedures *MakeHstructure*, *MakeVstructure*, *MakePGDfromH*, *MakePGD fromV*, are the basic tools for performing geometric modifications of diagrams of both kinds without destroying their topological structure and maintaining all of the constraints in AUTOMATIC mode at the same time.

Overlappings of non-nested block rectangles are eliminated by means of a special procedure *NormalizeBlockDistances* which is based on *Shift* and works with arbitrarily placed block rectangles. As the procedure is useful also for adjusting specially calculated allocations of blocks, we can represent the given diagram by an undirected graph and use its drawing as a sketch for the layout.

We have implemented two graph drawing algorithms: one for 2-connected graphs and another for graphs with cutting vertices. In the case of a 2-connected graph, first, we uniformly lay out its vertices, and, second, improve their positions by well known barycentric method.

Graphs with cutting vertices are processed in a more complicated way as follows. First, we build a special cutting vertex-bicomponent tree (CVB-tree). The vertex set of a CVB-tree consists of the cutting vertices of the given graph, and vertices associated with its bicomponents that are not single edges. The set of edges of a CVB-tree consists of all pairs (c, b), where c is a cutting vertex and b is a vertex associated with the bicomponent containing c. In addition, each vertex of the CVB-tree is associated with a disk. The area of the disk is taken proportional to the total area of all block rectangles represented by the same CVB-vertex.

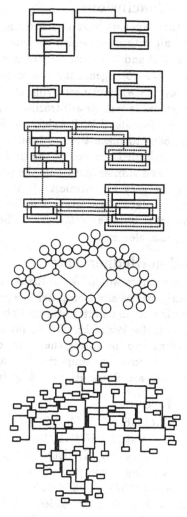

Fig. 3, 4, 5, 6

The next stage is to lay out the disks in non-overlapping way in order to get some initial planar representation of the CVB-tree. After this is done we compact iteratively the initial layout by translating both subtrees of each edge of the CVB-tree each towards the other. The disks are not allowed to intersect any region between any two adjacent disks bounded by their common outer tangents.

Finally, we place the vertices of the given graph into corresponding disks and adjust locations of the non-cutting vertices by barycentric method. Examples of drawings of both a CVB-tree and a corresponding PGD-structure with raw connection lines are given in Fig. 5, 6.

3 Conclusions

In this paper we have sketched the most important details of our data structures and algorithms on which various layout procedures for graph-like diagrams are based. H and V-structures represent diagram blocks, connections and superscriptions as a homogeneous set of rectangles. This ensures automatic transfiguration of layouts without destroying the diagram readability for relatively simple CD diagrams as well as arbitrarily nested ER diagrams.

K.Miriyala, S.W.Hornick and R.Tamassia [4] ask "what do we do if there is no space to place the arc label?" and "what happens if there is no space for the expanded version of the node?" Our approach gives a natural answer to these and similar questions: the procedures *Shift* and *NormalizeBlockDistances* are the very algoritmic tools which via the H and V-structures take care of freeing place of necessary size for the new object and eliminate overlappings of the other objects performing sufficiently small number of automatic moving of diagram elements.

Acknowledgements

The authors would like to thank the GRADE development group at Institute of Mathematics and Computer Science of University of Latvia for enabling them to participate in an interesting collaborative project, and particularly Ilona Etmane and Kārlis Podnieks. We would also like to thank Alvis Brāzma for carefully reading and discussing the manuscript.

The work was supported by the grant 93.601 of Latvian Council of Science, by Software House Riga, and by Infologistik GmbH, Munich.

References

1. GRADE V1.0: Modeling and Development Environment for GRAPES-86 and GRAPES/ 4GL, User Guide. – Siemens Nixdorf, 1993.
2. K.Sugiyama, K.Misue. Visualization of structural information: automatic drawing of compound digraphs. – *IEEE Trans. Syst. Man, Cybern.*, vol. 21, no. 4, 1991, pp. 876–892.
3. G.Di Batista, P.Eades, R.Tamassia, I.G.Tollis. Algorithms for drawing graphs: an annotated bibliography – *Computational Geometry: Theory and Applications*, vol. 4, no. 5, 1994, pp. 235–282.
4. K.Miriyala, S.W.Hornick, R.Tamassia. An incremental approach to aesthetic graph layout, – *Proc. Int. Workshop on Computer-Aided Software Engineering (CASE '93)*, 1993.
5. A.Papakostas, I.G.Tollis. Issues in interactive orthogonal graph drawing, – *Proc. Graph Drawing '95*, in *Lecture Notes in Computer Science*, this volume.

Grid Intersection and Box Intersection Graphs on Surfaces (Extended Abstract)

Jan Kratochvíl[1]* and Teresa Przytycka[2]

[1] Charles University, Prague, Czech Republic
[2] Odense University, Odense, Denmark

Abstract. As analogs to grid intersection graphs and rectangle intersection graphs in the plane, we consider grid intersection graphs, grid contact graphs and box intersection graphs on the other two euclidean surfaces – the annulus and the torus. Our first results concern the inclusions among these classes, and the main result is negative – there are bipartite box intersection graphs on annulus (torus), which are not grid intersection graphs on the particular surfaces (in contrast to the planar case, where the two classes are equal, cf. Bellantoni, Hartman, Przytycka, Whitesides: Grid intersection graphs and boxicity, Discrete Math. **114** (1993), 41-49). We also consider the question of computational complexity of recognizing these classes. Among other results, we show that recognition of grid intersection graphs on annulus and torus are both polynomial time solvable, provided orderings of both vertical and horizontal segments are specified.

1 Motivation

Intersection graphs of different types of geometric objects in the plane gained a lot of attention in the past years. One may consider interval graphs, circle graphs, circular arc graphs, permutation graphs and string graphs, to mention just a few of such classes of graphs. At this point, we want to pay closer attention to *grid intersection graphs* (intersection graphs of vertical and horizontal straight line segments in the plane, such that no two parallel segments of the representation overlap cf. [1], [4]), *grid contact graphs* (grid intersection graphs that have a representation in which no two segments cross each other cf. [3]) and *box intersection graphs* (graphs of boxicity two, i.e. intersection graphs of isothetic rectangles in the plane cf. [7], [10], [4], [1]).

The concept of vertical and horizontal directions translates naturally to the other two orientable euclidean surfaces – the annulus and the torus, and our aim

---

* This research was started when the first author was visiting Odense University in May 1994. The paper was written when the first author visited University of Oregon in Eugene, Oregon in the academic year 1994/5. The first author also acknowledges partial support from Czech Research Grants GAUK 361 and GAČR 2167.

is to study analogous classes of graphs on these surfaces. Our motivation stems from two results about the planar case, whose generalizations we question.

First, Bellantoni *et al.* [1] proved that every bipartite graph of boxicity two is in fact a grid intersection graph. In other words, given a set of rectangles in the plane such that its intersection graph is bipartite, one can shrink the rectangles corresponding to the vertices of one color class into vertical segments, and the rectangles corresponding to the other color class into horizontal segments, while keeping the intersection graph itself unchanged.

The second source of our motivation is the result of de Frayssiex *et al.* [3], which says that every planar bipartite graph is a grid contact graph. This result, which is based on visibility representations of general planar graphs [8, 9], was actually generalized to the torus by Mohar and Rosenstiehl [6].

The paper is organized as follows. In Section 2, we give exact definitions of the classes of graphs under consideration. Then, in Section 3, we present the inclusions among the classes (and reason about their strictness). In Section 4, we aim at the question of the computational complexity of recognizing the considered classes of graphs, in particular in the case when the coordinates of the segments are preordered. The last section contains final remarks and open problems.

2   Definitions

It is usual to view the torus as a rectangle whose vertical and horizontal sides are unified. In this model, segments of horizontal and vertical lines correspond to segments of geodetic circles on the torus. Having this model in mind, we will work with a rectangle whose vertical sides are unified as a model of the annulus, and with ordinary rectangle as a model of the plane. (Since we will only consider finite graphs, restriction to bounded parts of annulus and plane are irrelevant.) We will refer to this rectangle as the *base rectangle*. We will use the abreviations *pl*, *an* and *to* for the plane, the annulus and the torus, respectively. For each of these three surfaces, we consider the following classes of graphs (note that in view of the theorems whose generalizations we question, we only consider bipartite graphs):

GI(surface) = *grid intersection graphs* on the surface = graphs that have a representation by vertical and horizontal line segments on the surface, such that any two parallel segments are disjoint and two segments share a point if and only if the corresponding vertices are adjacent;

GC(surface) = *grid contact graphs* on the surface = graphs which have a representation by vertical and horizontal line segments on the surface, such that any two parallel segments are disjoint, no two segments cross and two segments share a *contact point* if and only if the corresponding vertices are adjacent;

BI(surface) = *box interesection graphs* on the surface = bipartite graphs that have a representation by isothetic rectangles on the surface such that two rectangles are disjoint if and only if the corresponding vertices are nonadjacent;

EM(surface) = *graphs embeddable* in the surface = bipartite graphs that allow a noncrossing drawing on the surface.

As an illustrative example, a grid contact representation of $K_{4,4}$ on the torus is shown in Figure 1.

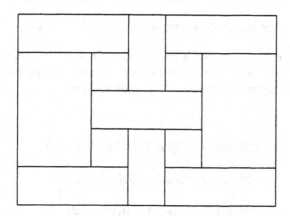

Fig. 1. A grid contact representation of $K_{4,4}$ on the torus.

We will view rectangles (boxes) as two-dimensional intervals, i.e., as products of intervals in the coordinates. In the case of the torus, both horizontal and vertical intervals that determine a box may be wrapped around the sides of the base rectangle, i.e., the coordinate of the left endpoint may be greater than the coordinate of the right one. In the case of the annulus, only the horizontal intervals may be of this type, and in the planar case, all intervals must be proper. For the case of grid intersection and grid contact graphs, the segments will be viewed as boxes with one side of length zero. Since we only deal with finite graphs, we may assume without loss of generality that the coordinates of the endpoints of the intervals that determine the boxes of a representation are mutually distinct integers.

It is well known that a graph is a box intersection graph in the plane if and only if it is the intersection of two interval graphs, i.e., if it has interval dimension at most 2 [2]. Similarly, a graph is a box intersection graph on the annulus if and only if it is the intersection of an interval graph and a circular arc graph. Finally, a graph is a box intersection graph on the torus if and only if it is the intersection of two circular arc graphs, i.e., if it has circular arc dimension at most 2.

3   Inclusions among the classes

The following three propositions are straightforward:

Proposition 1. *For any $X \in \{EM, GI, GC, BI\}$, $X(pl) \subseteq X(an) \subseteq X(to)$.*

Proposition 2. *For any $X \in \{pl, an, to\}$, $GC(X) \subseteq GI(X) \subseteq BI(X)$.*

Proposition 3. *For any $X \in \{pl, an, to\}$, $GC(X) \subseteq EM(X)$.*

The last proposition is in fact a special case of a more general theorem which says that a contact graph of curves on any surface with at most two curves sharing a contact point is embeddable on that surface.

$$
\begin{array}{cccc}
EM(to) \supseteq & GC(to) \subseteq & GI(to) \subseteq & BI(to) \\
\cup| & \cup| & \cup| & \cup| \\
EM(an) \supseteq & GC(an) \subseteq & GI(an) \subseteq & BI(an) \\
\cup| & \cup| & \cup| & \cup| \\
EM(pl) \supseteq & GC(pl) \subseteq & GI(pl) \subseteq & BI(pl)
\end{array}
$$

Fig. 2. Inclusions among the classes.

All these inclusions are depicted in Figure 2. It follows from the results of Bellantoni *et al.* [1] that $GI(pl) = BI(pl)$. Another inclusion in the bottom line reduces to equality due to the result of de Frayssiex *et al.* [3], namely $EM(pl) = GC(pl)$. But every graph embeddable on annulus is planar (the cylinder is topologically equivalent to the plane with two points removed), and hence $EM(pl) = EM(an) = GC(an) = GC(pl)$. It is well known that e.g. $K_{3,3}$ is embeddable on torus and not in the plane, hence $EM(an) \subset EM(to)$. Concerning the other inclusions, we have the following theorem:

Theorem 4. *All other inclusions are strict, as depicted in Figure 3.*

$$
\begin{array}{cccc}
EM(to) \supset & GC(to) \subset & GI(to) \subset & BI(to) \\
\cup & \cup & \cup & \cup \\
EM(an) = & GC(an) \subset & GI(an) \subset & BI(an) \\
\| & \| & \cup & \cup \\
EM(pl) = & GC(pl) \subset & GI(pl) = & BI(pl)
\end{array}
$$

Fig. 3. Inclusions among the classes.

Let us remark at this point that the equality EM(to) = GC(to) "almost" holds – Mohar and Rosenstiehl proved that every bipartite graph embeddable on the torus is indeed a grid contact graph on a *skew* torus (i.e., a torus whose horizontal and vertical grid directions are not parallel with the sides of the base rectangle). However, there are graphs embeddable on the torus that are not grid contact graphs on the torus in our sense [5]. The strict inclusion GC(an) $\subset$ GC(to) follows from the complete bipartite graph $K_{3,3}$, the inclusions GC(X) $\subset$ GI(X), $X \in$ {pl,an,to} follow from complete bipartite graphs as well. For the remaining inclusions, we use slightly more involved constructions of graphs that have topologically unique intersection representations. We will give the corresponding constructions in the full version of the paper.

4   Recognition of the classes

The first author has proved in [4] that recognition of grid intersection graphs in the plane (and hence also of graphs of boxicity 2) is NP-complete. By a straightforward reduction we can show the following theorem.

Theorem 5. *Recognition of grid intersection and box intersection graphs on the annulus and torus are NP-complete problems.*

Sketch of the proof. Let G be a graph whose membership to GI(pl) we want to test. Take a nonplanar graph, say H, embeddable on the torus and let H^* be the graph obtained from H by subdividing its edges by sufficiently many vertices. Then H^* can be represented as a grid intersection graph on the torus, but every face of any such representation is homeomorphic to a disk. Hence $H^* + G$, the disjoint union of H^* and G, is in GI(to) if and only if $G \in$ GI(pl).

A slightly more involved construction works for the annulus. Here we start with a 3-connected planar graph H which has at least 3 faces. Again, we take the subdivided graph H^*. To obtain a graph G', we place a copy of G in each of three chosen faces of H^*, connected to the vertices of the faces by paths of suitable length. In any grid intersection representation of H^* on the annulus, at least one of the three chosen faces bounds a region homeomorphic to a disk. Hence $G' \in$ GI(an) if and only if $G \in$ GI(pl). $\square$

Other interesting questions arise if we consider grid intersection graphs with preordered color classes. To be precise, we consider the following problem: Let G be a bipartite graph with color classes V and H and let the vertices be ordered $V = \{v_1, v_2, \ldots, v_n\}, H = \{h_1, h_2, \ldots, h_m\}$. We want to find a grid intersection representation for G in which each vertex v_i will be represented by a vertical segment with endpoint coordinates $[i, b_i], [i, u_i]$, and each vertex h_j will be represented by a horizontal segment with endpoints $[l_j, j], [r_j, j]$. (If $b_i < u_i$ then the segment is proper, otherwise it wraps around the horizontal sides of the base rectangle. This is allowed only in the case of the torus. Similarly, if $l_j > r_j$, then the corresponding horizontal segment wraps around the vertical sides of the base rectangle. This is allowed both in the case of the annulus and the torus.) In the planar case, there is a very straightforward way to decide if such an assignment exists. For each "horizontal" vertex h_j, obviously l_j must be $\leq$ $\min\{i|v_ih_j \in E(G)\}$ and r_j must be $\geq \max\{i|v_ih_j \in E(G)\}$. Hence if an assignment exists, then assigning $l_j = \min\{i|v_ih_j \in E(G)\}$, $r_j = \max\{i|v_ih_j \in E(G)\}$, $b_i = \min\{j|v_ih_j \in E(G)\}$, $u_i = \max\{j|v_ih_j \in E(G)\}$ yields an intersection representation, which is if and only if there are no indices $i_1 < i < i_2, j_1 < j < j_2$ such that $v_{i_1}h_j, v_{i_2}h_j, v_ih_{j_1}, v_ih_{j_2}$ are edges of G and v_ih_j is not. This condition is actually much better seen from the bipartite adjacency matrix of G. This matrix A_G is a matrix of type $n \times m$, and $(A_G)_{ij} = 1$ if $v_ih_j \in E(G)$ and $(A_G)_{ij} = 0$ otherwise. The necessary and sufficient condition for the existence of ordered grid intersection representation descirebed above is restated as "the bipartite adjacency matrix A_G does not contain a submatrix of the following form:"

$$\begin{pmatrix} & 1 & \\ 1 & 0 & 1 \\ & 1 & \end{pmatrix}$$

This description straightforwardly yields a polynomial recognition algorithm for ordered grid intersection graphs in the plane.

The situation is more interesting for grid intersection graphs on the annulus and the torus. Here we have:

Theorem 6. *A bipartite graph with preordered color classes has an ordered grid intersection representation on the annulus if and only if its bipartite adjacency matrix does not contain a submatrix of the following type*

$$\begin{pmatrix} & & 1 \\ & 1 & \\ 1 & 0 & 1 & 0 \\ & 1 & \\ & & 1 \end{pmatrix}$$

(where the first two and/or the last two rows may be swapped or may coincide). Consequently, there is a polynomial algorithm that decides if a preordered bipartite graph has an ordered grid intersection representation on the annulus.

Obviously, a finite number of forbidden submatrices can be read out from the pattern given in the theorem. On the other hand, we can prove that in the case of ordered grid intersection representations on the torus, the number of minimal forbidden submatrices is infinite. Despite of this fact, we can prove:

Theorem 7. *There is a polynomial algorithm that decides if a given bipartite preordered graph has an ordered grid intersection representation on the torus.*

Proof. Given a preordered bipartite graph $G = (V \cup H, E)$, we will consider its adjacency matrix A. Every row (and column) of the matrix consists of continuous intervals of zeros separated by intervals of ones. Let $I_{i,1}, I_{i,2}, \ldots, I_{i,n_i}$ (resp. $Y_{j,1}, Y_{j,2}, \ldots, Y_{j,m_j}$) be the continuous intervals of coordinates of zeros in the i-th row (resp. j-th column). (Note that again, one of these intervals may be non-trivial, i.e. may have the left endpoint placed to the right of the right endpoint.) We say that the horizontal segment $\overline{[l_i, i][r_i, i]}$ representing a vertex h_i *misses* an interval $I_{i,j}$ if $I_{i,j} \cap [l_i, r_i] = \emptyset$, and we say that this segment *covers* $I_{i,j}$ if $I_{i,j} \subset [l_i, r_i]$. The analogous notions for vertical segments and column-intervals are defined in obvious way.

Note that a general grid intersection representation may contain segments that neither miss nor cover a particular interval of zeros. However, it is easy to argue that if G has an ordered grid intersection representation, then it has a representation whose every horizontal (vertical) segment misses exactly one interval of zeros in the particular row (resp. column) and covers all the others. Such special representations can be easily described as instances of 2-SATISFIABILITY, a problem which is notoriously known to be solvable in polynomial time.

For every interval $I_{i,j}$, we introduce a variable $x_{i,j}$, and for every interval $Y_{r,s}$, a variable $y_{r,s}$. The idea is to construct a formula which would be satisfied if and only if G has an ordered grid intersection representation, under a truth valuation which encodes $x_{i,j}$ = true iff the segment representing h_i misses the interval $I_{i,j}$ (resp. $y_{r,s}$ = true iff the segment representing v_r misses the interval $Y_{r,s}$). The requirement that the segment representing h_i misses at most one interval is then expressed by a subformula

$$\Phi_i = \bigwedge_{1 \leq j < k \leq n_i} (\neg x_{i,j} \vee \neg x_{i,k}).$$

The analogous requirement for vertical segment v_r is expressed by

$$\Psi_r = \bigwedge_{1 \leq s < t \leq m_r} (\neg y_{r,s} \vee \neg y_{r,t}).$$

Any truth assignment that satisfies all these subformulas will correspond to a grid intersection representation which will realize all necessary crossings of the segments (i.e., all ones of the adjacency matrix). To avoid undesirable crossings, we have to ensure that for every zero in the adjacency matrix, at least one of the two intervals (vertical and horizontal) is missed by the coresponding segments of the representation. For a particular position i, r such that $A_{ir} = 0$, $r \in I_{i,j}$ and $i \in Y_{r,s}$, this requirement is expressed by a subformula

$$\Phi_{i,r} = x_{i,j} \vee y_{r,s}.$$

It is now clear that G has an ordered grid intersection representation if and only if the following formula Φ is satisfiable

$$\Phi = \bigwedge_{i=1}^{n} \Phi_i \wedge \bigwedge_{r=1}^{m} \Psi_r \wedge \bigwedge_{i,r:A_{ir}=0} \Phi_{i,r}.$$

All clauses that appear in Φ have size 2 and hence the ordered grid intersection representability problem is reduced to 2-SATISFIABILITY. $\quad\square$

5 Concluding remarks

The computational complexity of the ordered grid intersection problem is open if only one of the color classes comes preordered (and the vertices in the other color class can be permuted). For the planar case, this problem (which was first considered by the first author and J. Nešetřil) appears in [4]. If one does not believe in existence of a polynomial solution to this problem, then the analogous problems for annulus and torus should be easier to be proved NP-complete.

In the view of the nonequivalence of EM(to) and GC(to) [5], it would be very interesting to know a description of the class of grid contact graphs on the torus. In particular, is this class recognizable in polynomial time?

References

1. Bellantoni, S., Ben-Arroyo Hartman, I., Przytycka, T., Whitesides, S.: Grid intersection graphs and boxicity, Discrete Math. **114** (1993), 41-49
2. Cozzens, M.B., Roberts, F.S.: On dimensional properties of graphs, Graphs and Combinatorics **5** (1989), 29–46
3. de Fraysseix, H., de Mendez, P.O., Pach, J.: Representation of planar graphs by segments, Intuitive Geometry, Colloquia Mathematica Societatos Janos Bolyai **63** (1991), 109-117
4. Kratochvíl, J.: A special planar satisfiability problem and a consequence of its NP-completeness, Discrete Appl. Math. **52** (1994) 233–252
5. Mohar, B. (private communication)
6. Mohar, B., Rosenstiehl, P.: Tessellation and visibility representations of maps on the torus (preprint)
7. Roberts, F.S.: On the boxicity and cubicity of a graph, In: W.T. Tutte, ed., Recent Progress in Combinatorics, Academic Press, New York, 1969, pp. 301-310
8. Rosenstiehl, P., Tarjan, R.E.: Rectilinear planar layouts and bipolar orientations of graphs, Discrete Comput. Geometry **1** (1986), 343-353
9. Tamassia, R., Tollis, I.G.: A unified approach to visibility representations of planar graphs, Discrete Comput. Geometry **1** (1986), 321-341
10. Wood, D.: The riches of rectangles, In: Proceedings 5th International Meeting of Young Computer Scientists, Smolenice (1988), 67-75

How to Draw Outerplanar Minimum Weight Triangulations * (extended abstract)

William Lenhart[1] and Giuseppe Liotta[2]

[1] Department of Computer Science, Williams College, Williamstown, MA 01267.
lenhart@cs.williams.edu
[2] Department of Computer Science, Brown University, 115 Waterman Street,
Providence, RI 02912.
gl@cs.brown.edu

Abstract. In this paper we consider the problem of characterizing those graphs that can be drawn as minimum weight triangulations and answer the question for maximal outerplanar graphs. We provide a complete characterization of minimum weight triangulations of regular polygons by studying the combinatorial properties of their dual trees. We exploit this characterization to devise a linear time (real RAM) algorithm that receives as input a maximal outerplanar graph G and produces as output a straight-line drawing of G that is a minimum weight triangulation of the set of points representing the vertices of G.

1   Introduction

A widely used graph drawing standard represents vertices as points on the plane and edges as straight-line segments between points. Drawings that follow such a standard are called *straight-line drawings* and the design of algorithms to produce such drawings is a field of growing interest. An extensive survey of results on straight-line drawings as well as on other graphic standards is provided by Di Battista et al. [3]. Recently, attention has been devoted to a special type of straight-line drawings, called *minimum weight drawings*, which have applications in areas including computational geometry and numerical analysis.

Let C be a class of graphs, let P be a set of points in the plane. Let G be a graph such that

1. G has vertex set P,
2. the edges of G are straight-line segments connecting pairs of points of P,

* Research supported in part by the National Science Foundation under grant CCR-9423847, by the U.S. Army Research Office under grant 34990–MA–MUR, by Progetto Finalizzato Sistemi Informatici e Calcolo Parallelo of the Italian National Research Council (CNR), and by N.A.T.O.- CNR Advanced Fellowships Programme.

3. $G \in C$, and

4. the sum of the lengths of the edges of G is minimized over all graphs satisfying 1–3.

We call such a graph G a *minimum weight representative of* C. Given a graph $G \in C$, we say that G has a *minimum weight drawing* if there exists a set P of points in the plane such that G is a minimum weight representative of C. For example, a *minimum spanning tree* of a set P of points is a connected, straight-line drawing that has P as vertex set and minimizes the total edge length. So, letting C be the class of all trees, a tree G has a minimum weight drawing if there exists a set P of points in the plane such that G is isomorphic to a minimum spanning tree of P. A *minimum weight triangulation* of a set P is a triangulation of P having minimum total edge length. Letting C be the class of all planar triangulations, a planar triangulation G has a minimum weight drawing if there exists a set P of points in the plane such that G is isomorphic to a minimum weight triangulation of P.

The problem of testing whether a tree admits a minimum weight drawing is essentially solved. Monma and Suri [20] proved that each tree with maximum vertex degree at most five can be drawn as a minimum spanning tree of some set of vertices by providing a linear time (real RAM) algorithm. In the same paper it is shown that no tree having at least one vertex with degree greater than six can be drawn as a minimum spanning tree. As for trees having maximum degree equal to six, Eades and Whitesides [5] showed that it is NP-hard to decide whether such trees can be drawn as minimum spanning trees.

Surprisingly, nothing seems to be known about the problem of constructing a minimum weight drawing of a planar triangulation. Moreover, it is still not known whether computing a minimum weight triangulation of a set of points in the plane is an NP-complete problem (see Garey and Johnson [6]). Several papers have been published on this last problem, either providing partial solutions, or giving efficient approximation heuristics. A limited list of results includes the work by Meijer and Rappaport [19], Lingas [14, 16], Kirkpatrick [10], Keil [9], Dickerson et al. [4], and Aichholzer et al. [1].

In this paper we examine the problem of characterizing those triangulations admitting a minimum weight drawing and answer the question for maximal outerplanar graphs. The contribution is twofold:

1. Minimum weight triangulations for points that are vertices of a regular polygons are characterized. The characterization is based on the combinatorial structure of the dual tree of the minimum weight triangulations of such point sets. A consequence of the characterization is an optimal time algorithm for computing a minimum weight triangulation of a regular n-gon for any given n. Interestingly, it is not necessary to supply the algorithm with the actual points, only the size (number of vertices) of the polygon is needed. When arbitrary convex polygons are considered, the fastest known algorithms require $O(n^3)$ time, where n is the number of vertices of the polygon (see Gilbert [7], Klincsek [11], and Heath and Pemmaraju [8]).

The triangulation produced by our algorithm turns out to be that which would result from the application of the greedy algorithm described by Levcopoulos and Lingas [13]. Lloyd [17] has shown that the greedy triangulation of a convex polygon is not necessarily of minimum weight; lower bounds for the nonoptimality of the greedy triangulation are given by Manacher and Zobrist [18] and by Levcopoulos [12]. Lingas [15] shows that on average the greedy triangulation approximates the optimum by an $O(\log n)$ factor.

2. We show that every maximal outerplanar graph G has a minimum weight drawing. This is done by exhibiting an algorithm that computes a minimum weight drawing of G in time proportional to the number of vertices of G, within the real RAM model of computation. The drawing algorithm exploits the combinatorial properties of the minimum weight triangulations of regular polygons and is based on a decomposition rule of minimum weight triangulations.

2 Preliminaries

We assume familiarity with the basic terminology of graph theory and computational geometry (see also Bondy and Murty [2], and Preparata and Shamos [21]). A graph G is *outerplanar* if it has a planar embedding such that all vertices lie on a single face. G is *maximal outerplanar* if G is outerplanar, but the addition of any new edge results in a non-outerplanar graph. In geometric applications graphs often arise as the result of selecting a set S of points in the plane and then connecting certain pairs to be joined by straight line segments which form the edge set of the graph. In this paper, we will be interested in *triangulations* of a point set S: planar graphs obtained from S by taking as edge set a maximal number of mutually non-crossing straight line segments connecting pairs of points in S. In particular, a *triangulation of a regular n-gon P* is a triangulation obtained by adding $n - 3$ mutually non-intersecting diagonals connecting pairs of vertices of P. Every triangulation G of a regular n-gon gives a straight-line drawing of a maximal outerplanar graph on n vertices, such that the outer face forms a regular n-gon; similarly, every maximal outerplanar graph on n vertices gives rise to a unique triangulation of a regular n-gon. We will refer to any graph as a triangulation if the graph is isomorphic to a triangulation of some point set S.

Given an embedded planar graph G, the *extended dual tree (or e-dual)* of G is a planar graph G' defined as follows. G' has a vertex for each internal face of G and a vertex for each of the edges on the external face of G. Two vertices u, v of G' are adjacent either if they correspond to two internal faces of G that share an edge, or if u corresponds to an edge e of G and v to a face of G containing e.

Note that if G is a triangulation of an n-gon, then G' is a tree with $2n - 2$ vertices, with the property that every non-leaf vertex has degree 3. In the rest of the paper we assume that the e-dual of the triangulation of an n-gon is rooted at a non leaf-vertex r.

Let G be a triangulation of an n-gon, and let T be the extended dual of of G having root r. Observe that there is a natural way to associate the edges of G to all but vertices of T other than r: Each leaf of T corresponds to an external edge of the triangulation; each non-leaf $v \neq r$ of T will correspond to the third edge of the triangular face of G formed by v and its two children. Only r has no corresponding edge in G.

A (non minimum weight) triangulation of a regular 15-gon and its e-dual are shown in Figure 1. The vertex of the dual labeled r is the root of the tree; vertices labeled x, y and z will be of use in the rest of the paper.

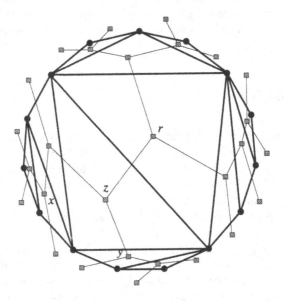

Fig. 1. A triangulation and its e-dual.

We will frequently be transforming rooted trees on the same number of vertices. It will be convenient to think of a given set V of vertices which has different trees defined on it. Each of these trees will be used to define several functions both on V and on subsets of V.

A rooted tree T defined on a set V is a *feasible* tree if every non-leaf has degree 3 and each subtree of the root has at most $|V|/2$ vertices. Clearly any e-dual of a triangulation of an n-gon is a feasible tree on $2n - 2$ vertices. Conversely, any feasible tree on $2n - 2$ vertices is an e-dual of some triangulation of a n-gon.

Let T be a rooted tree defined on a set V. Two vertices x and y are *incomparable* if neither x nor y is an ancestor of the other.

If T is a rooted tree defined on V, the subtree of T with root v is denoted by $T(v)$; the number of leaves in $T(v)$ is denoted by $l_T(v)$. If $x \in V$ and z is an ancestor of x in T, then the set of interior vertices on the path in T from x to z (i.e., all vertices except x and z) is denoted by $\pi(x, z)$. Note that if z is the parent of x then $\pi(x, z) = \emptyset$.

Let x and y be two vertices of a regular n-gon of radius 1. Let l be the number of edges on a path between x and y. Then the length of the diagonal from x to y is given by $2\sin(\frac{\pi}{n}l)$. Note that this formula holds regardless of whether we used the short or the long path between x and y.

Let T be a rooted tree defined on V. For any vertex $v \in V$, the *weight* of v, $w_T(v)$, is defined to be $2\sin(\frac{\pi}{n}l_T(v))$. Observe that if T is the (rooted) e-dual tree of a triangulation G of an n-gon, then $w_T(v)$ denotes the length of the edge of G corresponding to v. In particular, $w_T(r) = 2\sin(\frac{\pi}{n}n) = 0$, which conveniently agrees with the earlier observation that r does not correspond to any edge of G. Thus the weight of the triangulation G is given by the sum of the weights of the vertices of T. Moreover, for any $X \subseteq V$, we define the *weight* of X, $W_T(X)$, to be $\sum_{v \in X} w_T(v)$. Thus, for example, $W_T(V)$ is the weight of the triangulation. We will also denote this value by $W(T)$ and refer to it as the *weight of T*. Finally, we will also need to consider the related functions $\overline{w}_T(v) = 2\cos(\frac{\pi}{n}l_T(v))$ and $\overline{W}_T(X) = \sum_{v \in X} \overline{w}_T(v)$.

The problem of finding a minimum weight triangulation of a regular n-gon can be reformulated as follows: Given a set of $2n - 2$ vertices, minimize $W(T)$ over all feasible trees T.

In the next section, we solve the minimum weight triangulation problem just mentioned. In Section 4, we exploit our results to prove that every maximal outerplanar graph G admits a *minimum weight drawing*; that is, a drawing of G such that the edges of the drawing are those of some minimum weight triangulation of the vertices of the drawing. The last section summarizes our conclusions and mentions some open problems.

3 Minimum Weight Triangulations of Regular Polygons

We begin by defining an operation on a feasible tree T defined on V, which transforms T to another feasible tree T'' on V. Let x and y be incomparable vertices of T and let z be the lowest common ancestor of x and y. A *swap* consists of two steps:

1. The subtrees $T(x)$ and $T(y)$ are exchanged, resulting in a tree denoted by T',

2. If T' is feasible, then $T'' = T'$, otherwise T'' is constructed by choosing a new root for T' to make it feasible.

Notice that $W(T') = W(T'')$. A swap is an *improvement* (or *improving swap*) if $W(T') < W(T)$.

Observe that the only vertices of T' whose weights change as a result of Step 1 are those vertices in $\pi(x, z) \cup \pi(y, z)$, since these are the only vertices v

such that $l_T(v) \neq l_{T'}(v)$. Thus, $W(T') - W(T) = (W_{T'}(\pi(x, z)) - W_T(\pi(x, z))) + (W_{T'}(\pi(y, z)) - W_T(\pi(y, z)))$.

Lemma 1. *Let T be a feasible tree and let x and y be incomparable vertices of T such that $l_T(x) \neq l_T(y)$ and having lowest common ancestor z. The swap of $T(x)$ and $T(y)$ is an improvement if either $\overline{W}_T(\pi(x, z)) - \overline{W}_T(\pi(y, z))) = 0$, or $(l_T(x) - l_T(y)) \times (\overline{W}_T(\pi(x, z)) - \overline{W}_T(\pi(y, z))) > 0$.*

Proof. Let $x_i \in \pi(x, z)$, $y_j \in \pi(y, z)$ and $\Delta l = l_T(y) - l_T(x)$. Observe that $l_{T'}(x_i) = l_T(x_i) + \Delta l$, and $l_{T'}(y_j) = l_T(y_j) - \Delta l$.

Now, $W(T') - W(T) = \sum_{i=1}^{k}(w_{T'}(x_i) - w_T(x_i)) + \sum_{j=1}^{m}(w_{T'}(y_j) - w_T(y_j))$. Using the definition of $w_T()$ and properties of the sine function, we have,

$$w_{T'}(x_i) = 2sin(\frac{\pi}{n}l_{T'}(x_i))$$
$$= 2sin(\frac{\pi}{n}(l_T(x_i) + \Delta l))$$
$$= 2sin(\frac{\pi}{n}l_T(x_i))cos(\frac{\pi}{n}\Delta l) + 2cos(\frac{\pi}{n}l_T(x_i))sin(\frac{\pi}{n}\Delta l)$$
$$= w_T(x_i)cos(\frac{\pi}{n}\Delta l) + \overline{w}_T(x_i)sin(\frac{\pi}{n}\Delta l).$$

Similarly, $w_{T'}(y_j) = w_T(y_j)cos(\frac{\pi}{n}\Delta l) - \overline{w}_T(y_j)sin(\frac{\pi}{n}\Delta l)$.
Thus,

$$W(T') - W(T) = \sum_{i=1}^{k}((cos(\frac{\pi}{n}\Delta l) - 1)w_T(x_i) + sin(\frac{\pi}{n}\Delta l)\overline{w}_T(x_i)) +$$
$$\sum_{j=1}^{m}((cos(\frac{\pi}{n}\Delta l) - 1)w_T(y_j) - sin(\frac{\pi}{n}\Delta l)\overline{w}_T(y_j))$$
$$= (cos(\frac{\pi}{n}\Delta l) - 1)(W_T(\pi(x, z)) + W_T(\pi(y, z))) +$$
$$sin(\frac{\pi}{n}\Delta l)(\overline{W}_T(\pi(x, z)) - \overline{W}_T(\pi(y, z)))$$

The first term is negative since $cos(\frac{\pi}{n}\Delta l) - 1 < 0$. The second term is nonpositive since $sin(\frac{\pi}{n}\Delta l)$ has the same sign as Δl, and since, by assumption, either $\overline{W}_T(\pi(x, z)) = \overline{W}_T(\pi(y, z))$ or Δl has the opposite sign of $\overline{W}_T(\pi(x, z)) - \overline{W}_T(\pi(y, z))$. Thus $W(T') - W(T) < 0$ and the swap is improving. $\square$

The preceding result is used to establish the following two corollaries.

Corollary 2. *Let T be a feasible tree and let x and y be incomparable non-leaf vertices of T having children x', x'' and y', y'' respectively. If $l_T(x') > l_T(y')$ and $l_T(x'') < l_T(y'')$, then T admits an improving swap.*

Proof. Let z be the lowest common ancestor of x and y. The hypotheses of the corollary imply that $l_T(x') - l_T(y')$ and $l_T(x'') - l_T(y'')$ have opposite signs. Therefore one of these two differences has the same sign as $\overline{W}_T(\pi(x', z)) -$

$\overline{W}_T(\pi(y', z))$. Since $\pi(x', z) = \pi(x'', z)$ and $\pi(y', z) = \pi(y'', z)$, we can apply Lemma 1, to see that either swapping Tx' with Ty' or swapping Tx'' with Ty'' must decrease the weight of the tree. Thus, we have made an improving swap. □

As an example, vertices x and y of Figure 1 are two incomparable non-leaf vertices for which there is an improving swap.

Corollary 3. *Let T be a feasible tree and let z be a vertex having children x and y such that y has children y', y''. If $l_T(x) < l_T(y')$, then T admits an improving swap.*

Proof. Since the path from x to z is empty, $\overline{W}_T(\pi(x, z)) = 0$; also $\overline{W}_T(\pi(y', z)) = \overline{W}_T(\pi(y'', z)) = \overline{w}_T(y) \geq 0$, since the subtree rooted at y has at most half of the leaves in the tree. Thus $\overline{W}_T(\pi(x, z)) - \overline{W}_T(\pi(y', z)) \leq 0$. Therefore, if $l_T(x) - l_T(y') < 0$, there is an improving swap by Lemma 1. □

We now have two operations which can be applied to feasible trees in order to decrease their weight. It turns out that these two operations suffice to transform any feasible tree to a minimum weight tree.

A *weight-balanced tree* is a feasible tree which admits no improving swaps.

Clearly every minimum-weight tree is weight-balanced. The rest of this section consists in showing that every weight-balanced tree is minimum-weight.

To accomplish this, we show that every weight-balanced tree can be put into a standard form by repeatedly swapping left and right subtrees of vertices of the tree. Clearly swapping the left and right subtrees of a given vertex does not change the weight of the tree, since it does not change the weight of the vertex. Since every weight-balanced tree can be put into this form, all weight-balanced trees on V must have the same weight. This weight must be minimum since all minimum weight trees are themselves weight-balanced.

A weight-balanced tree T is *sorted* if, in a breadth-first, left-to-right traversal of T, the function $l_T()$ is not increasing.

Lemma 4. *There is only one sorted weight-balanced tree T having $2n - 2$ vertices.*

Proof. The vertices on each level are sorted by decreasing $l_T()$-value; if a level contains a leaf v, all vertices to the right of v on that level are also leaves. Since all vertices on lower levels have $l_T()$-values no greater than $l_T(v)$, all vertices on the next level down are also leaves. Because all feasible trees on $2n - 2$ vertices have exactly n leaves, this uniquely determines the structure of the tree. □

Let T be a weight-balanced tree. It is clear that we can arrange T so that each vertex has its children sorted from left to right by decreasing $l_T()$-value. It turns out that, once this is accomplished, the resulting tree is sorted. We show this in two steps.

Lemma 5. *Let T be a weight-balanced tree, and let x and y be incomparable vertices of T having children x_L, x_R and y_L, y_R respectively, such that $l_T(x_L) \geq l_T(x_R)$ and $l_T(y_L) \geq l_T(y_R)$. If $l_T(x) \geq l_T(y)$, then $l_T(x_R) \geq l_T(y_L)$.*

Proof. By assumption, $l_T(x_L) + l_T(x_R) = l_T(x) \geq l_T(y) = l_T(y_L) + l_T(y_R)$. So, if $l_T(x_R) < l_T(y_L)$, then it must be that $l_T(x_L) > l_T(y_R)$. Therefore, by Corollary 2, T admits an improving swap, contradicting the assumption that T is weight-balanced. □

Lemma 6. *If T is a weight-balanced tree, such that the children of each vertex are sorted left-to-right by decreasing $l_T()$-value, then T is sorted.*

Proof. The proof consists of two parts.

1. We first show that vertices at same depth have $l_T()$-values decreasing from left to right.
2. We then show that the right-most vertex at depth k has $l_T()$-values at least as large as that of the left-most vertex at depth $k + 1$.

We prove the first claim by induction on the depth of vertices in the tree. The base case is simple, the vertices at depth one are simply the children of the root, and so are sorted by decreasing $l_T()$-value. Suppose now, that all vertices at depth $k - 1$ are sorted by decreasing $l_T()$-value, and consider now the vertices at depth k. Let x and y be two consecutive vertices, in left-to-right order, at depth $k - 1$. By induction, $l_T(x) \geq l_T(y)$; by hypothesis, the children of x are in decreasing $l_T()$-value order, as are the children of y. Thus, by Lemma 5, the children of x and y together are in decreasing $l_T()$-value order. Since this holds for all pairs of consecutive vertices at depth $k - 1$, all vertices at depth k must also be sorted.

We also prove the second claim by induction on the depth of vertices in the tree. The base case is easy to establish (by contradiction using Corollary 3), so we show only the induction step. Let x_i and y_i denote the left-most and right-most vertices at depth i respectively. Assume that, for all $i < k$, y_i has $l_T()$-value at least as great as that of x_{i+1}. Thus $\overline{w}_T(x_{i+1}) \geq \overline{w}_T(y_i)$. Let $y = y_k$ be the right-most vertex at depth k, let $x = x_{k+1}$ be the left-most vertex at depth $k+1$, and let r be the root of T. Then $\overline{W}_T(\pi(x, r)) > \overline{W}_T(\pi(y, r))$, since $\overline{w}_T(x_1) > 0$ and $\overline{w}_T(x_{i+1}) \geq \overline{w}_T(y_i)$, for each $i < k$. Since T admits no improving swaps, by Lemma 1, it must be that $l_T(x) \leq l_T(y)$. □

Lemmas 4–6 immediately yield the following result.

Theorem 7. *A triangulation G of a regular n-gon is minimum weight if and only if its e-dual can be rooted so that it is weight-balanced.*

Figure 2 shows a minimum weight triangulation of a regular 15-gon; r designates the root of the e-dual.

We can use our understanding of the structure of weight-balanced trees to design an optimal algorithm for computing a minimum weight triangulation of

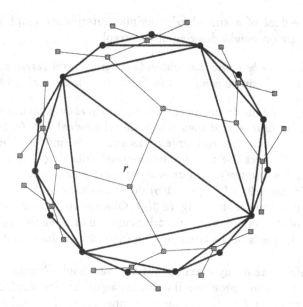

Fig. 2. A minimum weight triangulation of a regular 15-gon.

a regular n-gon. Let T be the sorted, weight-balanced tree on $2n-2$ vertices. T induces a triangulation of the regular n-gon which, by Theorem 7 is a minimum weight triangulation. Observe that the triangulation is the same as the one which would be obtained by an application of the greedy algorithm (always selecting the smallest segment which does not cross any of the segments selected so far). Moreover, this triangulation can be computed in time proportional to the size of T. We summarize this in the following:

Theorem 8. *A minimum weight triangulation of a regular n-gon can be computed in time proportional to the size of the triangulation produced. Furthermore, the triangulation obtained is the same as that obtained by the greedy algorithm.*

4 Minimum Weight Drawings of Maximal Outerplanar Graphs

Before proving the main result of this section, we note a feature of minimum weight triangulations.

Lemma 9. *Let G be a minimum weight triangulation of a set P of points in the plane. Let $\triangle abc$ be an interior face of G such that $\overline{ab}$ and $\overline{ac}$ are on the outer face of G. Then $E(G) - \{\overline{ab}, \overline{ac}\}$ is a minimum weight triangulation of $P - \{a\}$, where $E(G)$ denotes the set of edges of G.*

Using the e-dual of a embedded maximal outerplanar graph allows us to compute a minimum weight drawing of the graph.

Theorem 10. *Let G be a maximal outerplanar graph on n vertices. A minimum weight drawing of G can be computed in $O(n)$ time in the real RAM model.*

Proof. We begin by considering an embedded maximal outerplanar graph G with all vertices the outer face, and then computing the e-dual T of G. Let T' be any feasible weight-balanced tree containing T as a sub-tree, and let N be the number of leaves of T'. T' gives rise to a minimum-weight drawing Γ of a triangulated regular N-gon. A minimum weight drawing of G can now be obtained by repeated application of Lemma 9 to Γ. Figure 3(b) shows an example of such construction, when the input is the graph of Figure 3(a). Observe that the drawing is a sub-triangulation of the minimum weight triangulation of a regular 15-gon. Dotted lines describe the parts of the triangulation (and of its e-dual) that are not part of the drawing.

This establishes that any maximal outerplanar graph G admits a minimum-weight drawing. We now provide a linear-time algorithm for constructing a minimum weight drawing Γ of G. We use the approach of the previous paragraph to choose a particular feasible tree T', namely, a complete feasible tree T' which contains T as a subtree. Let k be the height of T'. Thus T' is the e-dual of a regular $3 \cdot 2^{k-1}$-gon. Clearly, the construction of the regular $3 \cdot 2^{k-1}$-gon corresponding to T' must be avoided if the algorithm is to be linear time. Observe that the vertices of the regular $3 \cdot 2^{k-1}$-gon are evenly distributed around the circumference of a disk. thus the vertices of Γ are also distributed around the disk. We need only give a method for computing the location of these vertices. Recall that each edge of G corresponds to a non-root vertex of its e-dual T. Now, the length in Γ of an external edge e of G having e-dual vertex x is given by $w_{T'}(x')$, where x' is the vertex of T' corresponding to x. Note that x' is not necessarily a leaf of T', even though x is a leaf of T. Observe also that the length of e is completely determined by $l_{T'}(x')$, and that $l_{T'}(x') = 2^{k-d}$, where d is the depth of x (or x'). Thus for each leaf x of T, the edge corresponding to x is a chord of length $2sin(\frac{\pi}{n}2^{k-d})$ connecting 2 vertices on the disk.

So, to construct Γ, do an inorder traversal of the leaves of T, drawing chords of the corresponding lengths on the disk. This gives a drawing of the outer face of G, which completely determines the location of all vertices; now draw all remaining edges as straight line segments. □

Note that if G happens to have an e-dual which is weight-balanced, then this e-dual can be used to directly produce a minimum weight drawing of G.

5 Conclusions and Open Problems

In this paper we have proved that every maximal outerplanar graph admits a straight-line drawing that is a minimum weight triangulation of the set of points representing the vertices in the drawing. We have also provided a complete

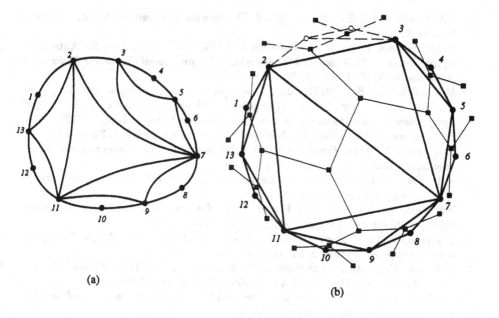

Fig. 3. (a) A maximal outerplanar graph and (b) its minimum weight drawing.

characterization of the minimum weight triangulations of regular polygons. The general problem of determining which triangulations are drawable as minimum weight triangulations is still far from solved. As an intermediate step toward answering the question, we think it might be worth investigating the minimum weight drawability of special classes of graphs, like the 4-connected planar triangulated graphs or the maximal k-outerplanar graphs (a graph is k-outerplanar when it has an embedding such that all vertices are on disjoint cycles properly nested at most k deep).

Acknowledgement

The authors wish to thank Andrzej Lingas for the information he provided on the status of the minimum weight triangulation problem and Roberto Tamassia for his encouragement during the preparation of this paper.

References

1. O. Aichholzer, F. Aurenhammer, G. Rote, M. Taschwer. Triangulations Intersect Nicely. *Proceedings 11th ACM Symposium on Computational Geometry*, Vancouver, Canada, 1995, pp. 220-229.

2. J.A. Bondy and U.S.R. Murty. *Graph Theory with Applications*. Elsevier Science, New York, 1976.

3. G. Di Battista, P. Eades, R. Tamassia and I.G. Tollis. Algorithms for Automatic Graph Drawing: An Annotated Bibliography. *Computational Geometry: Theory and Applications*, 4, 1994, pp. 235-282.

4. M.T. Dickerson, S.A. McElfresh, M. Montague. New Algorithms and Empirical Findings on Minimum Weight Triangulations Heuristics. *Proceedings 11th ACM Symposium on Computational Geometry*, Vancouver, Canada, 1995, pp. 238-247.

5. P. Eades and S. Whitesides. The Realization Problem for Euclidean Minimum Spanning Tree is NP-hard. *Proceedings 10th ACM Symposium on Computational Geometry*, Stony Brook NY, USA, 1994, pp. 49-56.

6. M.R. Garey and D.S. Johnson. *Computers and Intractability-A Guide to the Theory of NP-Completeness*. Freeman, New York, 1979.

7. P.N. Gilbert. New Results in Planar Triangulations. M.Sc. Thesis, Coordinated Science Laboratory, University of Illinois, Urbana, IL, 1979.

8. L.S. Heath and S.V. Pemmaraju. New Results for the Minimum Weight Triangulation Problem. *Algorithmica*, 12, 1994 pp. 533-552.

9. J.M. Keil. Computing a Subgraph of the Minimum Weight Triangulation. *Computational Geometry: Theory and Applications*, 4, 1994, pp. 13-26.

10. D. G. Kirkpatrick. A Note on Delaunay and Optimal Triangulations. *information Processing Letters*, 10, 1980, pp. 127-128.

11. G.T. Klincsek. Minimal triangulations of Polygonal Domains. *Ann. Discrete Math.*, 9, 1980, pp. 121-123.

12. C. Levcopoulos. An $\Omega(\sqrt{n})$ Lower Bound for the Nonoptimality of the Greedy Triangulation. *Information Processing Letters*, 25, 1987, pp. 247-251.

13. C. Levcopoulos and A. Lingas. On Approximation Behavior of the Greedy Triangulation for Convex Polygons. *Algorithmica*, 2, 1987, pp. 175-193.

14. A. Lingas. A Linear-Time Heuristic for Minimum Weight Triangulation of Convex Polygons. *Proc. 23rd Allerton Conference on Computing, Communication, and Control*, Urbana, 1985, pp. 480 -485.

15. A. Lingas. The Greedy and Delaunay Triangulations are not Bad in the Average Case. *Information Processing Letters*, 22, 1986, pp. 25-31.

16. A.Lingas. A New Heuristic for Minimum Weight Triangulation. *SIAM J. Algebraic Discrete Methods*, 8, 1987, pp. 646-658.

17. E.L. Lloyd. On Triangulations of a Set of Points in the Plane. *Proceedings of the 18th Conference on the Foundations of Computer Science*, Providence, RI, 1977, pp. 228-240.

18. K. Manacher, L. Zobrist. Neither the Greedy nor the Delaunay Triangulation of a Planar Point Set Approximates the Optimal Triangulation. *Information Processing Letters*, 9, 1979, pp. 31-34.

19. H. Meijer and D. Rappaport. Computing the Minimum Weight Triangulation of a Set of Linearly Ordered Points. *Information Processing Letters*, 42, 1992, pp. 35-38.

20. C. Monma and S. Suri. Transitions in Geometric Minimum Spanning Trees. *Proc. 7th ACM Symposium on Computational Geometry*, 1991, pp. 239-249.

21. F. P. Preparata and M. I. Shamos, *Computational Geometry – an Introduction*. Springer-Verlag, New York, 1985.

Portable Graph Layout and Editing

Brendan Madden, Patrick Madden, Steve Powers, and Michael Himsolt

Tom Sawyer Software, 1828 Fourth Street, Berkeley, CA 94710
info@TomSawyer.COM

Abstract. The *Graph Layout Toolkit* and the *Graph Layout Toolkit* are portable, flexible toolkits for graph layout and graph editing systems. The Graph Layout Toolkit contains four highly customizable layout algorithms, and supports hierarchical graphs. The Graph Editor Toolkit is a tightly coupled interactive front end to the Graph Layout Toolkit.

1 Introduction

The visualization of graphs has become very important over the last few years. Some examples include project management, compiler and software development tools, work-flow, and reverse engineering applications. Other applications include those for network management, CAD and CASE, diagramming, and database design.

Many graph layout and editing systems have been developed during the last ten years, examples include *D-ABDUCTOR*, *dag*, *dot*, *daVinci*, *Diagram Server*, *EDGE*, *GEM*, *grab*, *GraphEd* and *vcg* (for an overview, see [DETT95]). Research projects generally have different design, documentation, packaging, and testing goals than those of commercial software.

In addition, research systems usually have the freedom to experiment with novel ideas with uncertain commercial value. However, our goals are often different as a commercial vendor. Our approach is generally to seek very high quality layout algorithms that have general applicability to groups of markets.

In a commercial environment, it is usually better to avoid techniques that have limited applicability. A novel approach to a graph drawing problem may confuse, and sell poorly, if the solution does not match the traditions or expectations of the particular customer. We have produced a family of layout styles to try to anticipate the needs of diverse markets with varying requirements.

In production applications, it is unlikely that any restrictions on input graphs such as planarity, maximum degree or biconnectivity will hold in general. In spite of that, we try to adopt many useful techniques from graph drawing literature into our framework when those techniques can be made generally applicable. However, it seems to be very difficult, and perhaps impossible, to satisfy all of the requirements of commercial applications.

Furthermore, the Graph Layout Toolkit and the Graph Editor Toolkit support a generalized framework to visualize information that spans across many linked graphs. A programmer or user may navigate easily from one graph to another, or many graphs may be interactively nested in the same plane if desired. Each graph also maintains its own layout tailoring properties.

Additionally, it is very important that layout and editing systems are portable. Ideally, they should work with a number of different compilers, operating systems, industry standard graphics class libraries, rapid application development tools, and be easily embedded into other applications.

Tom Sawyer Software's Graph Layout Toolkit and Graph Editor Toolkit are mature software packages that provide many of the above-mentioned features.

2 The Graph Layout Toolkit

The Graph Layout Toolkit [GLT95a, GLT95b] currently supports four different layout styles: circular, hierarchical, orthogonal, and symmetric. Each layout style derives from a virtual function driven layout class hierarchy that is loosely coupled with a graph management class hierarchy. This design allows the user to flexibly switch between the installed layout styles at any time.

The graph class hierarchy supports directed and undirected graphs with multiedges and reflexive edges. There are no imposed implementation limits on the size of a graph, its maximum degree, and no restrictions on the classes of graphs that can be specified. All layout algorithms also support multiple connected components, navigation, variable node dimensions, edges with bends, and reading and writing of graphs and their drawings from and to disk.

Edges may either be drawn as straight lines or may be represented as a sequence of lines. An edge with bends owns a graph, i.e. a path, that is a sorted chain of dummy nodes and edges. This technique ensures that paths can be managed through standard graph operations. Each new layout algorithm extends the generic graph and layout services that are provided by the framework. The virtual function interface and separate name spaces ensure that one layout algorithm does not adversely affect another.

The Graph Layout Toolkit provides operations that make it easy to integrate graph drawing techniques into various applications. It supports a number of features for portable interactive editing such as continuously up-to-date drawing, and cut, copy, paste, duplicate, and clear functionality. Users just plug in their domain-specific graphics calls. It further supports graph and layout replication features and subject/view relations where several objects may be attached to a node or an edge and get notified of changes. Graphical queries can be applied to graphs and displayed in different windows with minimal programming overhead.

The Graph Layout Toolkit also supports highly customizable Postscript™ and Encapsulated Postscript™ output. It generates gray scale or color output and can distribute a drawing over multiple pages of arbitrary size. It can also optionally scale and rotate, insert crop marks, and insert page labeling and numbering detail into the final drawing. It has an extensible design that enables application developers to load icons into nodes, and to write application-specific PostScript procedures.

The Graph Layout Toolkit is implemented in C++, and provides more than six hundred functions for each of the C++ and ANSI C application programmer interfaces. It has been ported to more than fifteen different compilers and runs on all standard operating systems.

2.1 Circular Layout

The circular layout algorithm is designed for the layout of ring and star network topologies. It is an advanced version of the one developed by Kar, Madden, and Gilbert [KMG89]. It functions by partitioning nodes into logical groups based on a number of flexible node grouping models. Each group of nodes is placed on radiating circles based on their logical interconnection. The partitioning is either performed with a pre-defined method, such as biconnectivity of the graph or the degree of nodes, IP addresses, IP subnet masks or another domain specific technique. It also supports manually configured clustering.

The algorithm minimizes cluster to cluster crossings as well as crossings within each cluster. In additions, it also employs tree balancing routines, and has ring and star detection and placement techniques within each cluster. Figure 1 shows two sample drawings.

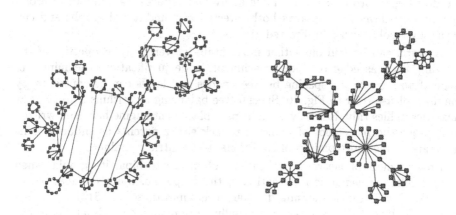

Fig. 1. Examples of circular layout.

2.2 Hierarchical Layout

The hierarchical layout algorithm is designed to lay out directed graphs. It has applications in project management software, compiler and software development tools, information management applications, work-flow, reverse engineering applications, database schemata and network management applications. Figure 2 shows two examples of this style.

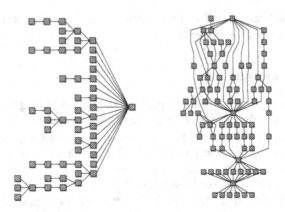

Fig. 2. Examples of hierarchical layout.

The hierarchical layout algorithm is a heavily optimized and extended version of the standard algorithm by Sugiyama, Tagawa and Toda [STT81] that organizes the nodes of a graph into levels, adds dummy nodes, as necessary, into the drawing to produce a proper hierarchy and reduces the number of crossings between levels. It supports both directed and undirected graphs and can optimally resolve cycles in directed graphs.

A graph can be laid out either horizontally or vertically. Properties of the drawing, such as edge to node attachment, node justification, the minimum slope of an edge or the spacing between nodes and levels can be controlled by the user, please refer to Fig. 10. Special tree balancing algorithms help to draw class hierarchies tidy. Recently, a multi-pass placement engine has been written to improve the placement of nodes that lack either parents or children. This eliminates some shortcomings of traditional barycentric techniques.

Programmers of network applications often create bipartite graphs when there is inherent semantic information in the graph, i.e. routers connecting to networks. If one uses the standard layout techniques from [STT81] for directed graphs, the layout that will result is usually an unreadable two-level hierarchy.

Therefore, we have extended the hierarchical layout style to unfold undirected network structures in a number of novel ways. One method employs a bipolar acyclic ordering for the biconnected components. Ethernet nodes typically have very high degree and variable width. Subtrees and end-nodes, i.e. terminals and hosts, are selectively inverted around these possibly variable-width Ethernet nodes, as shown in Figure 3.

These techniques have been applied to the Network Layout Assistant [NLA95] which is an add-on for a network management software package from Sun Microsystems named Solstice SunNet Manager™. In this system, Solstice SunNet Manager™ discovers devices on a network from which the Network Layout Assistant generates an intelligent drawing.

2.3 Orthogonal Layout

The orthogonal layout algorithm is based on papers by Biedl and Kant [BK94] and Papakostas and Tollis [PT94]. Orthogonal layouts are specifically suited for database design, object oriented analysis and design, and CAD and CASE diagrams. Figures 4 and 5 show examples of drawings of this style.

The algorithm produces a constant number of bends per edge and has relatively high performance in practice. It has further been extended by Biedl and Papakostas at Tom Sawyer Software to produce more aesthetically pleasing drawings than those of the currently published B/K and P/T algorithms. With this algorithm, planar graphs can be drawn planar, and nonplanar graphs are also easily supported. Support has been recently added for nodes with degree greater than four and for row and column reuse. For high degree nodes, the user has the choice to keep node dimensions fixed to their specified input which implies decreasing the separation of grid lines. Alternatively, the user can allow the node area to increase to allow high degree while keeping the separation of grid lines fixed.

2.4 Symmetric Layout

The symmetric layout algorithm is generally designed to display networks. The drawings stress symmetric and isomorphic substructures, and its uniform distribution of nodes and edges, in general, yields aesthetically pleasing drawings. Both directed and undirected graphs are supported. Figure 6 shows two examples. The algorithm is originally based on work by Kamada and Kawai [KK89].

Several parameters of this algorithm are available in order to fine-tune the heuristics and the termination conditions. These include the strength of the spring constant and the numbers of iterations that are performed. Generally, it produces relatively few edge crossings. We have also been able to more than quadruple the performance over subsequent Graph Layout Toolkit releases. We will try to continue this trend over coming releases to quickly lay out very large graphs.

2.5 Hierarchical Graphs

Hierarchical graphs become important when information becomes too complex to be modeled with a single graph, or when information is more naturally modelled hierarchically. The *navigation manager* realizes hierarchical relations among graphs:

1. *Node to graph* navigation supports hierarchical relations from a node to a graph. Figure 7 shows several examples of node to graph relationships. These relations can be visualized by expanding the parent node so that it visually contains the child graph.

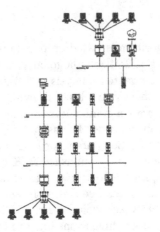

Fig. 3. A network drawing generated by the Network Layout Assistant. The Network Layout Assistant is a plug-in module for Solstice SunNet Manager™ that draws networks with the Graph Layout Toolkit.

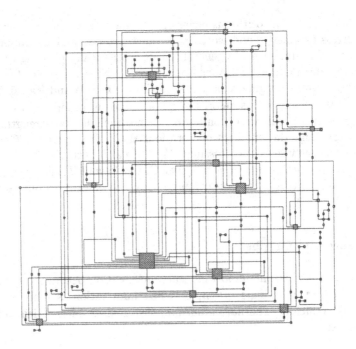

Fig. 4. Example of orthogonal layout.

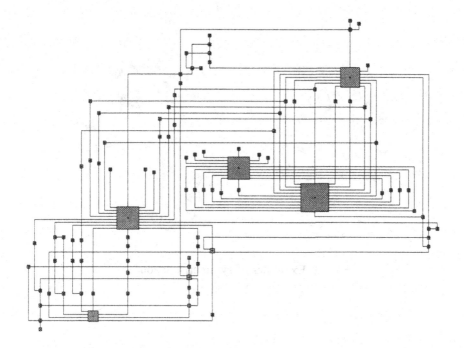

Fig. 5. Example of orthogonal layout.

2. *Edge to graph* navigation supports hierarchical relations from an edge to a graph. An edge (or node) to graph relation can be also be visualized by displaying the target graph in with another window or another application specific technique.

A navigation manager is organized as several graphs which model the hierarchical relation. One graph manages node to graph and edge to graph relations. Another recalculation graph manages which graphs are dependent on which in order to perform layouts in the correct order when graphs are nested. Arbitrarily many graphs can be recursively nested within a navigation manager to create large hypertext-like structures. Individual nodes and edges may also navigate to many graphs; further, many nodes and edges may navigate to the same graph.

The Graph Layout Toolkit supports interactive collapse and expand operations with continuously up-to-date drawings to show and hide detail in large structures. Nodes are scaled to fit around their contained graphs based on a post-order traversal of the nodes in the graph that maintains the navigation recalculation information. Recalculation operations are extremely fast under small changes to drawings as each graph maintains information that indicates whether it needs to be laid out again.

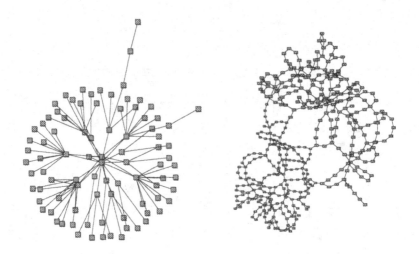

Fig. 6. Examples of symmetric layout.

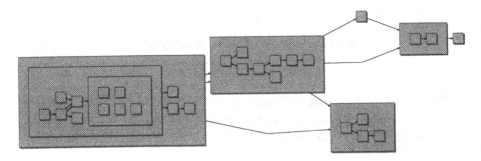

Fig. 7. An example of a hierarchical graph with several *node to graph* relations. Note the graphical visualization where the parent nodes contain the children graphs.

Each graph is laid out in an arbitrary world coordinate system and transformed dynamically into the coordinate system of the bottom-most graph in the drawing. This design helps to keep children graphs and their layouts invariant under collapse and expand operations.

Since the layout algorithms process each graph separately, layout works faster on hierarchically organized graphs than on flat graphs, and the structure is preserved. Navigation operations are efficient enough to manage hundreds of small graphs at the same time with an interactive graph editor.

3 The Graph Editor Toolkit

The Graph Editor Toolkit [GET95] is a portable front end for the Graph Layout Toolkit. It provides a user interface to create, display and edit graphs. Figure 8 shows a screen capture of the editor. Figures 9 and 10 show two dialogs of the editor. It is designed to be extended at the back end to tie it to various data sources and to be customized at the front end to add application semantics.

The Graph Editor Toolkit derives its classes from Graph Layout Toolkit classes and extends them with methods to handle user interface operations such as dispatching user input and drawing nodes and edges. It consists of four independent subsystems:

- The *Engine* is a portable module that determines how the editor reacts to input. The engine extends the Graph Layout Toolkit's classes into user interface aware graph, node, and edge classes. Potential users of the editor may extend these classes to adapt them for their own need.
- The *DocView manager* implements a document/view model that allows one graph to be displayed in more than one window, but with different visual representations.
- The *Graphics manager* can display nodes and edges as arbitrary graphic widgets. Each node is represented by a matrix of graphic primitives. In the standard cases, this is a 1×2 matrix that consists of a bitmap and a text label. Databases may use $2 \times n$ tables to display data entries.
- The *Dialog manager* interacts with the windowing system to generate windows, menus, and dialogs.

Naturally, the Graphics and the Dialog manager depend on the underlying toolkit and window system, and are less portable than the Engine and DocView managers.

The Graph Editor Toolkit is currently available as a stand-alone application or as a library for Borland ObjectWindows™ based programs for Microsoft Windows 3.1, Windows '95, and Windows NT.

4 Outlook

There are many improvements and extensions planned for the Graph Layout Toolkit and the Graph Editor Toolkit. Extensions will include incremental and interactive algorithms, constraint systems, enhanced navigation systems, and labeling systems. A Graph Editor Toolkit portable editor framework is being developed and the graphics portions are being ported to the Microsoft Foundation Classes for Windows, Microsoft OLE, and OSF/Motif™ for UNIX platforms.

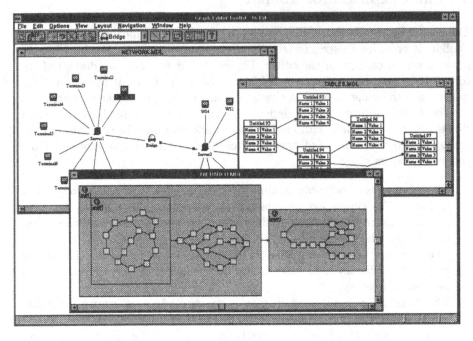

Fig. 8. The Graph Editor Toolkit for Microsoft Windows™ 3.1.

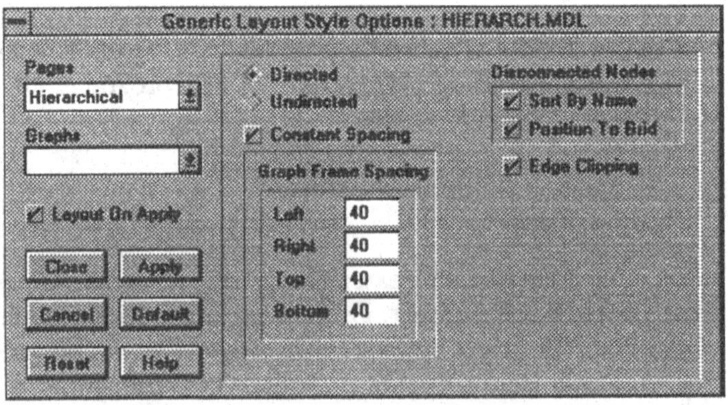

Fig. 9. This Graph Editor Toolkit dialog is used to set the basic layout options.

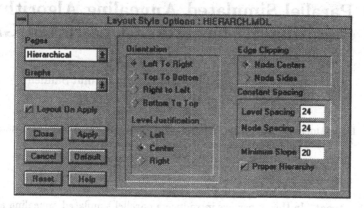

Fig. 10. This Graph Editor Toolkit dialog is used to set the options for the hierarchical layout style.

References

[BK94]   Biedl, T., Kant, G.: *A Better Heuristic for Orthogonal Graph Drawings*, In: Proc. of the 2nd European Symp. on Algorithms (ESA 94), Lecture Notes in Computer Science **855** Springer-Verlag (1994) 124–135.

[DETT95]   Di Battista, G., Eades, P., Tamassia, R., Tollis, I.G.: *Algorithms for drawing graphs: An annotated bibliography*. In Computational Geometry: Theory and Applications 4, (1994), 235–282.

[GET95]   Tom Sawyer Software: *Graph Editor Toolkit Manuals*, Berkeley, CA (1995) (to appear).

[GLT95a]   Tom Sawyer Software: *Graph Layout Toolkit User's Guide*, Berkeley, CA (1992 - 1995).

[GLT95b]   Tom Sawyer Software: *Graph Layout Toolkit Reference Manual*, Berkeley, CA (1992 - 1995).

[KK89]   Kamada, T., Kawai, S.: *An Algorithm for Drawing General Undirected Graphs*, Information Processing Letters **31** (1989) 7–15.

[KMG89]   Kar, G., Madden, B.P., Gilbert, R.S.: *Heuristic Layout Algorithms for Network Management Presentation Services*, IEEE Network November (1988) 29–36.

[NLA95]   Tom Sawyer Software: *Network Layout Assistant User's Guide* (1993 - 1995).

[PT94]   Papakostas, A., Tollis, I.G.: *Improved algorithms and bounds for orthogonal drawings*, In: Proc. Graph Drawing '94, Lecture Notes in Computer Science **894**, Springer-Verlag (1994) 40–51. (a revised version is in progress).

[STT81]   Sugiyama, K., Tagawa, S., Toda, M.: *Methods for visual understanding of hierarchical systems*. IEEE Transactions on Systems, Man and Cybernetics **11** (1981) 109–125.

A Parallel Simulated Annealing Algorithm for Generating 3D Layouts of Undirected Graphs*

Burkhard Monien,[1] Friedhelm Ramme,[2] Helmut Salmen[1]

[bm | ram | hlmut] @ uni-paderborn.de

[1] Department of Computer Science, University of Paderborn, Germany
[2] Paderborn Center for Parallel Computing (PC²),
University of Paderborn, Germany **

Abstract. In this paper, we introduce a parallel simulated annealing algorithm for generating aesthetically pleasing straight-line drawings. The proposed algorithm calculates high quality 3D layouts of arbitrary undirected graphs. Due to the 3D layouts, structure information is presented to the human viewer at a glance. The computing time of the algorithm is reduced by a new parallel method for exploiting promising intermediate configurations. As the algorithm avoids running into a local minimum of the cost function, it is applicable for the animation of graphs of reasonably larger size than it was possible before.

Subsequent to the discussion of the algorithm, empirical data for the performance of the algorithm and the quality of the generated layouts are presented.

Keywords: 3D graph layout, straight-line drawing, parallel simulated annealing

1 Motivation

During the last few years the problem of generating aesthetically pleasing layouts of a given graph $G = (V, E)$ has received an increasing amount of attention [BETT94]. After transforming an abstract graph description into an appealing drawing, a human viewer can, at a glance, derive additional information about the properties of G which are inherent in the abstract description. Due to the high complexity of the problem, which is $\mathcal{NP}$-hard and the small computational power of the desktop workstations of the past, current layout programs are mostly computing two dimensional straight-line drawings. Only a few authors [FR91, CELR94, FPS94] have considered the challenge of calculating 3D layouts. These kinds of drawings, however, are best suited to present inside information to human viewers.

Modern workstations are equipped with fast graphic interfaces which allow the user to rotate a 3D representation without visible disturbance and to create

* This work was partially supported by the ESPRIT Basic Research Action No. 7141 (ALCOM II)

** WWW: http://www.uni-paderborn.de/fachbereich/AG/monien/index.html
WWW: http://www.uni-paderborn.de/pcpc/pcpc.html

virtual walk-through animations. However, the calculation of 3D layouts requires a reasonable amount of computational power. For reducing the computing time we will take advantage of a parallel computer system. Criteria like:

- reflect inherent symmetries
- maximize angle resolution
- distribute vertices evenly
- avoid edge crossings
- keep edge lengths uniform

should be optimized in order to get aesthetically pleasing drawings [KK89, FR91, Tu93]. Unfortunately, some of them are competitive and there is no reasonable way to assign absolute weights to them. Most of the known algorithms for solving this problem can be classified into one of three groups:

The *combined heuristics* make use of heavy-duty preprocessing techniques to determine graph characteristics like strongly connected components, planar subgraphs or minimal height breadth-first spanning trees [HS93, Tu93]. The final layout is gradually obtained afterwards.

The *spring embedder model* for drawing undirected graphs was introduced by Eades [Ea84]. Using an analogy to physics, vertices are treated as mutually repulsive charges and edges as springs connecting and attracting the charges. Starting from an initial placement, the spring system is moved to a state with minimal energy. Kamada and Kawai [KK89] refined the model of Eades. They introduced the desirable length, a combined value of the shortest path between two vertices and the desired length of a single edge in the display plane. There, a partial differential equation system must be solved for each vertex in each iteration to determine its new location. Fruchterman and Reingold [FR91], like Eades, made only vertices that are neighbors attract each other, while all other vertices repelled each other. Their layouts were determined during a fixed number of iterations. The proposed algorithm calculates the effect of the attractive and repulsive forces on each vertex and limits the total displacement by a temperature value, afterwards. The temperature and its cooling schedule were borrowed from the simulated annealing method. While all these spring embedders are quite fast, they perform a gradual descent and converge at a local minimum of the energy function. The algorithm presented in [FLM94] avoids this drawback by introducing a local temperature value to each vertex. It terminates after a fixed number of rounds or if all local temperatures are below a certain threshold. Using a new heuristic to detect rotations and oscillations during the computations it was possible to outperform the speed of the algorithms presented in [KK89, FR91]. However, no statements could be made about the convergence behavior.

The third group of algorithms uses randomness to overcome the problem of ending up in local minima. Its model, called *simulated annealing* (SA), is a flexible optimization method, suited for large-scale combinatorial optimization problems [DLS93, JAMS89-91]. To our knowledge, Davidson and Harel [DH93] were the first who used an SA approach for calculating aesthetically pleasing graph layouts. The graph drawing problem, especially, is characterized by large configurations as well as a non-trivial neighborhood relation and a complex cost-function. These properties seem to be highly responsible for the long computing times required by the SA algorithm proposed. However, the resulting quality was comparable to or even better than that of the spring embedders [FR91]. Due to the considerable time requirement, even for graphs of small and medium sizes and the large number of parameters, the SA algorithm of [DH93] is not suited to draw larger graphs. A first parallel approach (implemented at a CM-2) for tackling the drawing problem was presented in [KMS94]. However, it was reported that the convergence rate was slow and that the network diagrams generated had only small sizes.

Within this paper, we will present a parallel SA algorithm for generating 3D layouts of arbitrary undirected graphs. The required time is significantly reduced compared to [DH93], by using the computing facilities of a modern MPP system. For our algorithm, the assignment of four parameters is sufficient. All others are either fixed or were made self-adjusting during runtime. This property is essential for being applicable in practise. In Section 2 of this paper, we will present the PARallel Simulated Annealing algorithm (PARSA) in details. Its properties and some layout results will be discussed in Section 3. Additionally, empirical performance data will be given.

2 The PARSA-Algorithm

SA has been applied successfully to many problems characterized by a large discrete configuration space, too large for an exhaustive search, over which a cost-function is to be minimized (or maximized). After picking some initial configuration, most of the iterative methods continue by choosing a new configuration at each step, evaluating it, and possibly replacing the previous one with it. This action is repeated until some termination condition is satisfied. In general, the procedure ends up in a local minimum, rather than the desired global one. SA tries to prevent this by allowing uphill moves with a certain probability. A sequential SA algorithm[3] consist of two nested loops. Within the interior loop, new configurations are continuously evaluated until an equilibrium of the corresponding cost-function $\mathcal{M}$ is reached. A new configuration C_k is derived from the previous one by choosing an arbitrary vertex of C_{k-1} and moving it to a randomly chosen position. The maximum displacement is limited by a value which is proportional to the temperature ϑ. If an improvement is reached, the new configuration is accepted. In case of deterioration, acceptance is probabilistic. After an equilibrium is detected, the virtual temperature is reduced according to a certain cooling-schedule. If a termination condition is fulfilled the final graph layout is returned.

At the beginning of the SA algorithm, the configurations resembles a muddle. This situation has to be resolved step by step afterwards. To attain a satisfactory convergence behavior, it is important that a single vertex (or only a small number of vertices) are moved at a time, and that each displacement is limited by a temperature dependent factor. When using SA to generate 2D layouts the second statement prevents single vertices to be placed at distant but reasonably better positions if the temperature is already lowered and a potential (or

---

[3] We assume that the reader is familiar with the basic annealing techniques. Otherwise, a good introduction can be found in [HRS86, JAMS89-91, OG89, DLS93] and the references given.

cost-) barrier must be overcome. Therefore, quite often drawings of unpleasing quality were returned. This effect can be reduced to a certain degree by using a sophisticated cost-function [DH93]. The PARSA-algorithm, however, abolishes this by generating 3D configurations directly. The additional dimension permits single vertices to circumnavigate obstacles and to travel to a distant destination at low temperature values and in situations where the intermediate configuration is relatively stable. The resulting advantage is twofold. Firstly, a simple cost-function speeds up the sequential algorithms. Secondly, the solution-space allows for smooth transitions which results in a convinced convergence behavior with high quality layouts, even if the graphs are going to become large.

To simplify reading, we kept the notation of the PARSA-algorithm (Fig. 1) close to the sequential case. Thus, for the discussion of the common parts one should ignore the lines 10-11, 13-15, 17, 21-23, 25, set $I = 0$, and assume that the remaining algorithm is executed at a single processor.

2.1 Cost-function

The cost-function is one of the most critical parts of any SA algorithm. The quality of a desired solution must be well complemented by the values returned for intermediate solutions rsp. configurations. As this function forms the compute-intensive kernel of the algorithm, it should be possible to calculate its new values $\mathcal{M}(C_k)$ very quickly and if possible incrementally from its previous values $\mathcal{M}(C_{k-1})$. The cost-function we have chosen consists of only three parts: $\mathcal{M} = \mathcal{M}_a + \mathcal{M}_e + \mathcal{M}_p$ with $\mathcal{M}_a$ the *angle-costs*, $\mathcal{M}_e$ the *edge-costs*, and $\mathcal{M}_p$ the *pseudo-edge-costs*. The new values of $\mathcal{M}_i(C_k)$ ($i \in \{a, e, p\}$) are determined by adding a fast computable term $\Delta_i(v)$ to $\mathcal{M}_i(C_{k-1})$, with v being the vertex moved within the transition $C_{k-1} \vdash C_k$.

Angle-costs: Each angle between neighboring edges of a vertex ($\sphericalangle$) in the 3D configuration space is associated with a cost value. The smaller an angle, the higher its cost. An angle α larger than π has costs equivalent to $2\pi - \alpha$. Let λ_a be a scaling factor and k_a a constant. Then, the angle-costs are determined by:

$$\mathcal{M}_a = \lambda_a \cdot \sum_{v \in V} \sum_{a \in A_v} \frac{c_a(a)}{deg(v)} \quad \text{with} \quad c_a(\alpha) = \frac{\pi - \alpha}{k_a \cdot \alpha + \pi}$$

$$\text{and} \quad A_v = \{\sphericalangle(k, l) \mid k = \{v, v_i\}, l = \{v, v_j\}, k, l \in E, i < j\}$$

Edge-costs: This term evaluates the difference between the desired length L_e of an edge and its Euclidean length Δ in the 3D space. Best results were achieved by using a twofolded function: A hyperbola if the length is below L_e, and a straight line otherwise. Let λ_e be a scaling factor and k_{e1}, k_{e2} some constants. Then, the edge-costs are determined by:

$$\mathcal{M}_e = \lambda_e \sum_{\substack{e = \{u, v\} \\ e \in E}} \frac{c_e(\Delta)}{deg(u) + deg(v)} \quad \text{with} \quad c_e(\Delta) = \begin{cases} L_e \cdot \frac{L_e - \Delta}{k_{e1} \cdot \Delta + L_e} & \text{if } \Delta < L_e \\ \frac{\Delta - L_e}{k_{e2}} & \text{otherwise} \end{cases}$$

```
-- k        : number of accepted configurations
-- C_k      : configuration under negotiation
-- M        : value of the cost-function
-- Ē        : mean value of M
-- V̄        : mean deviation of M
-- ϑ        : temperature value
-- N[0..h]  : with N[i] = (M, ϑ, V̄) the current
--          : cost-function, temperature, mean deviation   at PE i
--          : with i = 0 being this PE and 1 ≤ i ≤ h the neighbors
--          : initially M = ∞,  ϑ₀ = 3d,  V̄ = ∞
```

$PARSA\ (G = (V, E))$

```
 1:  BEGIN
 2:     C_0 := random_initial_configuration () ;
 3:     call_init (N[0], Ē);
 4:     k := 1;
 5:       REPEAT
 6:         REPEAT
 7:             C_k := create_new_conf (C_{k-1}, N[0].ϑ);
 8:             m := N[0].M;
 9:             N[0].M := M(C_k);
10:             I := i ∈ {0..h} : N[i].M is minimal ;
11:             IF short_subchain (C_k) THEN I := 0; FI
12:             IF (N[I].M < m) THEN
13:                IF (I > 0) THEN C_k := get_conf_from_neighbor (I);
14:                                N[0].M := N[I].M;
15:                FI
16:                calc_new (Ē, N[0].V̄);
17:                send_to_all_neighbors (N[0]);
18:                k := k + 1;
                                        ELSE
19:
20:                IF ( e^{-\frac{m-N[I].M}{N[0].ϑ}} ≥ rnd (0..1) ) THEN
21:                   IF (I > 0) THEN C_k := get_conf_from_neighbor (I);
22:                                   N[0].M := N[I].M;
23:                   FI
24:                   calc_new (Ē, N[0].V̄);
25:                   send_to_all_neighbors (N[0]);
26:                   k := k + 1;
27:                FI
28:             FI
29:           UNTIL equilibrium ();
30:           N[0].ϑ := temp_reduction ();
31:       UNTIL terminate () ;
32:     RETURN (C_{k-1});
33:  END ;
```

Fig. 1. The PARSA algorithm

Pseudo-edge-costs: Up to now, only those vertices were considered which are directly connected by an edge. To get aesthetically pleasing layouts, we have to consider all pairs of vertices. Instead of taking the original graph G we take the square-graph $G^2 = (V, E^2)$ with $E^2 = E \cup \{\{i,j\} \mid i,j \in V, \exists k \in V : \{i,k\}, \{k,j\} \in E\}$. Now, if all edges of its complementary graph $\overline{G^2} = (V, \overline{E^2})$ with $\overline{E^2} = \{\{i,j\} \mid i,j \in V, \{i,j\} \notin E^2\}$ are imposed by additional costs, non-adjacent vertices are placed distantly. When using $\overline{G^2}$ instead of G a double evaluation with the angle-costs at a vertex can be avoided. Especially when drawing large graphs, the muddle resulting from the initial random layout is rectified very fast by using the pseudo-edge-costs. Let λ_p be a scaling factor, k_{p1}, k_{p2} some constants and d the expected diameter of the final graph layout (see 2.2). Then, the pseudo-edge-costs are determined by:

$$M_p = \lambda_p \cdot \sum_{p \in \overline{E^2}} c_p(\Delta) \quad \text{with} \quad c_p(\Delta) = \begin{cases} L_e \cdot \frac{k_{p1} \cdot d - \Delta}{k_{p2} \cdot L_e + k_{p1} \cdot d} & \text{if} \quad \Delta < 3d \\ 0 & \text{otherwise} \end{cases}$$

2.2 Cooling-schedule

The number of intermediate configurations and the decrease of the temperature are controlled according to a cooling-schedule together with the function *equilibrium* () (line 29 in Fig. 1). The choice of the cooling-schedule has great influence on the convergence-behavior of the algorithm and its efficiency. To be applicable in practice, it is essential that the schedule is *self-adapting* to the parameters of a given problem instance [HRS86, DLS93].

Initial temperature: The initial temperature should be chosen depending on the problem instance. It should be large enough for potentially spying out each corner of the solution space. Taking three times the estimated diameter d of the 3D configuration space was found to be sufficient. In order to estimate d, a maximum plane graph with $|V|$ vertices and f faces each having three corners is considered. Using Eulers formula for plane graphs we get $f = 2(|V| - 2)$. With L_e being the desired edge length, the area of a face can be approximated by $\frac{1}{2}L_e^2$. If all faces are placed equally on the surface of a sphere, we get $f \frac{L_e^2}{2} = \pi d^2$ and thus $d = L_e \cdot \sqrt{\frac{|V|-2}{\pi}}$. Now, we can set $\vartheta_0 = 3d$ to get the initial value of the virtual temperature.

Temperature reduction: The schedule proposed in [HRS86] reduces the temperature according to the length of the Markov chains, whereby the length of a chain is redetermined at each temperature level. To keep the variations between two consecutive levels lower than the standard deviation σ of the cost-function, the new value ϑ_{k+1} is set to $\vartheta_k \cdot e^{-\frac{\lambda \cdot \vartheta_k}{\sigma}}$ with $0 < \lambda \leq 1$. As the reduction ratio depends on σ, large changes in the standard deviation resulting from short Markov chains will have significant impact on the current schedule. This influence can be alleviated by using the smoothing technique of [OG89] to approximate σ by $\sigma_{k+1}^s = (1 - w) \cdot \sigma_{k+1} + w \cdot \sigma_k^s \cdot \frac{\vartheta_{k+1}}{\vartheta_k}$ with $0 \leq w \leq 1$.

Equilibrium detection: The equilibrium detection is performed depending on the number of accepted configurations and on the standard deviation of the cost-function. It can be assumed [DLS93] that an equilibrium is reached, if the ratio between the number of accepted configurations with cost values in a range of $\pm\delta = \frac{1}{2}\sigma$ around the expected average value of the costs $\overline{E}$, and if the total number of generated configurations is convergent [HRS86].

Termination detection: In addition to the physical network of the MPP systems the PARSA algorithm makes use of a virtual tree network [Parix] for initialization and termination detection. As each Processing Element (PE) of the system computes its own SA algorithm, local annealing temperatures ϑ and local cost-values are provided at each node. Let $\overline{E}$ ($\overline{V}$) be the approximated mean value (mean deviation) of the cost-function. Then, the triple $(\overline{\vartheta}, \overline{E}, \overline{V})$ is determined successively from the leaves to the root of the virtual tree. Initial values are gathered from short local pre-runs. The PARSA algorithm terminates if $\overline{\vartheta}$ falls below of a given threshold or if $\overline{V}$ is less than three percent of $\overline{E}$. The second condition indicates that the single cost-values are in a relatively small range around $\overline{E}$ so that a solution approaching the global minimum is reached with high probability.

2.3 Parallelization

The PARSA algorithm was evaluated on two partitionable MPP systems. A GCPP consisting of 64 nodes with 64 MByte memory and two MPC 601 processors, and a GCel consisting of 1024 nodes with 4 MByte memory and single T8 processors. While T8 nodes can communicate with 20 Mbps per link, the communication speed of the MPC 601 nodes is four times higher. Each node performs the algorithm of Fig. 1. The processors communicate solely by message-passing.

At the beginning, each PE starts with its own random layout configuration. As the mean deviation and the annealing temperature at each PE is high, the different SA runs are nearly independent. Thus, up to # PEs areas of the solution space are explored simultaneously. Later on, this behavior is changed. If one PE detects a promising intermediate configuration now, its neighbors can decide to adopt this configuration for themselves and start an independent Markov subchain on it. Thus, once a promising configuration is found, more and more PEs can join the group to explore the surrounding solution space while others still continue on their local chains. In this way, only little effort is wasted with configurations of unpleasant quality. At the end of the computation, the only Markov chain left is the one which had returned the final layout. This relationship is illustrated in Fig. 2.

Talking about promising layouts: Each PE maintains a local data structure $N[0..h]$ with $N[i] = (\mathcal{M}, \vartheta, \overline{V})$. $N[i]$: $1 \leq i \leq h$ stores the cost-values, temperature, and deviation of the i-th neighboring processor of the physical MPP network. The local values are stored in $N[0]$ and m is used to back-up the values of the previous configuration. In line 10, the neighborhood of a PE is investigated. If $I > 0$ but the local Markov subchain is still too short (line 11), or

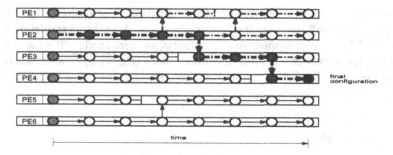

Fig. 2. A sample run with six PEs

the local configuration has currently minimal costs, then the configuration C_k is retained. Otherwise, the configuration from the neighbor PE I is ordered. After some housekeeping, the local data ($N[0]$) are send to the neighborhood. In case of increased costs (line 19) a configuration is accepted with the probability determined in line 20.

Communication vs. computation: The messages send by the PARSA algorithm can be divided into two classes: Configuration-transfer messages and others, whereby the latter consist of only a small number of bytes. These short messages are send to the physical neighbors if a transition at a PE is completed (line 17, 25). If an equilibrium is reached (line 29), or if the termination condition is fulfilled (line 31), they are send via a virtual tree network (see the virtual topology library of [Parix]). The configuration transfer messages (replies to line 13, 21) depend on G and have a much larger size. In order to reduce the computations necessary when a neighboring configuration is accepted, each vertex in the configuration-transfer message is associated with its partial cost-values. Thus, at the receiving PE, $\mathcal{M}(C_k)$ can be redetermined very fast. To prevent PEs from continuously altering their configurations, a new external one can be accepted only if the local Markov chain is of sufficient length (line 11). If few configurations from neighboring PEs are accepted, many PEs may waste their time by exploring inferior areas of the solution space. On the other hand, if too many configurations from the neighborhood are adopted, the communication overhead becomes dominating, and, because most PEs assume that their neighbors have more promising configurations, significant computation and communication effort is of less value. This behavior is controlled by the procedure *short_subchain()* which checks for the minimum length of the local Markov chain. On our MPP systems, good experiences were made with at least $|V|$ transitions.

3 Properties and Layout Results

Many SA algorithms are characterized by a large number of parameters. Finding the set of values suited best is a highly time-consuming task, even for experts. Thus, a central goal in the design of our algorithm was to keep the number of user-controlled parameters as little as possible. At the end, we were left with only four of those parameters: $\lambda_a, \lambda_e, \lambda_p$ of the cost-function and another one to control the speed of the temperature reduction. The desired edge length L_e has only weak influence on the layout structure. All other parameters were either fixed by a large number of experiments, or were made self-adapting to the problem instance (see 2.2). Initial control values (e.g. $\bar{\vartheta}, \overline{E}, \overline{V}$) are gathered from short pre-runs. This costs additional time, nevertheless, it is indispensable for practical applications.

3.1 Properties

A well designed cost-function is essential for each SA algorithm. It must be calculable very quickly, define a smooth solution space, and approximate the desired aesthetics well. The latter aspect was studied by observing layout sequences of different test-graphs with 'known' optimal layouts. For such graphs, the cost-values of their optimal layouts can be compared to the values of the generated layouts. Fig. 3 shows the result of a sample sequence. The 2% difference from optimum of this sequence, and others, indicates that the algorithm is able to approximate the minimum of the cost-function and the desired aesthetics quite well. This result was confirmed by several other test-sequences, not shown in this report.

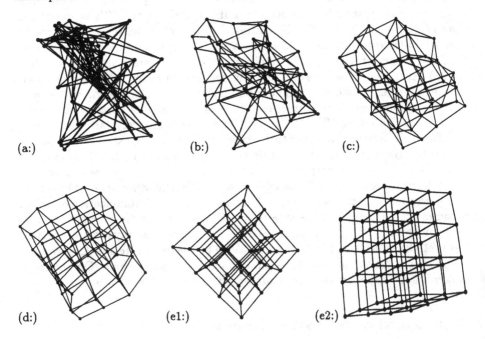

Fig. 3.   (a-d:)   3D intermediate states of a 4x4x4-Cube
          (e1, e2:)   final layout from different angles (2% from optimum)

The evolution of $\mathcal{M}$ and the temperature reduction gives further information on the behavior of the algorithm. To study the former, an 'average' graph

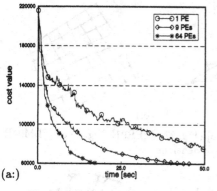 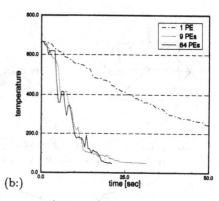

(a:)   (b:)

Fig. 4. Evolution of the cost-function   Temperature reduction

was chosen (see Fig. 5a) and $\mathcal{M}$ was investigated on the GCel system on different network sizes (Fig. 4). During the first 5 seconds, the cost-function drops rapidly. Thus quite often an early impression of the layouts can be obtained very quickly, while it takes much longer to get the required quality. As more PEs can explore a broader area of the solution space simultaneously (Fig. 2), the curves of the larger MPP networks are dropping below the curves of the smaller ones right from the beginning (Fig. 4a). For the test-graph above, a cost-value of about 5000 is reached when the termination condition is fulfilled. Thus, the advantage of the parallel approach, compared to the sequential case, is obvious. Nevertheless, a limitation on the network size for this graph is clearly indicated. By taking graphs of reasonable larger sizes (e.g. Fig. 8), up to 256 PEs can be exploited successfully.

The limitation for the graph discussed is also confirmed by the temperature reduction curves in Fig. 4b. While the temperature falls slowly at the one PE computation it drops rapidly when exploiting the MPP system. The peaks in the 9 and 64 PE curves result from adoptions of neighboring configurations with a significant difference in the temperature level. These differences, however, become negligible if the cost-function tends to converge.

3.2 Layout results

Due to few objective criteria, it is difficult to evaluate graph drawing algorithms in general. However, a number of test graphs currently substitute as a benchmark suite. They are used to 'measure' the aesthetics and the computation times [KK89, FR91, DH93, Tu93, FLM94, Sa95]. Applying the PARSA algorithm to these graphs, drawings similar to those presented by other authors (Fig. 1, 8, 12 of [DH93], Fig. 17, 18, 47-50, 64, 72 of [FR91] or Fig. 10, 11 of [FLM94]) were obtained. Moreover, the generated 3D layouts of dense graphs can be impressively animated on a computer screen. Our drawing of the common test-graph, mentioned above, is shown in Fig. 5a. Due to the competing aesthetic criteria, vertices were assigned to different levels of the third dimension (e.g. vertices which belong to the inner square were assigned to the top-most level of Fig. 5). Figure 5b shows a 3D layout of a two dimensional grid. These 2D-types of graphs are difficult to draw for any 3D layout algorithm.

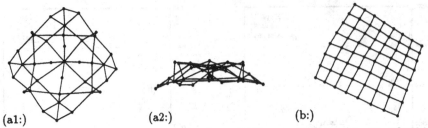

(a1:) (a2:) (b:)

Fig. 5. (a1,a2:) Drawing of Fig. 12 of [DH93] (Fig. 27 of [FR91], Fig. 19 of [FLM94])
(b:) 2D grid, layed out in three dimensions

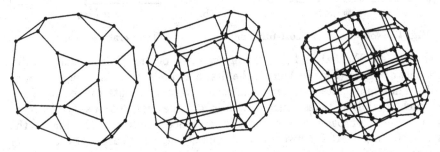

Fig. 6. CCCs of dimension 3, 4 and 5

Fig. 6 shows 3D drawings of Cube-Connected-Circles (CCCs) of dimension three, four and five. These drawings were generated in 7.6, 67 and 620 seconds respectively (on a GCPP with 64 nodes). Within the second figure, the encapsulated cubes with the circles at their corners are easy to recognize. However, for the animation of very complex graphs structures, such as the CCC 5 ($|V| = 160; |E| = 240$) the third dimension is indispensable. The CCCs and other more irregular graphs (like the graph 'C' of the GD'95 competition), show the ability of the PARSA algorithm to generate high quality layouts of large graphs with different symmetry-regions. These types of graphs are a challenge for any spring embedder.

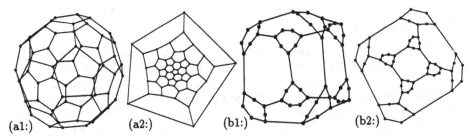

(a1:) (a2:) (b1:) (b2:)

Fig. 7. 3D layouts of soccer ball (a1) and a modified CCC (b1);
and its plane projections (a2) and (b2), respectively

Other graphs with inherent 3D structure are the soccer ball (Fig. 7a) with $|V| = 60; |E| = 90$ and the modified CCC (Fig. 7b). The former was computed in 58 seconds on average, the latter requires 24 seconds. Its plane projections (Fig. 7 (a2, b2)) were obtained by placing the view-point very close to a face of the 3D layouts of Fig. 7 (a1, b1). Thus, getting a fish-eyes view of the drawing from inside. Due to the 3D layout technique, plane representations for many cubic drawings were attained without any additional effort.

For drawing significantly larger graphs than those presented so far, it is essential that vertices with large distances in G are also placed at distant positions of the layout area. This was achieved by the pseudo-edge-costs of $\mathcal{M}$ (see 2.1). Therefore, it was possible to generate much larger 3D drawings than before. An impression of the aesthetic quality of such large graphs can be obtained from the generated layouts shown in Fig. 8 which was calculated in only 6 minutes. Depending on the characteristics of G speed-up values of up to 20 were measured. This, however, is not the final run-time result. Further potential is expected by moving to the subchain method [DLS93], for the final phase of the parallel SA algorithm.

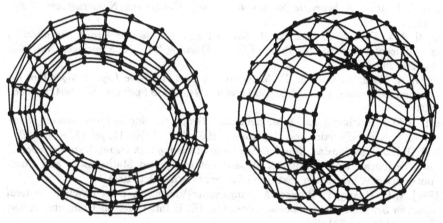

Fig. 8. 3D layout of a 10x20-Torus, computed in 6 minutes ($|V| = 200$; $|E| = 400$)

4 Conclusion

The problem of drawing arbitrary undirected graphs has received an increasing amount of attention during the last few years. However, only few of the published algorithms seem to be applicable in practise. To our knowledge, all of them are either restricted to special graph classes or limited to graphs of relatively small size. In this paper, we have presented a parallel SA algorithm for generating 3D straight-line drawings of arbitrary undirected graphs. Additional structure information is given to human viewers via a simple front-end for 3D graph animations. Due to the cost-function developed and the extended possibilities using a 3D configuration space, it was possible to generate drawings of very different graphs and of much larger size than it was possible before. Because the cooling-schedule and most parameters of the PARSA algorithm were made self-adapting, the presented approach is of high interest for practical applications.

References

[BETT94] G.D. Battista, P. Eades, R. Tamassia, I.G. Tollis : *Algorithms for drawing graphs: An annotated bibliography*, Report, Brown University, June 1994

[CELR94] R.F. Cohen, P. Eades, T. Lin, F. Ruskey : *Three-Dimensional Graph Drawing*, Proc. of Graph Drawing '94, LNCS Springer, Vol. 894, pp. 1-11

[DH93] R. Davidson, D, Harel : *Drawing graphs nicely using simulated annealing*, Technical Report CS89-13, Department of Applied Mathematics and Computer Science, Weizmann Institute of Science, Israel 1989, revised July 1993, to appear in Communications of the ACM

[DLS93] R. Diekmann, R. Lüling, J. Simon : *Problem Independent Distributed Simulated Annealing and its Applications*, in: R.V.V. Vidal (ed.): *Applied Simulated Annealing*, Lecture Notes in Economics and Mathematical Systems, Springer 1993, No. 396, pp. 17-44,

[Ea84] P. Eades : *A heuristic for graph drawing*, Congressus Numerantium, 1984, Vol. 42, pp. 149-160

[FPS94] P.W. Fowler, T. Pisanski, J. Shawe-Taylor : *Molecular Graph Eigenvectors for Molecular Coordinates*, Proc. of Graph Drawing '94, LNCS Springer, Vol. 894, pp. 282-285

[FLM94] A. Frick, A. Ludwig, H. Mehldau : *A Fast Adaptive Layout Algorithm for Undirected Graphs*, Proc. of Graph Drawing '94, LNCS Springer, Vol. 894, pp. 388-403

[FR91] T.M.J. Fruchtermann, E.M. Reingold : *Graph drawing by force-directed placement*, Software-Practice and Experience, 1991, Vol. 21, No. 11, pp. 1129-1164

[HS93] D. Harel, M. Sardas : *Randomized Graph Drawing with Heavy-Duty Preprocessing*, Technical Report CS93-16, Department of Applied Mathematics and Computer Science, Weizmann Institute of Science, Israel Oct. 1993

[HRS86] M.D. Huang, F. Romeo, A. Sangiovanni-Vincentelli : *An Efficient General Cooling Schedule for Simulated Annealing*, IEEE Int. Conf. on Computer Aided Design 1986, pp. 381-384

[JAMS89-91] D.S. Johnson, C.R. Aragon, L. A. McGeoch, C. Schevon : *Optimization by Simulated Annealing: An Experimental Evaluation*, Part I, "Graph Partitioning", Operations Research Vol. 37, No. 6, pp. 865-892, 1989; *Optimization by Simulated Annealing: An Experimental Evaluation*, Part II, "Graph Coloring and Number Partitioning", Operations Research Vol. 39, No. 3, pp. 378-406, 1991

[KK89] T. Kamada, S. Kawai : *An algorithm for drawing general undirected graphs*, Information Processing Letters, North-Holland 1989, Vol. 31, pp. 7-15

[KMS94] C. Kosak, J. Marks, S. Shieber : *Automating the Layout of Network Diagrams with Specified Visual Organization*, IEEE Trans. on Systems, Man, and Cybernetics, Vol. 24, No. 3, pp. 440-454

[OG89] O.E. Otten, L. van Ginneken : *The Annealing Algorithm*, Kluwer Academic Publishers 1989

[Parix] Parsytec Computer Ltd. : *Parix 1.3: Software Documentation*, Aachen

[Sa95] H. Salmen : *Dreidimensionale Auslegung beliebiger Graphen mittels parallelem Simulated Annealing Methoden*, Master Thesis, Univ. of Paderborn, 1995

[Tu93] D. Tunkelang : *A layout algorithm for undirected graphs*, In Graph Drawing '93, ALCOM Int. Workshop, Paris 1993

Incremental Layout in DynaDAG

Stephen C. North
north@research.att.com

Software and Systems Research Center
AT&T Bell Laboratories
Murray Hill, N.J. 07974 U.S.A.

Abstract. Graph drawings are a basic component of user interfaces that display relationships between objects. Generating incrementally stable layouts is important for many applications. This paper describes *DynaDAG*, a new heuristic for incremental layout of directed acyclic graphs drawn as hierarchies, and its application in the *DynaGraph* system.

1 Introduction

Effective techniques have been developed for some important families of graph layouts, such as hierarchies, planar embeddings, orthogonal grids and forced-directed (spring) models [1]. These techniques have been incorporated in practical user interfaces that display static diagrams of relationships between objects [19, 18, 17].

Static diagrams are not completely satisfactory because in many situations, the displayed graphs can change. Three common scenarios are:

Manual editing. Most interactive graph drawing systems allow users to manually insert and delete nodes and edges. Layouts must be updated dynamically to reflect such changes.

Browsing large graphs. When only static layout is available, browsing large graphs usually means drawing the entire graph and then viewing portions in a window with pan and zoom controls, fisheye lenses, etc. The problem is that the section in the current window may not be very informative. For example it may contain edge segments whose endpoint nodes are outside the window, or nodes whose placement can be rationalized globally but not locally. Incremental layout offers the alternative of directly adjusting the set of displayed objects to make informative displays.

Visualizing dynamic graphs. Often, data being visualized is subject to change. In our experience with the dotty system, we found many applications for graph animations:

- *CIAO* is a program database that displays dependencies between the types, data, functions and files in a C or *C++* program [6]. Programs change throughout their life-cycle as they are debugged, maintained, and improved, so graph views should reflect such changes.
- *Improvise* is a multimedia viewer for software process models [11]. These models are incrementally corrected and refined. Users report that stable incremental layout and manual editing of diagrams are essential.
- *LDBX* is a prototype graphical debugger that runs on an unmodified dbx text-based debugger [17]. It displays data structure graphs. Records are drawn as

nested boxes containing primitive data or pointer fields. Pointers may be traced interactively, or automatically by the system, yielding incremental graph updates.

- *VPM* displays distributed programs as graphs [4]. Processes and resources are drawn as nodes. Edges represent dependence and communication. Subgraphs (that is, *zones* [10] or *clusters* [16]) show distribution across hosts. *VPM* would benefit greatly from stable incremental graph layout. Furthermore, graph updates are issued at an exceptionally high rate (the rate of system call issue) so efficiency is critical.

Most graph drawing algorithms to date are not incrementally stable. They usually apply batch techniques to optimize objectives such as reducing total edge crossings or edge length. A small change in the input set, even just its ordering, may yield unpredictable, instable changes between successive layouts. This may occur even if a previous layout is taken as a starting configuration. The results can be confusing when viewing a sequence of layouts. An example is shown in fig. 1 made by an extension of the algorithm of Sugiyama et al [20]. Graphs (a) and (b) differ by only one edge. Although the drawing could be updated by moving a subgraph downward, as shown in (c), the layout system makes more drastic, unnecessary changes.

Most of our applications involve software engineering diagrams drawn hierarchically, so our immediate goal is to "incrementalize" our variant of Sugiyama's hierarchical drawing algorithm. Some similar issues, though, are encountered in making other kinds of layouts incrementally.

2 Previous Work

Significant progress has been made in drawing dynamic trees [15] (using subtree contours), planar graphs [2] and series-parallel graphs [7]. Although these are useful techniques for these restricted classes of graphs, they are not directly applicable to general graph drawing.

Hornick, Miriyala and Tamassia describe a practical incremental edge router for orthogonal drawings, such as entity-relation diagrams [14]. Nodes are placed externally (typically, by the user); the system routes edges incrementally by a shortest-path technique that accounts for edge length, crossings, and number of bends. The technique would have to be extended to handle automatic node placement and adjusting layouts to make space for new objects.

Lyons describes a way of incrementally improving layouts of undirected graphs such as those made by force directed modeling with unconstrained optimization [13]. The idea is to adjust regions where nodes are too close by computing Voronoi sets around each node and moving each node that has a conflict to a better place within its region. Because each node moves to a point that is closer to its old position than to any other node, faces are preserved. This technique is claimed to give better and more stable results than alternatives, such as re-solving a spring model with adjustments in springs intended both to force overlapping nodes apart, and to anchor nodes to their original positions. Lyons' algorithm is effective for this problem, but does not address how to maintain stability if the underlying graphs change. Also, virtual physical modeling is somewhat limited by the characteristic that all edges tend toward unit length.

Newbery and Bohringer show how to add stability constraints to graphs drawn by a batch technique in the *Edge* system using Sugiyama's directed graph drawing algorithm [20, 3]. After making an initial layout, intra-rank node ordering constraints are appended. If u, v are neighbors on rank r, they will appear in the same order in succeeding layouts as long as they stay on r. Adding such constraints is a good step toward making stable layouts of directed graphs, but preserving this ordering is not difficult if one assumes an on-line layout algorithm. The more fundamental problem is how to adjust layouts when groups of nodes must change ranks, and how to maintain geometric stability.

3 Incremental layout

3.1 Goals

The basic problem is, given a sequence of graphs

$$G_0, G_1, G_2, \ldots, G_n$$

interpreted as successive versions of G, find a "good" sequence of layouts

$$L_0, L_1, L_2, \ldots, L_n$$

where each L_i is a drawing of G_i. An update $G_i \rightarrow G_{i+1}$ may be written $U_i = (V+, V-, E+, E-)$ where these are sets of nodes and edges inserted and deleted. There are important advantages if L_{i+1} resembles L_i:

- Users can retain a persistent "mental map" [8].
- Graphical updates reflect actual changes in the data.
- Large layouts can potentially be updated quickly.

The first two properties concern effective data visualization. In any context, visualization should help reveal meaningful patterns in data while avoiding irrelevancy and display artifacts. Efficiency is also desirable but not particularly important at this stage until the right problem has been identified.

While stability is important, it is not the only desirable characteristic for incremental layouts. We propose the following, in order of importance:

- consistency
- stability
- readability

Consistency or adherence to layout style rules is most important because the displayed diagram should always reveal properties of interest. Otherwise, visualization is pointless. For example, if the purpose of visualization is to demonstrate that a given graph is a tree, DAG, planar, or embeddedable in a grid, the diagram should always capture this property. If consistency is relaxed or abandoned, successive incremental layouts could quickly become obscure, ambiguous, or even incorrect. The second property, stability, refers to a principle of least change between successive diagrams, subject to consistency. Readability refers to other properties to make diagrams pleasing and easy to read.

Assuming that consistency is more important than stability, then edits that cause fundamental changes in graph structure may be expected to cause large changes in layouts in order to maintain consistency. As a simple example, if a display shows two trees in a conventional downward drawing, and an edge is inserted that makes one tree become a subtree of the other, consistency requires moving the entire subtree, no matter how large. Thus, stability implies weak constraints on node and edge placement. This is a useful view because weak constraints may also reflect user-specified object placement requests.

3.2 Graph updates

An important question is what updates U_i to allow. Some possibilities include: arbitrary updates, single node or edge operations, append-only updates, homeomorphic expansion and subgraph abstraction (collapsing a subgraph into a node or restoring the subgraph [5, 18]).

3.3 Look-ahead

Often the entire sequence $G_0 \ldots G_i$ is known in advance. This should be important information and it is available whenever an animation is made off-line. In other situations, some look-ahead may be available, such as when updates are batched.

3.4 Stability

A key question is how to characterize stability between layouts. The answer depends on how people perceive and remember the structure of diagrams. For now we assume that important factors include:

- position (geometry)
- order (topology)

Geometry and order can be considered absolute properties, or relative to a neighborhood, such as the set of logically or geometrically adjacent nodes. An interesting proposition is that node stability is more crucial than edge stability. The rationale is that nodes are sites that users learn and return to in a diagram, while edges are generally traced on the fly to discover connections, and consequently their routing is less important. (Some researchers have suggested that interactive displays of dense layouts can be improved by making only a small subset of edges visible at a time.) If this is true, it seems advantageous to adjust edge routes aggressively to improve layouts, but move nodes more conservatively.

Locality (spatial and temporal) is often relevant in designing user interfaces. For graph layout, if a node is in the geometric or logical neighborhood of another that was recently updated, it may be a better candidate for update than a node that is not in any such neighborhood. Likewise, a node that was recently moved may be a better candidate for another update than one that was not recently adjusted. This suggests using "age" or "memory" to control stability over time.

3.5 Display update

Smooth animation is often easier to understand than instantaneously switching images in a display. This means extending

$$L_i \rightarrow L_{i_0}, L_{i_1}, \ldots, L_{i_k}$$

to perform in-betweening. For smooth animation, some L_{i_j} may be inconsistent with the layout rules. Some nodes may overlap other nodes or edges as they are moving. Because it is cumbersome to support this directly in a layout system that assumes consistent diagram structure, we propose to separate the logical layout and physical display. The physical update layer is also an appropriate place to implement cues that emphasize updates, such as blinking or changing color.

4 *DynaDAG* Heuristic

4.1 Overview

DynaDAG is an incremental heuristic for drawing ranked digraphs, based on previous refinements to Sugiyama's heuristic [21, 9]. In the following discussion, we assume graphs are drawn in levels numbered from top to bottom, and that long edges are broken into chains of virtual nodes on adjacent ranks. *DynaDAG* preserves stability geometrically (exactly) and topologically (heuristically). It does not yet incorporate temporal information nor a separate display update module, but it is a suitable testbed for such future experiments.

As a simplifying assumption, *DynaDAG* supports only these operations:

$$\{ \text{ insert } | \text{ optimize } | \text{ delete } \} \times \{\text{node} | \text{edge}\}$$

More complex updates must be decomposed into these primitives. Our underlying hypothesis was that this decomposition yields good incremental layouts. This turned out to be partially correct, but an unnecessary restriction. The internal primitives of the heuristic do operate on only one node or edge at a time. On the other hand, applications often require performing a number of updates at once. In retrospect, it would make more sense to collect the updates and perform them together. This would not involve many changes to the heuristic.

The procedure *insert_node* (with an optional edge set) is most interesting because it may potentially involves moving many pre-existing nodes and edges. There are several phases. A rank assignment is determined for the incoming node. Pre-existing nodes and edges are adjusted to be consistent with this assignment. As a simplifying assumption, we compute new rank assignments by DFS and only move nodes downward. A reasonable enhancement would be to re-solve the global rank assignment problem and allow moving some nodes upward, symmetric to the downward case. Finally, the new node is installed with local optimization of its position and that of adjacent edges.

make_feasible moves an individual node to a different rank. First, the node is moved to the same X coordinate in the adjacent rank. Second, it is shifted right or left so that its label is locally consistent with the median sort order of nodes in the new rank. This process is iterated until the node reaches its destination. Finally, the node's adjacent edges are updated by adjusting (shrinking, moving, or stretching) the virtual node chain. Virtual nodes are moved by a similar heuristic. Informally, when a node moves upward or downward, it follows a "valley" in the median function.

DynaDAG contains two heuristics that find edge routes. The first applies a variant of the median sort heuristic to any nodes and edges of the graph marked as movable with the rest of the configuration held fixed. The second heuristic routes individual edges by exhaustive search using limited backtracking [12].

update_geometry employs a form of linear programming for node coordinate assignment. This technique was previously introduced in dot. As illustrated in fig. 3, given an initial ranking, there is a way of adding a variables and two constraint edges to impose a linear penalty for moving a node from its old assignment. The construction involves creating an additional node as an anchor or reference point for the layout. The minimum lengths of the edges marked λ reflect the stable coordinate. The stable coordinate is 2.0 for u and 4.0 for v and w. In this figure, an additional constraint has been introduced between v and w that forces at least one of them to move away from its old position. The cost of this adjustment is set by the weights of the auxiliary edges. (A coarse approximation of a non-linear penalty could thus be simulated by summing linear terms.)

```
procedure insert_node(view, user_node, edge_set, hint_coord)
{
    // map node to its layout representative
    v = layout_node(view, user_node);

    range = feasible_ranks(v, edge_set);
    if feasible(range)
        rank = choose_rank(v,hint_coord.y);
    else
        { rank = low(range); make_feasible(v,rank); }

    if is_valid_point(hint_coord) pos = hint_coord.x;
    else pos = mean_x(adjacent(v,edge_set));

    install_node(view,v,pos);
    install_edges(edge_set);
    opt_neighborhood(v,is_valid_point(hint_coord));
    update_geometry(view);
}

procedure insert_edge(view, orig_edge)
{
    u = layout_node(view, tail(orig_edge));
    v = layout_node(view, head(orig_edge));

    if path_exists(v,u) {temp = u; u = v; v = temp;}
    e = layout_edge(view, u, v);

    if rankof(u) + minlength(e) > rankof(v)
```

```
            make_feasible(v, rankof(u) + minlength(e));

    route_edge(e);
    update_geometry(view);
}

procedure make_feasible(v, v_rank)
{
    by DFS, find new ranks of nodes in G w.r.t. v on v_rank
    MS = { u in G : newrank(u) != oldrank(u) }
    for u in MS
        move_node_down(u,newrank(u))
    for u in MS
        for e in adjacent_edges(u)
            adjust_edge(e)
}

procedure move_node_down(v,newrank)
{
    x = position(v).x;
    for i = oldrank(v) + 1 to newrank(v) {
        set_medians(G,i);
        // place and reopt makes a new leaf under v at
        // the same x coordinate, moves it to a locally
        // optimal position, and replaces the leaf with v.
        x = place_and_reopt(v,i);
    }
}

procedure adjust_edge(e)
{
    compute shrink, same, stretch segment sizes of e
    r = oldrank(tail(e)) + 1;
    for i = 1 to shrink
        { delete_vnode(e,r); r = r + 1; }
    for i = 1 to same
        { move_node_down(vnode(e,r),r+1); r = r + 1; }
    for i = 1 to stretch
        { copy_node_down(vnode(e,r),r+1); r = r + 1; }
}

procedure opt_neighborhood(node, node_is_movable)
{
    if node_is_movable
        set_movable(node,TRUE);
    for e in edges(node)
        set_movable(e,TRUE);          // vnode chain

    range = movable_region(node);
    for iter = 1 to MAXITER {
        for r = range.low to range.high
```

```
            optimize_rank(r,r-1);    // use in-edges
        for r = range.high downto range.low
            optimize_rank(r,r+1);    // use out-edges
    }
    set_movable(n,FALSE);
    for e in edges(node)
        set_movable(e,FALSE);
}

procedure update_geometry(view)
{
    // update the auxiliary constraint graph
    r = low_rank(view);
    while (r <= high_rank(view)) {
        v_left = leftmost_node(view, r);
        while (v_left) {
            constrain_prevposition(v_left);
            constrain_outedge_cost(v_left);
            v_right = right_neighbor(v_left);
            if v_right
              constrain_separation(v_left,v_right);
            v_right = v_left;
            v_left = right_neighbor(v_left);
        }
        r = r + 1;
    }
    // invoke network simplex coordinate solver
    ns_solve(view);
    // node and edge callbacks can be done here
}
```

This heuristic has some useful properties. Its internal primitives are not difficult to program. Some can be applied individually, opening the way to explore higher-level incremental update strategies, for example, allowing a heuristic or the user to identify specific nodes and edges to be re-optimized. Further, using linear constraints and weights to solve coordinates gives precise control of tradeoffs between consistency, stability, and readability and yields predictable results.

Figure 4 shows frames from an animation created by incrementally inserting all the nodes and edges of an example graph. Most updates are apparently stable; an exception can be found between frames 26 and 27, where there is a larger adjustment, but other parts of the layout still closely resemble previous frames.

4.2   Incremental Layout Systems

We implemented *DynaDAG* and several other algorithms to provide that provide an incremental layout service through a library interface (*DynaGraph*). The other algorithms include an implementation of the Hornick-Miriyala-Tamassia orthogonal embedder, and a spring embedder with a constraint enforcement heuristic (following Lyons [13]). In *DynaGraph*'s model, an abstract graph may have one or more independent views, each containing a subset of the base graph. *DynaGraph* deals only with graphs and coordinates, and is window-system independent.

We created several compatible graph viewers on top of this interface. *DynaGraph* is an OLE-compliant Microsoft Windows graph viewer.[1] It can act as a server, managing active diagrams embedded in other documents, and as a client allowing external objects to be embedded as nodes. This greatly simplifies integration of hypertext documents or multimedia clips in graph diagrams.

dged is a programmable front end implemented in the Unix TCL/tk environment.[2] Fig. 2 is a sample of some output frames in a sequence that was created by inserting new nodes and edges. (In this execution of *DynaDAG*, stability of absolute coordinates was intentionally disabled on nodes that change ranks to see how this affects displays.) *dged* supports multiple views maintained in synchrony by different layout engines. Its intended use is to construct prototype network management utilities.

5 Conclusion

DynaDAG is a heuristic for hierarchical layout of directed graphs that incorporates geometric and topological stability. It incorporates a heuristic to move nodes between adjacent ranks, based on median sort. The heuristic is effective for viewing incremental layouts of graphs of at least several dozen nodes, though further tuning is needed. Experience with a working implementation in real applications offers invaluable guidance. The heuristic does not use look-ahead, temporal information, or adaptive update strategies; there is a good opportunity for further work here.

There are many factors that may affect how users perceive stability in graph drawings. More work is needed to understand what properties are most important, and to find efficient incremental layout algorithms for general graphs.

References

1. G. Di Battista, P. Eades, R. Tamassia, and I.G. Tollis. Algorithms for drawing graphs: An annotated bibliography. *Computation Geometry: Theory and Applications*, 4(5):235–282, 1994. Available at ftp.cs.brown.edu in /pub/compgeo/gdbiblio.tex.Z.

2. G. Di Battista and R. Tamassia. Incremental planarity testing. In *Proc. 30th IEEE Symp. on Foundations of Computer Science*, pages 436–441, 1989.

3. K. Bohringer and F. Newbery Paulisch. Using constraints to acheive stability in automatic graph layout algorithms. In *Proceedings of ACM CHI 90*, pages 43–51., 1990.

4. Yih-Farn Chen, Glenn S. Fowler, David G. Korn, Eleftherios Koutsofios, Stephen C. North, David S. Rosenblum, and Kiem-Phong Vo. Intertool connections. In B. Krishnamurthy, editor, *Practical Reusable UNIX Software*, chapter 11. Wiley, 1995. To appear January 1995.

5. Yih-Farn Chen, Leftheris Koutsofios, and David Rosenblum. Intertool connnections. In Balachander Krishnamurthy, editor, *Practical Reusable UNIX Software*, chapter 11. John Wiley & Sons, 1995.

---

[1] Written by Giampiero Sierra, Princeton University.
[2] Written by John Ellson, AT&T Bell Laboratories.

6. Yih-Farn Chen, Michael Nishimoto, and C. V. Ramamoorthy. The C Information Abstraction System. *IEEE Transactions on Software Engineering*, 16(3):325–334, March 1990.

7. Robert F. Cohen, Giuseppe Di Battista, Roberto Tamassia, and Ionnis G. Tollis. Dynamic graph drawings: Trees, series-parallel digraphs, and planar st-digraphs. In *Proc. Symposium on Computational Geometry*, pages 261–270, 1992. to appear in SIAM J. Computing.

8. P. Eades, W. Lai, K. Misue, and K. Sugiyama. Preserving the mental map of a diagram. In *Proceedings of Compugraphics 91*, pages 24–33, 1991.

9. E.R.Gansner, E. Koutsofios, S.C. North, and K.-P. Vo. A technique for drawing directed graphs. *IEEE Trans. on Soft. Eng.*, 19(3):214–230, 1993.

10. C. Kosak, J. Marks, and S. Shieber. Automatic the layout of network diagrams with specific visual organization. *IEEE Transactions on Systems, Man and Cybernetics*, SMC-24(3):440–454, 1994.

11. B. Krishnamurthy and N. Barghouti. Provence: A Process Visualization and Enactment Environment. In *Proc. of the Fourth European Conference on Software Engineering*, pages 151–160, Garmisch-Partenkirchen, Germany, September 1993. Springer-Verlag. Published as *Lecture Notes in Computer Science* no. 717.

12. Panagiotis Linos, Vaclav Rajlich, and Bogdan Korel. Layout heuristics for graphical representations of programs. In *Proc. IEEE Conf. on Systems, Man and Cybernetics*, pages 1127–1131, 1991.

13. K. Lyons. Cluster busting in anchored graph drawing. In *Proceedings of the 1992 CAS Conference*, pages 7–16, 1992.

14. Kanth Miriyala, Scot W. Hornick, and Roberto Tamassia. An incremental approach to aesthetic graph layout. In *Proc. Sixth International Workshop on Computer-Aided Software Engineering*, pages 297–308. IEEE Computer Society, July 1993.

15. S. Moen. Drawing dynamic trees. *IEEE Software*, 7:21–8, 1990.

16. Stephen C. North. Drawing ranked digraphs with recursive clusters. In *Proc. ALCOM Workshop on Graph Drawing '93*, September 1993. submitted.

17. Stephen C. North and Eleftherios Koutsofios. Applications of Graph Visualization. In *Graphics Interface '94*, pages 235–245, 1994.

18. F. Newbery Paulish and W.F. Tichy. Edge: An extendible graph editor. *Software - Practice and Experience*, 20(S1):1/63–S1/88, 1990. also as Technical Report 8/88, Fakultat fur Informatik, Univ. of Karlsruhe, 1988.

19. L.A. Rowe, M. Davis, E. Messinger, C. Meyer, C. Spirakis, and A. Tuan. A browser for directed graphs. *Software - Practice and Experience*, 17(1):61–76, 1987.

20. K. Sugiyama, S. Tagawa, and M. Toda. Methods for visual understanding of hierarchical systems. *IEEE Transactions on Systems, Man and Cybernetics*, SMC-11(2):109–125, 1981.

21. K. Sugiyama, S. Tagawa, and M. Toda. Methods for visual understanding of hierarchical systems. *IEEE Transactions on Systems, Man and Cybernetics*, SMC-11(2):109–125, 1981.

Issues in Interactive Orthogonal Graph Drawing (Preliminary Version)

Achilleas Papakostas and Ioannis G. Tollis

Department of Computer Science
The University of Texas at Dallas
Richardson, TX 75083-0688
email: papakost@utdallas.edu, tollis@utdallas.edu

Abstract. Several applications require human interaction during the design process. The user is given the ability to alter the graph as the design progresses. *Interactive Graph Drawing* gives the user the ability to dynamically interact with the drawing. In this paper we discuss features that are essential for an interactive drawing system. We also describe some possible interactive drawing scenaria and present results on two of them. In these results we assume that the underline drawing is always orthogonal and the maximum degree of any vertex is at most four at the end of any update operation.

1 Introduction

Graphs have been extensively used to represent various important concepts or objects. Examples of such objects include parallel computer architectures, networks, state graphs, entity-relationship diagrams, subroutine call graphs, automata, data-flow graphs, Petri nets, VLSI circuits, etc. In all of these cases, we require that the graph be represented (or drawn) in the plane so that we can understand and study its structure and properties. It is for that reason that, typically, drawing of a graph is accompanied by optimizing some cost function such as area, number of bends, number of edge crossings, uniformity in the placement of vertices, minimum angle, etc. For a survey of graph drawing algorithms and other related results see the annotated bibliography of Di Battista, Eades, Tamassia and Tollis [4]. An *orthogonal* drawing is a drawing in which vertices are represented by points of integer coordinates and edges are represented by polygonal chains consisting of horizontal and vertical line segments. In this paper we focus our attention on interactive orthogonal graph drawing.

In [17] and [19] it is shown that every biconnected planar graph of maximum degree four can be drawn in the grid with $2n + 4$ bends. If the graph is not biconnected then the total number of bends rises to $2.4n + 2$. In all cases, no more than four bends per edge are required. The algorithms of [19] take linear time and produce drawings, such that at most one edge may have four bends. Kant [9] shows that if the graph is triconnected of maximum degree four, then it can be drawn on an $n \times n$ grid with at most three bends per edge. The total number of bends is no more than $\lfloor \frac{3}{2}n \rfloor + 3$. For planar graphs of maximum degree three it is shown in the same paper that a gridsize of $(\frac{n}{2} - 1) \times \frac{n}{2}$ is sufficient and no more than $\frac{n}{2}$ bends are required totally. In this case, no edge bends more than twice. Even and Granot [6] present an algorithm for obtaining an orthogonal drawing of a 4-planar graph with at most three bends per edge. If the embedding of a planar graph is fixed, then an orthogonal drawing with the minimum number of bends can be computed in $O(n^2 \log n)$ time [18]. If the planar embedding is not given, the problem is polynomially solvable for 3-planar graphs [5], and NP-hard for 4-planar graphs [8]. There is a lower bound of $2n - 2$ bends for biconnected planar graphs [20].

Upper and lower bounds have been proved in the case when the orthogonal drawing of the graph is not necessarily planar. Leighton [10] presented an infinite family of planar graphs which require area $\Omega(n \log n)$. Independently, Leiserson [11] and Valiant [21] showed that every planar graph of degree three or four has an orthogonal drawing with area $O(n \log^2 n)$. Valiant [21] showed that the orthogonal drawing of a general (nonplanar) graph of degree three or four requires area no more than $9n^2$, and described families of graphs that require area $\Omega(n^2)$. Schäffter [16] presented an algorithm which constructs orthogonal drawings of graphs with at most two bends per edge. The area required is $2n \times 2n$. A better algorithm is presented in [1] and [2], which draws the graph within an $n \times n$ grid with no more than two bends per edge. This algorithm introduces at most $2n + 2$ bends.

Recently, we presented an algorithm that produces an orthogonal drawing of a graph of maximum degree four that requires area no more than $0.76n^2$ [14]. This algorithm introduces at most $2n + 2$ bends, while the number of bends that appear on each edge is no more than two. If the maximum degree is three, then there is another algorithm which produces an orthogonal drawing which needs maximum area $(\frac{n}{2} + 1) \times \frac{n}{2}$ and $\frac{n}{2} + 3$ bends [14, 15]. In this drawing, no more than one bend appears on each edge except for one edge, which may have at most two bends.

In all of the above, the drawing algorithm is given a graph as an input and it produces a drawing of this graph. If an insertion (or deletion) is performed on the graph, then we have a "new" graph. Running the drawing algorithm again will result in a new drawing, which might be vastly different from the previous one. This is a waste of time and resources from two points of view: (a) the time to run the algorithm on the new graph, and (b) the user had probably spent a significant amount of time in order to understand and analyze the previous drawing. We investigate techniques that run efficiently and introduce minimal changes to the drawing.

The first systematic approach to dynamic graph drawing appeared in [3]. There the target was to perform queries and updates on an implicit representation of the drawing. The algorithms presented were for straight line, polyline and visibility representations of trees, series-parallel graphs, and planar graphs. Most updates of the data structures require $O(\log n)$ time. The algorithms maintain the planarity of the drawing. The insertion of a single edge however, might cause a planar graph to drastically change embedding, or even to become non planar. An incremental approach to orthogonal graph drawing is presented in [13], where the focus is on routing edges efficiently without disturbing existing nodes or edges.

In this paper we investigate issues in interactive graph drawing. We introduce four scenaria for interactive graph drawing, and we analyze two of them. These scenaria are based on the assumption that the underlying drawing is orthogonal and the maximum degree of any vertex is four at the end of an update operation.

We show that in one scenario (Relative-Coordinates), the general shape of the current drawing remains unchanged after an update is performed. The coordinates of some vertices of the current drawing may increase or decrease by at most 3 units along the x and y axes. In another scenario (No-Change), we discuss an interactive algorithm for building an orthogonal drawing of a graph from scratch, so that any update inserts a new vertex and routes new edges in the drawing without disturbing the current drawing. Our algorithm guarantees that the area of the drawing at any time t is no more than $(n(t)+n(t)_4)^2$, where $n(t)$ is the total number of vertices at time t, and $n(t)_4$ is the total number of vertices up to time t, that had degree four *when they were inserted*. Apart from the area, our interactive algorithm has a good performance in terms of the total number of bends which are no more than $2.66n(t)+2$, while introducing at most 3 bends per edge.

In Sect. 2 we give an example of some features that an interactive drawing system should have. In Sects. 3 and 4 we present preliminary results on two interactive drawing scenaria. Section 5 presents conclusions and open problems.

2   Interactive Scenaria

First, the software which supports interactive graph drawing features should be able to create a drawing of the given graph under some layout standard (e.g., orthogonal, straight line, etc.). Secondly, the software should give the user the ability to interact with the drawing in the following ways (other interactive features are possible too):

- move a vertex around the drawing,
- move a block of vertices and edges around the drawing,
- insert an edge between two specified vertices,
- insert a vertex along with its incident edges,
- delete edges, vertices or blocks.

The drawing of the graph that we have at hand at some time moment t is called *current drawing*, and the graph is called *current graph*. The drawing resulting after the user request is satisfied is called *new* drawing. There are various factors which affect the decisions that an interactive drawing system takes at each moment a user request is posted and before the next drawing is displayed. Some of these factors are the following:

- The amount of control the user has upon the position of a newly inserted vertex.
- The amount of control the user has on how a new edge will be routed in the current drawing connecting two vertices of the current graph.
- How different the new drawing is, when compared with the current drawing.

Keeping these factors in mind, in this section we propose four different scenaria for interactive graph drawing. They are the following:

1. The *Full-Control* scenario. The user has full control over the position of the new vertex in the current drawing. The control can range from specifying lower and upper bounds on the x and y coordinates that the new vertex will have, up to providing the exact desired coordinates to the system. The edges can be routed by the user or by the system.
2. The *Draw-From-Scratch* scenario which is based on a very simple idea: every time a user request is posted, take the new graph and draw it using one of the popular drawing techniques. Apart from the fact that this scenario gives very slow drawing systems, the new drawing might be completely different compared to the current one.
3. The *Relative-Coordinates* scenario. The general shape of the current drawing remains the same. The coordinates of some vertices and/or edges may change by a small constant because of the insertion of the new vertex (somewhere in the middle of the current drawing) and the insertion of a constant number of rows and columns.
4. The *No-Change* scenario. In this approach, the coordinates of the already embedded vertices, bends and edges *do not change at all.* In order to achieve such a property, we need to maintain some invariants after each insertion.

There is a close connection between the Full-Control scenario and global routing in VLSI layout [12]. The reason is that this approach deals with (re)location of vertices and (re)routing of edges using the free space in the current drawing. Also, the technique presented in [13] computes routes for new edges inserted in the graph without disturbing any of the existing vertices and edges. The Draw-From-Scratch scenario is not interesting since every time an update is requested by the user, the drawing system ignores all the work that it did up to that point. The major disadvantage here is that the user has to "relearn" the drawing. In the rest of the paper, we discuss the other two scenaria and present preliminary results.

3   The Relative-Coordinates Scenario

In this scenario, every time a new vertex is about to be inserted into the current drawing, the system makes a decision about the coordinates of the vertex and the routing of its incident edges. New rows and columns are inserted anywhere in the current drawing in order for this routing to be feasible. The coordinates of the new vertex (say v) as well as the locations of the new rows and/or columns will depend on the following:

- v's degree (at the time of insertion).
- How many of v's adjacent vertices allow the insertion of a new incident edge towards the same direction (i.e., up, down, right, or left of the vertex).
- How many of v's adjacent vertices allow a new incident edge towards opposite directions.
- Whether or not the required routing of edges can be done utilizing segments of existing rows or columns that are free (not covered by an edge).
- Our optimization criteria.

Of course, there are many different cases because there are many possible combinations. It is relatively easy to come up with various cases for each insertion. Instead of enumerating all of them in this section, we will give examples of some of the best and worst cases one might encounter. In the example shown in Fig. 1a vertices u_2 and u_1 have a free direction (edge) to the right and up respectively. In this case no new rows/columns are needed for the insertion of vertex v and no new bends are introduced. On the other hand however, in the example shown in Fig. 1b all four vertices u_1, u_2, u_3 and u_4 have pairwise opposite free directions. The insertion of new vertex v requires the insertion of 3 new rows and 3 new columns in the current drawing. Additionally, eight bends are introduced. As discussed above, single edge insertions can be handled using techniques from global routing [12] or the technique of [13]. The easiest way to handle deletions is to delete vertices/edges from the data structures without changing the coordinates of the rest of the drawing. Occasionaly, or on demand, the system can perform a linear-time compaction similar to the one described in [19], and refresh the screen.

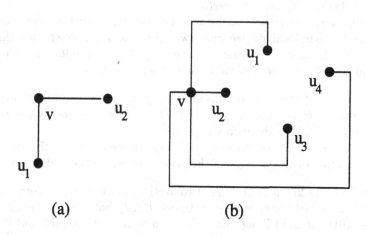

(a)   (b)

Fig. 1. Insertion of v: (a) no new row or column is required, (b) three new rows and three new columns are required.

The scenario that is described in this section maintains the general shape of the current drawing after an update (vertex/edge insertion/deletion) takes place. The coordinates of the vertices of the current drawing increase or decrease by at most 3 units along the x and y axes. This scenario works well when we build a graph from scratch, or we are presented with a drawing (which was produced somehow, perhaps by a different system) and we want our interactive system to update it. In order to refresh the drawing after each update, the coordinates of every vertex/edge affected must be recalculated.

4 The No-Change Scenario

In this scenario, the drawing system *never* changes the positions of vertices and edges of the current drawing. It just increments the drawing by adding the new elements. This is useful in many cases where the user has already spent a lot of time studying a particular drawing and he/she does not want to have to deal with something completely different after each update.

There is no work known which gives satisfactory answers to the above described scenario. In this section we present a simple yet effective scheme for allowing the insertion of vertices in an orthogonal drawing so that the maximum degree of any vertex in the drawing at any time is less than or equal to four. In the description of our interactive drawing scheme, we assume that we build a graph from scratch. If a whole subgraph needs to be drawn initially, we can draw it by simulating the above scenario, inserting one vertex at a time. We assume that the graph is always connected.

An embedded vertex v has a *free direction* to the right (bottom) when the grid edge that is adjacent to the right (bottom) of v is not covered by any graph edge. Let v be the next vertex to be inserted in the current graph. The number of vertices in the current graph that v is connected to, is called the *local degree* of v, and is denoted by $local\_degree(v)$.

Since the graph is always connected, we only consider the case where an inserted vertex has local degree one, two, three or four, except for the first vertex inserted in an empty graph. In order to prove our results, as vertices are inserted in the drawing, we maintain the following invariants:

- Every vertex of the current drawing of degree one or two has at least one free direction to the bottom and at least one free direction to the right of the grid point where the vertex is placed.
- Every vertex of the current drawing of degree three has a free direction either to the bottom or to the right of the grid point where the vertex is placed.

Figures 2a and 2b show the first two vertices inserted in an empty graph. Notice that after vertices v_1 and v_2 are inserted, they both satisfy the invariants set above. Different embeddings of the first two vertices are possible but the edge that connects them always has to have one bend in the way shown in Fig. 2b. If a straight no-bend line is used to connect v_1 and v_2, at least one of these two vertices will not satisfy the first invariant.

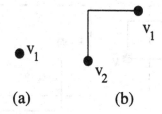

Fig. 2. Inserting the first two vertices in an empty graph.

Let us assume that v_i is the next vertex to be inserted in the current drawing. We distinguish the following cases:

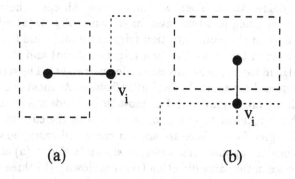

Fig. 3. The insertion of local degree one vertex v_i requires one one column and one row.

1. v_i has local degree one. There are two cases which are shown in Figs. 3a and 3b. At most one new column and one new row are required, and at most one bend is introduced. Notice that this bend is introduced along an edge which is incident to v_i and whose other end is open. In Fig. 3a the vertex will have one free direction to the bottom and two to the right. The second free direction to the right (which is responsible for introducing an extra row and bend to the drawing) will be inserted in the drawing later and only if v_i turns out to be a full blown degree four vertex. We take a similar approach for the second downward free direction of v_i of Fig. 3b.

2. v_i has local degree two. There are four cases. We have shown two cases in Figs. 4a and 4b (the other two are symmetric and are treated in a similar fashion). At most one new row and one new column is required, and at most two bends are introduced along edges which are incident to v_i and connect v_i with the current drawing.

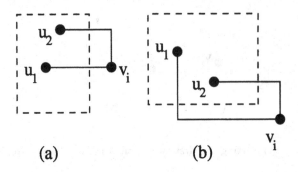

Fig. 4. (a) The insertion of local degree two vertex v_i requires one column; (b) the insertion of v_i now requires one column and one row.

3. v_i has local degree three. There are nine cases. All cases, however, can be treated by considering just two cases, as shown in Fig. 5: (a) all the vertices have a free edge in the same direction (right or down), and (b) two vertices have a free edge in the same direction (right or down) and the other vertex has a free edge in the opposite direction (down or right). The rest of the cases are symmetric and are treated in a similar fashion. At most one new row and one new column are required, and at most three bends are introduced along edges which are incident to v_i and connect v_i with the current drawing.

4. v_i has local degree four. There are sixteen cases. All cases, however, can be treated by considering just three cases, as shown in Fig. 6: (a) all the vertices have a free edge in the same direction (right or down), (b) three vertices have a free edge in the same direction (right or down) and one vertex has a free edge in the other direction (down or right), and (c) two vertices have a free edge in the same direction (right or down) and the other two vertices have a free edge in the other direction (down or right). The symmetric cases are treated in a similar fashion. At most two new rows and two new columns are required, and at most six bends are introduced along edges which are incident to v_i and connect v_i with the current drawing.

As we described above, the easiest way to handle deletions is to delete vertices/edges from the data structures without changing the coordinates of the rest of the drawing. Occasionaly, or on demand, the system can perform a linear-time compaction similar to the one described in [19], and refresh the screen.

Lemma 1. *The total number of bends introduced by the algorithm for the "No-Change scenario" up to time t is at most $2.66n(t) + 2$, where $n(t)$ is the number of vertices at time t.*

Theorem 2. *There exists a simple interactive orthogonal graph drawing scheme for the "No-Change scenario" with the following properties:*

1. every insertion operation takes constant time,

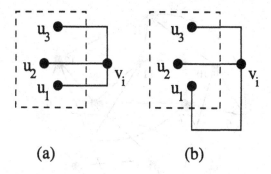

Fig. 5. (a) The insertion of local degree three vertex v_i requires one column; (b) the insertion of v_i now requires one row and one column.

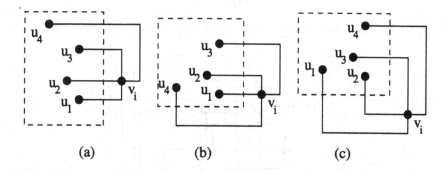

Fig. 6. (a) The insertion of local degree four vertex v_i requires two columns; (b) the insertion of v_i requires two columns and one row; (c) the insertion of v_i requires two columns and two rows.

2. every edge has at most three bends,
3. the total number of bends at any time t is at most $2.66n(t) + 2$, where $n(t)$ is the number of vertices of the drawing at time t, and
4. the area of the drawing at any time t is no more than $(n(t) + n(t)_4)^2$, where $n(t)_4$ is the number of vertices of local degree 4 which have been inserted up to time t.

The interactive scheme we just described is simple and efficient. The area and bend bounds are higher than the best known [1, 2, 14, 15], but we have to consider that this is a scheme that gives the user a lot of flexibility in inserting any node at any time. Moreover, any insertion takes place without disturbing the current drawing, since the insertion is built around it. Besides, $n(t)_4$ is the number of vertices of local degree four which have been inserted up to time t, and not the total number of vertices of degree four in the graph. The area will be much smaller if the user chooses an insertion strategy which keeps $n(t)_4$ low.

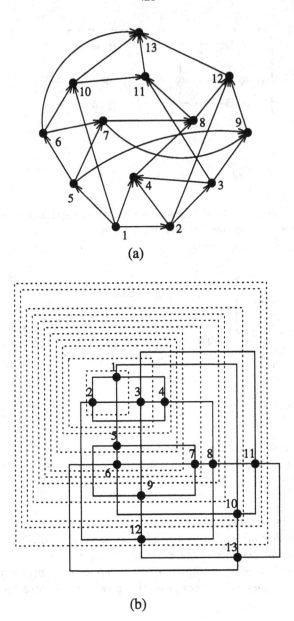

Fig. 7. (a) A regular graph of degree 4 with 13 vertices, (b) drawing the graph with the algorithm of the No-Change scenario

Notice that it is possible to reuse rows and columns on which other vertices have been placed before. Although we cannot guarantee that this will always happen, the interactive drawing program should be able to see if a reuse is possible during an insertion, and take advantage of it.

Depending on the situation, we can make slight changes to the invariants to improve the quality of the drawing. In Fig. 7 we show an example of our technique when applied on a graph that we know in advance. This is a regular degree 4 graph, has 13 vertices and is shown in Fig. 7a, together with an st-numbering for it. We simulate our algorithm for the No-Change scenario and we insert the vertices following the st-numbering, starting with an empty drawing. The final drawing has width 10 and height 11 and is demonstrated in Fig. 7b. Notice that the insertion of vertex 3 allowed vertex 2 to have two free directions to the bottom. In this way, we had a more efficient placement for vertex 4. Finally, the dotted boxes denote the current drawing at all intermediate steps, and we can see that it always remains unaltered.

5 Conclusions and Open Problems

We presented some preliminary results on interactive orthogonal graph drawing. Our algorithm for the No-Change scenario guarantees that the area of the drawing at any time t is no more than $(n(t) + n(t)_4)^2$, where $n(t)$ is the total number of vertices at time t, and $n(t)_4$ is the total number of vertices up to time t, that had degree four *when they were inserted*. In the same time, our algorithm guarantees no more than 3 bends per edge, while keeping the total number of bends at low levels (at most $2.66n(t) + 2$).

It would be interesting to analyze the Relative-Coordinates scenario and compare its performance with that of the No-Change scenario. In the future we will study algorithms that allow the degree to increase arbitrarily. Also, techniques for interactive graph drawing in other standards (straight line, polyline, etc.) are needed. Since it is counterproductive for the user to spend a significant amount of time to "relearn" the new drawing, the main target is to produce a drawing that is as close to the drawing before the update as possible.

Acknowledgement

We would like to thank Brendan Madden and Roberto Tamassia for helpful discussions.

References

1. Therese Biedl, *Embedding Nonplanar Graphs in the Rectangular Grid*, Rutcor Research Report 27-93, 1993.
2. T. Biedl and G. Kant, *A Better Heuristic for Orthogonal Graph Drawings*, Proc. 2nd Ann. European Symposium on Algorithms (ESA '94), Lecture Notes in Computer Science, vol. 855, pp. 24-35, Springer-Verlag, 1994.

3. R. Cohen, G. DiBattista, R. Tamassia, and I. G. Tollis, *Dynamic Graph Drawing* to appear in SIAM Journal on Computing.

4. G. DiBattista, P. Eades, R. Tamassia and I. Tollis, *Algorithms for Drawing Graphs: An Annotated Bibliography*, Computational Geometry: Theory and Applications, vol. 4, no 5, 1994, pp. 235-282. Also available via anonymous ftp from ftp.cs.brown.edu, gdbiblio.tex.Z and gdbiblio.ps.Z in /pub/papers/compgeo.

5. G. DiBattista, G. Liotta and F. Vargiu, *Spirality of orthogonal representations and optimal drawings of series-parallel graphs and 3-planar graphs*, Proc. Workshop on Algorithms and Data Structures, Lecture Notes in Computer Science 709, Springer-Verlag, 1993, pp. 151-162.

6. S. Even and G. Granot, *Rectilinear Planar Drawings with Few Bends in Each Edge*. Tech. Report 797, Comp. Science Dept., Technion, Israel Inst. of Tech., 1994.

7. S. Even and R.E. Tarjan, *Computing an st-numbering*, Theor. Comp. Sci. 2 (1976). pp. 339-344.

8. A. Garg and R. Tamassia, *On the Computational Complexity of Upward and Rectilinear Planarity Testing*, Proc. DIMACS Workshop GD '94, Lecture Notes in Comp. Sci. 894, Springer-Verlag, 1994, pp. 286-297.

9. Goos Kant, *Drawing planar graphs using the lmc-ordering*, Proc. 33th Ann. IEEE Symp. on Found. of Comp. Science, 1992, pp. 101-110.

10. F. T. Leighton, *New lower bound techniques for VLSI*, Proc. 22nd Ann. IEEE Symp. on Found. of Comp. Science, 1981, pp. 1-12.

11. Charles E. Leiserson, *Area-Efficient Graph Layouts (for VLSI)*, Proc. 21st Ann. IEEE Symp. on Found. of Comp. Science, 1980, pp. 270-281.

12. Thomas Lengauer, *Combinatorial Algorithms for Integrated Circuit Layout*, John Wiley and Sons, 1990.

13. K. Miriyala, S. W. Hornick and R. Tamassia, *An Incremental Approach to Aesthetic Graph Layout*, Proc. Int. Workshop on Computer-Aided Software Engineering (Case '93), 1993.

14. A. Papakostas and I. G. Tollis, *Algorithms for Area-Efficient Orthogonal Drawings*, journal version, in preparation.

15. A. Papakostas and I. G. Tollis, *Improved Algorithms and Bounds for Orthogonal Drawings*, Proc. DIMACS Workshop GD '94, Lecture Notes in Comp. Sci. 894, Springer-Verlag, 1994, pp. 40-51.

16. Markus Schäffter, *Drawing Graphs on Rectangular Grids*, Discr. Appl. Math. (to appear).

17. J. Storer, *On minimal node-cost planar embeddings*, Networks 14 (1984), pp. 181-212.

18. R. Tamassia, *On embedding a graph in the grid with the minimum number of bends*, SIAM J. Comput. 16 (1987), pp. 421-444.

19. R. Tamassia and I. Tollis, *Planar Grid Embeddings in Linear Time*, IEEE Trans. on Circuits and Systems CAS-36 (1989), pp. 1230-1234.

20. R. Tamassia, I. Tollis and J. Vitter, *Lower Bounds for Planar Orthogonal Drawings of Graphs*, Information Processing Letters 39 (1991), pp. 35-40.

21. L. Valiant, *Universality Considerations in VLSI Circuits*, IEEE Trans. on Comp. vol. C-30, no 2, (1981), pp. 135-140.

Automatic Drawing of Compound Digraphs for a Real-Time Power System Simulator

Gilles Paris

IREQ (Institut de recherche d'Hydro-Québec)

1800, montée Ste-Julie, Varennes (Québec) Canada J3X 1S1

Abstract. Two implemented versions of a compound digraph algorithm using an object-oriented approach are shown for use in the field of automatic drawing of electrical circuit diagrams. One is a test version with minimal user interface, the other a full-fledged editor part of a graphical user interface project currently under development. Orthogonal drawing of edges and interactive implosion and explosion of compound vertices are key features of the application.

1 Introduction

This demonstration shows two implementations of the algorithm published in [1] for the automatic drawing of *compound digraphs* (directed graphs that have both *inclusive* and *adjacent* relations among vertices). This algorithm is to be included in IGSIM [2], a graphical user interface for IREQ's real-time power system simulator. The work presented here is still preliminary and will evolve along with IGSIM.

2 Application Context and Justification

2.1 The Hybrid Real-Time Power System Simulator

The existing simulator is constructed with both digital and analog components, and so has hardware control panels, but the next version, still a prototype, will be fully digital, and will rely mostly on software control panels and graphical user interface.

Although well equipped with control, database and plotting software, the simulator is lacking a schematic-based user interface, so that general-purpose text-processing software has to be used to produce diagrams as documentation only.

2.2 Single-Line Electrical Diagrams

The main type of diagram used in power studies is the single-line schematic diagram where vertices are either electrical symbols or stretchable bars, while edges are usually represented orthogonally. More general electronic circuit schematics may also be used at a lower level, but they are not considered as part of the automatic drawing, and are represented as a fixed-content compound vertex. This facility is supported but not yet sufficiently developed to be discussed further.

2.3 Other Types of Diagrams

Block diagrams are used frequently in power system analysis, so representations of a type similar to [1] are also possible with this automatic drawing facility. The implementation currently is in one of two modes: *block diagram* or *single-line diagram*.

2.4 Use of Circuit Analysis Programs

An important part of power system design is done with highly specialized software tools such as the ElectroMagnetic Transients Program (EMTP), MatLab, and Stability Study Program, some of which have low-level interface, so there is a need for an automatically generated schematic derived from the many existing input files.

2.5 Why Directed Graphs?

Electrical circuits are undirected by nature, but there is a tendency, because of the topographical characteristics of our network, to order components from generating to consuming centers. This feature may be used to specify the layout in a mostly directed way. Remaining cycles in the graph are automatically resolved by the algorithm itself.

2.6 High Performance Requirements

Any interactive change to a compound digraph forces a total redrawing, so care was taken to implement the algorithm efficiently, using inline functions, pointers and integer arithmetic as much as possible, as well as minimizing structure traversals and recomputations. Iteration limits may also be set as compromise between speed and readability.

3 Reusable Object-Oriented Implementation Using C++

3.1 Independence of Graphical User Interface Development

One goal of this approach was to separate this work from IGSIM development.

3.2 Independence of Drawing Primitives

An alternative goal was to isolate in a separate C++ class the drawing primitives, so that specific classes may be derived for integration in any program written in C++. This is shown by the two different versions of the algorithm in this demonstration.

3.3 Independence of Input Formats

A third goal was to remain independent of any specific input format for graph description. This goal was met by designing a simple tree language using overloaded C++ operators. This feature, along with C++ stream operators, means that a graph (or compound vertex) specification may reside in memory, be directly coded as an expression in the application, or be input from a file.

4 Extensions to the Original Algorithm

The direction of edges may be vertical or horizontal. In the following discussion, vertical direction is assumed. The metrical layout portion of [1] has been modified as follows to support the single-line style.

4.1 Separation of Multiple Input and Output Connections to Vertices

In single-line mode, a vertex having multiple input and/or output edges connected to it is stretched horizontally and edges are ordered according to the x position of connected vertices.

4.2 Orthogonal Edge Drawing

Oblique edges generated by [1] are drawn with three orthogonal segments. A separate horizontal corridor is allocated for each vertex at the same compound level and edges connected to a vertex are laid out in it. This also stretches the drawing vertically.

5 The First Version: An RPC-Based Test Implementation

This version draws directly on an event-driven canvas and receives interactive commands from a separate process using Remote Procedure Calls (RPC). Vertices are drawn as simple labeled rectangles. Positions are in pixels.

5.1 Goal

The goal was to have an independent test-bench for the algorithm along with a method of exercising interactive operations without having to write a proper user interface. Fast unstructured graphics isolate the performance of the algorithm from the drawing itself and thus ease subjective performance evaluation with interactive operations.

5.2 Interactive Commands

Commands consist of a single letter followed by arguments which may be simple integers, strings, or arbitrary length constructor expressions of the input language. Possibilities are *creation/destruction* of lists of edges and of sub-trees or single vertices, *implosion/explosion* and *visibility* of compound vertices, *structural changes* of the inclusion relation, *direction* and *mode* of drawing. A number of *built-in test graphs* may also be constructed by invoking the corresponding function name.

Commands may be used to change the size and labeling of rectangles, vary iteration limits and distances separating rectangles in and between compound levels.

6 The Second Version: Integration with an Unidraw-Based Editor

6.1 A Snapshot of Work in Progress for IGSIM

This editor is based on Unidraw [3]. Because IGSIM is undergoing changes, a snapshot version was taken for this implementation, and does not include control panels. It is a generic schematic editor.

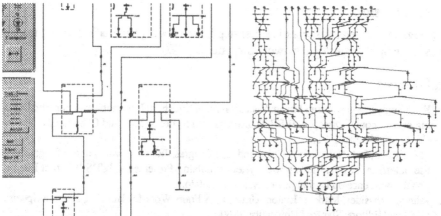

Fig. 1. Two views of Hydro-Québec's 735 kV network automatically drawn in the editor

6.2 Integration of Automatic Drawing Classes

All classes implementing the automatic drawing algorithm were first included without changing the editor. Code was then added to by the author to reimplement the drawing class by repositioning or modifying elements created with the editor.

A special menu was added to control the automatic drawing function. These exercise the same functions as the RPC-based version, except that selection is done directly on the drawing instead of referencing elements by their number.

Edges which are conceptually made of multiple line segments have to be reimplemented as a single multiline in the editor. Unidraw's concept of a generic multiline modified by a transformation matrix poses some distortion problems with orthogonal edge redrawing when moving connected objects. A temporary fix was included.

6.3 Structured Versus Unstructured Graphics

The use of hierarchical structured graphics is essential for picking objects, and Unidraw's implementation pays great attention to optimization of redrawing, but it is quite apparent that a loss of performance appears when repositioning large graphs.

7 Further Work

7.1 Filter Programs

A filter program transforming EMTP files in the input language is the next step.

7.2 Better Orthogonal Routing of Edges

Orthogonal drawing of edges adds too much vertical stretch to the drawing. Better use of each corridor must be made without impairing performance. The barycenter heuristic also creates ambiguities in some vertical paths for complex graphs by the alignment of unrelated vertices. These ambiguities do not appear with oblique edges.

7.3 Bidirectional Edge Drawing

A more difficult task is automatic drawing of edges in both directions. A simple way to approach this is by having the direction local to a compound vertex, but direction changing poses non-trivial problems which must first be addressed theoretically.

8 Conclusion

Although incomplete, this work is a first step in introducing automatic drawing techniques in the field of electrical network diagrams at IREQ.

References

1. Kozo Sugiyama and Kazuo Misue: Visualization of Structural Information: Automatic Drawing of Compound Digraphs. IEEE Transactions on Systems, Man and Cybernetics, Vol 21, No. 4, July/August 1991.
2. Marie Rochefort, Nathalie De Guise and Luc Gingras, IREQ: Development of a graphical user interface for a real-time power system simulator. Presented at ICDS '95, first International Conference on Digital power system Simulators.
3. John M. Vlissides, Mark A. Linton: Unidraw: A FrameWork for Building Domain-Specific Graphical Editors, Stanford University, 1989.

Validating Graph Drawing Aesthetics

Helen C. Purchase[1] and Robert F. Cohen[2] and Murray James[1]

[1] Department of Computer Science, University of Queensland,
St. Lucia, Queensland, 4072, Australia.
[2] Department of Computer Science, University of Newcastle,
Callaghan, New South Wales, 2308, Australia

Abstract. Designers of graph drawing algorithms and systems claim to illuminate application data by producing layouts that optimize measurable aesthetic qualities. Examples of these aesthetics include *symmetry* (where possible, a symmetrical view of the graph should be displayed), *minimize edge crossings* (the number of edge crossings in the display should be minimized), and *minimize bends* (the total number of bends in polyline edges should be minimized).

The aim of this paper is to describe our work to validate these claims by performing empirical studies of human understanding of graphs drawn using various layout aesthetics. This work is important since it helps indicate to algorithm and system designers what are the aesthetic qualities most important to aid understanding, and consequently to build more effective systems.

1   Introduction

The visualization produced by a graph drawing subsystem should *illuminate* application data. That is, it should help the user to understand and remember the information being visualized. A good layout can be a picture worth a thousand words; a poor layout can confuse or mislead. Designers of graph drawing algorithms and systems claim to illuminate application data by producing layouts that optimize measurable aesthetic qualities. Examples of these aesthetics include *symmetry* (where possible, a symmetrical view of the graph should be displayed, see e.g. [4, 7]), *minimize edge crossings* (the number of edge crossings in the display should be minimized, see e.g. [5]), and *minimize bends* (the total number of bends in polyline edges should be minimized, see e.g. [10, 11]).

The aim of this paper is to describe our work to validate these claims by performing empirical studies of human understanding of graphs drawn using various layout aesthetics. This work is important since it helps indicate to algorithm and system designers what are the aesthetic qualities most important to aid understanding, and consequently to build more effective systems.

There has been very little work in this area. This is the first study to address the effect of aesthetics on general understanding of graphs. Prior work has focused on specific applications [1, 3] and comparisons of two- and three-dimensional layouts [12].

Although many algorithms consider more than one aesthetic in their attempt to create an illuminating graph display, this initial study focuses on individual aesthetic principles, leaving the testing of the effectiveness of the various algorithms which embody them for a further project. We study the following popular graph drawing aesthetics: minimize edge crossings, minimize bends, and maximize symmetries. Our initial results are quite encouraging. Our results indicate that minimizing edge bends and minimizing edge crossings are both important aids to human understanding. However, our study is inconclusive as to the importance of maximizing symmetry. This final non-result is surprising to us, since we believed that maximizing symmetry would have a high impact on the understandablity of drawings.

The rest of the paper is organized as follows. Section 2 describes the scope of our experiment, section 3 explains our methodology, section 4 presents our results, and section 5 describes our analysis. Finally, in section 6 we give conclusions and directions for future work.

2   Scope

2.1   Hypotheses

Three common graph-drawing aesthetics were chosen for this study: *symmetry* (where possible, a symmetrical view of the graph should be displayed), *minimize edge crossings* (the number of edge crossings in the display should be minimized), and *minimize bends* (the total number of bends in polyline edges should be minimized). We restricted our study to undirected graphs. Other aesthetics that were considered were the direction of flow for directed graphs, and maximising the minimum angle between arcs from one node: these have been left for a later study.

Thus, our hypotheses were:

- Increasing the number of arc bends in a graph drawing decreases the understandability of the graph
- Increasing the number of arc crossings in a graph drawing decreases the understandability of the graph
- Increasing the local symmetry displayed in a graph drawing increases the understandability of the graph

To measure the understandability of a graph drawing, three graph-theoretic questions were chosen. The performance of a person in answering these questions about a graph drawing was considered an appropriate indicator of their understanding of the drawing. The three questions chosen were:

- How long is the shortest path between two given nodes?
- What is the minimum number of nodes that must be removed in order to disconnect two given nodes such that there is no path between them?
- What is the minimum number of arcs that must be removed in order to disconnect two given nodes such that there is no path between them?

2.2 Graphs

A dense graph and a sparse graph were defined. The sparse graph has 16 nodes and 18 arcs. The dense graph was created by adding 10 extra arcs onto the sparse graph.

Nine drawings of each graph were created, with the number of bends, crossings and the amount of perceived symmetry varied appropriately. Metrics for the number of crossings and bends were easy: we merely counted them; determining a metric to measure the amount of symmetry in a graph drawing was more difficult. We needed a metric that could give a numerical value to perceived symmetry. The method used in this study is described in the appendix (see also [2]).

For each aesthetic, three drawings of each graph were produced, one with a small aesthetic measurement (*few*), one with an interim aesthetic measurement (*some*), and one with a large aesthetic measurement (*many*). The values of these measurements for the dense graph drawings were as follows:

dense graph drawing		bends	crossings	symmetry metric
dense-bends-few	(**dbf**)	6	0	0
dense-bends-some	(**dbs**)	18	0	0
dense-bends-many	(**dbm**)	30	0	0
dense-crossings-few	(**dcf**)	0	6	0
dense-crossings-some	(**dcs**)	0	24	0
dense-crossings-many	(**dcm**)	0	42	0
dense-symmetry-few	(**dsf**)	0	0	4.57
dense-symmetry-some	(**dss**)	0	0	25.7
dense-symmetry-many	(**dsm**)	0	0	51.0

The nine sparse graph drawings had a similar variation in the values of the aesthetics, maintaining the same *few:some:many* ratios as used in the dense graph drawings as much as possible (see Figs. 1 and 2).

3 Methodology

Experiments were conducted: 49 subjects worked on the dense graph, and 35 of these same subjects worked on the sparse graph a week later. The subjects were all second year computer science students, as representatives of a more general target population who use graph drawings in their work. They were each given an experimental booklet of the following form:

1. A brief description of graphs, and definitions of the terms *node, arc, path, path length*.
2. Explanation of the three graph-theoretic questions that they would be required to answer about the experimental graphs.

aesthetic	variation		
	few	some	many
bends	dbf	dbs	dbm
crossings	dcf	dcs	dcm
symmetry	dsf	dss	dsm

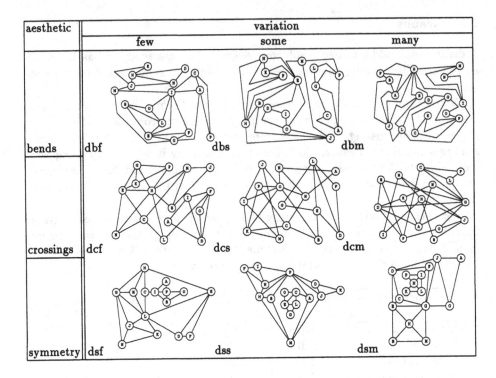

Fig. 1. The Experimental Graph Drawings: dense graph

3. A simple example graph drawing (including some bends, crossings and symmetry), with three representative questions and their correct answers.

   At this stage in the experiment, the subjects were asked if they had any questions about graphs in general, or about the experiment. It was important to ensure that all the subjects knew what was going to be expected of them.

4. Six drawings of two "practise" graphs (sparse: 20 nodes and 21 arcs; dense: 20 nodes and 32 arcs). Each graph was drawn three times: once with crossings, once with bends and once with some symmetry. The three graph theoretic questions were asked of each of these drawings. The purpose of this section was to get the subjects used to the nature of graph drawings and the time constraints, and to ensure that by the end of this section they were comfortable with the task, before tackling the nine experimental graphs. The subjects were unaware that these graph drawings were not experimental.

5. A "filler" task which engaged the subjects' mind on a small problem unrelated to graphs. This was to ensure that their performance on the subsequent experimental graphs was not affected by any follow-on effect from the practise graphs. We used a simple word puzzle.

6. The nine experimental graph drawings (all sparse, or all dense), in a random order for each subject.

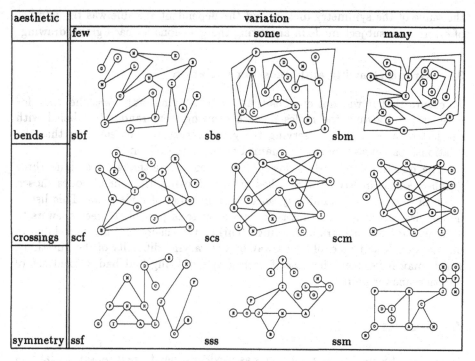

Fig. 2. The Experimental Graph Drawings: sparse graph

7. A questionnaire, asking the subjects about their current degree, how far they were into their degree, any previous experience of reading graphs, and if there were any particular features about the graphs that they could identify as assisting or hindering their performance (this information was for our own interest only: it was not used in the quantitative analysis of the results).

The time allowed for each component of the experiment was as follows:

Task	Time: dense graph	Time: sparse graph
Introduction and example	As necessary	As necessary
Practise graph 1	60 seconds	45 seconds
Practise graph 2	60 seconds	45 seconds
Practise graph 3	55 seconds	40 seconds
Practise graph 4	55 seconds	40 seconds
Practise graph 5	50 seconds	35 seconds
Practise graph 6	50 seconds	35 seconds
Filler task	60 seconds	60 seconds
Experimental graphs 1–9	45 seconds	30 seconds

The experiments were thus controlled for time, the graph structure and the questions; the independent variables were the number of bends, crossings and the value of the symmetry metric, and the dependent variable was the number of errors each subject made in answering the questions for each graph drawing.

3.1 Graph Drawing and Question Variation

To ensure that it was not obvious that the underlying graph was the same for each graph drawing, the nodes in each drawing were randomly labeled with the letters A – P, thus preserving the graph structure, but ensuring that the relationships between particular named nodes were different.

The questions themselves were randomised too: although the same three graph-theoretic questions were asked of each drawing, the pair of nodes chosen for each question was randomly selected from a list of node-pairs. This list of node-pairs was defined so that there was a range of possible correct answers to each question for each graph drawing.[3] This was to ensure that any variability in our data could not be explained away by the varying difficulty of the questions, which may have been the case if each graph drawing had had a fixed set of question-answer pairs.

4 Results

A within-subjects analysis method was used in order to reduce any variability that may have been attributable to the difference between the subjects (age, experience *etc*). Any learning effect was minimised by of the large number of graphs used in the experiment, the inclusion of the practise graphs, and the randomisation of the ordering of the graph drawings.

For each graph, the results for each aesthetic were collated and ranked. As an example, Table 1 shows the results for the *dense-bends* drawings, and the associated ranking for each subject.

The *Friedman analysis of variance by ranks* test was used to determine the χ^2 value for each aesthetic, and the probability that the ranks were produced by chance (p). The results were:

graph	aesthetic	χ^2	p
dense	bends	17.48	< 0.001
	crossings	23.09	< 0.001
	symmetry	2.3	< 0.3
sparse	bends	10.99	< 0.01
	crossings	10.56	< 0.01
	symmetry	0.27	< 0.9

[3] Thus, for each graph drawing, four node-pairs were defined that would give correct answers to the first question of either 2, 3, 4 or 5, two node-pairs were defined that would give correct answers to the second question of either 1 or 2, and three node-pairs were defined that would give correct answers to the third question of either 1, 2 or 3.

With a level of significance set at $\alpha = 0.05$, these results are therefore significant for bends and crossings in both the dense and sparse graphs, as $p < \alpha$ in both these cases.

Table 1. subject scores and rankings for *dense-bends* drawings

Subject	Scores			Rankings			Subject	Scores			Rankings		
	dbf	dbs	dbm	dbf	dbs	dbm		dbf	dbs	dbm	dbf	dbs	dbm
1	3.0	3.0	3.0	2.0	2.0	2.0	26	3.0	2.0	2.0	1.0	2.5	2.5
2	3.0	3.0	0.0	1.5	1.5	3.0	27	3.0	3.0	2.0	1.5	1.5	3.0
3	3.0	1.0	1.0	1.0	2.5	2.5	28	2.0	2.0	2.0	2.0	2.0	2.0
4	2.0	2.0	1.0	1.5	1.5	3.0	29	2.0	1.0	2.0	1.5	3.0	1.5
5	0.0	0.0	0.0	2.0	2.0	2.0	30	3.0	3.0	1.0	1.5	1.5	3.0
6	2.0	2.0	1.0	1.5	1.5	3.0	31	2.0	2.0	1.0	1.5	1.5	3.0
7	1.0	0.0	1.0	1.5	3.0	1.5	32	2.0	2.0	0.0	1.5	1.5	3.0
8	2.0	1.0	1.0	1.0	2.5	2.5	33	2.0	3.0	3.0	1.0	2.5	2.5
9	0.0	1.0	0.0	2.5	1.0	2.5	34	3.0	3.0	3.0	2.0	2.0	2.0
10	0.0	1.0	1.0	3.0	1.5	1.5	35	3.0	3.0	3.0	2.0	2.0	2.0
11	3.0	3.0	1.0	1.5	1.5	3.0	36	2.0	2.0	2.0	2.0	2.0	2.0
12	3.0	3.0	3.0	2.0	2.0	2.0	37	2.0	3.0	2.0	2.5	1.0	2.5
13	3.0	3.0	1.0	1.5	1.5	3.0	38	2.0	3.0	1.0	2.0	1.0	3.0
14	3.0	2.0	1.0	1.0	2.0	3.0	39	3.0	3.0	2.0	1.5	1.5	3.0
15	2.0	2.0	1.0	1.5	1.5	3.0	40	3.0	2.0	1.0	1.0	2.0	3.0
16	2.0	3.0	0.0	2.0	1.0	3.0	41	3.0	2.0	2.0	1.0	2.5	2.5
17	2.0	1.0	1.0	1.0	2.5	2.5	42	2.0	2.0	1.0	1.5	1.5	3.0
18	1.0	2.0	2.0	3.0	1.5	1.5	43	1.0	2.0	2.0	3.0	1.5	1.5
19	1.0	1.0	0.0	1.5	1.5	3.0	44	2.0	1.0	1.0	1.0	2.5	2.5
10	3.0	2.0	2.0	1.0	2.5	2.5	45	1.0	3.0	2.0	3.0	1.0	2.0
21	2.0	1.0	2.0	1.5	3.0	1.5	46	1.0	2.0	2.0	3.0	1.5	1.5
22	3.0	3.0	2.0	1.5	1.5	3.0	47	3.0	3.0	3.0	2.0	2.0	2.0
23	2.0	2.0	1.0	1.5	1.5	3.0	48	3.0	3.0	2.0	1.5	1.5	3.0
24	3.0	2.0	1.0	1.0	2.0	3.0	49	2.0	3.0	0.0	2.0	1.0	3.0
25	3.0	0.0	1.0	1.0	3.0	2.0							

Figs. 3 and 4 show the trends in the average number of errors made by each subject, when plotted against the variation in each aesthetic.

One of the questions we asked the subjects at the end of the questionnaire was: "Can you identify any characteristics of the graph drawings that made them more difficult to read?" Responses to this question included:

- "Many interwoven arcs like spiderwebs"
- "Lines crossing over"
- "Crooked lines"
- "The zig-zag lines and lack of efficient structure"
- "Disorganised, unstructured"

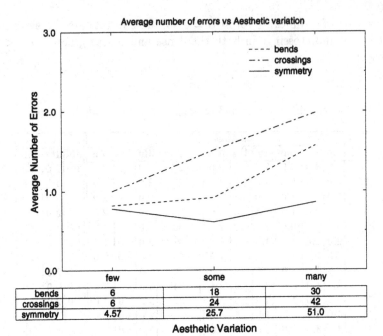

Fig. 3. Results for the dense graph

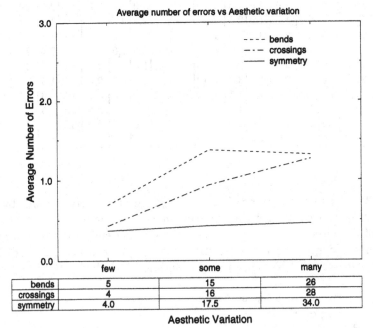

Fig. 4. Results for the sparse graph

5 Analysis

Two of our hypotheses are confirmed in both sparse and dense graphs (crossings and bends), but the symmetry hypothesis is inconclusive. We believe that this is due to a 'ceiling effect'. The length of time for the subjects to answer the questions (45 seconds for dense, 30 seconds for sparse), was too generous: most of the subjects managed to get all the questions right for the symmetrical graphs (even those which had a low measure of symmetry). This meant that there was no variability in the symmetry data, and thus the results of this analysis are not significant. In pilot tests, when we used times as high as 75 seconds for dense graphs and 60 seconds for sparse graphs, we observed a similar ceiling effect on the graph drawings for all three aesthetics.

Additionally, symmetry is very different in nature to the other two aesthetics considered here. Other significant points which may explain why the symmetry hypothesis was not as conclusive as the other two hypotheses are:

- The symmetry metric we defined may not adequately quantify perceptual symmetry. In future work we will examine weighting the metric by the proportion of nodes involved in symmetrical relations as well as the proportion of arcs. We also only considered local symmetry: it is possible that running similar experiments with global symmetries might produce better results.

- Secondly, these experiments tested the ability of the subjects to understand the structure of graph drawings. It is possible that maximising the amount of perceptual symmetry in a graph drawing is of more benefit when the symmetry is used for the perceptual organisation of semantic concepts instead of merely the organisation of the graph structure. Future experiments will test subjects' ability to understand the meaning of graph drawings from specific application areas.

- In controlling the experiment, we avoided introducing potential interaction between the aesthetics by having no variation of two aeshetics while varying the third. For example, all the graph drawings that varied the number of bends had no crossings and a zero value for symmetry. There is a potential conflict here between the nature of our three hypotheses: with bends and crossings, the hypothesis was that *increasing* their number would result in increased errors; with symmetry, the hypothesis was that *decreasing* the value of the metric would result in increased errors. By keeping the number of crossings and bends at zero for the symmetrical graph drawings, the drawings were being made simpler. This is a contrast to the bends and crossings diagrams, where keeping the symmetry metric at zero made the drawings more complex. Therefore, perhaps it is no surprise that a ceiling effect was observed with the reading of the symmetry drawings, when the same time period was used for reading all the graphs of the three aesthetics.

It is possible that, although we cannot state conclusively that the symmetry hypothesis is confirmed, it may indeed be the case that increasing the number of crossings and bends is more detrimental to the understandability of a graph drawing than a reduction in symmetry.

6   Conclusions and Further Work

In this paper, we have described our study to validate important graph drawing aesthetics. Our study confirms two of our hypotheses:

- Increasing the number of arc bends in a graph decreases the understandability of the graph.
- Increasing the number of arc crossings in a graph decreases the understandability of the graph.

These results indicate that the general intuition that edge bends and edge crossings should be minimized is correct.

Our third hypothesis remains unconfirmed:

- Increasing the local symmetry displayed in a graph increases the understandability of the graph.

We plan to conduct a further study to investigate this hypothesis further. We are building an interactive system to conduct experiments which will allow us to get more valid and conclusive results, and to expand our studies. It will enable us to:

- measure the time taken to answer the questions correctly as the dependent variable, rather than the number of errors made within a set time
- study the effect of aesthetics on human understanding of graphs drawn on a display device
- compare the understandability of drawings on paper and on a display
- design and perform other experiments

Future work includes the following:

- studying other graph drawing aesthetics (for example, orthogonality, maximizing the minimum angle, etc)
- studying the effectiveness of various graph drawing algorithms
- studying the effectiveness of graph drawing aesthetics when used in application domians (for example, software enginnering design diagrams)
- studying the effect of aesthetics in three-dimensional drawings.

Appendix: A Metric for Symmetry

A measurement of the amount of symmetry in a graph drawing was an essential part of the study: we needed to be able to vary the symmetry displayed in a graph drawing, and to have some metric by which we could measure it. For the purposes of this experiment, symmetry is quantified by [2]:

1. counting all the symmetrical relationships about each axis of symmetry.
2. adding the symmetry values from all the axes together.

3. multiplying this total by the proportion of the graph drawings' arcs which are symmetrical objects.

This resulted in the following formula for measuring symmetry:

$$\sigma = \left(\frac{A_o}{A_{total}}\right) \times \sum_{i=1}^{M} \left(\frac{A_{di}}{2} + \frac{N_i}{2} + A_{si}\right)$$

where:

- σ is the measure of symmetry; the higher the value of σ, the more symmetrical the drawing
- A_o is the number of arcs in the graph which are symmetrical objects
- A_{total} is the total number of arcs in the graph
- M is number of axes of symmetry
- A_{di} is the number of double symmetrical arcs about axis i
- N_i is number of symmetrical nodes about axis i
- A_{si} is the number of single symmetrical arcs about axis i

It appears that this metric gives an appropriate measure of the perceived symmetry of a graph drawing [2].

Definitions

An *axis of symmetry* is defined as a straight line that has a minimum of three symmetrical objects (nodes (N), singly symmetrical arcs (A_s) and doubly symmetrical arcs (A_d)) mirrored on either side. The minimum of three symmetrical objects prevents an axis of symmetry being drawn at right angles, midway through every arc in a graph. Thus preventing every arc and pair of directly connected nodes contributing to symmetry, regardless of the overall appearance of the graph.

Around any axis of symmetry is a group of symmetrical objects. Collectively these objects make a *locality of symmetry*. While a *locality of symmetry* is defined by one axis of symmetry, the same group of symmetrical objects can have numerous axes. There is no restriction on the direction of these axes, but for each *locality of symmetry*, no two axes can be parallel. This prevents the same symmetrical relationship being counted over multiple axes.

The following objects are considered *symmetrical objects*:

- A double symmetrical arc (A_d) is an arc that is mirrored by another arc about the axis of symmetry.
- A single symmetrical arc (A_s) is a single arc that is bisected at right angles by the axis of symmetry.
- An axially symmetrical arc (A_a) is an arc that runs along the axis of symmetry.
- A symmetrical node (N) is a node that is mirrored by another node about the axis of symmetry.

The following three relationships are considered *symmetrical relationships*:

- Two mirrored double symmetrical arcs (A_d).
- Both halves of a single symmetrical arc (A_s).
- Two mirrored symmetrical nodes (N).

The final metric value is obtained by multiplying the number of symmetrical relationships in the graph by the *proportion of arcs which are symmetrical*. Thus, axially symmetrical arcs (A_a) contribute to this proportion even though they are not involved in a symmetrical relationship. If all the arcs in a graph are symmetrical objects then σ is simply the total number of symmetrical relationships over all axes of symmetry.

Acknowledgments

The authors would like to thank the following people and organizations for their helpful advice and participation: Shelina Bhanji, Julie McCreddon, Tim Mansfield, and the staff and students of the University of Queensland Computer Science Department. This work is funded by the Australian Research Council.

References

1. C. Batini, L. Furlani, and E. Nardelli. What is a good diagram? a pragmatic approach. In *Proc. 4th Int. Conf. on the Entity Relationship Approach*, 1985.
2. S. Bhanji, H.C. Purchase, R.F. Cohen, and M. James. Validating graph drawing aesthetics: A pilot study. Technical Report 336, University of Queensland Department of Computer Science, 1995.
3. C. Ding and P. Mateti. A framework for the automated drawing of data structure diagrams. *IEEE Transactions on Software Engineering*, SE-16(5):543–557, 1990.
4. P. Eades. A heuristic for graph drawing. *Congressus Numerantium*, 42:149–160, 1984.
5. D. Ferrari and L. Mezzalira. On drawing a graph with the minimum number of crossings. Technical Report 69-11, Istituto di Elettrotecnica ed Elettronica, Politecnico di Milano, 1969.
6. R. Gottsdanker. *Experimenting in Psychology*. Prentice-Hall, 1978.
7. R. Lipton, S. North, and J. Sandberg. A method for drawing graphs. In *Proc. ACM Symp. on Computational Geometry*, pages 153–160, 1985.
8. G.L. Lohse, K. Biolsi, N. Walker, and H.H. Rueter. A classification of visual representations. *Communications of the ACM*, 37(12):36–49, December 1994.
9. S. Siegel. *Nonparametric Statistics for the Behavioral Sciences*. McGraw-Hill, 1956.
10. R. Tamassia. On embedding a graph in the grid with the minimum number of bends. *SIAM J. Computing*, 16(3):421–444, 1987.
11. H. Trickey. Drag: A graph drawing system. In *Proc. Int. Conf. on Electronic Publishing*, pages 171–182. Cambridge University Press, 1988.
12. C. Ware, D. Hui, and G. Franck. Visualizing object oriented software in three dimensions. In *CASCON 1993 Proceedings*, 1993.

A Fast Heuristic for Hierarchical Manhattan Layout

G. Sander (sander@cs.uni-sb.de)

Universität des Saarlandes, FB 14 Informatik, 66041 Saarbrücken

Abstract. A fast heuristic for the layout of directed graphs according to Manhattan convention is presented. Nodes are placed into layers. Edges consist of sequences of vertical and horizontal segments. Sharing of segments is allowed in certain situations. The algorithm is an extension of the hierarchical layout method [11, 15] that includes crossing reduction and emphasis on a uniform edge orientation. Compared to the original algorithm, the time overhead is $O(n + ek)$ where n, e and k are the number of nodes, of edges, and the maximal number of line rows between two layers of nodes. It produces drawings where each edge has at most four bends.

1 Introduction

We address the problem of constructing drawings of graphs from the viewpoint of compiler construction where these drawings are used for demonstration, debugging and documentation of algorithms and data structures. Typically, these data structures are large and dense, while the construction of the drawing should be fast enough to support online debugging and algorithm animation.

In Manhattan drawings all edges consist of sequences of line segments which have a strict horizontal or vertical orientation. Manhattan layout (also called orthogonal layout) is widely used in VLSI design [1, 16]. However, the focus of VLSI design is different to our applications: First, in VLSI design, normally the nodes (terminals) have a fixed place, while only the edges (wires) need to be routed. Secondly, the cost, correctness and usability of the VLSI circuit (e.g., the minimizing of wire length [3]) has much more importance than the speed of the layout process. Typical VLSI routing times for the wires range up to 1h, while in interactive data structure visualization, the user should never wait more than a minute for the positioning of both nodes and edges.

There are good solutions for subclasses of graphs, mostly for grid embeddings. Vijayan and Wigderson [14] discuss the straight line embedding of graphs with n nodes in a square grid without edge crossings in time complexity $O(n^2)$. They also give a classification, which planar graphs can be handled. Tamassia's algorithm [12] allows edge bendings, thus it is suitable for all planar graphs with nodes of maximal degree 4. It runs in $O(n^2 log\, n)$ and produces a layout with minimal number of bends. A linear algorithm for 4-ary planar graphs is presented by Tamassia and Tollis [13], where the number of bends is not minimal but restricted by the upper bound $2.4n + 2$. Biedl and Kant [4] describe a linear time method to draw an undirected graph without loops or multiple edges, but with maximal node degree 4, in an $n \times n$ area with at most $2n + 2$ bends. This method is based on st-numbering. Nodes are drawn as points, i.e. have size zero. Even and Granot [5] present a similar approach for the case that nodes are structures of nonzero size. Orthogonal layouts for general graphs can be obtained by the layout method of Batini et al. [2]. Here, the graph is embedded in an orthogonal grid according to several aesthetics like minimization of edge crossings, of edge bends, of edge lengths, of the area, and placement of specified nodes on the external boundary. Since some of the aesthetics are contradicting, and some of the subproblems in this algorithm are NP-hard (e.g., the planarization problem to minimize edge crossings [8]), the layout method is a heuristic. This approach is perfectly suitable for undirected graphs. It is also applied to directed graphs, however it does not emphasize a uniform edge orientation. Protsko et al. [9] present a similar approach. They further sketch the idea how the uniform edge orientation can be specified by additional constraints.

Our preconditions are very special: We have large directed graphs (not necessarily planar or 4-ary). Nodes have nonzero size and different shapes. The layout should emphasize a uniform edge orientation and allow the grouping of nodes into the levels of a hierarchy. This implies a strong restriction of the y-position of the nodes to discrete coordinates, but there are only weak constraints to the x-position of the nodes. The layout algorithm should be very fast. Reducing the number of edge crossings, the number of bends in edges or the area of the drawing is useful, but has not the first priority.

Why do we prefer orthogonal edge segments: Manhattan layout gives a very uniform aesthetics which is in particular of interest for annotated control flow graphs (Fig. 1). Most of hand drawn control flow diagrams in the literature use Manhattan convention, thus programmers are familiar with this aesthetics. Edges tend to be bundled like busses such that it is easy to follow chains of long edges. The visual resolution is very high even if the graphs are very dense.

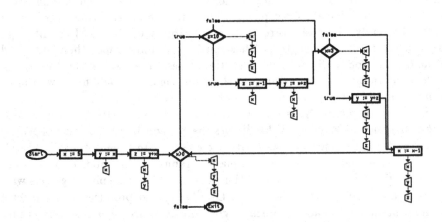

Fig. 1. Annotated CFG by VCG Manhattan Layout

Our algorithm is implemented in the VCG tool [10]. It is an extension of our variant of the hierarchical layout methods of Warfield [15] and Sugiyama et al. [11] which are known to be suitable for large directed graphs. It runs fast: we show that the overhead of Manhattan layout compared to normal hierarchical layout is $O(n + ek)$ where n is the number of nodes, e is the number of edges in the proper hierarchy, and k is the maximal number of line rows between two layers of nodes. Note that k is very small for most layouts. It produces drawings where each edge has maximally four bends.

In the next section, we explain the drawing rules for our variant of the Manhattan layout. Section 3 shortly sketches well known results about edge crossing reduction. Section 4 presents the segment ordering graph needed for the calculation of node positions. Section 5 deals with the adjustment of positions to a grid. Then, we describe the algorithm for the routing of edges, and finally, we show some statistics and experiences.

2 Manhattan Drawing Conventions

We define the drawing conventions of our applications:

1) Nodes are placed in a hierarchy of layers and never overlap.
2) Edges are drawn as a sequence of horizontal or vertical line segments.
3) An edge segment never crosses a node.
4) A horizontal segment of one edge and a vertical segment of another edge may share a point (i.e., we have an edge crossing, Fig. 3, Sit. b).
5a) Two horizontal segments of different edges never share a point.
6) Vertical segments of different edges may share subsegments, if and only if the segments are adjacent to a node (i.e., they must be the first or the last segment of the edge, Fig. 2, Sit. b).
7) Horizontal segments should have a minimal vertical distance d_h. Vertical segments should have a minimal horizontal distance d_v.

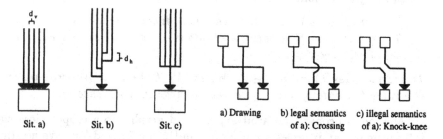

Fig. 2. Diff. Segment Sharing Conv.   **Fig. 3.** Crossing and Knock-knee

In a sequence of edge segments belonging to the same edge two adjacent segments always share an endpoint. Convention 4 allows edge crossings, thus the convention is applicable for nonplanar graphs, but the aesthetics requires that the number of edge crossings should be minimized. Convention 5a and 6 contribute to the visual readability: The route of two different edges is clearly distinguishable if they never share common equally-oriented segments (Fig. 2, Sit. a). However, we relax this constraint and allow the sharing on the first and the last segment of edges. The reason: If the last segments of edges are shared, they also share the arrowhead, which increases the resolution (Fig. 2, Sit. b). Of course, an edge segment with arrowhead never shares an edge segment without arrowhead. Otherwise, it would not be recognizable which segments have arrowheads. In particular, convention 5a forbids knock-knees which would not be distinguishable from edge crossings (Fig. 3).

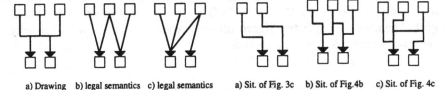

a) Drawing   b) legal semantics   c) legal semantics   a) Sit. of Fig. 3c   b) Sit. of Fig.4b   c) Sit. of Fig. 4c

Fig. 4. Confusing Sit. with Conv. 5b   **Fig. 5.** Drawings with Conv. 5a

In tree layouts, convention 5a is often relaxed (see later Fig. 10):

5b) Two horizontal segments of different edges may share a subsegment, if they also share the vertical segment adjacent to the subsegment (Fig. 2, Sit. c).

However in general graphs, drawings according to convention 5b are sometimes misleading. For instance, the drawing Fig. 4a can represent the semantics sketched by Fig. 4b and Fig. 4c. Note that the corresponding drawings according to convention 5a are more readable (Fig. 5).

3   Hierarchy Mapping

Definition 1 *A* **proper** *n*-**level hierarchy** *is a directed graph* $G = (V, E)$ *which satisfies the following conditions:*

- *The set of nodes V is partitioned into n disjoint sequences $V_1, \ldots, V_n$.*
- *The set of edges E is partitioned into $n - 1$ disjoint subsets $E_1, \ldots, E_{n-1}$ with $E_i \subseteq V_i \times V_{i+1}$.*

Furthermore $\mathbf{lev}(v) = i$ *iff* $v \in V_i$. *We call V_i the* **layer** *i. We write* $\mathbf{pos}(v) = k$ *if v is the kth node of a layer, i.e. the kth node in a sequence V_i.*

Methods to convert an arbitrary directed graph into a proper hierarchy by introducing dummy nodes are discussed in [15, 11, 6], and [10]. We use the notions predecessor $\mathbf{pred}(v)$ of a node v, successor $\mathbf{succ}(v)$, $\mathbf{indeg}(v) = |\mathbf{pred}(v)|$, $\mathbf{outdeg}(v) = |\mathbf{succ}(v)|$ with respect to the proper hierarchy. Note: a predecessor of v in a proper hierarchy need not to be a predecessor of v in the original graph. For instance, an edge of the original graph may occur reverted in the proper hierarchy.

Definition 2 *A linear segment S is a maximal sequence of nodes $w_1, \ldots, w_m$ with*

(1) $\text{lev}(w_i) = \text{lev}(w_{i+1}) - 1$

(2) $\text{indeg}(w_1) \leq 1, \text{outdeg}(w_m) \leq 1, \text{succ}(w_1) = \{w_2\}, \text{pred}(w_m) = \{w_{m-1}\}$

(3) $\text{pred}(w_i) = \{w_{i-1}\}$ *and* $\text{succ}(w_i) = \{w_{i+1}\}$ *for* $i \in \{2, \ldots, m-1\}$.

A linear segment is a sequence of nodes that should be drawn as a straight vertical line (Fig. 6). Each node belongs to at most one linear segment. We can easily partition the proper hierarchy into disjoint linear segments and remaining nodes in time $O(|V| + |E|)$.

Edge crossing reduction in the proper hierarchy is done as follows: We traverse the hierarchy from layer 1 to layer n; for each layer, we calculate a weight W_p for each node and change the positions $\text{pos}(v)$ by sorting the nodes of the layer according to this weight. Then, we traverse the hierarchy from layer n to layer 1 and reorder according to a weight W_s. This is iterated several times [11].

Proposition 1 *Assume that*

(1) $(\forall(v', w') \in \text{pred}(v) \times \text{pred}(w) \ \text{pos}(v') < \text{pos}(w')) \Rightarrow W_p(v) < W_p(w)$

(2) $(\forall(v', w') \in \text{succ}(v) \times \text{succ}(w) \ \text{pos}(v') < \text{pos}(w')) \Rightarrow W_s(v) < W_s(w)$

Then after a crossing reduction traversal, for two linear segments $S_a \neq S_b$ $\nexists v_i, v_{i+1} \in S_a, w_j, w_{j+1} \in S_b$ *with* $\text{lev}(v_i) + 1 = \text{lev}(w_j) + 1 = \text{lev}(v_{i+1}) = \text{lev}(w_{j+1})$ *and* $\text{pos}(v_i) < \text{pos}(w_j) \wedge \text{pos}(v_{i+1}) > \text{pos}(w_{j+1})$

Proposition 1 implies that after crossing reduction, two linear segments will never cross. Assume that the proposition does not hold, i.e. that such v_i, v_{i+1}, w_j, w_{j+1} would exist. Obviously, v_{i+1} and w_{j+1} would be reordered in a top down traversal of the layers, or v_i and w_j would be reordered in a bottom up traversal, respectively. Good selections of weights W_p and W_s that satisfy assumptions (1) and (2) can be found in [11] and [6].

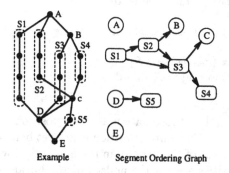

Example   Segment Ordering Graph

Fig. 6. Linear Segments

4 Positioning of Nodes

Now, we try to find positions of nodes such that (a) layers 1 to n are positioned top down, (b) nodes with $\mathbf{pos} = 1 \ldots m$ are positioned left to right, (c) all nodes of a layer are on a horizontal line, (d) the layout is balanced, (e) all nodes of a linear segment are on a vertical line. In particular, this implies that the edge segments and dummy nodes of a linear segment that formerly represented an edge are now drawn in a long vertical line without bends. First, we calculate the x-coordinates of the nodes. To simplify the explanation, each node v that does not belong to a linear segment forms a trivial linear segment with one element $\{v\}$. This implies that the whole graph is now partitioned into disjoint linear segments. The linear segments are ordered according to the relation 'is left of' ($\sqsubseteq$) by using the segment ordering graph SG (Fig. 6):

(1) Nodes(SG) := all linear segments; Edges(SG) := $\emptyset$
(2) **for** each layer V_i **do**
(3)   **for** $j := 2$ **to** $|V_i|$ **do**
(4)   let $v_{j-1} \in S_a$ and $v_j \in S_b$
(5)   /* it holds $\mathbf{lev}(v_{j-1}) = \mathbf{lev}(v_j)$ and $\mathbf{pos}(v_{j-1}) = \mathbf{pos}(v_j) - 1$ */
(6)   Edges(SG) := $\{(S_a, S_b)\} \cup$ Edges(SG)
(7)   **od**
(8) **od**
(9) topological_sort(SG)

SG is acyclic because of proposition 1. Thus topological sorting is possible, and each linear segment S has an ordering number $\mathbf{spos}(S)$ afterwards. The calculation of SG needs time $O(|V|)$ if the layers are represented by ordered, doubly linked lists. Topological sorting needs time $O(|SG|)$, but because the number of linear segments does not exceed the number of nodes, and each node introduces only one edge in SG, the whole calculation of ordering numbers can be done in time $O(|V|)$. Now, we can produce an initial position $x(v)$ for each node that satisfies (b) and (e):

(10) **for all** S in increasing order of **spos do**
(11)   **for all** $v \in S$ **do**
(12) $x_{min}(v) :=$ leftmost possible x-coordinate
(13)   **od**
(14) $X_{min} := \max \{x_{min}(v) \mid v \in S\}$
(15)   **for all** $v \in S$ **do** $x(v) := X_{min}$ **od**
(16) **od**

The leftmost possible x-coordinate must be a position right to the segments that are already placed. This initial position is not yet balanced. We balance the graph by using a variant of the pendulum method [10]. However, different from [10], the movable entities are not the nodes but the linear segments, thus linear segments remain straight vertical lines. We sketch the pendulum method again: The linear segments are the ball and the edges between the linear segments are the strings of the pendulum. If the uppermost segments are fixed on a ceiling, the balls on the strings swing to a balanced layout driven by their gravity. Balls attached by several strings swing into the middle position of their predecessor balls. Neighbored balls may influence each other. For instance, if the left ball is pulled to the right and the right ball is pulled to the left, the balls form a region which is positioned such that the sum of pulling forces of the region becomes zero. We simulate this simplified physical model.

Definition 3 *The* **predecessor deflection** *of an edge* $e = (s, t)$, *of a node* v, *of a linear segment* S *and of a region of linear segments* $\{S_1, \ldots, S_k\}$ *is defined as*

$$\mathbf{D}_p(e) = x(s) - x(t) \qquad \mathbf{D}_p(v) = \frac{\sum_{(w,v) \in E} \mathbf{D}_p((w, v))}{\mathrm{indeg}(v)}$$

$$\mathbf{D}_p(S) = \sum_{v \in S} \mathbf{D}_p(v) \qquad \mathbf{D}_p(\{S_1, \ldots, S_k\}) = \frac{\sum_{i \in \{1, \ldots, k\}} \mathbf{D}_p(S_i)}{k}$$

The **successor deflection** $\mathbf{D}_s$ *can be defined similarly.*
Two segments S_a *and* S_b *are* **touching** *if there is a node* $v \in S_a$ *and a directly neighbored node* $w \in S_b$ *in the same layer and the distance between both nodes is the minimal allowed distance.*

If $\mathbf{D}_p(S) > 0$, the segment is pulled to the right, and if $\mathbf{D}_p(S) < 0$, it is pulled to the left. Because each linear segment S is always positioned as a straight vertical line, $\mathbf{D}_p(S) = \mathbf{D}_p(v_1)$ holds where v_1 is the first (topmost) node of S, because $\mathbf{D}_p(v_i) = 0$ for $v_i \in S$ with $v_i \neq v_1$. We start with a trivial partition PR of segments into regions $R_1 \ldots, R_m$ and subsequently replace regions that influence each other by their union.

(17) $PR := \{R_i \mid R_i = \{S_i\} \text{ for each } S_i\}$
(18) **repeat**
(19)   **if** $S_a \sqsubseteq S_b$ are touching **then let** $S_a \in R_a$ and $S_b \in R_b$
(20)   **if** $\mathbf{D}_p(R_a) \geq \mathbf{D}_p(R_b)$ **then** $PR := PR + \mathrm{union}(R_a, R_b) - R_a - R_b$ **fi**
(21)   **fi**
(22) **until** the regions don't change anymore.

Finally, we try to correct the position of each segment S of a region R by moving all nodes of S horizontally by the minimum of $|\mathbf{D}_p(R)|$ and the space between S and its neighboring segments. We move to the left if $\mathbf{D}_p(R) < 0$, otherwise to the right. In this way, the nodes never overlap. The whole process is repeated iteratively with the predecessor deflection $\mathbf{D}_p$ and the successor deflection $\mathbf{D}_s$ until the layout is balanced.

The time overhead of this algorithm with respect to the original algorithm in [10] is constant:[1] The lines (19)-(21) are in fact identical to the original algorithm.

---

[1] This holds for the worst case (this is, if we have $n - 1$ unions for all n nodes of the graph, i.e. if the result is one large region containing all nodes/segments). For the normal case, a comparison of the complexity is difficult: Working on segments instead of nodes might require more unions of regions, i.e. more iterations of (18)-(22). On the other hand, our experience is, that the pendulum method comes faster to a balanced situation, if it works on segments instead of nodes.

We store for each node, which segment it belongs to, and for each segment, which region it belongs to. In order to check whether two segments are touching, we check whether two nodes are touching (we did this in the original algorithm, too) and get the segment and the region from the nodes in constant time (this is the overhead).

5 Adjustment to the Horizontal Grid

Before we calculate the relative positions of the edges, which will also give the positions $y(v)$, we adjust the positions $x(v)$ to a grid with raster d_v: We traverse the segments in increasing order of **spos** (i.e. from the leftmost to the rightmost segment) and move each segment to the nearest possible grid point. This might reduce the balance by a small degree, but it ensures that vertical edge segments will have a minimal distance d_v.

Edges of the proper hierarchy connecting nodes of the same linear segment can be drawn as straight vertical lines. Edges connecting nodes of different linear segments have two potential bend points. Since an original edge of the initial directed graph may consist of a linear segment of dummy nodes and two edges that connect the linear segment with the remaining graph, the maximal number of bends of the original edge is four. In this section, we calculate the end points of the edges. In the next section, we derive the bend points from the end points.

The end points of the edges depend on the port points where the edge is adjacent to the node. Ports are points at the border of the node. A node v needs at most **indeg**(v) incoming ports and **outdeg**(v) outgoing ports. The calculation of the incoming ports starts with the ordered sequence $SP(v)$ of trivial ports and subsequently unifies neighboring ports if all end points of the corresponding edges have the same arrowhead style. The arrowhead style indicates the existence of an arrowhead, it might further indicate different colors, sizes or drawing modes of arrowheads. Edges of the same port share a vertical edge segment.

(23) $SP(v) := (P_1(v), \ldots, P_{indeg(v)}(v))$ with $P_j(v) = \{(w, v) \in E \mid \mathbf{pos}(w) = j\}$
(24) **for** j from 2 to **indeg**(v) **do**   /* $SP(v) = (\ldots, P_{j-1}(v), P_j(v), \ldots)$ */
(25)   **if** for some $e_a \in P_{j-1}(v), e_b \in P_j(v)$ is ahead_style$_{in}(e_a) = $ ahead_style$_{in}(e_b)$
(26)   **then** $SP(v) := (\ldots, P_{j-1}(v) \cup P_j(v), \ldots)$
(27)   **fi**
(28) **od**

Note that all edges in $P(v)$ always have the same arrowhead style. Thus in line (25), we need only one element $e_a \in P_{j-1}(v)$ and another element $e_b \in P_j(v)$ in order to check whether for all $e \in P_{j-1}(v), e' \in P_j(v)$ holds ahead_style$_{in}(e)$ = ahead_style$_{in}(e')$. This algorithm needs time $O(\mathbf{indeg}(v))$. The calculation of the outgoing ports is symmetrical. It is done for all nodes. Thus, we get all ports of the whole graph in time $O(|V| + |E|)$. Note: Since the predecessor nodes w are already sorted according to **pos**(w) after the crossing reduction, the creation of the ordered sequence $SP(v)$ is possible without sorting the $P_j(v)$. Traverse the predecessor nodes s in that order, create for each outgoing edge (s, t) a set

$P = \{(s,t)\}$ and append P to the sequence of ports $SP(t)$. In this way, the sequence of ports will automatically be ordered.

The ports are now assigned to x-coordinates. The corresponding end points of all edges of a port get the same x-coordinate. The x-coordinate of a port must be on the grid with raster d_v. If a node needs more ports than available on its border, this can be corrected by reducing the raster d_v or by increasing the size of the node. However, both solutions require a recalculation of the previous steps. For simplicity, our current solution combines ports in this case until the number of ports fits the number of available raster points on the border of the node. Unfortunately, this fallback solution decreases the readability, but these cases are rare: they occur only if the degree of a node is very high, there are nearly as much backward edges as forward edges, and the schedule of forward and backward edges at the node happens to be very unfavorable.

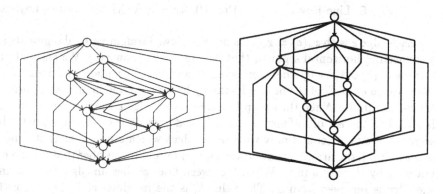

Fig. 7. K8 by Original Sugiyama Layout   **Fig. 8.** K8 by Nearly Manhattan Layout

Is is important to note that until this step the layout process is not strictly related to Manhattan layout. The algorithm enforces long edges to be drawn straight vertical, which is a necessary precondition for the Manhattan convention. But even without the following final step, the algorithm would produce a nice, balanced drawing, that produces more straight vertical lines than the original algorithm of Sugiyama et al. [11] (compare Fig.7, made by GraphEd [7], with Fig.8. Figure 7 contains much more zig-zags).

6 Positioning of Edges

We have the following situation: Each node v of the proper hierarchy has a position $x(v)$ and a level $\text{lev}(v)$ but not yet a position $y(v)$. Each edge e between two layers has a start position $x_s(e)$ and an end position $x_e(e)$ derived from the positions of the incoming and outgoing ports e belongs to. The start position of e is in the upper layer and the end position in the lower layer. Next, the free vertical space between layers V_i and V_{i+1} is divided into k_i line rows each containing horizontal line segments (Fig. 9). The rows have the vertical distance d_h. Thus, the distance between two layers must be $(k_i + 1)\, d_h$. It holds $k_i \leq |E_i|$, but in order to avoid wasting of vertical space, we calculate a minimal k_i. From the k_i the y-coordinates of the nodes and edges can be derived directly.

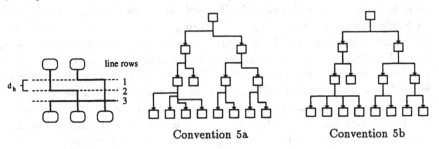

Convention 5a   Convention 5b

Fig. 9. Line Rows   **Fig. 10.** Tree by VCG Manhattan Layout

Each edge e with $x_s(e) \neq x_e(e)$ is drawn by two straight vertical segments and a connecting horizontal segment that belongs to line row $r(e)$. We calculate the line rows $r(e)$ of the edges between layers V_i and V_{i+1} by a plain sweep algorithm: We traverse each E_i of the proper hierarchy by a sweep line in increasing order of the x-coordinate. When the sweep line touches an edge e the first time, the edge is added to the set of unfinished edges U and gets its $r(e)$. The horizontal line segment of e is in conflict only with those edges which are in U at that time. In order to fulfill the drawing convention 5a, $r(e)$ must be a number that is currently not used by the edges in U. When the sweep line touches an edge the last time, the edge is removed from U. The value k_i is the maximal $r(e)$. The algorithm requires a list C of pairs (e, x), sorted according to x. For each edge $e \in E_i$, there are two pairs $(e, x_s(e))$ and $(e, x_e(e))$. Similar as the ordered sequence of port sets, the list C can be constructed in time $O(|V| + |E|)$ without the call of a sorting procedure, since the nodes of the layers are already sorted after crossing reduction and the edges are already sorted after the port calculation.

```
(29)  U := ∅;   k_i := 0
(30)  for each (e, x) ∈ C in increasing order do
(31)      if x_s(e) ≠ x_e(e) then
(32)          if x = min{x_s(e), x_e(e)} then
(33)              r(e) := 1 + max{r(e')|e' ∈ U}
(34)              if k_i < r(e) then k_i = r(e) fi
(35)              U := U + e
(36)          else/* x = max{x_s(e), x_e(e)} */
(37)              U := U - e
(38)          fi
(39)      fi
(40)  od
```

The algorithm can easily be adapted to drawing convention 5b by a modification of line (33). Further, with some time overhead, line (33) can be changed to search for smaller values for $r(e)$ that currently are not in U. This avoids wasting of vertical space. The loop (30)-(40) is executed $2|E_i|$ times. We implement U as doubly linked linear list, i.e. each edge e has a reference to its occurrence in U. Thus, insertion and deletion can be done in constant time. The maximal $r(e)$ in U is found in time $O(k_i)$, i.e. the whole algorithm needs time $O(|V| + |E_i|k_i)$.

At last, the y-coordinates of nodes and edges are derived. For nodes $v \in V_0$ we set $y(v) = 0$. For nodes $v \in V_i$ we set $y(v) = (k_i + 1) * d_h + max\{y(w) + height(w) \mid w \in V_{i-1}\}$. Assume that for instance the shape of the nodes are boxes. Then, for edges $e = (s, t)$ we set $y_s(e) = y(s) + height(s)$ and $y_e(e) = y(t)$. For edges e with $x_s(e) \neq x_e(e)$ we set further the two bend points $(x_s(e), y_s(e) + r(e)d_h)$ and $(x_e(e), y_s(e) + r(e)d_h)$. This completes the calculation of the Manhattan layout.

7 Experiences

Table 1 shows the time and space of some examples laid out by the normal algorithm [15, 11, 10] and by our extended Manhattan layout algorithm. It is difficult to analyze the time complexity of the algorithms. It depends heavily on the number of iterations made during crossing reduction and node positioning. However, it is known that the normal algorithm is very fast even for large graphs. Compared to it, we need additional time $O(|V| + |E| \max\{k_i\})$ in the worst case. In some cases (graph 2), Manhattan layout is even faster than normal layout, because less iteration are needed for node positioning.

Example	Nodes	Edges	Crossings	t_{norm}	W_{norm}	t_{manh}	W_{manh}	Bendings$_{manh}$
Graph 1	33	38	0	2.3	490	2.4	620	16
Graph 2	20	190	2251	14.3	960	12.7	1370	628
Graph 3	40	131	585	3.1	1780	3.3	1830	418
Graph 4	50	151	190	5.8	270	6.1	340	474
Graph 5	615	1310	15640	61.5	15420	65.0	26970	2261

Times (sum of user time and system time in secs) for parsing, layout and drawing (Sun Sparc 10/30, 32 MB mem., X11R6). t_{norm}/W_{norm} are time/width of normal layout [10], and t_{manh}/W_{manh} are time/width of Manhattan layout. The number of crossings is independent of the layout method, because in both cases the same crossing reduction method is used. The number of bendings is measured for Manhattan layout.

Table 1. Statistics

In dense graphs (graph 2, 3, 4) the number of bend points tends to the maximal number $4|E|$. Manhattan layout needs more space (W of graph 1, 2, 5) because long linear segments block close node positions. With very long linear segments the plane tend to be separated into a grid with large meshes. Most nodes are placed on the grid points while only few nodes are inside the meshes. This might be not satisfying because of the waste of space. In this case, the reduction of bend points of edges has minor importance than the usage of space. As solution the size of the linear segments can be restricted. In this variant of our algorithm edges may have more than four bends but the nodes are closer together.

8 Acknowledgment

We like to thank R. Wilhelm, C. Fecht, and R. Heckmann for their comments on the presentation of the algorithm.

References

1. Baker, B.S.; Bhatt, S.N.; Leighton, F.T.: An Approximation Algorithm for Manhattan Routing, in Preparata, F.P., ed.: Advances in Computing Research, Vol. 2, pp. 205-229, JAI Press, Greenwich, Connecticut, 1984.
2. Batini, C.; Nardelli, E.; Tamassia, R.: A Layout Algorithm for Data Flow Diagrams, IEEE Trans. on Software Engineering, SE-12(4), pp. 538-546, 1986,
3. Bhatt, S.N.; Cosmadakis, S.S.: The Complexity of Minimizing Wire Lengths in VLSI Layouts, Information Processing Letters, 25, pp. 263-267, 1987.
4. Biedl, T.; Kant, G.: A Better Heuristic for Orthogonal Graph Drawings, Technical Report UU-CS-1995-04, Utrecht University, 1995, also in Proc. 2nd Ann. European Symposium on Algorithms (ESA '94), LNCS 855, pp. 24-35, Springer-Verlag, 1994.
5. Even, S.; Granot, G.: Grid Layouts of Block Diagrams – Bounding the Number of Bends in Each Connection, in Tamassia, R.; Tollis, I.G., eds.: Graph Drawing, Proc. DIMACS Intern. Workshop GD'94, LNCS 894, pp. 64-75, Springer-Verlag, 1995.
6. Gansner, E.R.; Koutsofios, E.; North, S.C.; Vo, K.-P.: A Technique for Drawing Directed Graphs, IEEE Trans. on Software Engineering, 19(3), pp. 214-230, 1993.
7. Himsolt, M.: GraphEd – A Graphical Platform for the Implementation of Graph Algorithms, in Tamassia, R.; Tollis, I.G., eds.: Graph Drawing, Proc. DIMACS Intern. Workshop GD'94, LNCS 894, pp. 182-193, Springer-Verlag, 1995.
8. Johnson, D.: The NP-completeness column: An ongoing guide, Journal on Algorithms, 3(1), pp. 215-218, 1982
9. Protsko, L.B.; Sorenson, P.G.; Tremblay, J.P.; Schaefer, D.A.: Towards the Automatic Generation of Software Diagrams, IEEE Trans. on Software Engeneering, 17(1), pp. 10-21, 1991,
10. Sander, G.: Graph Layout Through the VCG Tool, in Tamassia, R.; Tollis, I.G., eds.: Graph Drawing, Proc. DIMACS Intern. Workshop GD'94, LNCS 894, pp. 194-205, Springer-Verlag, 1995. The VCG tool is publicly available via http://www.cs.uni-sb.de:80/RW/users/sander/html/gsvcg1.html.
11. Sugiyama, K., Tagawa, S., Toda, M.: Methods for Visual Understanding of Hierarchical Systems, IEEE Trans. Sys., Man, and Cybernetics, SMC 11(2), pp. 109-125, 1981.
12. Tamassia, R.: On Embedding a Graph in the Grid with the Minimum Number of Bends, SIAM Journal of Computing, 16(3), pp. 421-444, 1987.
13. Tamassia, R., Tollis, I.G.: Planar Grid Embedding in Linear Time, IEEE Trans. on Circuits and Systems, 36(9), pp. 1230-1234, 1989.
14. Vijayan G.; Wigderson A.: Rectilinear Graphs and Their Embeddings, SIAM Journal of Computing, 14(2), pp. 355-372, 1985.
15. Warfield, N.J.: Crossing Theory and Hierarchy Mapping, IEEE Trans. Sys., Man, and Cybernetics, SMC 7(7), pp. 505-523, 1977.
16. Wieners-Lummer, C.: Manhattan Channel Routing with Good Theoretical and Practical Performance, ACM SIAM Symp. on Disc. Alg., pp. 465-474, 1990.

CLAX – A Visualized Compiler

G. Sander*, M. Alt*, C. Ferdinand*, R. Wilhelm
(sander|alt|ferdi|wilhelm@cs.uni-sb.de)

Universität des Saarlandes, FB 14 Informatik, 66041 Saarbrücken

Abstract. The CLAX compiler was developed in the project COM-PARE as reconfigurable demonstration compiler. Its various optimization phases are visualized and animated by the graph layout tool VCG. Visualization allows to understand and to improve the behavior of the algorithms of the compiler phases. We present a tour through the CLAX compiler and demonstrate the newest extensions of the VCG tool, that help to explore large compiler data structures.

1 Introduction

The Esprit project #5399 COMPARE (COMpiler Generation for PARallel MachinEs) aims at reconfigurable, optimizing compilers for different source-target combinations. A new compiler organization CoSY [1] has been designed and implemented. It is based on a generic data base approach with modularization of compilers into phases, where all phases work concurrently on a common intermediate representation and each phase solves a certain subtask of the compilation.

Before developing production quality compilers the feasibility of CoSY was evaluated by the prototype compiler CLAX [3]. The CLAX compiler was used to demonstrate the benefits of distributed compiler phase supervision. The compiler data structures are visualized and animated. This allows to understand and to improve the behavior of the algorithms of the phases.

The VCG tool [4], also developed in the project COMPARE, was used for the layout of the compiler graphs. This tool is specialized for the requirements of compiler data structures which are usually very large. It provides different folding mechanism to focus on certain aspects of the graphs, and allows to reduce the amount of visible information to the points of interest. It provides browsing facilities like tracing of chains of edges or searching of nodes, unlimited scaling and different views onto the graph. The layout methods are a fast variant of the hierarchical layout algorithm of Sugiyama et al. [6] for directed graphs, and a fast force directed placement for undirected graphs (the spring embedder algorithm of Frick et al. [2]). Both layout algorithms can be controlled in a wide range.

The VCG tool and the CLAX compiler run concurrently while the CLAX compiler controls the VCG tool via an animation interface. The VCG tool produces good drawings and runs reasonably fast even on very large graphs.

* This work was partially supported by the ESPRIT Project #5399 COMPARE

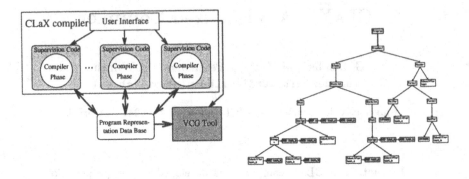

Fig. 1. CLaX Compiler Organization   **Fig. 2.** Annotated Syntax Tree

2 Compiler Phases

In traditional compiler construction, the phases of a compiler are executed in a fixed sequential order. Each phase translates from one intermediate representation of the program into another one. In this sequence, the representations change from source dependent information to target dependent information. Since each phase requires a different format of the representation as input, a reorganization of the compiler or a reuse of a phase is difficult or impossible.

The problem with the traditional model is, that a fixed, optimal order of compiler phases in the middle end does not exist. One optimization may influence another optimization, while the latter might calculate a good starting position for the former. A fixed order of the compiler phase would not produce an optimal result, even if each individual phase would calculate the optimum. Thus, the CLaX compiler works demand driven by the phase dependence graph on a data base of program representations (Fig. 1). This graph identifies which phases are required as precondition of an optimization, and which phases optionally support the optimization. In CoSy, there is no fixed order, but all sequences of phases that do not violate the dependences are allowed. The dependence graph further shows the potential phase parallelism of the compiler. The data base is kept consistent, i.e. if the precondition of a phase is destroyed, the phase is automatically restarted. The supervision code is distributed among the phases. Cyclic execution of optimization is possible, and the progress is visualized by the phase dependence graph.

3 Data Structure Visualization and Animation

The CLaX compiler is demand driven and can be controlled interactively: Sending a request to a data structure results in the generation of this data structure. Sending a optimization request starts all optimization phases that can contribute, until a fixed point is reached. In each step a new instance of the program representation is calculated. The representation is transformed towards their optimum. The compiler uses syntax trees with different annotations (Fig. 2), control flow graphs (Fig. 3), data dependence graphs (Fig. 4), dominator trees, etc. [7]. These graphs are visualized by the VCG tool and can be examined interactively after each step to inspect details. Changes after each phase can be animated automatically. This shows an overview over the compiling process. It is also possible to examine one phase by animating the internal steps.

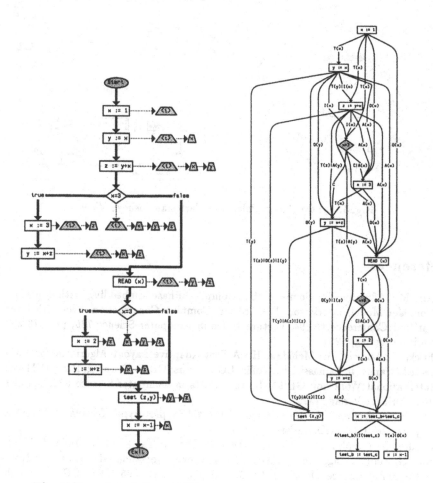

Fig. 3. Annotated Control Flow Graph (Manhattan Layout)

Fig. 4. Data Dependence Graph (Spline Layout)

Each kind of compiler graph needs different visualization methods: Attributed syntax trees are visualized by a specialized tree layout (Fig. 2). It is possible to interactively hide or expose the attributes. Control flow graphs have a uniform appearance if a manhattan layout is used, where all edges consist of horizontal or vertical line segments (Fig. 3). Very large graphs cause the problem that the user may loose the orientation when only a small part of the graph is displayed

462

in the VCG tool window. Thus, fisheye views [5] can be used which distort the graph picture such that the whole graph (or a large part of it) is visible while only the focus point is displayed in all details at the same time. By moving the focus point through the graph, details can be inspected without loosing the orientation (Fig. 5).

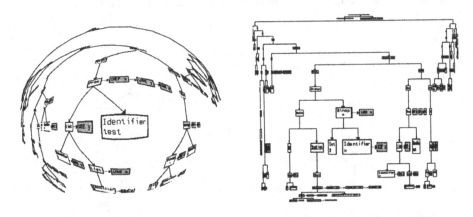

Fig. 5. Syntax Trees, Polar and Cartesian Fisheye View

References

1. Alt, M.; Aßmann, U.; Someren, H.: Compiler Phase Embedding with the CoSy Compiler Model, *in* Fritzson, P.A., editor: Compiler Construction, Proc. 5th International Conference CC'94, Lecture Notes in Computer Science 786, pp. 278-293 Springer-Verlag, 1994
2. Frick, A.; Ludwig, A.; Mehldau, H.: A Fast Adaptive Layout Algorithm for Undirected Graphs, *in* Tamassia, R.; Tollis, I.G., editors: Graph Drawing, Proc. DIMACS International Workshop GD'94, Lecture Notes in Computer Science 894, pp. 388-403, Springer-Verlag, 1995.
3. Müller, T.; Vollmer, J.: Description of the CoSY prototype, COMPARE Technical Report, Rel 1.3, GMD Karlsruhe, 1991
4. Sander, G.: Graph Layout Through the VCG Tool, *in* Tamassia, R.; Tollis, I.G., editors: Graph Drawing, Proc. DIMACS International Workshop GD'94, Lecture Notes in Computer Science 894, pp. 194-205, Springer-Verlag, 1995. The VCG tool is publicly available via http://www.cs.uni-sb.de:80/RW/users/sander/html/gsvcg1.html.
5. Sarkar, M.; Brown, M.H.: Graphical Fisheye Views, Commun. of the ACM 37(12), pp. 74-84, 1994.
6. Sugiyama, K., Tagawa, S., Toda, M.: Methods for Visual Understanding of Hierarchical Systems, IEEE Trans. Sys., Man, and Cybernetics, SMC 11(2), pp. 109-125, 1981.
7. Wilhelm, R.; Maurer, D.: Compiler Design, Addison-Wesley, 1995.

Crossing Numbers of Meshes

Farhad Shahrokhi[1], Ondrej Sýkora[2,*], László A. Székely[3], Imrich Vrt'o[2]

[1] Department of Computer Science, University of North Texas
P.O.Box 13886, Denton, TX, USA
[2] Institute for Informatics, Slovak Academy of Sciences
P.O.Box 56, 840 00 Bratislava, Slovak Republic
[3] Department of Computer Science, Eötvös University
Múzeum krt. 6-8, 1088 Budapest, Hungary

Abstract. We prove that the crossing number of the cartesian product of 2 cycles, $C_m \times C_n$, $m \leq n$, is of order $\Omega(mn)$, improving the best known lower bound. In particular we show that the crossing number of $C_m \times C_n$ is at least $mn/90$, and for $n = m, m+1$ we reduce the constant 90 to 6. This partially answers a 20-years old question of Harary, Kainen and Schwenk [3] who gave the lower bound m and the upper bound $(m-2)n$ and conjectured that the upper bound is the actual value of the crossing number for $C_m \times C_n$. Moreover, we extend this result to $k \geq 3$ cycles and paths, and obtain such lower and upper bounds on the crossing numbers of the corresponding meshes, which differ by a small constant only.

1 Introduction

The crossing number of a graph G, denoted by $cr(G)$, is the minimum number of crossings of its edges over all drawings of G in the plane, such that no more than two edges intersect in any point and no edge passes through a vertex. Computing $cr(G)$ is NP-hard and there have been only few results concerning the exact value of crossing number for very special and restricted classes of graphs. Besides Kleitman's exact result [6] on the crossing number of $K_{m,n}$, for $m \leq 6$, most effort has been devoted to crossing numbers of some cartesian product graphs [4, 7, 8, 22]. For a detailed exposition of the crossing number problem see our survey [19].

For $G_1 = (V_1, V_2)$ and $G_2 = (V_2, E_2)$, let $G_1 \times G_2$ denote the cartesian product of G_1 and G_2. Thus $G_1 \times G_2$ is a graph with the vertex set $V_1 \times V_2$ in which $(i, j)(r, s)$ are adjacent iff either $i = r$ and $js \in E_2$ or $j = s$ and $ir \in E_1$. Let P_n and C_n denote the n-vertex path and the n-vertex cycle , respectively. For $2 \leq n_1 \leq n_2 \leq ... \leq n_k$, let $M_k = \prod_{i=1}^{k} P_{n_i}$ and $TM_k = \prod_{i=1}^{k} C_{n_i}$. We will call M_k and TM_k the k-dimensional mesh and the k-dimensional toroidal mesh, respectively.

* Research of the 2nd and the 4th author was partially supported by grant No. 2/1138/94 of Slovak Grant Agency and Alexander von Humboldt Foundation.

Clearly, M_2 is planar and has crossing number 0, but estimating the crossing number of TM_2 has been an open problem. Harary et al. [3] provided a simple drawing of $C_m \times C_n$, $m \leq n$ with $(m-2)n$ crossings. They also derived a weak lower bound m on $cr(C_m \times C_n)$ and conjectured that $cr(C_m \times C_n) = (m-2)n$, for $3 \leq m \leq n$. Beineke and Ringeisen [1, 16] proved the conjecture for $m \leq 4$. Richter and Thomassen [15] proved the conjecture for $m = n = 5$. Finally, very recently Klešč [9] and Richter and Stobert [14] announced their proof for $m = 5$ and arbitrary $n \geq 5$. It is instructive to mention that all existing standard methods for estimating lower bounds on the crossing number fail to give good lower bounds for $cr(C_m \times C_n)$. In particular two very powerful methods developed by VLSI community [11] – the bisection method and the embedding method – give a weak lower bound for $cr(C_m \times C_n)$. We suspect that the reason is that $C_m \times C_n$ has genus 1 and very much resembles the planar 2-dimensional mesh.

In this paper we take a major step to prove the conjecture and show that for $6 \leq m \leq n$, $cr(C_m \times C_n) \geq mn/90$. For $n = m, m+1$ we improve the constant 90 to 6. It is worth mentioning that the method used here to prove our main result employs Dilworth' chain decomposition theorem which was shown to be very effective when dealing with problems in combinatorial geometry [10, 13].

Moreover for $k \geq 3$, we derive upper and lower bounds within a constant multiplicative factor for the crossing number of M_k and TM_k. We indicate that, since bisection width of M_k and TM_k are known, see for instance [12], one can use the relationship between the crossing number and the bisection and derive lower bounds for $cr(M_k)$ and $cr(TM_k)$, $k \geq 3$ which are of the same order magnitude as the lower bounds we have derived here. Nevertheless, such an approach does not provide constants, whereas, our method identifies relatively large constants associated with our lower bounds. Moreover, for $k \geq 3$ we have been able to provide (new) drawings of M_k and TM_k, with number of crossings which are within relatively small constant factors from the lower bounds, under reasonable conditions.

2 Crossing Number of $C_m \times C_n$

One finds in $C_m \times C_n$ in a natural way n vertex disjoint row cycles and m vertex disjoint column cycles. We will call them r-cycles and c-cycles. Deleting all edges of any r-cycle (c-cycle) yields a graph which is a subdivision of $C_m \times C_{n-1}$ (of $C_{m-1} \times C_n$) and therefore has a crossing number less than or equal to that of $C_m \times C_n$. It immediately follows, that $cr(C_m \times C_n)$ is monotone nondecreasing in both parameters. We will use this fact implicitly throughout our proof many times.

Theorem 2.1 For $6 \leq m \leq n$,

$$cr(C_m \times C_n) \geq \frac{mn}{90}.$$

Proof. If $6 \le m \le n \le 180$ then by monotonicity and a result of Beineke and Ringeisen [1]: $cr(C_4 \times C_n) = 2n$ we have

$$cr(C_m \times C_n) \ge cr(C_4 \times C_n) = 2n \ge \frac{mn}{90}.$$

Assume $n \ge 181$. For a closed curve C in the plane R^2, define its body $B(C)$ as the closure of the union of the bounded components of $R^2 \setminus C$. See Fig. 1. Define the exterior of C as the complement of $B(C)$ (the white part of Fig. 1).

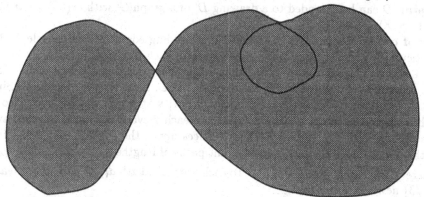

Fig. 1 : The body of a closed curve

Let us be given a drawing D of $C_m \times C_n$ in the plane. We may assume without loss of generality that (1) crossing edges really cross, i.e. no touching situation occurs, (2) the number of crossings in D is finite. Define a partial order on the set of its r-cycles by $C < Z$ if C lies in one of the bounded connected components of $R^2 \setminus Z$. Dilworth' Theorem [2] p. 62 applies to this poset. If k is the size of the largest antichain, then the poset may be decomposed into chains of length $a_1, a_2, ..., a_k$ such that $a_1 + a_2 + ... + a_k = n$. In such a chain each member contains the next member of the chain in one of the bounded connected components of the plane that it defines. Each c-cycle has exactly one vertex on every member cycle in the i-th chain. Hence, by Jordan's Curve Theorem, every c-cycle has at least $a_i - 2$ crossings with the r-cycles of the i-th chain. Therefore, $cr(D) \ge m(a_1 + a_2 + ... + a_k - 2k) = m(n - 2k) \ge mn/50$, if $k < 49n/100$.

Assume that $k \ge 49n/100$. We have an antichain of size at least $49n/100$, i.e. at least $49n/100$ r-cycles such that any two cross or they lie in the exterior of each other. Ignoring the edges of the other r-cycles, we consider our drawing as a drawing of a $C_m \times C_{\lceil 49n/100 \rceil}$. For any r-cycle C let $a(C)$ denote the number of c-cycles whose bodies lie in the body of C. Observe that for two r-cycles C, Z taken from an antichain, $a(C) > 0$ implies 2 crossing of C and Z, since every c-cycle was supposed to have a common vertex with every r-cycle and if two r-cycles cross then they must cross at least twice. Either we have at least $(1/24)(49n/100)$ r-cycles C with $a(C) > 0$, and the total number of crossings is $2(49n/2400) \cdot (49n/100 - 1) > mn/90$ and we are at home, or we have at least $(23/24)(49n/100)$ r-cycles C with $a(C) = 0$. Reduce our drawing further to the drawing of the corresponding $C_m \times C_{\lceil 1127n/2400 \rceil}$.

Now every c-cycle either crosses an r-cycle or is in the exterior of it. Either we have at least $(1/3)(1127n/2400)$ r-cycles, such that each of them is crossed by the c-cycles at least $m/14$ times, or we have at least $(2/3)(1127n/2400)$ r-cycles, such that the c-cycles cross them at most $m/14$ times. In the first case $cr(D) \geq mn/90$, and the Theorem holds. In the second case reduce our interest to a drawing D of $G = C_m \times C_{\lceil 1127n/3600 \rceil}$ so that for every r-cycle C, $a(C) = 0$ and at most $m/14$ c-cycles cross C. Let $cr(D)$ denote the number of crossings in D. The proof is completed employing the following claim.

Claim: D can be extended to a drawing D' of a graph G' with $cr(G') \geq mn/15$ and $cr(D') \leq 6cr(D)$.

Proof of the Claim. We construct D' by adding edges to each r-cycle C. In particular we add at least $3m/14$ new edges to each r-cycle C.

Let S be the set of all c-cycles that cross C. By deleting all edges of C that are crossed by c-cycles from S, we divide C to at most $\lfloor m/14 \rfloor$ vertex-disjoint paths. On these paths, there are at most $\lfloor m/14 \rfloor$ vertices, in which c-cycles from S have a vertex in common with C, since each r-cycle and each c-cycle have exactly one vertex in common. These vertices divide the $\lfloor m/14 \rfloor$ vertex disjoint paths into at most $2\lfloor m/14 \rfloor$ edge disjoint paths of lengths $d_1, d_2, d_3, ..., d_{2\lfloor m/14 \rfloor}$, where $\sum_{i=1}^{2\lfloor m/14 \rfloor} d_i \geq m - \lfloor m/14 \rfloor$. To each path of length d_i, we can add at least $\lfloor d_i/3 \rfloor$ new edges so that

(i) in each path above we join vertices of distance 3 in a greedy way, such that the new "long edges" do not overlap

(ii) the new edges are drawn very close to C and inside $B(C)$, as far as it is possible

(iii) the new edges do not cross each other unless the corresponding part of the drawing of C is self-intersecting

(iv) the new edges do not cross c-cycles.

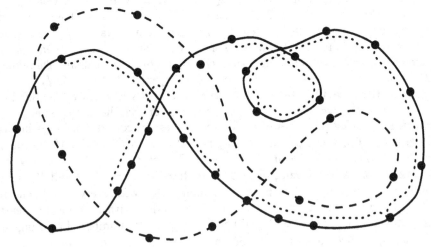

Fig. 2

Fig. 2 shows details of a drawing of $C_{23} \times C_{11}$ with one r-cycle (solid line) and one c-cycle (dashed line). The new edges are drawn by dotted lines. Hence the total number of added edges is

$$\sum_{i=1}^{2\lfloor \frac{m}{14} \rfloor} \left\lfloor \frac{d_i}{3} \right\rfloor \geq \sum_{i=1}^{2\lfloor \frac{m}{14} \rfloor} \frac{d_i - 2}{3} \geq \frac{1}{3}\left(m - \left\lfloor \frac{m}{14} \right\rfloor - 4\left\lfloor \frac{m}{14} \right\rfloor\right) \geq \frac{3m}{14}.$$

The drawing D' obtained in this way corresponds to a graph G' which has $m\lceil 1127n/3600 \rceil$ vertices, at least $2m\lceil 1127n/3600 \rceil + 161mn/2400$ edges, and of girth 4. Due to the construction $cr(D') \leq 6cr(D)$, since a crossing of two distinct r-cycles in D may result in at most 4 crossings in D', a self-crossing of an r-cycle in D may result in 6 crossings in D' (see Fig. 2). Crossings of r-cycles and c-cycles do not multiply. Finally, Kainen's lower bound [5] applies to G':

$$cr(G') \geq |E(G')| - \frac{girth(G')}{girth(G') - 2}(|V(G')| - 2).$$

This formula immediately finishes the proof. $\qquad\square$

In special cases, we can improve the lower bound to the following:

Theorem 2.2 For $3 \leq m \leq n$, where $n = m$ or $m + 1$

$$cr(C_m \times C_n) \geq \frac{(m-2)n}{6}.$$

Proof. We inductively show that if the claim holds for $cr(C_m \times C_m)$, then it also holds for $cr(C_m \times C_{m+1})$, and use this to show that the claim holds for $cr(C_{m+1} \times C_{m+1})$. We call the crossing of two different r-cycles as rr-crossing. Similarly define the cc-crossing and the rc-crossing. The claim is true for $C_m \times C_m$ when $m \leq 10$, as $cr(C_m \times C_m) = (m-2)m$, for $m \leq 5$, [1, 16, 15] and $cr(C_m \times C_m) \geq cr(C_5 \times C_5) = 15 \geq (m-2)m/6$, for $m = 6, 7, ..., 10$. Let the claim hold for $C_m \times C_m$, $m \geq 10$. We first show it holds for $C_m \times C_{m+1}$. Consider any drawing of $C_m \times C_{m+1}$. Suppose that there exists an r-cycle in $C_m \times C_{m+1}$ containing at least $m/6$ crossings. Deleting the edges of this r-cycle we get a subdivision of $C_m \times C_m$. Hence

$$cr(C_m \times C_{m+1}) \geq cr(C_m \times C_m) + \frac{m}{6} \geq \frac{(m-2)m}{6} + \frac{m}{6} \geq \frac{(m-2)(m+1)}{6}.$$

Suppose that each r-cycle in $C_m \times C_{m+1}$ contains at most $\lfloor m/6 \rfloor$ crossings. We may assume that there exist 3 distinct r-cycles, so that no two of them have a crossing. Otherwise each triplet of r-cycles would determine at least 2 rr-crossings and a simple counting argument shows that

$$cr(C_m \times C_{m+1}) \geq \frac{2\binom{m+1}{3}}{m-1} > \frac{(m-2)(m+1)}{6}.$$

Consider the 3 distinct r-cycles. Since for each of them, there exist at least $m - \lfloor m/6 \rfloor$ c-cycles, which do not cross it, there are at least $m - 3\lfloor m/6 \rfloor$ c-cycles, none of them crosses any of the 3 distinct r-cycles. Now every triplet of these $m - 3\lfloor m/6 \rfloor$ c-cycles together with the 3 distinct r-cycles determine a subdivision of $C_3 \times C_3$. If the subdivision contains a selfcrossing of a cycle we can redraw it without the selfcrossing and without producing a new crossing. Hence we may assume that the subdivision is without selfcrossings. It has 3 cc-crossings, since $cr(C_3 \times C_3) = 3$, but it must contain an even number of cc-crossings, since c-cycles are vertex disjoint and if they cross, they must cross an even number of times. Thus, this subdivision contains at least 4 cc-crossing. Further, a counting argument shows that,

$$cr(C_m \times C_{m+1}) \geq \frac{4\binom{m-3\lfloor m/6 \rfloor}{3}}{m - 3\lfloor m/6 \rfloor - 2} \geq \frac{(m-2)(m+1)}{6}.$$

We may conclude that if the claim holds for $C_m \times C_m$ it also holds for $C_m \times C_{m+1}$. Now we use this fact to prove a lower bound for $C_{m+1} \times C_{m+1}$. Suppose that there exists an r-cycle in $C_{m+1} \times C_{m+1}$ containing at least $(m+1)/6$ crossings. Deleting the edges of the r-cycle we get a subdivision of $C_{m+1} \times C_m$. Hence

$$cr(C_{m+1} \times C_{m+1}) \geq cr(C_{m+1} \times C_m) + \frac{m+1}{6} \geq \frac{(m-2)(m+1)}{6} + \frac{m+1}{6}$$
$$\geq \frac{(m-1)(m+1)}{6}.$$

Suppose that each r-cycle in $C_{m+1} \times C_{m+1}$ contains at most $(m+1)/6$ crossings. The rest of the proof is the same as the proof of the lower bound for $C_m \times C_{m+1}$. $\square$

3 Crossing Number of M_k and TM_k

It is straightforward to show, using Theorem 2.1 that, for $k \geq 3$, $cr(TM_k) = \Omega(\Pi_{i=1}^k n_i)$ which is a weak lower bound in some cases. In this section we prove near-optimal lower and upper bounds on the crossing number of TM_k and M_k, for $k \geq 3$. In particular, when $n_i = n, i = 1, 2, .., k$, the upper and lower bounds differ by a multiplicative factor only. Let $G_1 = (V_1, E_1)$ and $G_2 = (V_2, E_2)$ be graphs such that $|V_1| \leq |V_2|$.

An embedding ω of G_1 in G_2 is a pair of injections (ϕ, ψ)

$$\phi : V_1 \rightarrow V_2, \qquad \psi : E_1 \rightarrow \{\text{all paths in } G_2\},$$

such that if $uv \in E_1$ then $\psi(uv)$ is a path between $\phi(u)$ and $\phi(v)$. Define the congestion of ω

$$\mu_\omega = \max_{e \in E_2}\{|\{f \in E_1 : e \in \psi(f)\}|\}.$$

Leighton [11] invented a lower bound technique for crossing numbers, based on an embedding of the complete graph in the given graph. Several authors [18, 20, 22] realized that the method can be generalized for arbitrary graphs in the following form:

Lemma 3.1 *Let ω be an embedding of $G_1 = (V_1, E_1)$ into $G_2 = (V_2, E_2)$, $|V_1| \leq |V_2|$. Let D_2 be a drawing of G_2 with $cr(D_2)$ crossings, then there is a drawing D_1 of G_1 with $cr(D_1)$ crossings so that*

$$cr(D_2) \geq \frac{cr(D_1)}{\mu_\omega^2} - \frac{|V_2|\Delta_2^2}{2},$$

where Δ_2 is the maximum degree of G_2. □

For $2 \leq n_1 \leq n_2 \ldots \leq n_k$, let $N_i = n_1 n_2 \ldots n_i$, for $i = 1, 2, \ldots, k$.

Theorem 3.1 *For $k \geq 3$*

$$\frac{N_{k-1}^2}{5} - \frac{(5k^2 + 8)N_k}{2} \leq cr(M_k) \leq 4N_{k-2}N_k,$$

$$\frac{4N_{k-1}^2}{5} - 2(k^2 + 2)N_k \leq cr(TM_k) \leq 16N_{k-2}N_k + 8k^2 N_k.$$

Proof. We first prove the lower bound for $cr(TM_k)$ and construct an upper bound for $cr(M_k)$. Set $G_2 = TM_k$ and $G_1 = 2K_{N_k}$, where $2K_{N_k}$ denote the complete multigraph on N_k vertices, obtained from K_{N_k} by replacing every edge by two new edges. Shahrokhi and Székely [17] constructed an embedding ω of G_1 into G_2 with

$$\mu_\omega \leq \frac{N_k n_k}{4}.$$

Substituting this into Lemma 3.1, recalling from [21] the following formula:

$$cr(K_{N_k}) \geq \frac{N_k^4 - 7N_k^3}{80},$$

and noting that $cr(2K_{N_k}) = 4cr(K_{N_k})$ we get the claimed lower bound for TM_k.

Now we prove an upper bound for $cr(M_k)$. We construct a recursive drawing L_k for M_k in which all vertices are placed along a straight line in the plane. For $M_1 = P_{n_1}$ we place successively the vertices of P_{n_1} on a line and obtain a drawing with no crossings. Assume that we have constructed a drawing L_{k-1} of M_{k-1} with $cr(L_{k-1})$ crossings. The drawing L_k of $M_k = M_{k-1} \times P_{n_k}$ is constructed in the following way. Place n_k copies of the drawings L_{k-1} successively on a line such that 2 neighboring copies are symmetric according to a perpendicular line between the 2 copies. Join the corresponding vertices of the first and the second copy by edges drawn as half-circles above the line. Similarly, join the corresponding vertices of the second and the third copy by edges drawn as half-circles below the line and continue in this fashion until a drawing L_k of M_k is obtained. Let us call the inserted edges the edges of the dimension k and denote the number of crossings in L_k by $cr(L_k)$. Clearly

$$cr(L_k) \leq n_k cr(L_{k-1}) + l_k,$$

where l_k denotes the number of crossings of the edges of the k-th dimension with edges of n_k copies of the drawing L_{k-1}, i.e. the number of crossings of edges of the k-th dimension with edges of smaller dimensions. A counting analysis shows that there are at most $N_k N_{i-1}$ crossings of the edges of the k-th dimension with edges of the i-th dimension, where $N_0 = 1$. Hence

$$l_k \leq N_k \sum_{i=1}^{k-1} N_{i-1}$$

and

$$cr(L_k) \leq n_k cr(L_{k-1}) + N_k \sum_{i=0}^{k-2} N_i.$$

The solution is

$$cr(L_k) \leq N_k \sum_{i=0}^{k-2} (k - i - 1) N_i$$

$$\leq N_k N_{k-2} \left(1 + \frac{2}{n_{k-2}} + \frac{3}{n_{k-2} n_{k-3}} + ... + \frac{k-1}{n_{k-2} ... n_1} \right) \leq 4 N_k N_{k-2}.$$

Finally, we use Lemma 3.1 with $G_1 = TM_k$ and $G_2 = M_k$ and note that there is an embedding ω of G_1 into G_2 with $\mu_\omega = 2$. To get the upper bound for $cr(TM_k)$, we take D_2 to be L_k, then D_1 is a desirable drawing of TM_k. To get the lower bound for $cr(M_k)$, we substitute the term $cr(D_1)$ in the Lemma, by our lower bound for $cr(TM_k)$. $\square$

Corollary 3.1 If $n_k = O(n_{k-1})$ then the bounds in Theorem 3.1 are optimal within a constant multiplicative factor and in addition, if $n_1 = n_2 = ... = n_k = n$ then

$$\frac{n^{2k-2}}{5} - \frac{(5k^2 + 8)n^k}{2} \leq cr(M_k) \leq 4n^{2k-2},$$

$$\frac{4n^{2k-2}}{5} - 2(k^2 + 2)n^k \leq cr(TM_k) \leq 16n^{2k-2} + 8k^2 n^k.$$

$\square$

Corollary 3.2 Consider a general k-dimensional mesh $GM_k = \prod_{i=1}^{k} A_{n_i}$, where A_{n_i} equals either P_{n_i} or C_{n_i}. Then

$$cr(M_k) \leq cr(GM_k) \leq cr(TM_k).$$

$\square$

References

1. L. W. Beineke, R. D. Ringeisen, On the crossing number of product of cycles and graphs of order four, *J. Graph Theory* **4** (1980), 145–155.
2. M. Hall, Jr., *Combinatorial Theory*, Blaisdell Publ. Co., Waltham, 1967.
3. F. Harary, P. C. Kainen, A. Schwenk, Toroidal graphs with arbitrary high crossing numbers, *Nanta Mathematica* **6** (1973), 58–67.
4. S. Jendrol', M. Ščerbová, On the crossing numbers of $S_m \times P_n$ and $S_m \times C_n$, *Časopis pro Pestování Matematiky* **107** (1982), 225–230.
5. P. C. Kainen, A lower bound for crossing number of graphs with applications to K_n, $K_{p,q}$, and $Q(d)$, *J. Combinatorial Theory, Series B* **12** (1972), 287–298.
6. D. J. Kleitman, The crossing number of $K_{5,n}$, *J. Combinatorial Theory* **9** (1970), 315–323.
7. M. Klešč, On the crossing number of the cartesian product of stars and paths or cycles, *Mathematica Slovaca* **41** (1991), 113–120.
8. M. Klešč, The crossing number of product of path and stars with 4-vertex graphs, *J. Graph Theory* **18** (1994), 605–614.
9. M. Klešč, On the crossing numbers of products of cycles, preprint.
10. D. Larman, J. Matoušek, J. Pach, J. Töröcsik, A Ramsey-type result for planar convex sets, to appear.
11. F. T. Leighton, Complexity Issues in VLSI, MIT Press, Cambridge, 1983.
12. F. T. Leighton, Introduction to Parallel Algorithms and Architectures: Arrays.Trees.Hypercubes, Morgan Kaufmann, San Mateo, 1992.
13. J. Pach, J. Töröcsik, Some geometric applications of Dilworth' theorem, *Discrete Computational Geometry*, **21**(1994), 1–7.
14. R. B. Richter, I. Stobert, The crossing number of $C_5 \times C_n$, preprint.
15. R. B. Richter, C. Thomassen, Intersection of curve systems and the crossing number of $C_5 \times C_5$, *Discrete and Computational Geometry* **13** (1995), 149–159.
16. R. D. Ringeisen, L. W. Beineke, The crossing number of $C_3 \times C_n$, *J. Combinnatorial Theory, Series B* **24** (1978), 134–144.
17. F. Shahrokhi, L. A. Székely, An algebraic approach to the uniform concurrent multicommodity flow problem: theory and applications, *Technical Report CRPDC-91-4*, Department of Computer Science, University of North Texas, Denton, 1991.
18. F. Shahrokhi, L. A. Székely, Effective lower bounds for crossing number, bisection width and balanced vertex separators in terms of symmetry, in: *Proc. 2-nd IPCO Conference*, Pittsburgh, 1992, 102–113, also in *Combinatorics, Probability and Computing* **3** (1994), 523–543.
19. F. Shahrokhi, L. A. Székely, I. Vrťo, Crossing numbers of graphs, lower bound techniques and algorithms: a survey, in: *Proc. DIMACS Workshop on Graph Drawing'94*, Lecture Notes in Computer Science 894, Springer Verlag, Berlin, 1995, 131–142.
20. O. Sýkora and I. Vrťo, On the crossing number of hypercubes and cube connected cycles, *BIT* **33** (1993), 232–237.
21. A. T. White, L. W. Beineke, Topological graph theory, in: *Selected Topics in Graph Theory*, (L.W. Beineke R.J. Wilson, eds.), Academic Press, N.Y., 1978, 15–50.
22. K. Wada, K. Kawaguchi, H. Suzuki, Optimal bounds of the crossing number and the bisection width for generalized hypercube graphs, in: *Proc. 16th Biennial Symposium on Communications*, 1992, 323–326.

Directed Graphs Drawing by Clan-Based Decomposition

Fwu-Shan Shieh and Carolyn L. McCreary
Department of Computer Science and Engineering
Auburn University, Alabama, USA
{fwushan, mccreary}@eng.auburn.edu

Abstract. This paper presents a system for automatically drawing directed graphs by using a graphanalysis that decomposes a graph into modules we call clans. Our system, CG (Clan-based Graph Drawing Tool), uses a unique clan-based graph decomposition to determine intrinsic subgraphs (clans) in the original graph and to produce a parse tree. The tree is given attributes that specify the node layout. CG then uses tree properties with the addition of "routing nodes" to route the edges. The objective of the system is to provide, automatically, an aesthetically pleasing visual layout for arbitrary directed graphs. Using the clan-based decomposition, CG's drawings are unique in several ways: (1)The node layout can be balanced both vertically and horizontally; (2) Nodes within a clan, a subgraph of nodes that have a common relationship with the rest of the nodes in the graph, are placed close to each other in the drawing; (3) Nodes are grouped according to a two-dimensional affinity rather than a single dimension such as level or rank [13]; (4) The users can contract a clan into a single node and later expand the node to show the subgraph in its original clan; and (5) Crossings reduction processing by clan-based graph decomposition is faster than Sugiyama, Tagawa, and Toda [20, 21] barycentric ordering algorithm.

In addition to the capabilities of the old drawing system [16], several features have been added: (1) The modified barycentric technique is used to reduce crossings. In the modified technique, the components of matrix representation are clans instead of nodes. The users can (2) specify a node's size, shape, and label, and a edge's label in the textual input file; (3) contract a clan (subgraph) into a single node; (4) extract a clan (subgraph) and hide the rest of the graph; and (5) save the drawing into a file in Postscript format.

1. Introduction

Directed graphs, or digraphs, are an excellent means of conveying the structure and operation of many types of systems. They are capable of representing not only the overall structure of such a system, but also the smallest details in a simple and effective way. However, drawing digraphs by hand can be tedious and time consuming, because much time can be spent just trying to plan how the graph should be organized on the page, especially if the number of nodes and edges is large. In addition, it is difficult for a user to draw a graph when the data is generated by applications (e.g., compiler-generated parse trees[1] and dialogue state diagrams generated by reverse engineering [3]). We have developed an automated system capable of converting a textual description of a digraph into a well organized and readable drawing of the digraph.

Many researchers have studied this problem and many graph drawing systems have been developed [see 6 for complete list]. The aesthetic criteria of the systems vary. The objectives may include requirements of uniform edge length, minimum number of edge crossings, straight edges, grid drawings (edges are either horizontal or vertical), minimal bends in the edges, etc. Cruz and Tamassia [4], Tamassia, Batini, and Battista [22], and Messinger [17] have lists of aesthetic criteria of drawings. Some criteria limit the input graphs to a particular class such as planar graphs, trees, graphs with maximum degree of four, or some application-specific graphs such as Petri nets, data-flow diagrams, DBMS models diagrams, digital system schematic diagrams, PERT diagrams, flowcharts, etc. Originally, CG is designed for program dependency graphs of parallel computation, and has been adopted by social networks, automatic graphical user interface design, reverse engineering graphical representations [3]. Like dot [13, 14] and its predecessor DAG [12], CG takes a textual description of an arbitrary directed graph (digraph) and produces a visual representation of it.

The remainder of the paper is divided into several sections which are related to CG: (section 2) clan-based graph decomposition; (section 3) node layout; (section 4) edge routing; (section 5) crossings reduction; (sections 6) subgraphs contract / extract; (section 7) cyclic directed graphs; and (section 8) example of applications.

2. Clan-Based Graph Decomposition

In general, a graph can be decomposed in two ways: (1) application-specific decompositions suggested by the semantics of the input graph; and (2) graph-theoretic decompositions based on syntactic decomposing algorithms [18]. CG uses a new method called clan-based parsing [15, 16]. Clan-based graph decomposition is a parse of a directed acyclic graph (DAG) into a hierarchy of subgraphs. These new subgraphs generated by the decomposition are called clans [9, 10].

Let G be a DAG. A subset $X \subseteq G$ is a clan iff for all x, y $\in$ X and all z $\in$ G - X, (a) z is an ancestor of x iff z is an ancestor of y, and (b) z is a descendant of x iff z is a descendant of y. A simple clan C, with more than three vertices, is classified as one of three types. It is (i) primitive if the only clans in C are the trivial clans; (ii) parallel if every subgraph of C is a clan; or (iii) series if for every pair of vertices x and y in C, x is an ancestor or descendant of y. Any graph can be constructed from these simple clans. Applying clan-based graph decomposition algorithms, any DAG can be decomposed into a tree of subgraphs (clans) whose leaves are trivial clans (graph nodes) and whose internal nodes are complex clans (series or parallel) built from their descendants [15, 16]. The primitive clans are decomposed into series and parallel clans by augmenting edges from all the source nodes of the primitive to the union of the children of the sources [15, 16]. After adding edges (2, 7), (2, 10), (3, 7), (3, 9), (3, 10), and (4, 5) into figure 1(a), sets {2, 3, 4}, {8, 9}, {7, 10}, {7, 8, 9, 10}, and {5, 6, 7, 8, 9, 10} are some of the nontrivial clans. Figure 1(b) is the parse tree of the figure 1(a) graph. After graph decomposition, the series clan are displayed vertically and connected by inter-clan edges, and the parallel clans are displayed horizontally and there are no edges connected between them.

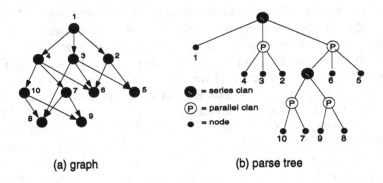

(a) graph   (b) parse tree

Figure 1. Graph and Its Parse Tree

3. Node Layout

The parse tree of the graph is used to preserve node attributes (such as shape, and label) and to provide geometric interpretations to the graph. A node's shape and label are used to determine the size of its bounding box. The default shape for a node is a circle and default label is a number. The user can specify the shape and label for each node in the textual description input file. Figure 2 shows two parse trees with the node shape attribute specified for figure 1. In figure 2(a), the nodes shapes are from system default values. In figure 2(b), the shapes are user specified.

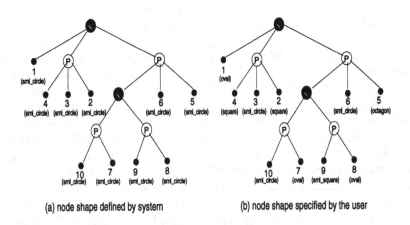

(a) node shape defined by system   (b) node shape specified by the user

Figure 2. Parse Trees with Node Shape Attribute

A "bounding box" with computed dimension is associated with each clan and the nodes in the clan are assigned locations within the bounding box. A bounding box is used to specify the allotted area for a subgraph. To calculate the bounding box for the parse tree, the bounding boxes of leaves are computed first. For a leaf node N, the bounding box length (N.l) and width (N.w) can be determined by N's shape, label size, application-specific settings, or user-specific settings. A series clan is bounded by a rectangle whose length is the sum of the lengths of the component clans and whose width is the maximum width of the component clans. An parallel clan is placed in an area whose width is the sum of the widths of the component clans and whose length is the maximum of the lengths of the component clans. After all the bounding boxes have been computed, CG uses the parse tree with the computed bounding boxes to map the graph onto coordinates in the planar window [15, 16]. Figure 3(a) and 3(c) show the parse trees with bounding boxes computed from figure 2's defined shapes. The bounding boxes of figure 3(a) are computed from figure 2(a)'s shape settings and figure 3(c) are from figure 2(b)'s. Figure 3(b) and 3(d) are node layouts for 3(a) and 3(c) respectively.

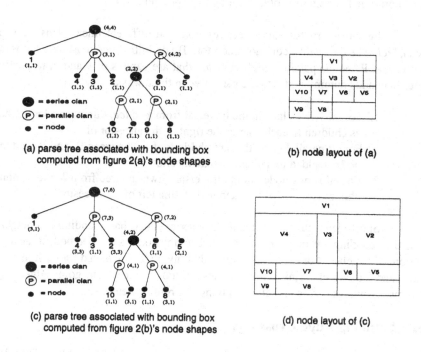

(a) parse tree associated with bounding box computed from figure 2(a)'s node shapes

(b) node layout of (a)

(c) parse tree associated with bounding box computed from figure 2(b)'s node shapes

(d) node layout of (c)

Figure 3. Parse Tree with Bounding Box and Node Layout

4. Edge Routing

Using the parse tree to place nodes is simple and elegant and provides for an aesthetically pleasing balanced placement. If adjacent nodes are connected by straight edges, several unacceptable visualizations may occur when the nodes are placed according to the location attributes.

- Edges could pass through nodes on their path.
- Edges might be superimposed upon other edges
- Unnecessarily long edges may be drawn.
- There may be an unnecessary number of edge crossings.

The first 3 problems listed above are caused by "long" edges, i.e. edges connecting nodes whose levels (y-values) differ by more than one. The traditional solution is to place dummy nodes at each intermediate level and route the long edge through the intermediate nodes [21]. One of the problems with this approach is that the long edges, by passing through nodes placed at arbitrary horizontal displacements, may contain unnecessary bends and may cross other edges unnecessarily. CG provides inter-clan and short-clan heuristics to solve the long edge problems [16].

Inter-clan routes edges between nodes in different linear clans. For edge (x,y), let lca be the nearest common ancestor. By definition, lca must be a series node. Let P_x and P_y be the parallel children of lca that are parents of x and y, respectively. Inter-clan adds dummy nodes to the parse tree in three ways.

1. For all series clans in the traversal from x to lca, dummy nodes are added as children in each clan to the right of the ancestor of x.
2. For all parallel clans that are children of lca between P_x and P_y, a dummy node is added in the appropriate location.
3. For all series nodes in the traversal down the tree from lca to y, dummy children are added for each node to the left of y's ancestor.

Short-clan is invoked when node or series clan C has bounding box height less than the bounding box height of its parent. Dummy nodes are added both at the top and bottom of the clan. For each clan source, dummy nodes are added for each in-coming edge, and for each clan sink, dummy nodes are added for each out-going edge. Figure 4 shows drawings before and after applying CG routing heuristics.

5. Reducing Edge Crossings

Minimizing the number of edge crossing is a NP-hard problem [7, 8]. Warfield [23] developed a heuristic method, barycentric ordering, for two-level graphs. A value called barycenter is computed for each of the vertices in the two levels. For each vertex, this value is a weighted average of the horizontal positions of the vertices in the adjacent level to which the vertex is connected. The barycenters of each of the vertices in a level are computed and the vertices are sorted according to their barycenters. Carpano [2], and Sugiyama, Tagawa, and Toda [20, 21] generalized the two-level

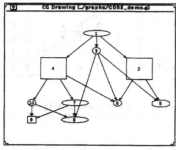

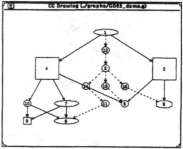

(a) original drawing (edge (3,5) & (3,8) pass node 2 & 7 respectively)

(b) dummy nodes 11 added by inter-clan heuristic, and 12, 15, & 16 added by short-clan heuristic

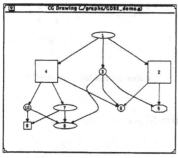

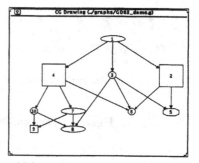

(c) new drawing with dummy nodes added

(d) final drawing with unnecessary bends removed

Figure 4. Drawings Before and After Applying Inter-clan and Short-clan Heuristics

barycentric method to reduce edge crossings for k-level hierarchical graphs. CG adopts the concept of barycentering to be used in conjunction with clans. A matrix is used to describe the connections of subgraphs (clans) instead of nodes. The value [i, j] of a matrix is defined as the number of the connecting edges between $clan_i$ and $clan_j$.

In CG, matrices are formed for the component clans of series clans not of parallel clans, because, by definition, there are no connections between parallel clans. There are fewer matrices used in CG and each is no larger than a matrix of individual nodes, because they are formed by subgraphs. As a result, CG's crossings reduction processing is faster than the original barycentric method. Figure 5 shows the matrix representation of the original barycentric technique and CG. After adding edges (c, m) and (d, m) into figure 5(a), the parse is produced as figure 5(c). According to the parse tree, only two matrices are required, M_1 for series clan C_0 and M_2 for C_2. There is no matrix needed for series clans C_1, C_5, C_4, and C_3, because each of them contains two components and at least one of the two components is a single node. In the matrix M_1, the value of $[C_2, C_3]$ is 2, because there are two edges between clan C_2 and C_3.

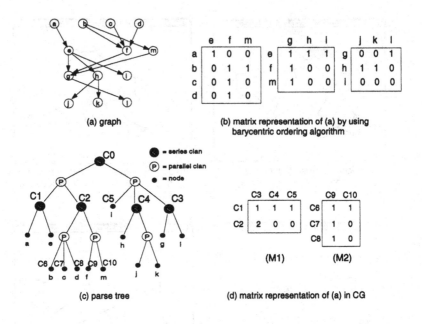

(a) graph

(b) matrix representation of (a) by using barycentric ordering algorithm

(c) parse tree

(d) matrix representation of (a) in CG

Figure 5. Graph and Matrix Representation

6. Subgraphs (Clans) Contract/Extract

Because the graph layout is based on a parse tree created by extracting subgraphs, it is possible to abstract those subgraphs and represent them by a single node, or just to display the selected subgraph. CG supports subgraph contraction, expansion, and extraction.

Since clans are defined as sets of nodes with identical ancestors and descendants within the rest of the graph, clans can easily be contracted to a single node. By selecting a single node, the user can contract the smallest non-trivial clan containing that node into a single node. Any node not in the clan that was connected to a clan source or sink will be connected to the contracted node. By allowing segments of the graph to be contacted, the user can simplify dense graphs for viewing by contracting those parts which are not relevant to the investigation. Contracted nodes can be expanded to show the original clan configuration. Similarly, CG can extract and display only the clan, ignoring the rest of the graph. In figure 6(b), the user contracts the clan containing node 6. In figure 6(c), the user displays the clan containing node 6. The parse tree of figure 6(a) is in figure 1(b).

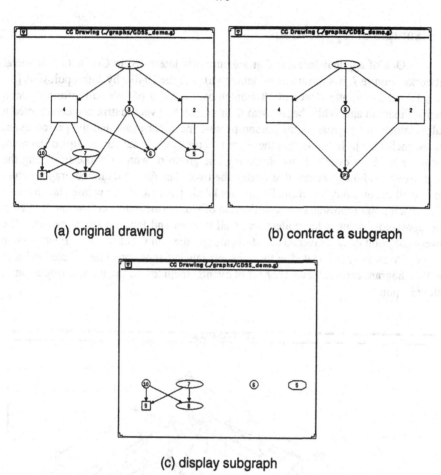

(a) original drawing

(b) contract a subgraph

(c) display subgraph

Figure 6. Subgraph Contract/Extract

7. Cyclic Directed Graphs

The clan-based graph decomposition can be used to draw cyclic directed graphs by reversing certain edges [19]. A simple transformation is required to apply the graph decomposition method to cyclic graphs. Cycles can be found in a depth-first graph traversal. To break a cycle, the edge that identifies the cycle is given the reverse orientation. When the layout is ready, its orientation will be corrected. This method of breaking cycles will show a cycle not as a circular arrangement of nodes, but as a vertical line of nodes with an edge connecting the bottom to the top. This view is consistent with some applications such as the visualization of program control flow graphs. The general philosophy of a top to bottom flow for directed graphs is supported by this layout, with only few edges reversing that direction.

8. Example of Applications

One of application areas that are currently interested in CG is that of social networks. Figure 7. is an example of data reported in the 1940's by anthropologists [5]. They tallied the co-attendence of 18 women over a series of 14 small informal social events. Freeman and White began with their person by event matrix and constructed a Galois lattice that represents the person-person, the event-event and the person-event dependencies [11]. It shows that there are two pretty clear-cut sub-groups of women, and three kinds of events: those involving one group of women, those involving the other group, and those events that bridge the two. The 65 points in the graph picture the overall dependency structure. Events are labeled as A to N and women labeled as 1 to 18. Each point represents some collection of women and some collection of events. The uppermost point is the collection of all women and the null set of events. The lowermost point is the universal set of events and the null set of women. Each woman (or set of women) participated in those events labeled at or above her labeled point in the line diagram and each event (or set of events) included all the women labeled at or below its point.

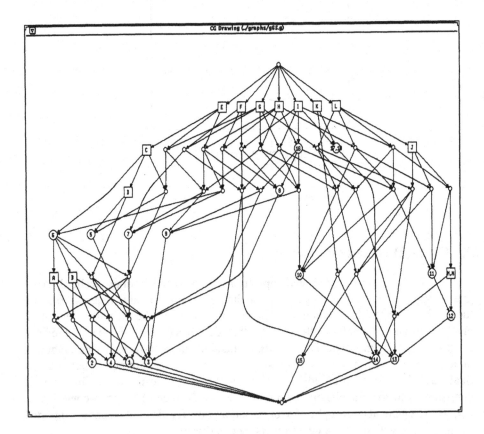

Figure 7. Social Networks Lattice

Acknowledgements

Thanks to Jeff Okerson for helping us to add the feature of saving the drawing to file in PostScript format.

References

1. K. Andrews, R. R. Henry, and W. K. Yamamoto, "Design and implementation of the UW illustrated compiler, " Univ. Washington, Dep. of Computer Science, Tech. Rep. 88-03-07, Mar. 1988.

2. M. J. Carpano, "Automatic Display of Hierarchized Graphs for Computer Aided Decision Analysis", IEEE Trans. on Systems Nab and Cybernetics, Vol. SMC-10, No. 11, 705-715, 1980.

3. J. H. Cross II and R. S. Dannelly, "Reverse Engineering Graphical Representations of X Source Code," will be appeared in International Journal of Software Engineering and Knowledge Engineering, Spring, 1996.

4. I. F. Cruz, and R. Tamassia, "How to Visualize a Graph: Specification and Algorithms, " Tutorial on Graph Drawing, available at http://www.cs.brown.edu/calendar/gd94, 1994.

5. A. Davis, B. Gardiner and M. Gardiner, 1941, Deep South, Chicago: U. of Chgo. Press

6. G. Di Battista, P. Eades, R. Tamassia, I. Tollis, "Algorithms for Drawing Graphs: an Annotated Bibliography", June, 1994, available via ftp from wilma.cs.brown (128.148.33.66) in file pub/papers/compgeo/gdbiblio.tex.Z.

7. P. Eades and D. Kelley, "Heuristics for drawing 2-layered graphs," Ars Combinatoria , Vol. 21-A, 89-98, 1986.

8. P. Eades and N. Wormald, "Edge Crossings in Drawings of Bipartite Graphs", Algorithmica, Vol. 11, No. 4, 379-403, April 1994.

9. A. Ehrenfeucht and G. Rozenberg, "Theory of 2-Structures, Part I: Clans, Basic Subclasses, and Morphisms," Theoretical Computer Science, Vol. 70, 277-303, 1990.

10. A. Ehrenfeucht and G. Rozenberg, "Theory of 2-Structures, Part II: Representation Through Labeled Tree Families," Theoretical Computer Science, Vol. 70, 305-342, 1990.

11. L. C. Freeman and D.R.White, 1993, "Using Gaiois Lattices to Represent Network Data," in P.V. Marsden, ed., Sociological Methodology 1993, Oxford: Blackwell, 127-146.

12. E. R. Gansner, S. C. North, and K. P. Vo, "DAG - A Program that Draws Directed Graphs," Software - Practice and Experience, Vol 18, No. 11, 1047-1062, Nov. 1988.

13. E. R. Gansner, E. Koutsofios, S. C. North, and K. P. Vo, "A Technique for Drawing directed Graphs," IEEE Transactions on Software Engineering, Vol. 19, No. 3, 214-230, 1993.

14. Koutsofios and S. North, "Drawing Graphs with dot", Dot User's Manual, AT&T Bell Labs, Murray Hill, N. J., 1994

15. C. L. McCreary and A. Reed"A Graph Parsing Algorithm and Implementation," Tech. Rpt. TR-93-04, Dept. of Comp. Sci and Eng., Auburn U. 1993.

16. C. McCreary, F. S. Shieh, and H. Gill, "CG: a Graph Drawing System Using Graph-Grammar Parsing," Lecture Notes in Computer Science, Vol. 894, 270-273, Springer-Verlag, 1995.

17. E. B. Messinger, "Automatic Layout of Large Directed Graphs," Univ. of Washington, Ph.D. dissertation, 1988.

18. E. B. Messinger, L. A. Rowe, and R. R. Henry, "A Divide-and-Conquer Algorithm for the Automatic Layout of Large Directed Graphs," IEEE Trans. on Sys. Man, and Cyb., Vol. 21 No. 1,, 1-12, 1991.

19. L. A. Rowe, M. Davis, E. Messinger, and C. Meyer, "A Browse for Directed Graphs," Software-Practice and Experience, Vol. 17(1), 61-76, January 1987.

20. K. Sugiyama, S. Tagawa, and M. Toda, "Effective Representation of Hierarchies," in Proc. IEEE Int. Conf. Cybernetics and Society, New York, Oct. 1979.

21. K. Sugiyama, S. Tagawa and M. Toda, "Methods for Understanding of Hierarchical system structures," IEEE Trans. on Sys. Man, and Cyb., SMC-11, 109-125, 1981.

22. R. Tamassia, G. D. Battista, and C. Batini, "Automatic Graph Drawing and Readability of Diagrams," IEEE Trans. on Sys. Man, and Cyb., Vol. 18, No. 1, 61-79, 1988.

23. J. N. Warfield, "Crossing Theory and Hierarchy Mapping," IEEE Trans. on Syst., Man, and Cybernetics, Vol. 7, No. 7, 505-523, Jul. 1977.

TOSCANA
Management System for Conceptual Data

Martin Skorsky

Technische Hochschule Darmstadt
D-64277 Darmstadt
skorsky@mathematik.th-darmstadt.de

TOSCANA is a software for the visualization of data with nested line diagrams. It is a tool for navigating in databases and for the retrieval of objects in a database. It enables the user to discover relationships and implications between attributes.

TOSCANA is based on Formal Concept Analysis as introduced in [Wi82]. This method formalizes a *relation* I between a set G of *objects* and a set M of *attributes* by the *formal context* (G, M, I). A formal context can be handled as a table or a view of a relational database. The formal context determines its *concept lattice* which is the structure of the relation subconcept–superconcept. The concept lattices are graphically represented as line diagrams. For the basic notions of Formal Concept Analysis see also [DP90, chapter 11] or in [GW95].

The line diagrams for practical applications get very large. TOSCANA uses *nested line diagrams* which were introduced in [Wi84] (see also [Wi89]). We split a given context into parts, draw a line diagram of each part and then nest these line diagrams into each other. This gives a simplified diagram because groups of parallel lines are replaced by one line and the ellipses around the small diagrams (see Figure 2). TOSCANA automatizes the nesting of the diagrams and allows a flexible choice of the diagrams. As mentioned above, the formal context can be kept as a table or view in a database. TOSCANA selects the information for the nested line diagram from such a database. Thus, TOSCANA helps to analyze large sets of data and enables the user to navigate through this database.

The ideas for TOSCANA are first described in [VWW91]. The further development of these ideas is reflected in [SSVWW93], [KSVW94], [VW95]. This last paper was presented to the Graph Drawing conference in 1994 at Princeton.

We will demonstrate one basic direction of applying TOSCANA. The example (see 1, 2, 3) shows how TOSCANA is used to structure a set of documents. The documents, which take the role of the objects of the formal context, are related to search keys, which are the attributes. The documents are German law documents and technical norms. The search keys are grouped into themes. Each theme is represented by a separate line diagram. Now a users sees the documents not only for one search key but for the whole group of keys for the selected theme. The documents were unfolded from the very specific documents to the general ones. The combination of themes allow complex selections and searches.

The screen dumps in figure 1, 2, 3 show the functionallity of TOSCANA for navigating within the diagrams and selecting objects. The functions to improve

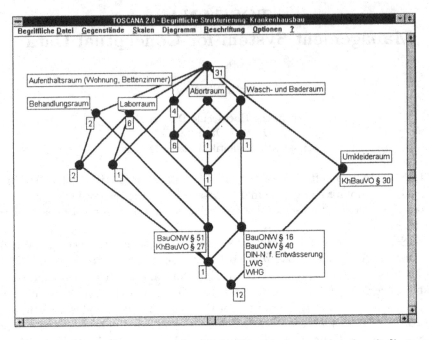

Fig. 1. line diagram about the theme 'functional rooms in a hospital'

the layout are not shown.

1. Figure 1: line diagram about the theme 'functional rooms in a hospital' ('Funktionsräume im Krankenhausbau')
2. Figure 2: nested line diagram of two layers; the first layer (ellipses) shows the theme 'functional rooms in a hospital' (same structure as in Figure 1), the second layer (small diagrams) shows the theme 'reliability in service and fire protection' ('Betriebs- und Brandsicherheit').
3. Figure 3: Zooming into the ellipse labelled with

| BauONW §16 |
| BauONW §40 |
| DIN-N. f. Entwässerung |
| LWG |
| WHG |

selects those five documents and shows how these documents concern 'reliability in service and fire protection' ('Betriebs- und Brandsicherheit')

Searching and Navigating is supported in two directions. First we can combine up to 4 attributes or themes to a nested line diagram. Nested diagrams with more than 4 attributes can not be handled in a practical way. The second direction is the process of zooming in and out. The user clicks with the mouse on an ellipse and gets a more detailed diagram of the situation specified by this ellipse. This is connected with a selection in the database. There is no limit to

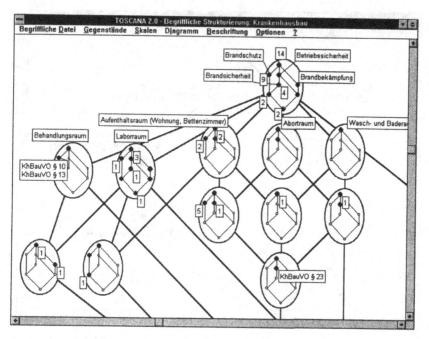

Fig. 2. nested line diagram of the themes 'functional rooms in a hospital' and 'reliability in service and fire protection'

the number of attributes or themes for the zoom process, thus we can refine the selection of the objects in the database.

The current implementation of TOSCANA runs on a PC with MS-Windows™ (486DX/50 processor or better, 8 MB RAM). It takes the data from databases managed by MS-Access™. The queries are done with SQL and transfered via DDE. The software is developed in C++ and uses a class library of concept analysis classes written by Frank Vogt.

References

[DP90] B. A. Davey and H. A. Priestley. *Introduction to Lattices and Order.* Cambridge University Press, Cambridge, 1990.

[GW95] B. Ganter and R. Wille. *Formale Begriffsanalyse.* B.I.–Wissenschaftsverlag, to appear.

[KSVW94] W. Kollewe, M. Skorsky, F. Vogt, and R. Wille. TOSCANA — Ein Werkzeug zur begrifflichen Analyse und Erkundung von Daten. In: R. Wille and M. Zickwolff, editors, *Begriffliche Wissensverarbeitung — Grundfragen und Aufgaben,* pages 267–288, Mannheim, 1994. B. I.–Wissenschaftsverlag.

[PC93] PC Magazine, Volume 12, number 2. Ziff-Davis Publishing Company, Boulder 1993, pages 110–247.

486

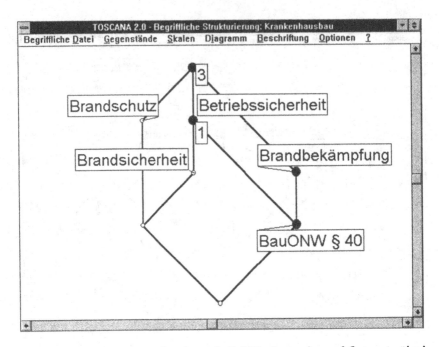

Fig. 3. line diagram about the theme 'reliability in service and fire protection'

[SSVWW93] P. Scheich, M. Skorsky, F. Vogt, C. Wachter, and R. Wille. Conceptual data systems. In: O. Opitz, B. Lausen, and R. Klar, editors, *Information and classification*, pages 72–84, Heidelberg, 1993. Springer–Verlag.

[VWW91] F. Vogt, C. Wachter, and R. Wille. Data analysis based on a conceptual file. In: H.-H. Bock and P. Ihm, editors, *Classification, data analysis, and knowledge organization*, pages 131–140, Berlin–Heidelberg, 1991. Springer–Verlag.

[VW95] F. Vogt and R. Wille. TOSCANA — A Graphical Tool for Analyzing and Exploring Data. In: R. Tamassia and I. G. Tollis, editors, *Graph Drawing*, pages 226–233, Heidelberg, 1995. Springer–Verlag.

[Wi82] R. Wille. Restructuring lattice theory: an approach based on hierarchies of concepts. In: I. Rival, editor, *Ordered sets*, pages 445–470, Dordrecht–Boston, 1982. Reidel.

[Wi84] R. Wille. Liniendiagramme hierarchischer Begriffssysteme. In: H.-H. Bock, editor, *Anwendungen der Klassifikation: Datenanalyse und numerische Klassifikation*, pages 32–51. Indeks–Verlag, Frankfurt, 1984. Line diagrams of hierarchical concept systems (engl. Translation). *Int. Classif.* 11 (1984), 77–86.

[Wi89] R. Wille. Lattices in data analysis: how to draw them with a computer. In: I. Rival, editor, *Algorithms and order*, pages 33–58, Dordrecht–Boston, 1989. Kluwer.

Graph Layout Adjustment Strategies

Margaret-Anne D. Storey[12] and Hausi A. Müller[2]

[1] School of Computing Science, Simon Fraser University, Burnaby, BC, Canada.
[2] Department of Computer Science, University of Victoria, Victoria, BC, Canada.
{mstorey,hausi}@csr.uvic.ca

Abstract. *When adjusting a graph layout, it is often desirable to preserve various properties of the original graph in the adjusted view. Pertinent properties may include straightness of lines, graph topology, orthogonalities and proximities. A layout adjustment algorithm which can be used to create fisheye views of nested graphs is introduced. The SHriMP (Simple Hierarchical Multi-Perspective) visualization technique uses this algorithm to create fisheye views of nested graphs. This algorithm preserves straightness of lines and uniformly resizes nodes when requests for more screen space are made. In contrast to other layout adjustment algorithms, this algorithm has several variants to preserve additional selected properties of the original graph. These variants use different layout strategies to reposition nodes when the graph is distorted. The SHriMP visualization technique is demonstrated through its application to visualizing structures in large software systems.*

1 Introduction

Although the computer screen is relatively small, it is easy to fill it with so much information and detail that it completely overwhelms the user. It is not the amount of information displayed that is relevant, but rather how it is displayed [17]. Frequently large knowledge bases are represented by graphs. Layout algorithms are often used to present graphs in a more meaningful format. Many visualization tools allow a user or other applications to interact with and adjust these graph layouts.

Misue *et al.* in [8] describe three properties which should be maintained in adjusted layouts to preserve the user's *mental map*: orthogonal ordering, clusters and topology. The orthogonal ordering between nodes is preserved if the horizontal and vertical ordering of points is maintained. Clusters are preserved by keeping nodes close in the distorted view if they were close in the original view. The topology is preserved if the distorted view of the graph is a homeomorphism of the original view. Other properties which are important to preserve for some applications include straightness of lines, orthogonality of lines parallel to the x and y axes [9], and relative sizes of nodes.

It is impossible to allocate more space to a portion of a graph constrained to fit on a fixed screen size without distorting one or more of the properties described above. The *type* of layout and its application should be considered when deciding which properties to preserve or distort. In a simple grid layout, it is preferable to preserve parallel and orthogonal relationships among nodes. This is important for the visualization of large circuit diagrams. For other layouts, such as subway routing maps, the proximity relationships among nodes is a more important property to preserve.

One layout adjustment problem is that of showing more detail (perhaps by increasing the size of nodes) without hiding the remainder of the graph. Approaches based on the fisheye lens paradigm seem well suited to this task. However, many of these techniques are non-trivial to implement and their distortion techniques often cannot be altered to suit different graph layouts. In addition, several techniques have the side effect of causing too much distortion in some areas of the graph. For some applications, it would be better to evenly distribute the distortion throughout the entire graph by uniformly scaling nodes outside the focal points.

This paper presents the SHriMP layout adjustment algorithm. This algorithm is suitable for creating fisheye views of nested graphs[3] by uniformly resizing nodes when requests for more screen space are made. It preserves straightness of lines and non-overlapping nodes in the original view will not overlap in the adjusted view. Moreover, this algorithm is flexible in its distortion technique as it can be altered to suit different graph layouts.

Several fisheye view methods are briefly discussed in Section 2. The subsequent section presents the SHriMP layout adjustment algorithm. Section 4 describes different layout strategies which are used to preserve important properties of various graph layouts. Finally, the SHriMP visualization technique is applied to the task of visualizing software structures.

2 Fisheye Views

Manipulating large graphs on a small screen can be problematic. Because of this, various methods have been proposed for displaying and manipulating large graphs. One approach partitions the graph into pieces, and then displays one piece at a time in a separate window. However, context is lost as detail is increased. Another approach makes the entire drawing of the graph smaller, thus preserving context, but the smaller details become difficult to read and interpret as the scale is reduced. A combination view can be given by providing context in one window and detail in another but this requires that the user mentally integrate the two—not always an easy task.

Techniques have been developed to view and navigate detailed information while providing the user with important contextual cues. Fisheye views, an approach proposed by Furnas [4], provide context and detail in one view. This display method is based on the fisheye lens metaphor where objects in the center of the view are magnified and objects further from the center are reduced in size. In Furnas' formulation, each point in the structure is assigned a *prior-*

---

[3] Nested graphs, in addition to nodes and arcs, contain composite nodes which are used to implicitly communicate the hierarchical nature of the graph [5].

ity that is calculated using a *degree of interest (DOI)* function. Objects with a priority below a certain threshold are filtered from the view.

In order to deemphasize information of lesser interest, several variations on this theme have been developed that use size, position, colour, or shading in addition to filtering. For example, *SemNet* uses three-dimensional point perspective that displays close objects larger than objects further away [3]. *Graphical Fisheye Views*, a technique developed by Sarkar and Brown [13], magnifies points of greater interest and correspondingly demagnifies vertices of lower interest by distorting the space surrounding the focal point. Therefore, nodes that are further away from the focal point appear smaller. The *Continuous Zoom Algorithm* by Ho *et al.* [2], suitable for interactively displaying hierarchically-organized, two-dimensional networks, allows users to view and navigate nested graphs by expanding and shrinking nodes. This algorithm uniformly resizes nodes to provide space for focal points. However, the zoom-out operation is not the reverse of the zoom-in operation.

Two methods based on a rubber sheet metaphor are described by Sarkar *et al.* in [14]. The first method, *Orthogonal Stretching*, uses handles to stretch an area of the graph in the x and y directions. Items which fall in these areas are stretched uniformly, while everything outside of these areas contract uniformly. The second method, *Polygonal Stretching*, allows a user to specify a polygonal region. Items inside the polygon are scaled as the polygon is stretched and the rest of the view is scaled smoothly to integrate it with the enlarged region. This method does not have an inverse mapping once a region is scaled.

Misue *et al.* describe three approaches in [9]. They are the *Biform Display Method* (BF), the *Fisheye Display Method* (FE) and the *Orthogonal Fisheye Display Method* (OFE). The BF method uses view areas, where items inside the view areas are uniformly magnified, and items outside of the areas are uniformly demagnified. The BF method is similar to the Continuous Zoom algorithm and the Orthogonal Stretching techniques; all three of these approaches preserve straightness of lines and orthogonal ordering in the distorted view. The FE uses an inverse tangent function to apply a fisheye lens to the view. Objects closer to the center of the lens appear increasingly larger. However, the orthogonal ordering of points is not maintained using this approach. OFE, a variant of FE, does maintain the orthogonal orderings, but this method (as well as FE) tends to have too much distortion for some parts of the graph. A survey of these approaches and others such as *Perspective Wall* and *Cone Trees* are described by Noik in [11].

3   The SHriMP Layout Adjustment Algorithm

The SHriMP layout adjustment algorithm is elegant in its simplicity. Nodes in the graph uniformly *give up* screen space to allow a node of interest to grow.

Figure 1 gives an example where one node is enlarged. Figure 1(a) shows the graph before the node of interest (the center node) is scaled by the desired factor.

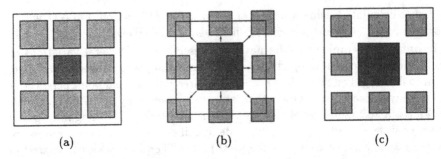

Fig. 1. (a) The graph before any scaling is done. (b) The node of interest (center node) grows by the desired scale factor and pushes its siblings outward. (c) Finally, the node and its siblings are scaled to fit inside the screen. This last step is the only step visible to the user of SHriMP, the other step is shown to describe the algorithm only.

The node grows by *pushing* its sibling[4] nodes outward as if there were infinite screen space, see Fig. 1(b). The node and its siblings are then scaled around the center of the screen so that they will fit inside the available space, see Fig. 1(c).

Each sibling is pushed outward by adding a translation vector $[\mathbf{T}_x, \mathbf{T}_y]$ to its coordinates. The scaling operation makes use of an equation which scales an object around an arbitrary fixed point [6]. In this case, the fixed point is the center of the screen, (x_p, y_p), and the scale factor, s, is equal to the size of the screen divided by the requested size of the screen. Equations (1) and (2) show the function applied to the coordinates (x, y) of the sibling nodes.

$$x' = x_p + s\,(x + \mathbf{T}_x - x_p) \tag{1}$$
$$y' = y_p + s\,(y + \mathbf{T}_y - y_p) \tag{2}$$

In a nested graph, the node of interest pushes the boundaries of its parent node outward also. The parent in turn pushes its siblings out and so on until the root is reached. As a final step, everything is scaled to fit inside the root.

To shrink a node that has previously been enlarged, the scale factor, s, will be < 1. The zoom-out operation will be the reverse operation of the zoom-in, and vice versa, when the scale factor is set appropriately. A simple extension allows for multiple focal points of varying scaling factors. To scale multiple nodes, each node in turn may grow (or shrink) pushing outward (or pulling inward) their siblings. Finally, nodes are scaled to fit inside the available space.

This algorithm is simple, fast and effective. When considering only one focal point, the algorithm is linear with respect to the number of nodes in the graph. When scaling multiple nodes, it is $O(k\,n)$ where k equals the number of focal points and n is the total number of nodes in the graph. In most applications, k is much smaller than n. The next section describes how different translation vectors may be used for repositioning siblings when a node is resized.

---

[4] Nodes which have the same parent in the nested graph are siblings.

4 Layout Strategies

When zooming a node in a graph layout it may be desirable to maintain pertinent properties of the layout such as orthogonality, proximity, straightness of lines and the overall topology of the graph. The layout strategies presented in this paper preserve straightness of lines and the graph topology of the nodes. However, it is difficult to preserve both orthogonality and proximity relations using a fixed screen size. Depending on the graph layout, it is often only necessary to preserve one of these properties.

In the SHriMP layout adjustment algorithm, a node grows (or shrinks) by pushing its sibling nodes outward along vectors. The translation vectors determine how the sibling nodes are repositioned after a request for more space is made. The following section describes three methods for setting the magnitude and direction of a vector. Figure 2 shows a simple grid layout of a graph. Figure 2(a) shows the grid before any scaling has been done. Parts (b),(c) and (d) show how different translation vectors can alter the appearance of the graph when it is distorted to allocate more space to the center node.

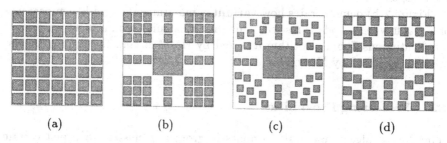

| (a) | (b) | (c) | (d) |

Fig. 2. (a) Grid before any scaling is done. (b) The center node is scaled using a layout strategy which preserves orthogonality in the graph. (c) and (d) The center node is scaled using layout strategies which are more suited to preserving proximities in the graph.

4.1 Preserving Orthogonality

One layout strategy, called **variant 1**, preserves the orthogonal relationships among nodes. The graph is partitioned into nine partitions by extending the edges of the scaled node. The translation vector for each sibling node is calculated according to the partition containing its center. Figure 3 shows the translation vectors for each of the nine partitions. For example, the translation vector for those nodes in partition 1 is $\mathbf{T} = [-d_x, -d_y]$, where d_x and d_y are the x and y differences between the new size of the scaled node and its previous size. All sibling nodes above (below) the scaled node are pushed upward (downward) by the same amount, thereby maintaining the orthogonality relationships of these nodes with respect to the y axis. Similarly, nodes to the right (left) are pushed

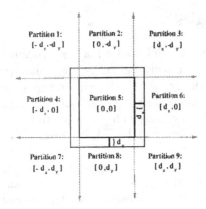

Fig. 3. In this layout strategy, the translation vector for each sibling node is determined by the partition containing its center.

right (left) by the same amount, maintaining the orthogonality relationships with respect to the x axis.

Figure 2(b) demonstrates how **variant 1** maintains the grid-like appearance of a graph when a node is resized. This variant is quite similar in appearance to the Continuous Zoom algorithm reported by Ho *et al.* and the Biform Display method described by Misue *et al.*

4.2   Preserving Proximities

Many layout algorithms position nodes in groups or clusters to depict certain relationships in the graph. For example, spring layout algorithms position nodes which are highly connected closer to one another [1]. Therefore, a layout strategy which keeps those nodes that are close in the original view close in the distorted view would be beneficial.

This subsection describes another variant, called **variant 2**, where proximity relationships are preserved by constraining each sibling node to stay on the line connecting its center to that of the node being resized. When a node is resized, it pushes a sibling node outward along this line. The direction of each sibling node's translation vector is equal to the direction of the line connecting the centers. The magnitude of this vector is equal to the distance that a corner point of the scaled node moves as it is enlarged.

In Fig. 4(a), the node A is enlarged. d_x and d_y are the x and y differences between the new size of A and its previous size. (x_a, y_a) is the center of A. (x_b, y_b) is the center of B, a sibling of A. The direction of B's translation vector is equal to the *direction* of the connecting line, and its magnitude is equal to μ as per (3) Equations (4) and (5) are used to calculate the translation vector $\mathbf{T} = [\mathbf{T}_x, \mathbf{T}_y]$ Note that μ is constant for all sibling nodes, and need only be calculated once.

493

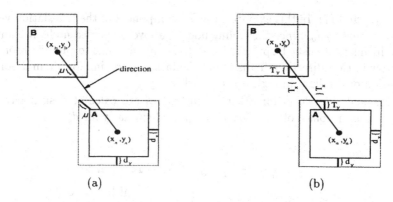

(a) (b)

Fig. 4. (a) Sibling node, B, is pushed outward along the line connecting its center and the center of A, the node being scaled. Each node is pushed out by the distance μ. (b) A sibling node, B, is pushed along the vector between its center and that of A, the node being scaled. The distance it is pushed along this vector is determined by the displacement of the intersecting node's edge as it moved along the vector.

$$\mu = \sqrt{d_x^2 + d_y^2} \tag{3}$$

$$T_x = \mu \frac{x_a - x_b}{\sqrt{(x_a - x_b)^2 + (y_a - y_b)^2}} \tag{4}$$

$$T_y = \mu \frac{y_a - y_b}{\sqrt{(x_a - x_b)^2 + (y_a - y_b)^2}} \tag{5}$$

This strategy has been applied to the grid in Fig. 2(c). This figure demonstrates that this strategy does not preserve all of the orthogonal relationships of **variant 1**, but that it does appear to keep those nodes which were close in the original view, close in the transformed view. However, the screen space is not being used effectively by this method. **Variant 3**, described in the next subsection, makes better use of screen space while maintaining similar proximity relations.

4.3 An Alternative Proximity Preservation Strategy

In **variant 3**, the direction of the translation vectors is the same as in **variant 2**, but the magnitude (μ) is not the same for all sibling nodes. Instead, a node pushes out sibling nodes according to the displacement of the scaled node's edge as it moved along the line connecting their centers. This strategy makes use of the fact that nodes are drawn as rectangles in SHriMP views. **Variant 3** is applied to a grid layout in Fig. 2(d).

In Fig. 4(b), node A is being enlarged. m is equal to the slope of the line connecting the centers of A and B. The $\mathbf{T}_x$ and $\mathbf{T}_y$ components are calculated using (6) and (7). In this example, the $\mathbf{T}_y$ component of the translation vector is simply equal to d_y. Since the sibling node is above the scaled node, it is intuitive that it must be pushed upwards by at least this amount to provide room for A to grow in that direction. $\mathbf{T}_x$ is then calculated by solving for $\mathbf{T}_x$ in a point-line equation of the line through (x_b, y_b) and (x_a, y_a).

Translation vectors for other sibling nodes are calculated similarly, where $-d_x$ is used in place of d_x when $x_b < x_a$, and similarly for d_y.

$$
\mathbf{T}_x = \begin{cases} \frac{1}{m}(y_b \pm d_y - y_a) + x_a - x_b & \text{if } |m| >= 1 \\ 0 & \text{if } |m| = \infty \\ \pm d_x & \text{otherwise} \end{cases} \tag{6}
$$

$$
\mathbf{T}_y = \begin{cases} m(x_b \pm d_x - x_a) + y_a - y_b & \text{if } 0 < |m| < 1 \\ 0 & \text{if } m = 0 \\ \pm d_y & \text{otherwise} \end{cases} \tag{7}
$$

Figure 5 shows a spring layout of a graph. Figure 5(b) shows the result of applying **variant 3** when scaling several nodes. The general appearance of the spring layout is maintained by retaining the proximity relationships between nodes in the adjusted view. Figure 5(c) shows the same nodes scaled using the orthogonality preservation layout strategy, **variant 1**, which distorts some of the clusters created by the spring layout and destroys the user's mental map in the process.

4.4 Hybrid Strategies

A graph layout may be composed from a variety of layout algorithms [7]. For example, the overall structure of the graph may be that of a tree, where subgraphs are laid out using a simple grid strategy. When zooming a node in any part of the graph, the overall layout as well as the subgraph layouts should be maintained. This is possible as the algorithm can apply different layout strategies to the sibling nodes when a request for more or less space is made. In other words, the method of calculating translation vectors need not be the same for all sibling nodes.

In a tree layout it may be preferable to preserve parallel relationships between levels in the tree hierarchy while repositioning children so that they remain close to their parent node. A hybrid strategy based on **variant 1** can retain both properties for these layouts. If we set the $\mathbf{T}_x$ component of the translation vector for children of the node of interest to 0, the children will be not be spread apart horizontally. Figure 6 shows the advantage of applying this hybrid strategy to a tree layout.

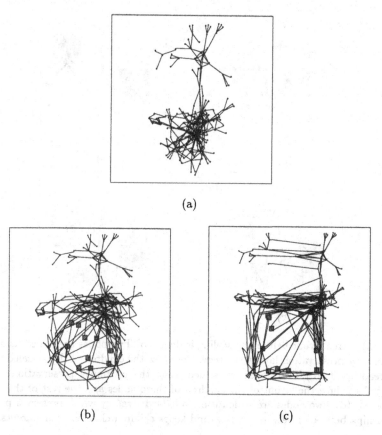

(a)

(b)   (c)

Fig. 5. (a) A spring layout of a graph before any scaling is done. (b) Several nodes are scaled using **variant 3**, which preserves proximities. The clusters are not distorted using this strategy. (c) The same nodes are scaled using the orthogonality preservation layout strategy, **variant 1**. Note how some of the clusters created by the spring layout are distorted.

5 Visualizing Software Structures Using SHriMP Views

The SHriMP visualization technique has been incorporated in the Rigi system for documenting and manipulating structures of large software systems [15]. The Rigi reverse engineering system is designed to analyze, summarize, and document the structure of large software systems [18]. The SHriMP visualization technique helps to alleviate problems of losing context while exploring the many relationships in a multi-million line legacy system.

For large software systems, understanding the structural aspects of a system's architecture is initially more important than understanding any single component [18]. The SHriMP visualization technique is particularly well suited to showing different levels of abstraction in a system's architecture concurrently. Nodes are used to represent artifacts in the software, such as functions or data

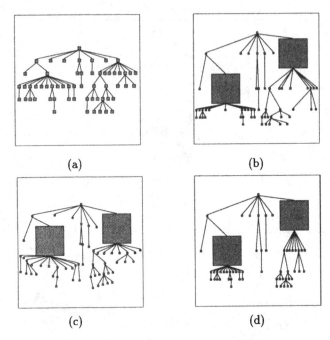

Fig. 6. (a) A tree layout before any scaling is done. (b) Two nodes are scaled using the orthogonality preservation layout strategy. Note how the children of these scaled nodes are spread apart. (c) Two nodes are scaled using the proximity preservation layout strategy. Note how the layout of the children of these nodes and the rest of the graph is distorted. (d) Two nodes are scaled using a hybrid strategy which preserves parallel relationships between levels in the tree, and keeps children close to their parents.

variables. Arcs represent dependencies among these artifacts, such as call dependencies. Composite nodes correspond to subsystems in the software. The nesting feature of nodes communicates the hierarchical structure of the software (e.g. subsystem or class hierarchies). The user may incrementally expose the structure of the software by magnifying subsystems of current interest. The SHriMP layout adjustment algorithm provides the ability to browse groups of nodes and arcs in large software systems. By zooming on different areas in a large graph, a software engineer can quickly identify important features such as highly connected nodes and candidate subsystems.

Reverse engineering a system involves information extraction and information abstraction [10]. One objective of a reverse engineer is to obtain a mental model of the structure of a subject software system and to communicate this model effectively. A reverse engineer uses visualization techniques to facilitate the identification of candidate subsystems and to assist in the visualization of structures and patterns in the graph. The application of graph layout algorithms play a key role in communicating the reverse engineer's mental model, and in the identification of structures and patterns in the software.

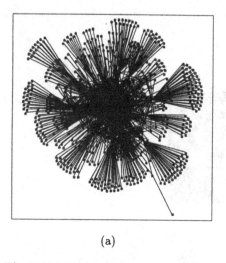 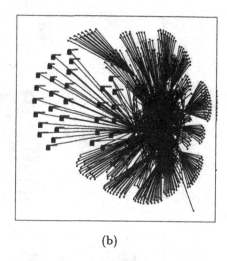

(a)   (b)

Fig. 7. (a) The spring layout algorithm has been applied to the SQL/DS software system. This algorithm helped to expose clusters of nodes on the fringe of the graph, which are candidates for subsystems. (b) One of the clusters of nodes is enlarged to show more detail.

Figure 7(a) shows the result of applying a spring layout algorithm to the graph representation of the SQL/DS software [18]. This layout algorithm assisted in the identification of several candidates for subsystems, by clustering groups of nodes around the fringe of the graph. In Fig. 7(b), the user has selected and zoomed one of these clusters, in order to see more detail. By using the SHriMP layout adjustment algorithm which preserves the proximity relations, this structure was emphasized without adversely affecting the general layout of the graph.

In the Rigi system, a variety of tree layout algorithms are used for visualizing call graphs, data dependency trees and other hierarchies. For example, Fig. 8 shows a call dependency tree routed at the main function in a small program written in the C language. This program implements a list data structure. One of these nodes, mylistprint has been expanded by the SHriMP layout adjustment algorithm using the hybrid layout strategy suitable for tree layouts. By zooming the node in this fashion, a software engineer can read the source code of the mylistprint function and at the same time maintain his mental map of the location of this function in the call dependency tree.

Rigi is end-user programmable [16] through the RCL (Rigi Command Language) which is based on the Tcl/Tk language [12]. The SHriMP visualization technique is implemented in the Tcl/Tk language and was therefore easily integrated in the Rigi system. Since SHriMP (through Rigi) is end-user programmable, the layout strategy can be dynamically changed for one or more nodes. The user can experiment with a variety of hybrid strategies based on the graph layout hierarchy.

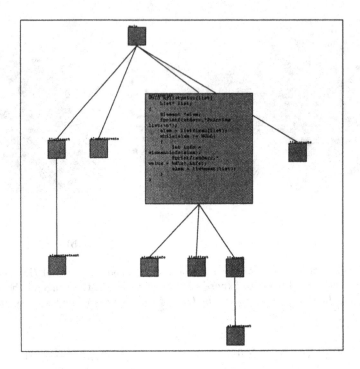

Fig. 8. Browsing software source code using SHriMP Views

6 Conclusions

This paper introduced the SHriMP layout adjustment algorithm suitable for uniformly resizing nodes when requests for more screen space are made. It preserves straightness of lines and graph topology in the adjusted views. Moreover, the SHriMP algorithm is flexible in its distortion technique and can be changed to suit different graph layouts. Several variants were presented for preserving orthogonal and proximity relationships. Hybrid strategies were also shown to be feasible, and are useful when trying to preserve the mental map of more sophisticated layouts.

This algorithm, due to its simplicity, can be easily integrated with existing graph drawing tools. This has been demonstrated through its integration with the Rigi system, where it was used for creating fisheye views of nested graphs. This approach can also be applied to the node disjointness problem.

We are currently investigating whether or not these methods will be useful in the design phase of graph layouts such as those found in hierarchical petri nets and user interfaces. Future work involves investigating other layout strategies and to analyze the strategies presented in this paper to determine which classes of proximity relations and orthogonal relations are preserved using the different strategies.

Acknowledgments The authors gratefully thank Bryan Gilbert, James McDaniel and Minou Bhargava for editing suggestions. This work was supported in part by the Natural Sciences and Engineering Research Council of Canada.

References

1. G. DI BATTISTA, P. EADES, R. TAMASSIA, AND I. TOLLIS, *Algorithms for graph drawing: An annotated bibliography*, Comput. Geom. Theory Appl., 4 (1994), pp. 235–282.

2. J. DILL, L. BARTRAM, A. HO, AND F. HENIGMAN, *A continuously variable zoom for navigating large hierarchical networks*, in Proceedings of the 1994 IEEE Conference on Systems, Man and Cybernetics, 1994.

3. K. M. FAIRCHILD, S. E. POLTROCK, AND G. W. FURNAS, *Semnet: Three-dimensional graphic representations of large knowledge bases*, in Cognitive Science and its Applications for Human-Computer Interaction, R. Guindon, ed., Lawrence Erlbaum Associates, 1988.

4. G. FURNAS, *Generalized fisheye views*, in Proceedings of ACM CHI'86, Boston, MA, April 1986, pp. 16–23.

5. D. HAREL, *On visual formalisms*, Communications of the ACM, 31(5) (May 1988).

6. D. HEARN AND M. P. BAKER, *Computer Graphics*, Prentice Hall, 1986.

7. T. R. HENRY AND S. E. HUDSON, *Interactive graph layout*, in UIST, Hilton Head, South Carolina, November 11-13, 1991, pp. 55–64.

8. K. MISUE, P. EADES, W. LAI, AND K. SUGIYAMA, *Layout adjustment and the mental map*, Journal of Visual Languages and Comput., 6(2) (1995), pp. 183–210.

9. K. MISUE AND K. SUGIYAMA, *Multi-viewpoint perspective display methods: Formulation and application to compound graphs*, in 4th Intl. Conf. on Human-Computer Interaction, Stuttgart, Germany, vol. 1, September 1991, pp. 834–838.

10. H. A. MÜLLER, M. A. ORGUN, S. R. TILLEY, AND J. S. UHL, *A reverse engineering approach to subsystem structure identification*, Journal of Software Maintenance: Research and Practice, 5(4) (December 1993), pp. 181–204.

11. E. NOIK, *A space of presentation emphasis techniques for visualizing graphs*, in Proceedings of Graphics Interface '94, (Banff, Alberta), May 1994, pp. 225–233.

12. J. K. OUSTERHOUT, *Tcl and the Tk Toolkit*, Addison-Wesley, 1994.

13. M. SARKAR AND M. BROWN, *Graphical fisheye views*, Communications of the ACM, 37(12) (December 1994).

14. M. SARKAR, S. SNIBBE, O. TVERSKY, AND S. REISS, *Stretching the rubber sheet: A metaphor for viewing large layouts on small screens*, in User Interface Software Technology, 1993, November 3-5, 1993, pp. 81–91.

15. M.-A. D. STOREY AND H. A. MÜLLER, *Manipulating and documenting software structures using shrimp views*. To appear in *Proceedings of the 1995 International Conference on Software Maintenance (ICSM '95), Opio (Nice), France*, October 16-20, 1995.

16. S. R. TILLEY, K. WONG, M.-A. D. STOREY, AND H. A. MÜLLER, *Programmable reverse engineering*, International Journal of Software Engineering and Knowledge Engineering, 4 (1994).

17. E. R. TUFTE, *Envisioning Information*, Graphics Press, 1990.

18. K. WONG, S. R. TILLEY, H. A. MÜLLER, AND M.-A. D. STOREY, *Structural redocumentation: A case study*, IEEE Software, 12 (1995), pp. 46–54.

A Generic Compound Graph Visualizer/Manipulator : D-ABDUCTOR

Kozo Sugiyama and Kazuo Misue

Institute for Social Information Science, Fujitsu Laboratories Limited
140 Miyamoto, Numazu, Shizuoka, 410-03, JAPAN
{sugi, misue}@iias.flab.fujitsu.co.jp

1 Introduction

We often exploit diagrams when we think something. One reason of using diagrams is that they are good media to represent and organize personal thought, and to share ideas among persons on group works. However, editing, reforming, and redrawing diagrams, which are necessary on thinking processes, are troublesome works. Moreover they might be unessential for thought. D-ABDUCTOR is a generic compound graph visualizer/manipulator which is implemented to support such the human thinking processes by using sophisticated techniques for diagram visualization/manipulation in a direct manipulation and animation environment. D-ABDUCTOR can effectively support the processes to manipulate the diagram interactively and to organize idea fragments into an integrated diagram. D-ABDUCTOR is still undergoing updates and technical evolution.

2 Features and facilities

Features of D-ABDUCTOR are summarized as:

- Fast automatic drawing algorithm for a higher class of graphs: compound graphs
- Advanced GUI based on menu, direct manipulation, animation and fisheye view environment
- Communication among multiple set of D-ABDUCTOR on a network and interface for applications
- Multi-media support.

Facilities of D-ABDUCTOR include[1]:

Compound graphs: D-ABDUCTOR provides an environment to deal with diagrams of a generic class of graphs called compound graphs which represent both adjacency and inclusion relationships among nodes[2].
Automatic graph drawing: D-ABDUCTOR provides a fast automatic layout facility of compound graphs utilizing a Sugiyama style extended drawing algorithm[2]. Certain operations chosen by the users can trigger the invocation of this facility for visual response.

Graph editing: D-ABDUCTOR provides various editing operations from standard operations(e.g. create/delete/move a node or link, change color/width/shape of a node or link) to group operations(e.g. create/resolve a group node) in a menu and direct manipulation environment.

Collapse and Expand operations: D-ABDUCTOR allows groups of nodes to be represented by one node(information hiding).

Fisheye view: Our fisheye view facility called Diagram Dressing changes size and visibility of nodes according to their importance. More important nodes become larger, and less important nodes become smaller or omitted. Structure of diagrams, semantics of nodes and the specific user's viewpoint influence the importance of nodes[3].

Display with Animation: D-ABDUCTOR can display the instantaneous visual change with an animation so that the user's mental map is preserved[3]

Communication: Multiple set of D-ABDUCTOR can communicate with each other on a network to share database called Card Base consisted of diagram data, text data and images. D-ABDUCTOR can communicate with other applications through the language 'Simple.'

Language 'Simple': The language Simple is designed to store the current graph and communicate with other processes. Graphs can be created, edited, loaded and saved through Simple. Furthermore, Simple has the capability to record working histories with D-ABDUCTOR and so we can replay the editing histories. D-ABDUCTOR works as an interpreter of Simple.

The system architecture and facilities of D-ABDUCTOR are shown in Fig. 1.

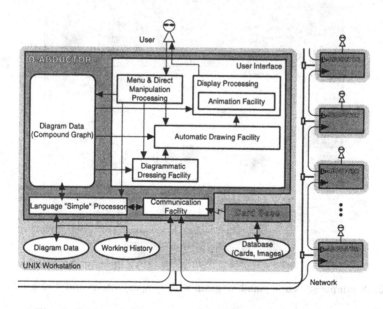

Fig. 1. System architecture and facilities of D-ABDUCTOR.

3 Sample Screens

Sample screens of D-ABDUCTOR are shown in Fig. 2 and 3. In Fig. 2 some of menus and dialog boxes are opened and a small diagram with images is shown in the Main Window. Text can be placed into nodes by using a text editor(Create Menu). Graphic images can also be attached to nodes instead of text. Any node (whether a group node or leaf node) can be connected together with links(Mouse Operation). Each node has several handles allowing links to be attached. Pre-defined actions such as adding a new node or link can be set up to trigger the automatic layout facility and redraw the diagram (Layout dialog box). Diagrams can be oriented north, east, south and west (Layout dialog box). In Fig. 3 larger examples are shown. The sizes of nodes in both diagrams are changed by the diagram dressing facility (Dressing dialog box).

D-ABDUCTOR has been implemented in C as an application of X Window system (V11R5) by using a GUI toolkit XView. D-ABDUCTOR does not have the facilities to edit text and therefore the user should prepare a text editor such as Emacs in the same environment. D-ABDUCTOR is delivered as a SPARC binary for Sun workstations (SunOS 4.1.x). The binary is available for non-profit use by anonymous ftp from ftp.fujitsu.co.jp in the directory /pub/misc/abd2.23.tar.gz.

The usage of D-ABDUCTOR other than idea organizations includes layout for documentation, outline processing, software engineering, WWW browsing, CAI design, database schema design and so on.

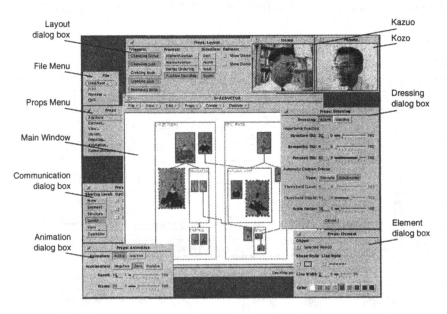

Fig. 2. A sample screen: a small diagram with images, menus and dialog boxes.

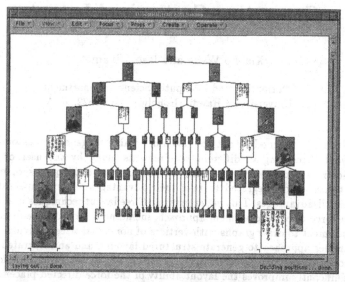

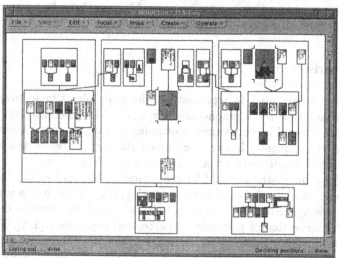

Fig. 3. Diagrams of a binary tree (upper) and a compound graph (lower). The number of leaf nodes in each diagram is 100. In both cases the diagram dressing facility is utilized. Two marked nodes are focused in the upper diagram and three in the lower.

References

1. Kazuo Misue: D-ABDUCTOR 2.0 user manual, Res. Rep. IIAS-RR-93-9E, Fujitsu Laboratories Ltd., IIAS, 1993.
2. Kozo Sugiyama and Kazuo Misue: Visualization of structural information: Automatic drawing of compound digraphs, IEEE Trans.SMC, 21(4): 876-893, 1991.
3. Kazuo Misue, Peter Eades, Wei Lai and Kozo Sugiyama: Layout adjustment and the mental map, J. of Visual Languages and Computing, 6(2): 183-210, 1995.

Generating Customized Layouts

Xiaobo Wang and Isao Miyamoto

Information and Computer Sciences Department
University of Hawaii, Honolulu, HI 96922, USA

Abstract. A good layout tool should be able to generate customized layouts according to different requirements given by the user or applications. To achieve this goal, existing layout techniques should be enhanced and integrated to take their advantages while compensating their disadvantages. This paper presents three layout techniques based on the force-directed placement approach, including a revised force-directed placement to draw graphs with vertices of nontrivial sizes, a divide-and-conquer approach to generate structured layouts, and an integrated approach to support constraints. The combination of the three techniques significantly improves the layout ability of the force-directed placement. They can be used to generate customized layouts that reflect semantics, preference, or principles of perceptual psychology.

1 Introduction

Graphs are widely used to represent relational problems in software systems. An important task for a graph-based application is generating the layouts of the graphs. Manual layout is time-consuming and error-prone. Automating the layout task consequently has received much attention in recent years [1].

A classical requirement on an automatic layout method is that the layout generated by the method should be syntactically valid and satisfy the aesthetic criteria, e.g., the layout should have no overlapping vertices and a small number of edge-crossings. Secondly, the layout should satisfy constraints derived from preference, semantics, or principles of perceptual psychology [3, 9]. This ensures that the layout conveys the correct message to the user. The third requirement is that the response time of the layout method should be reasonable for interactive application.

In the algorithmic approach, the layout of the graph is generated by optimizing the aesthetics of the graph. The algorithmic approach is computationally efficient and is very successful in generating layouts that are aesthetically pleasant to the eye. But the algorithmic approach draws the graph according to a set of pre-defined criteria. Most algorithmic methods do not support constraints and can not generate customized layout to reflect the semantics of the graph or the preference of the user. In the declarative approach, the layout of the graph is generated by searching a solution of a set of constraints. The power of constraints makes the declarative approach well-suited to express semantics or preference in the drawing of the graph. But it is difficult to specify global criteria such as aesthetics with constraints. It is also computationally inefficient to solve a large number of constraints. The drawbacks of the algorithmic and the declarative approaches make them inadequate to be adopted by current applications [10].

The work of Eades and Lin [10] has shown that an integrated approach can take the advantages of both algorithmic and declarative approaches while compensating their disadvantages. In this paper, we focus on the *force-directed placement* [6] and discuss how to enhance the algorithm to satisfy the requirements described above. We present a layout tool called LYCA. The tool employs the force-directed placement to improve aesthetics of undirected graphs. A constraint solver is integrated with the layout algorithm to satisfy constraints. LYCA has several features:

- The force-directed placement in [6] is modified to draw graphs that contain vertices of different sizes. The improved algorithm can distribute vertices evenly if sizes of vertices are considered. Overlaps between large vertices are eliminated.
- A divide-and-conquer approach is introduced to generate structured layouts that reflect zones, proximity, and symmetries of subgraphs.
- The constraint solver and the layout algorithm cooperate to resolve the blocking problem caused by the competition between the constraint solver and the layout algorithm. This improves the layout quality.

With those functions, LYCA can generate layouts that are nice-looking as well as functional.

The organization of this paper is the following. Section 2 briefly explains Fruchterman's algorithm and discusses how to improve the algorithm to draw graphs with vertices of different sizes. Section 3 presents the divide-and-conquer approach. Section 4 describes the integration of the constraint solver and the force-directed placement. The last section gives discussion and concluding remarks.

2 Drawing Graphs with Large Vertices

In the first half of this section, we explain the force-directed placement algorithm [6]. In the second half of the section, we discuss how to modify the algorithm to draw graphs with vertices of different sizes.

2.1 Force-directed Placement

LYCA uses the force-directed placement [6] to improve aesthetics of undirected graphs. The force-directed placement is a variation of the well-known *spring embedding* algorithm [3, 4, 8, 13]. The algorithm draws graphs by applying an analogy from a physical process. Briefly, vertices are represented as atomic particles that exert forces upon each other. All vertices repel each other with repulsive forces. Neighbor vertices attract each other with attractive forces. Given an initial layout, the vertices are moved by the forces until they reach a stable state, which is returned as the final layout.

In [6], the strengths of forces between vertices are defined as:

$$F_r = -k^2/d$$

$$F_a = d^2/k$$

where F_a is the attractive force, F_r is the repulsive force, d is the distance between a pair of vertices, k is the optimal distance between vertices.

2.2   The Revised Force-directed Placement

In the algorithm of [6], k represents the optimal distance between the centers of vertices. When drawing graphs which contain vertices of different sizes, k must be increased by the size of the largest vertex to avoid overlaps between vertices. This may yield layout with large area and uneven distribution of vertices if we consider sizes of vertices in the distribution, as shown in Figure 1 (a).

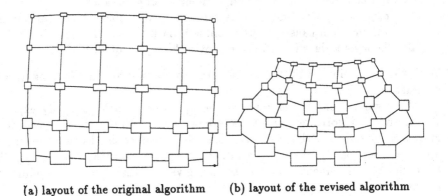

(a) layout of the original algorithm   (b) layout of the revised algorithm

Fig. 1. Example layouts of graph with vertices of different sizes

To overcome the problem, we modify the forces between a pair of vertices v and w as follows:

$$F_a = \begin{cases} 0 & \text{if } w, v \text{ overlap} \\ \frac{d_{out}^2}{k' + d_{in}} & \text{otherwise} \end{cases}$$

$$F_r = \begin{cases} C\frac{k'^2}{d} & \text{if } w, v \text{ overlap} \\ \frac{k'^2}{d} & \text{otherwise} \end{cases}$$

where F_a is the attractive force between v and w, F_r is the repulsive force between v and w, d is the distance between the centers of v and w, d_{out} is the

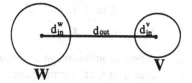

Fig. 2. Distance between vertices with areas

distance between the boundaries of v and w (see Figure 2), $d_{in} = d - d_{out}$, k' is the optimal distance between vertices, and C is a constant.

In the revised algorithm, the repulsive and attractive forces between a pair of vertices cancel each other when d_{out} between the vertices is equal to k'. Therefore, k' represents the optimal distance between the boundaries of vertices. Because the revised algorithm adjusts distances between vertices according to the actual sizes of the vertices, it distributes vertices evenly into a compact area. Overlaps between vertices are effectively eliminated since the attractive force between a pair of vertices becomes zero while the repulsive force is increased when the vertices overlap. An example layout created by the revised force-directed placement is shown in Figure 1 (b).

3 Generating Structured Layouts

Kosak el at. [9] pointed out that an important property of a layout is its perceptual organization. It was found that layouts that are organized according to the principles of perceptual psychology are easy to understand, while layouts that violate those principles are likely misleading [9].

We use a divide-and-conquer approach to generate visually structured layouts and display certain *Visual Organization Features* defined in [9], including zones, proximity, and symmetries of subgraphs.

3.1 Basic Notations

We first explain necessary concepts before presenting the divide-and-conquer algorithm.

Given a graph $G = (V, E)$, a *partition* P splits G into disjoint subgraphs, $P = \{G_1, \ldots, G_n\}$, such that

$$G_i = (V_i, E_i)$$
$$\bigcup_{i=1}^{n} V_i = V$$
$$V_i \cap V_j = \emptyset \text{ for } i \neq j$$
$$E_i = \{(v, w) \in E | v, w \in V_i\}$$

P also divides the edges of G into *intra-edges* and *inter-edges*. Intra-edges are edges between vertices in the same subgraph:

$$E_{intra} = \bigcup_{i=1}^{n} E_i$$

Inter-edges are edges between vertices in different subgraphs:

$$E_{inter} = E - E_{intra}$$

An undirected graph G_{meta} called a *meta-graph* is constructed by collapsing subgraphs of G into *meta-vertices* and transforming inter-edges of G into *meta-edges*. A layout of G_{meta} is called a *meta-layout* of G. A meta-layout can be obtained by the force-directed placement where the dimensions and center of each meta-vertex are set to the dimensions and center of the underlying sub-graph, respectively.

Forces in the force-directed placement [6] are also divide into two categories: a force between a pair of vertices in the same subgraph is called an *intra-force*, a force between a pair of vertices in different subgraphs is called an *inter-force*. For a meta layout constructed from a partition P and a layout obtained by the original force-directed placement [6], the improved force-directed placement described in the previous section is used to calculate forces between meta-vertices in the meta-layout. The net force on a meta-vertex is defined as the *meta-force* on all vertices contained by the subgraph that is represented by the meta-vertex.

In Figure 3, a layout and a partition are given on the left side, the corresponding meta-layout is shown on the right side. The intra-force on vertex C is the sum of the forces between C, A and C, B, the inter-force on C is the sum of forces between C and vertices in subgraphs S_2 and S_3, the meta-force on C is the net force on the meta vertex $S1$ in the meta-layout on the right side.

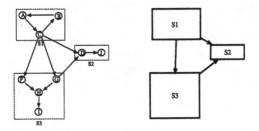

Fig. 3. Meta-graph and meta-layout

3.2 The Divide-and-conquer Approach

A divide-and-conquer approach draws a graph in three steps: 1) partition a graph into subgraphs, 2) draw subgraphs, 3) compose subgraph layouts together to form the resulting layout. A difficulty in divide-and-conquer layout is how to position inter-edges. If inter-edges are ignored in subgraph layouts, the resulting layout may have long-edges and edge-crossings. On the other hand, considering inter-edges in subgraph layouts leads to a circular dependency problem: inter-edges depend on subgraph layouts which in turn depend on inter-edges. In LYCA, this problem is solved as follows.

The divide-and-conquer approach uses a composite force to position a vertex:

$$F_{comp} = F_{intra} + S(t)F_{inter} + (1 - S(t))F_{meta}$$

where F_{comp} is the net force on a vertex, F_{intra} is the intra-force on a vertex, F_{inter} is the inter-force on a vertex, and F_{meta} is the meta force on a vertex. $S(t) \in [0, 1]$ is a function of layout time t such that $S(t)$ decreases as t increases after a threshold t' and reaches 0 at another threshold t'' ($> t'$).

The layout process consists of three phases. Between time 0 and time t', $S(t) = 1$, $F_{comp} = F_{inter} + F_{intra}$. The force-directed placement [6] is used to generate a layout with uniform edges and a small number of edge-crossings, as shown in Figure 4 (a). In the phase between time t' and t'', $S(t)$ decreases. This reduces the strengths of inter-forces. Meanwhile, the strengths of meta-forces are increased. At the time threshold t'', $S(t)$ reaches 0, $F_{comp} = F_{intra} + F_{meta}$. Since meta forces do not change the relative positions of vertices in a subgraph, vertices in subgraphs are positioned by intra-forces like in divide-and-conquer layout.

The divide-and-conquer approach displays zones of subgraphs because the meta-forces eliminate overlaps between subgraphs in the last phase of the layout process. Proximity can be reflected if vertices in subgraphs are placed closely and subgraphs are placed sparsely. This can be achieved by choosing a small optimal distance between vertices and a large optimal distance between meta-vertices. A problem of the force-directed placement is that it may not display symmetries of subgraphs if the entire graph is not symmetric. This problem is resolved in the divide-and-conquer approach since inter-forces are masked after the time threshold t''. In addition, because the resulting layout is evolved from the layout generated at time t' which has uniform edges and small number of edge-crossings, long edges and edge-crossings are avoided in the resulting layout. A structured layout created by the divide-and-conquer approach is shown in Figure 4 (b).

4 Integration of Constraint Solver and Layout Algorithm

A major limitation of the force-directed placement is that it does not support constraints. LYCA's solution to the problem is integrating the layout algorithm with a constraint solver. In this section, we focus on the issue of how to integrate the solver with the layout algorithm to ensure layout quality.

4.1 The Integrated Approach

LYCA takes an integrated approach to support constraints. A constraint solver is employed to solve three kinds of constraints: 1) an *absolute constraint* fixes a vertex at its current position, 2) a *relative constraint* constrains the position of a vertex in relation with others, 3) a *cluster constraint* clusters several vertices into a subgraph that can be processed as a whole. The solver uses a propagation style algorithm to satisfy constraints [15].

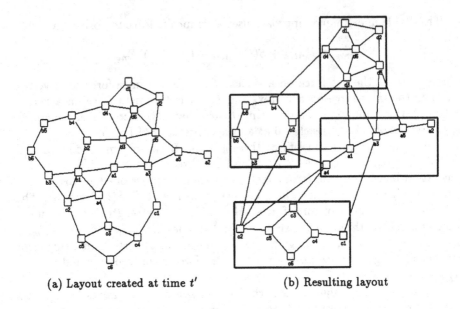

(a) Layout created at time t' (b) Resulting layout

Fig. 4. Structured Layout

(a) $A.x = B.x$ (b) $A.x < B.x$ (c) $A.x > B.x$

Fig. 5. Examples of barriers caused by solving constraints

During the layout process, the solver inputs coordinates of vertices from the layout algorithm and changes coordinates of constrained vertices to satisfy constraints. This integration may return poor layouts since solving constraints may make a vertex block others from reaching their optimal positions calculated by the force-directed placement. We define vertex B as a *barrier* for vertex A if solving a constraint between A and B prevents A from reaching its optimal position assigned by the force-directed placement. Figure 5 gives several examples of barriers. In the figure, the arrows on A and B indicate the movements of A and B assigned by the force-directed placement, respectively. Since the movements violate the constraint between A and B (shown below A and B), the solver has to change the position of either A or B to satisfy the constraint. If the solver chooses to change the position of A to satisfy the constraint, B becomes a barrier for A since it blocks A from being improved by the layout algorithm.

In Figure 5, if we move A and B together by the forces on them, we can reduce the total force on A and B. The solver will not change the position of A

to satisfy constraint because the constraint between A and B is not violated when A and B are moved together. Therefore, we can avoid barriers while improving the overall aesthetics of a graph with the principle of force-directed placement.

Based on the above observation, we introduce *rigid sticks* in the force-directed placement to represent constraints. If vertex v_1 becomes a barrier for vertex v_2, a rigid stick is introduced between v_1 and v_2 such that v_1 and v_2 must move together like one rigid object. The movements of v_1 and v_2 are determined by the weighted average of forces on them:

$$f = \frac{w_1 f_1 + w_2 f_2}{w_1 + w_2}$$

where f is the new force on v_1 and v_2, f_1 and f_2 are the original forces on v_1 and v_2, respectively, and w_1 and w_2 are weights of v_1 and v_2, respectively. During the layout process, the solver and the layout algorithm cooperate to remove barriers caused by constraints with following rules:

- If vertices v_1 and v_2 are aligned by an *"equal"* or a *"neighbor"* constraint, v_1 and v_2 are connected by rigid stick.
- If v_1 and v_2 are constrained by a *"less−than"* or *"greater−than"* constraint, a rigid stick is introduced between v_1 and v_2 only when one of the vertices becomes a barrier for the other one.
- If v is constrained as in the center of a set of vertices $v_1, \ldots, v_n$, the force on v is evenly distributed on $v_1, \ldots, v_n$.

With the cooperation between the solver and the layout algorithm, each layout iteration consists of four steps:

- Step 1. Calculate forces.
- Step 2. Introduce sticks and distribute forces.
- Step 3. Calculate new positions.
- Step 4. Satisfy constraints.

Step 1 and step 3 are performed by the layout algorithm. Step 2 and step 4 are performed by the solver. Our experiments showed that the heuristic works reasonably well to remove barriers and improve layout quality. As illustrated by the example in Figure 6, if the solver changes the positions of $n6$ and $n7$ to satisfy constraints "$n6.x > n15.x - 32$" and "$n7.x < n8.x + 32$", $n15$ and $n8$ become barriers for $n6$ and $n7$, respectively. If the barriers are not removed, the resulting layout has a long edge (see Figure 6 (a)). The problem is resolved when the solver and the layout algorithm cooperate to remove barriers, as shown in Figure 6 (b).

The integrated approach enhances the expressive capability of the force-directed placement significantly. Two example layouts generated with constraints are shown in Figure 7 and Figure 8. The constraints used to generated the layout in Figure 8 are given below:

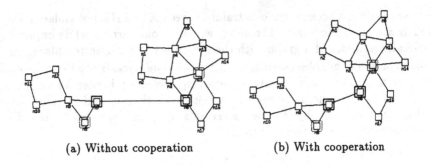

(a) Without cooperation   (b) With cooperation

Fig. 6. Remove barriers caused by solving constraints

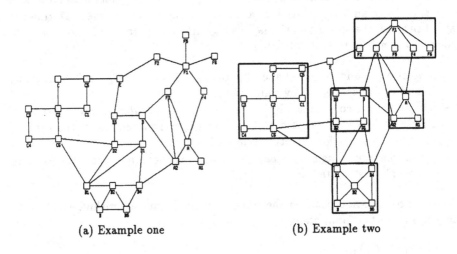

(a) Example one   (b) Example two

Fig. 7. Example layouts generated with constraints

```
abs(a4.x - a2.x) <= 120
abs(a1.x - a5.x) <= 70
a5.y = a1.y = a2.y - 64
a4.y = a3.y = a2.y
c2.y = c5.y = c4.y = c1.y
d3.y = d5.y
```

4.2 Analysis

LYCA uses a propagation style algorithm with linear time-complexity to solve constraints [15]. Internally constraints as represented as constraint graphs. Each edge in a constraint graph represents one constraint. To overcome barriers, the solver first marks the edges that are causing barriers as rigid sticks. The solver

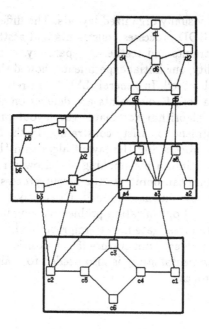

Fig. 8. Layout with constraints

then retrieves the vertices that are connected by rigid sticks and distributes forces on the retrieved vertices. The two steps also can be done in linear time. Therefore, the integrated approach is efficient. In our experiment, LYCA took less than 6 seconds on a Sparc 10 workstation to generate the layout shown in Figure 4. Since the current implementation of LYCA is not optimal, we expect that the time performance could be further improved.

5 Concluding Remarks

In the paper, we first discussed how to revise the force-directed placement in [6] to generate compact layouts for graphs with vertices of different sizes. A related work is the *force-scan* algorithm reported in [12]. The force-scan algorithm can keep the user's mental map on an existing layout while resolving overlaps between vertices in layout adjustment. In LYCA, the force-directed placement is mainly used for layout creation so it does not maintain the user's mental map on existing layouts.

LYCA's divide-and-conquer method draws a graph twice to create a structured layout: it first generates an aesthetically pleasing layout, it then transforms the first layout into a structured one. This may double the layout time. But the penalty is necessary to avoid long edges and edge-crossings in the resulting layout. Otherwise, manual modifications of the resulting layout or recursive adjustment of subgraph layouts have to be applied to ensure the layout quality in divide-and-conquer layout [7, 11]. Both LYCA and ANDD [3] use spring algorithm to generate visually organized layouts. The difference between LYCA and ANDD is that ANDD processes constraints and aesthetics together, while LYCA processes constraints and aesthetics separately.

About layout quality, our initial experiments showed that the integrated approach works reasonably well. In general, LYCA can return good layouts when a few constraints or a lot of constraints are defined on the graph. In the former case, the layout algorithm dominates the layout process and the solver performs minor adjustment to satisfy constraints. In the latter case, the solver dominates the layout process and the layout algorithm "beautifies" the layout generated by the solver into a nice-looking one. However, if neither the solver nor the layout algorithm can dominate the layout process, poor layout may be returned and manual adjustment is needed. Obviously there is no trivial solution since the constrained optimization problem is computationally intractable. It will be interesting to investigate how to improve the layout quality with some known techniques, e.g., local temperature [5] or positioning vertices in certain order [14]. More experimental study is also needed to evaluate the performance of the integrated approach.

Acknowledge

I would like to thank Peter Eades, Kozo Sugiyama, and Joe Marks for their helps and suggestions. I would also like to thank anonymous referees for their useful comments.

References

1. G. D. Battista, P. Eades, R. Tamassia and I. G. Tollis, "Algorithms for drawing graphs: An annotated bibliography," Tech. Report, Computer Science Dept., Brown Univ., June, 1993.

2. R. Davidson and D. Harel, "Drawing graphs nicely using simulated annealing," *Technical Report CS89-13*, Department of Applied Mathematics and Computer Science, The Weizmann Institute of Science, Rehovot, Israel, 1989.

3. E. Dengler, M. Friedell and J. Marks, "Constraint-driven diagram layout," *Proc. of Visual Language 93*, 1993.

4. P. Eades, "A heuristic for graph drawing," *Congress Numeratium*, Vol. 42, 1984.

5. A. Frick, A. Ludwing, and H. Mehldau, "A fast adaptive layout algorithm for undirected graphs," *Graph Drawing 94*, Princeton, New Jersey, October, 1994.

6. T. J. Fruchterman and E. M. Reingold, "Graph drawing by force-directed placement," *Software - Practice and Experience*, Vol. 21, No. 11, Nov. 1991, pp. 1129–1164.

7. T. R. Henry, "Interactive graph layout: The exploration of large graphs," Tech. Report 92–03, Computer Science Dept., Univ. of Arizona, Tucson, Arizona, 1992.

8. T. Kamada and S. Kawai, "An algorithm for drawing general undirected graphs," *Information Processing Letters*, Vol. 31, 1989.

9. C. Kosak, J. Marks and S. Shieber, "Automating the layout of network diagrams with specified visual organization," *IEEE Trans. on Syst., Man, and Cyb.*, Vol. 24., No. 3, March 1994.

10. T. Lin and P. Eades, "Integration of declarative and algorithmic approaches for layout creation," *Graph Drawing 94*, Princeton, New Jersey, October, 1994.

11. S. C. North, "Drawing ranked digraphs with recursive clusters," *Proc. of ALCOM Int'l Workshop on Graph Drawing*, Paris, France, Sept. 1993.

12. K. Misue, P. Eades, W. Lai and K. Sugiyama, "Layout adjustment and the mental map," Research Report ISIS-RR-94-6E, FUJITSU Lab. Ltd., Shizuoka, Japan, 1994.

13. K. Sugiyama and and K. Misue, "A simple and unified method for drawing graphs: magnetic-spring algorithm," *Graph Drawing 94*, Princeton, New Jersey, October, 1994.

14. D. Tunkelang, "An aesthetic layout algorithm for undirected graphs," Thesis for Master Degree, Computer Science and Engineering Department, M.I.T., 1992.

15. X. Wang, "Generating Customized Layouts Automatically," PhD thesis, Univ. of Hawaii at Manoa, August, 1995.

GOVE
Grammar-Oriented Visualisation Environment

Richard Webber and Aaron Scott

Department of Computer Science, University of Newcastle, Callaghan 2308, Australia

1 Introduction

Most Information Visualisations have been developed in an ad hoc manner. To overcome this, we have proposed an architecture that formalises the structure of Information Visualisations. By using this architecture, software developers will benefit from the use of tools that support this approach, and from the increased potential to reuse parts of visualisations built under the architecture. We have embodied this architecture in a software environment, GOVE (Grammar-Oriented Visualisation Environment) [7], that can be used to develop Relational Information Visualisations — visualisations that deal specifically with the entities and relationships found in the information source (which is currently limited to static, textual sources), usually presenting them as a graph. This form of Information Visualisation is very common, and accounts for much of the work in the field [1, 8].

2 Architecture

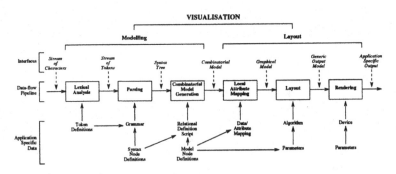

Fig. 1. Visualisation architecture

The traditional model of information visualisation consists of two parts [4, 5]: *Modelling* (extracting entities and relationships from an information source), and *Layout* (converting this into a graphical representation).

We propose an architecture which subdivides these parts into six modules, as outlined in Fig. 1. Together, these form a dataflow pipeline, which accepts an information source as input, and produces a visualisation as output. Unlike many systems, GOVE incorporates the process of obtaining relational information from the (textual) source.

Each module performs a specific task, and is made up of two logical components: *General Functionality* (the underlying engine for that module's part of the visualisation process), and *Application Specific Data* (the definitions that tailor a module to a specific visualisation). The interfaces between neighbouring modules are defined to allow modules from one visualisation to be inserted into the same position in another visualisation, greatly increasing the potential for reuse.

3 Modules

GOVE presents its user with an explicit visual representation of the Information Visualisation pipeline they are developing. A simple pipeline to visualise a graph description file is given in Fig. 2. The user interacts with this pipeline, supplying the application-specific data through a set of form-based interfaces (see Fig. 3), then instructing GOVE to build and execute the pipeline of modules to perform the visualisation.

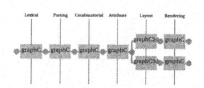

Fig. 2. Graph description pipeline

Fig. 3. Lexical Analysis interface

Lexical Analysis: divides a stream of characters (the textual source) into tokens. The application specific data for a Lexical Analysis module is a list of token names and definitions given as Lex [6] regular expressions.

Parsing: builds a syntax tree from the sequence of tokens it receives. The application specific data for a Parsing module is a Yacc [6] grammar.

Combinatorial Model Generation: analyses the syntax tree, and generates a Combinatorial Model (an Entity-Relationship Diagram). It does this by matching patterns in the syntax tree, and generating corresponding components in the Combinatorial Model. This process is similar to the construction of graphs through Graph-Grammars [3], and is shown in Fig. 4.

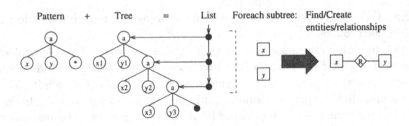

Fig. 4. Combinatorial Model Generation

Local Attribute Mapping: assigns graphical attributes to the entities and relationships of the Combinatorial Model. The application specific data consists of expressions that assign a value to each attribute for each *type* of entity and relationship, possibly using the data stored in the Combinatorial Model.

Layout (Global Attribute Mapping): applies the user's choice of layout algorithm [1] to assign positions to the nodes. GOVE has support for three-dimensional graph drawing, with an increasing set of layout algorithms.

Rendering: converts the resulting graph to the desired format for storage or viewing. The two visualisations resulting from the pipeline in Fig. 2 are shown in Figs. 5 and 6 (two- and three-dimensional spring layout respectively).

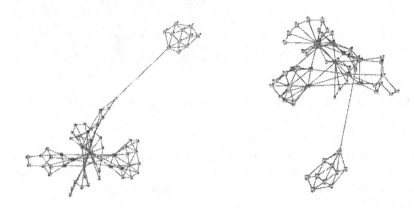

Fig. 5. Two-dimensional spring   **Fig. 6.** Three-dimensional spring

As shown in Fig. 2, GOVE can develop pipelines that branch to produce multiple layouts of the same Combinatorial Model. The same principle can be applied to any stage of the pipeline. Fig. 7 shows a pipeline that visualises the naming scope and procedure calls in a pseudo-pascal source file.

In addition to the spring algorithms shown in Figs. 5 and 6, GOVE currently supports a three-dimensional DAG algorithm [2] shown in Fig. 8.

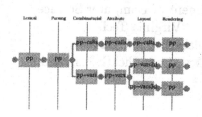

Fig. 7. Pseudo-pascal pipeline

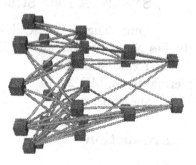

Fig. 8. Three-dimensional DAG

4 Future Improvements

GOVE is currently undergoing redevelopment. Some of the possible improvements include: the use of *diagrammatic* interfaces; interpreted modules to speed-up development; more specific/efficient Combinatorial Models; and many *useful* layouts algorithms. Allowing interaction with the resulting visualisations is also desirable — simply manipulating the output is easy, while supporting feedback to the information source is much more difficult.

References

1. G. Di Battista, P. Eades, R. Tamassia and I.G. Tollis, "Algorithms for Drawing Graphs: An Annotated Bibliography". Jun, 1994.
   ftp://wilma.cd.brown.edu/pub/papers/compgeo/gdbiblio.tex.Z
2. R.F. Cohen, D. Fogarty, P. Murphy and D.I. Ostry, "Animated Three-Dimensional Information Visualizations". Submitted to ACSC '96. Aug, 1995.
3. H. Ehrig, H.-J. Kreowski, G. Rozenberg, *Graph Grammars and their Applications to Computer Science*. Lect. Notes in Comp. Sc., vol. 532, Springer-Verlag. 1991.
4. T. Kamada, *Visualizing Abstract Objects and Relations: A Constraint Based Approach*. Series in Comp. Sc., vol. 5. 1989.
5. J. Mackinlay, "Automating the Design of Graphical Presentations of Relational Information" in *ACM Trans. Graphics*, vol. 5, no. 2, pp. 110–141. 1986.
6. T. Mason and D. Brown, *Lex and Yacc*, A Nutshell Handbook. 1990.
7. A. Scott and R. Webber, "Grammar Oriented Visualisation Environment (GOVE)". Hons. Proj., Dep. Comp. Sc., Uni. Newcastle, Callaghan 2308, Australia. Nov, 1993.
8. A. Scott, "A Survey of Graph Drawing Systems". Tech. Report 95-1, Dep. Comp. Sc., Uni. Newcastle, Callaghan 2308, Australia. Dec, 1994.

SWAN: A Data Structure Visualization System [1]

Jun Yang
Hughes Network Systems
11717 Exploration Lane
Germantown, MD 20876
jyang@hns.com

Clifford A. Shaffer and Lenwood S. Heath
Department of Computer Science
Virginia Tech
Blacksburg, VA 24061
shaffer@cs.vt.edu, heath@cs.vt.edu

1 Introduction

Swan is a data structure visualization system. It allows users to visualize the data structures and execution process of a C/C++ program. **Swan** views a data structure as a graph or collection of graphs. By "graph" we mean general directed and undirected graphs and special cases such as trees, lists and arrays.

As a part of Virginia Tech's NSF Educational Infrastructure Grant, **Swan** will be used in two ways: by instructors as a teaching tool for data structures and algorithms, and by students visualizing their own programs to understand how and why they do or do not work. To use **Swan**, a program must first be *annotated*, i.e., **Swan** calls are added to an existing program. The program is then compiled and linked with the **Swan** Annotation Interface Library (**SAIL**). The viewer then runs the annotated program.

Many program visualization systems exist. See [5, 4] for examples. These have been used for teaching, presentation, and debugging purposes. The main design goal for **Swan** was to create an easy-to-use annotation library combined with a simple, yet powerful, user interface for the resulting visualization. Several features distinguish **Swan** from most other program visualization systems:

1. **Swan** provides a compact annotation library. Fewer than 20 library functions are frequently used.
2. The viewer's user interface is simple and straightforward.
3. The annotator decides the view semantics, i.e., the association between program variables and graphical elements in the views. The annotator also controls the progress of the annotated program.
4. **Swan** provides automatic layout of a graph so the annotator need only concentrate on the logical structure of the graph.
5. **Swan** allows the viewer to modify the data structure.
6. **Swan** was built on the **GeoSim** Interface Library developed at Virginia Tech, which allows **Swan** to be easily ported to X Windows, MS-DOS and Macintosh computers. It is crucial for educational software to run on a variety of operating systems that are widely used in computer science classes.

Currently, two versions of **Swan** are available: one for the X Window system and one for MS-DOS. Information about **Swan** can be obtained through the World Wide Web at URL http://geosim.cs.vt.edu/Swan/Swan.html.

---

[1] The authors gratefully acknowledge the support of the National Science Foundation under Grant CDA-9312611.

Visualization can be applied either to the physical implementation for a data structure in a program or to the abstraction represented by that data structure. For example, two views of a graph can be provided as part of an annotated minimum spanning tree algorithm. In Figure 1, the view on the right is an adjacency list representation of the graph, i.e., the program's physical implementation. The view on the left shows the logical topology of the graph, an abstraction represented by the adjacency list. Separate views of the same data structure coexist in **Swan**, each as a separate graph.

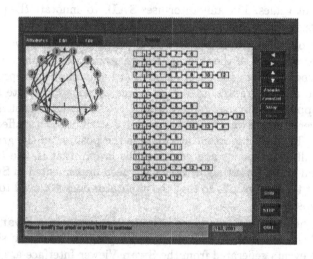

Fig. 1. Two views of a graph created in an annotated minimum spanning tree algorithm

In **Swan**, information can be passed from the annotator to the viewer in the form of graphical representation of data structures. Information can also be passed from the viewer to the annotator in the form of modification requests, providing a powerful mechanism to encourage the viewer to be more active in exploring the program and gaining new insights. This capability makes **Swan** different from most program visualization systems in which the viewer can only watch the animation passively. We believe the ability to modify the program's data structures not only makes **Swan** more suitable as an instructional tool, but also shows the potential for **Swan** to be used as a graphical debugging tool at the abstract level.

2 System Components

Swan has three main components: the **Swan** Annotation Interface Library (**SAIL**), the **Swan** Kernel, and the **Swan** Viewer Interface (**SVI**). **SAIL** is a small set of easy to use library functions that allow the annotator to design different views of a program. **SVI** allows a viewer to explore a **Swan** annotated program. The **Swan** Kernel is the main module in **Swan**. It is responsible for constructing, maintaining, and rendering all views generated through **SAIL** library functions. It accepts viewer's requests through **SVI**. It is also the medium through which the annotator communicates with the viewer.

All views in **Swan** are composed of **Swan** graphs. A **Swan** graph has a set of **nodes** and **edges**. A **Swan** graph is defined by the annotator via its nodes and edges. A **Swan** graph has default display attributes for its nodes and edges, which are used by **Swan** to render the corresponding graphical objects. Nodes and edges can have their own individual display attributes that can override the graph's default values. The annotator uses **SAIL** to annotate the program; the viewer investigates the annotated program through **SVI**.

The topology of a graph is stored in the **Swan** Logical Layer. The **Swan** Logical Layer contains all the internal representations of graphs created by the annotated program. For each graph, a standard adjacency list representation is used to store all the graph's nodes and edges. After appropriate layout algorithms are applied, a physical representation of the layout is kept in the **Swan** Physical Layer. Every **Swan** graph has *physical* attributes, that affect its graphical display. The most important attribute is the position of the graph and the positions of all of the nodes and edges in this graph, that is, the *layout* of the graph. Several graph layout algorithms have been implemented in **Swan** to deal with different types of graphs so that the annotator does not need to spend time on layout himself.

Events generated by interactions between the viewer and **Swan** are sent to the **Swan** Event Handler. A **Swan** annotated program runs as a single thread process. The events generated from the **Swan** Viewer Interface are stored in an event queue. Initially the annotated program has control of the process. Whenever a **SAIL** function is invoked, **Swan** will process all events in the event queue. At this point, **Swan's** Event Handler takes control. After the **SAIL** function completes, control is returned to the annotated program.

There are three basic states in **Swan** when it is active: **Run**, **Step** and **Pause**. Essentially, the process may run continuously (i.e., in **Run** state) or step by step (i.e., in **Step** state). "Step" here refers to the execution of a code segment ending at the next breakpoint set by the annotator. **Swan** lets the annotator decide the size of the step because it is difficult, if not impossible, for **Swan** to identify the interesting events in the annotated program.

The viewer interacts with an annotated program through the **Swan** Viewer Interface (**SVI**) as shown in Figure 1. The **SVI** main window contains a control panel and three child windows: the **display window**, **I/O window** and **location window**. The display window contains the graphs output by **Swan**. From here the viewer can get information about nodes and edges, or pan and zoom over the graph display area. The I/O window is used by the annotator and the **Swan** system to display one-line messages and get input from the viewer.

The **Swan** Annotation Interface Library (**SAIL**) is a set of easy to use functions for annotating a program so that its significant data structures and the manner in which the data structures change during the execution of the program can be visualized. Given an appropriate description of the data structures used in a program, **Swan** is able to display them using different graphical elements as specified by the annotator.

3 SwanGraph Layout Algorithms

A graph in **Swan** consists of a set of layout components. Each layout component has nodes and edges. When the graph is displayed, the layout of nodes and edges is determined by the type of the layout component they belong to. There are several algorithms implemented in **Swan** to lay out different kinds of graphs automatically. New graph layout algorithms can be integrated into **Swan** easily.

Linked lists and arrays are examples of layout components. Layout components allow an annotator to build a more complicated structure than the simple linked list or array. In **Swan**, a node in a layout component may be a parent node of another layout component. Therefore, a simple linked list can be recursively expanded to represent relatively complex structures.

Swan also contains an algorithm to draw rooted trees, based on the aesthetic criteria suggested by Bloesch [1]. These criteria include aligning sibling nodes horizontally; centering parent nodes between their leftmost and rightmost children; keeping edges from crossing; and good horizontal and vertical separation.

Swan implements two algorithms to draw general undirected graphs. The first distributes nodes along the circumference of a circle evenly (Figure 1). Edges are drawn as straight lines between its two end nodes. The second algorithm implements Kamada and Kawai's algorithm [3], which is a force-directed placement method. Here, the total balance of a layout is considered to be more important than simply reducing the number of edge crossings. **Swan** also includes a hierarchical layout algorithm for digraphs based on the procedures of Eades and Sugiyama [2].

References

1. A. Bloesch, "Aesthetic Layout of Generalized Trees", *SOFTWARE — Practice and Experience*, Vol. 23(8), August 1993, pp. 817-827.

2. P. Eades and K. Sugiyama, "How to Draw a Directed Graph", *Journal of Information Processing*, Vol. 13, No. 4, 1990, pp. 424-437.

3. T. Kamada and S. Kawai, "An Algorithm for Drawing General Undirected Graphs", *Information Processing Letters*, Vol. 31, April 1989, pp. 7-15.

4. G.-C. Roman and K.C. Cox, "A Taxonomy of Program Visualization Systems", *IEEE Computer*, Vol. 26, No. 12, 1993, pp. 11-24.

5. R. Tamassia and I.G. Tollis, Eds., *Graph Drawing'94*, Lecture Notes in Computer Science 894, Springer, Berlin, 1994.

Author Index

Springer-Verlag
and the Environment

We at Springer-Verlag firmly believe that an international science publisher has a special obligation to the environment, and our corporate policies consistently reflect this conviction.

We also expect our business partners – paper mills, printers, packaging manufacturers, etc. – to commit themselves to using environmentally friendly materials and production processes.

The paper in this book is made from low- or no-chlorine pulp and is acid free, in conformance with international standards for paper permanency.

Lecture Notes in Computer Science

For information about Vols. 1–949

please contact your bookseller or Springer-Verlag

Vol. 985: T. Sellis (Ed.), Rules in Database Systems. Proceedings, 1995. VIII, 373 pages. 1995.

Vol. 986: Henry G. Baker (Ed.), Memory Management. Proceedings, 1995. XII, 417 pages. 1995.

Vol. 987: P.E. Camurati, H. Eveking (Eds.), Correct Hardware Design and Verification Methods. Proceedings, 1995. VIII, 342 pages. 1995.

Vol. 988: A.U. Frank, W. Kuhn (Eds.), Spatial Information Theory. Proceedings, 1995. XIII, 571 pages. 1995.

Vol. 989: W. Schäfer, P. Botella (Eds.), Software Engineering — ESEC '95. Proceedings, 1995. XII, 519 pages. 1995.

Vol. 990: C. Pinto-Ferreira, N.J. Mamede (Eds.), Progress in Artificial Intelligence. Proceedings, 1995. XIV, 487 pages. 1995. (Subseries LNAI).

Vol. 991: J. Wainer, A. Carvalho (Eds.), Advances in Artificial Intelligence. Proceedings, 1995. XII, 342 pages. 1995. (Subseries LNAI).

Vol. 992: M. Gori, G. Soda (Eds.), Topics in Artificial Intelligence. Proceedings, 1995. XII, 451 pages. 1995. (Subseries LNAI).

Vol. 993: T.C. Fogarty (Ed.), Evolutionary Computing. Proceedings, 1995. VIII, 264 pages. 1995.

Vol. 994: M. Hebert, J. Ponce, T. Boult, A. Gross (Eds.), Object Representation in Computer Vision. Proceedings, 1994. VIII, 359 pages. 1995.

Vol. 995: S.M. Müller, W.J. Paul, The Complexity of Simple Computer Architectures. XII, 270 pages. 1995.

Vol. 996: P. Dybjer, B. Nordström, J. Smith (Eds.), Types for Proofs and Programs. Proceedings, 1994. X, 202 pages. 1995.

Vol. 997: K.P. Jantke, T. Shinohara, T. Zeugmann (Eds.), Algorithmic Learning Theory. Proceedings, 1995. XV, 319 pages. 1995.

Vol. 998: A. Clarke, M. Campolargo, N. Karatzas (Eds.), Bringing Telecommunication Services to the People – IS&N '95. Proceedings, 1995. XII, 510 pages. 1995.

Vol. 999: P. Antsaklis, W. Kohn, A. Nerode, S. Sastry (Eds.), Hybrid Systems II. VIII, 569 pages. 1995.

Vol. 1000: J. van Leeuwen (Ed.), Computer Science Today. XIV, 643 pages. 1995.

Vol. 1001: M. Sudan, Efficient Checking of Polynomials and Proofs and the Hardness of Approximation Problems. XIV, 87 pages. 1995.

Vol. 1002: J.J. Kistler, Disconnected Operation in a Distributed File System. XIX, 249 pages. 1995.

VOL. 1003: P. Pandurang Nayak, Automated Modeling of Physical Systems. XXI, 232 pages. 1995. (Subseries LNAI).

Vol. 1004: J. Staples, P. Eades, N. Katoh, A. Moffat (Eds.), Algorithms and Computation. Proceedings, 1995. XV, 440 pages. 1995.

Vol. 1005: J. Estublier (Ed.), Software Configuration Management. Proceedings, 1995. IX, 311 pages. 1995.

Vol. 1006: S. Bhalla (Ed.), Information Systems and Data Management. Proceedings, 1995. IX, 321 pages. 1995.

Vol. 1007: A. Bosselaers, B. Preneel (Eds.), Integrity Primitives for Secure Information Systems. VII, 239 pages. 1995.

Vol. 1008: B. Preneel (Ed.), Fast Software Encryption. Proceedings, 1994. VIII, 367 pages. 1995.

Vol. 1009: M. Broy, S. Jähnichen (Eds.), KORSO: Methods, Languages, and Tools for the Construction of Correct Software. X, 449 pages. 1995. Vol.

Vol. 1010: M. Veloso, A. Aamodt (Eds.), Case-Based Reasoning Research and Development. Proceedings, 1995. X, 576 pages. 1995. (Subseries LNAI).

Vol. 1011: T. Furuhashi (Ed.), Advances in Fuzzy Logic, Neural Networks and Genetic Algorithms. Proceedings, 1994. (Subseries LNAI).

Vol. 1012: M. Bartošek, J. Staudek, J. Wiedermann (Eds.), SOFSEM '95: Theory and Practice of Informatics. Proceedings, 1995. XI, 499 pages. 1995.

Vol. 1013: T.W. Ling, A.O. Mendelzon, L. Vieille (Eds.), Deductive and Object-Oriented Databases. Proceedings, 1995. XIV, 557 pages. 1995.

Vol. 1014: A.P. del Pobil, M.A. Serna, Spatial Representation and Motion Planning. XII, 242 pages. 1995.

Vol. 1015: B. Blumenthal, J. Gornostaev, C. Unger (Eds.), Human-Computer Interaction. Proceedings, 1995. VIII, 203 pages. 1995.

VOL. 1016: R. Cipolla, Active Visual Inference of Surface Shape. XII, 194 pages. 1995.

Vol. 1017: M. Nagl (Ed.), Graph-Theoretic Concepts in Computer Science. Proceedings, 1995. XI, 406 pages. 1995.

Vol. 1018: T.D.C. Little, R. Gusella (Eds.), Network and Operating Systems Support for Digital Audio and Video. Proceedings, 1995. XI, 357 pages. 1995.

Vol. 1019: E. Brinksma, W.R. Cleaveland, K.G. Larsen, T. Margaria, B. Steffen (Eds.), Tools and Algorithms for the Construction and Analysis of Systems. Selected Papers, 1995. VII, 291 pages. 1995.

Vol. 1020: I.D. Watson (Ed.), Progress in Case-Based Reasoning. Proceedings, 1995. VIII, 209 pages. 1995. (Subseries LNAI).

Vol. 1021: M.P. Papazoglou (Ed.), OOER '95: Object-Oriented and Entity-Relationship Modeling. Proceedings, 1995. XVII, 451 pages. 1995.

Vol. 1022: P.H. Hartel, R. Plasmeijer (Eds.), Functional Programming Languages in Education. Proceedings, 1995. X, 309 pages. 1995.

Vol. 1023: K. Kanchanasut, J.-J. Lévy (Eds.), Algorithms, Concurrency and Knowlwdge. Proceedings, 1995. X, 410 pages. 1995.

Vol. 1024: R.T. Chin, H.H.S. Ip, A.C. Naiman, T.-C. Pong (Eds.), Image Analysis Applications and Computer Graphics. Proceedings, 1995. XVI, 533 pages. 1995.

Vol. 1025: C. Boyd (Ed.), Cryptography and Coding. Proceedings, 1995. IX, 291 pages. 1995.

Vol. 1026: P.S. Thiagarajan (Ed.), Foundations of Software Technology and Theoretical Computer Science. Proceedings, 1995. XII, 515 pages. 1995.

Vol. 1027: F.J. Brandenburg (Ed.), Graph Drawing. Proceedings, 1995. XII, 526 pages. 1996.